Phytochemical Potentials for Dermatological Applications

Some wild, herbal and medicinal plants and their parts contain chemicals that provide a concise scientific evaluation of the effectiveness of foods, nutrients, herbs and dietary supplements in preventing dermal damage and cancer as well as improving skin health without disrupting the natural microflora. With climatic change, including increasing temperatures and longer periods of sun exposure, there is a growing concern among the population regarding skin protection. As the use of plant-based skin products becomes more popular, ***Phytochemical Potentials for Dermatological Applications*** describes biologically active components of plants and explains the numerous ways in which they support the treatment of different skin conditions.

FEATURES

- Explores the research in dermatology using bioactive compounds from plants representing the alternative treatments of the future
- Presents innovative methods of treatment in frequently encountered dermatological conditions
- Includes figures and tables that compress the information for the novice medical reader or non-medical reader

Phytochemical Potentials for Dermatological Applications summarizes studies on medicinal plants containing phytochemicals with skin-beneficial properties that have been tested *in vitro* or *in vivo* and describes appropriate and innovative treatment methods. This book is an essential resource for scientists and researchers interested in phytotherapy, with special emphasis on dermatology.

Phytochemical Potentials for Dermatological Applications

Edited by
Florina Miere (Groza), Simona Ioana Vicas,
and Amit Kumar Mandal

CRC Press
Taylor & Francis Group
Boca Raton London New York

CRC Press is an imprint of the
Taylor & Francis Group, an informa business

Designed cover image: Shutterstock

First edition published 2025
by CRC Press
2385 NW Executive Center Drive, Suite 320, Boca Raton FL 33431

and by CRC Press
4 Park Square, Milton Park, Abingdon, Oxon, OX14 4RN

CRC Press is an imprint of Taylor & Francis Group, LLC

ISBN: 978-1-032-49113-4 (hbk)
ISBN: 978-1-032-49657-3 (pbk)
ISBN: 978-1-003-39484-6 (ebk)

DOI: 10.1201/9781003394846

Typeset in Times
by Apex CoVantage, LLC

Contents

About the Editors

Florina Miere (Groza) took the first step in her biomedical career in 2017 when she graduated from the Faculty of Medicine and Pharmacy at the University of Oradea. She has been awarded for her exceptional results from the beginning of her career until now by people in the medical and pharmaceutical fields. She completed her PhD in Biology at the Doctoral School of Biomedical Sciences of the University of Oradea.

She currently works as a university assistant at the University of Oradea in the discipline of biochemistry. Since 2020, she has been able to work with a multidisciplinary team of researchers/teachers, resulting in multiple scientific articles published in renowned journals. It has articles that have been designated as the most cited articles of 2022.

Due to published articles, she has an H-index = 11 (Web of Science).

In the field of research and innovation, she continued to develop, being involved in two projects of the University of Oradea of the INO-TRANSFER-UO type (Scientific Research of Excellence Related to Priority Areas with Capitalization through Technological Transfer). In one project as a team member and one as a project manager.

She is also involved as a member of the editorial team in various journals in the medical field.

The major areas of interest for Florina Miere (Groza) are represented by bioactive compounds from plants, nanotechnologies in the medical and pharmaceutical fields and how they can be used in other fields such as cosmetics and food.

Simona Ioana Vicas completed her bachelor's degree in biochemistry and her master's degree in cellular and molecular biology from the University of Bucharest, Romania. She achieved her PhD in biotechnology from USAMV University in Cluj-Napoca, Romania. She currently holds the position of a full professor at the Department of Food Engineering at the University of Oradea, Romania. Additionally, she has the expertise to supervise a PhD program in the biology domain. With more than 25 years of research and teaching experience, she has published over 70 papers in various high-level journals (Hirsch Index = 20, Web of Science) in the fields of phytochemistry, cosmetics, functional foods, and bioeconomy. She was responsible for the project manager and was a member of multiple research projects. She is interested in bioactive compounds such as polyphenols and glucosinolates from different plant matrices, and how they can be used in the fields of food, cosmetics, pharmaceuticals, and medicine. Her other areas of interest include antioxidant capacity, antigenotoxicity, functional foods, nanoparticles and cell culture.

Dr. Amit Kumar Mandal is a distinguished academic and researcher with deep expertise in chemical biology and structural biology. He earned both his M.Sc. and PhD from the University of North Bengal. Before joining Raiganj University, he served as an assistant professor in the Department of Microbiology at Vidyasagar University under the UGC-Innovative Programme. Currently, Dr. Mandal leads the Centre for Nanotechnology Science (CeNS) at Raiganj University, where he has over a decade of experience in research, teaching, and administration.

Dr. Mandal has authored more than 80 peer-reviewed SCI publications in leading journals and holds three research patents, with a cumulative impact score exceeding 400, according to the journal citation report Web of Science. He is also actively involved in academic publishing, serving on the editorial boards and as a reviewer for several renowned publishers, including Nature Portfolio, RSC, ACS, Wiley, Elsevier, PLoS, and BioMed Central.

In recognition of his contributions, Dr. Mandal was awarded the title of MRSC by the Royal Society of Chemistry, UK, and ranked among the top 5% of highly cited authors by the Royal Society of Chemistry in 2020. He was also honoured with the Top Cited Papers India Award from IOP Publishing in 2022. Based on his H-index over the last six years, Dr. Mandal was ranked among the top 3% of scientists in biological sciences in India in 2024, as per the AD Scientific Index. His scholarly achievements also place him within the top 0.5% of all scholars globally, according to ScholarGPS. Additionally, he has secured research grants from DST-SERB to support his R&D efforts.

Contributors

Manjila Adhikari
Department of Biomedical Engineering, College of Life
Science and Technology, Huazhong University of Science and Technology
Wuhan, China

Seyfure Adıguzel
Experimental Medicine Research and Application Centre
University of Health Sciences
Istanbul, Turkey

Sevde Altuntas
Experimental Medicine Research and Application Centre
University of Health Sciences
Istanbul, Turkey

Merve Güdül Bacanlı
University of Health Sciences
Faculty of Pharmacy Department of Pharmaceutical Toxicology
Ankara, Turkey

Gamze Balci
University of Health Sciences Turkey
Experimental Medicine Research and Application Centre
Istanbul, Turkey

Saroj Ballav
Department of Human Physiology
Vidyasagar University
West Bengal, India

Sumitaksha Banerjee
Burdwan Medical College and Hospital
Barddhaman, India

Jatindra Nath Bhakta
Department of Food and Nutrition
University of Kalyani
Kalyani, West Bengal, India

Bakhtawar Bukhari
Gene Ed and Molecular Lab, FBRC, PCSIR Laboratories Complex
Lahore, Pakistan

Şevval Çelikten
University of Health Sciences Turkey
Gülhane Faculty of Pharmacy
Department of Pharmaceutical Toxicology
Ankara, Turkey

Biswajit Das
Centre for Life Sciences
Vidyasagar University
West Bengal, India

Kankana Das
Department of Physiology and Allied Sciences
Amity Institute of Health Allied Sciences
Amity University
Uttar Pradesh, India

Debankur Dey
Medical College and Hospital
Kolkata, India

Onur Erdem
University of Health Sciences Turkey
Gülhane Faculty of Pharmacy
Department of Pharmaceutical Toxicology
Ankara, Turkey

Ragip Ertas
Medical Palace Hospital
Department of Dermatology
Kayseri, Turkey

Sule Ketenci Ertas
Kayseri Acıbadem Hospital, Division of Internal Medicine
Department of Rheumatology
Kayseri, Turkey

Zozan Guleken
Department of Physiology
Faculty of Medicine
Gaziantep Islam Science and Technology University
Gaziantep, Turkey

Chandradipa Ghosh
Department of Human Physiology
Vidyasagar University
West Bengal, India

Jawad Hussain
Department of Biotechnology
Huazhong University of Science and Technology
Wuhan, China

Wajid Hussain
Advanced Biomaterials and Tissue Engineering Center
Huazhong University of Science and Technology
Wuhan, China

Debarati Jana
Department of Human Physiology
Vidyasagar University
West Bengal, India

Pradip Jana
Centre for Life Sciences
Vidyasagar University
West Bengal, India

Elif Ayse Kacar
Experimental Medicine Research and Application Centre
University of Health Sciences
Istanbul, Turkey

Miray Karamese
Experimental Medicine Research and Application Centre
University of Health Sciences
Istanbul, Turkey

Ahmet Kati
University of Health Sciences Turkey
Experimental Medicine Research and Application Centre
Istanbul, Turkey

Ayşe Baldemir Kılıç
Department of Pharmaceutical Botany
Faculty of Gulhane Pharmacy
University of Health Sciences
Ankara, Turkey

Harendra Kumar
Dow University of Health Sciences
Karachi, Pakistan

Prasenjit Maity
Department of Chemistry
Sabang Sajanikanta Mahavidyalaya
West Bengal, India

Amitava Mandal
Molecular Complexity Laboratory
Department of Chemistry
Raiganj University
Raiganj, India

Avrajit Manna
Department of Chemistry
Raja Narendra Lal Khan Women's College (Autonomous)
Gope Palace, India

Krishnendu Manna
Department of Food and Nutrition
University of Kalyani
Kalyani, West Bengal, India

Priyabrata Mitra
Molecular Complexity Laboratory
Department of Chemistry
Raiganj University
Raiganj, India

Dattatreya Mukherjee
Raiganj Government Medical College and Hospital
Raiganj, India

Soumyadip Mukhopadhayay
Department of Human Physiology
Vidyasagar University
West Bengal, India

Jannatun Nesa
Department of Zoology
Gangarampur College
Gangarampur, West Bengal, India

Mustafa Nisari
Department of Medical Biochemistry
Faculty of Dentistry
University of Nuh Naci Yazgan
Kayseri, Turkey

Ismail Ocsoy
Department of Analytical Chemistry
Faculty of Pharmacy
Erciyes University
Kayseri, Turkey

Bilge Olceroglu
University of Health Sciences Turkey
Experimental Medicine Research and Application Centre
Istanbul, Turkey

Amiya Kumar Panda
Department of Chemistry
Vidyasagar University
West Bengal, India

Anuttam Patra
Chemistry of Interfaces Group, Luleå University of Technology
Luleå, Sweden

Debjyoti Paul
Department of Dietetics and Applied Nutrition
Amity Institute of Applied Sciences Kolkata
Amity University Kolkata
West Bengal, India

Ishrat Perveen
Gene Ed and Molecular Lab, FBRC, PCSIR Laboratories Complex
Lahore, Pakistan

Dwija Poddar
Raiganj Government Medical College and Hospital
Raiganj, India

Sumita Rani Pradhan
Centre for Life Sciences
Vidyasagar University
West Bengal, India

Poulami Roy
North Bengal Medical College and Hospital
West Bengal, India

Saba Sabahat
Rashid Latif College of Pharmacy
Ferozepur Road, Lahore, Pakistan

Abdul Sadat
Insect Ecology and Conservation Biology Laboratory
Department of Sericulture
Raiganj University
Raiganj, India

Jülide Secerli
University of Health Sciences Turkey
Gülhane Faculty of Pharmacy
Department of Pharmaceutical Toxicology
Ankara, Turkey

Kishwar Sultana
Medicinal Botanical Centre
Pakistan Council for Scientific and Industrial Laboratories Complex
Peshawar, Pakistan

Sandipa Tripathi
Centre for Life Sciences
Vidyasagar University
West Bengal, India

Muhammad Wajid Ullah
Department of Pulp & Paper Engineering
College of Light Industry and Food Engineering, Nanjing Forestry University
Nanjing, China.

Huan Wang
Advanced Biomaterials and Tissue Engineering Centre
Huazhong University of Science and Technology
Wuhan, China

Shenqi Wang
Advanced Biomaterials and Tissue Engineering Centre
Huazhong University of Science and Technology
Wuhan, China

Xiaohan Yang
Advanced Biomaterials and Tissue Engineering Centre
Huazhong University of Science and Technology
Wuhan, China

Muhammed Burak Yucel
University of Health Sciences
Kayseri City Education and Research Hospital
Dermatology and Venereology Clinic
Turkey
Department of Pharmaceutical Botany
Faculty of Gulhane Pharmacy
University of Health Sciences
Ankara, Turkey

Sadi Yusufbeyoğlu
Department of Pharmaceutical Botany
Faculty of Gulhane Pharmacy
University of Health Sciences
Kayseri, Turkey

1 Traditional Medicine for Maintaining Skin Health

Debarati Jana, Sumita Rani Pradhan, Sandipa Tripathi, Biswajit Das, Pradip Jana, Saroj Ballav, Soumyadip Mukhopadhayay, Anuttam Patra, Amiya Kumar Panda and Chandradipa Ghosh

1.1 INTRODUCTION

Traditional medicines are a collection of all fundamental medical theories, indigenous knowledge, folklore, prehistoric theories, and basic techniques for using natural ingredients from plants, animals, minerals, and other sources in maintaining health including disease diagnosis, prevention, and treatment. Fossil proof suggests that people are using natural products as medicine for at least 60,000 years (Fabricant and Farnsworth, 2001; Shi et al., 2010). The utilization of natural products as medicines must have been extremely difficult for early people.

The oldest form of healthcare is traditional medicine worldwide (TM) and is used in the prohibition and therapy of physical and mental disorders. Traditionally, several communities have formed a collection of practical therapeutic techniques to fight against life-threatening diseases and to maintain well-being. Traditional medicine is also known as complementary or ethnic medicine and is still in widespread use in many countries today (WHO, 2000).

Natural products are used in many medical practices, including traditional Chinese medicine (TCM), Kampo (Japanese traditional medicine), Russian herbal medicine, traditional medicine in Africa, Ayurveda, and Unani (Indian traditional medicine), Egyptian traditional medicine, traditional Korean medicine (TKM), etc. all over the world for hundreds or even thousands of years and have established as well-organized medical systems. Even if they have certain weaknesses in different forms, they are nonetheless considered an important basis of human knowledge (Fabricant and Farnsworth, 2001; Alves and Rosa, 2007).

Minerals, plants, and animals found in nature have been the cornerstone of treating 15 human illnesses. According to current estimates, 80% of population from developing countries continue to have faith in conventional medicine, preferably relying on specific species of animals and plants for basic medical care. There are existing herbal medications along with their popularity and demands are rising daily. There are about 800 plants that have been extensively explored in ethnic medicinal practices, and around 500 plants with therapeutic relevance are mentioned in ancient literature. According to Chopra and Nayar (1956), India is home to countless medicinal plants that are in use in conventional medical practices.

The human skin helps maintain the homeostasis of the body through physiological functions meant for protection, sensation, thermoregulation, the defence system, and metabolic processes (Lai-Cheong and McGrath, 2013). It is primarily exposed to a variety of environmental elements, including infections, hazardous chemicals, and radiation (Celleno and Tamburi, 2009). Many skin conditions are brought on by sunlight's ultraviolet (UV) radiation. Constant UV-B exposure causes the skin to suffer from some negative consequences (Im et al., 2015).

The need for herbal skincare solutions has risen steeply as people's knowledge of the photoageing and carcinogenic consequences of UV radiation (UVR) has grown. Herbal remedies have

DOI: 10.1201/9781003394846-1

been used for ages to treat a variety of skin diseases and demonstrate many therapeutic characteristics (Afnan et al., 2016). Numerous herbal items received a lot of attention in the market as skincare products when it was discovered that they have medicinal potential (Afaq and Katiyar, 2011). Research has been carried out on a global scale using a variety of *in vivo* and *in vitro* models to discover the medicinal potential of diverse herbal compounds. However, new methods and long-term research are needed in this field (Katiyar et al., 1999).

Herbal remedies have been utilized to treat skin disorders, and various plant chemicals are currently employed as topical treatments (Jarić et al., 2018). Wounds, haemorrhoids, boils, and eczema are the most common conditions for which medicinal plants and their preparations are administered; also, their antibacterial and anti-inflammatory properties aid in the healing of skin. For example, *Urtica dioica L* is primarily suggested for topical use against eczemas, but *Plantago major L* is the species most frequently listed in treating injuries, wounds, and boils.

Since ancient ages herbs, plant materials, and products have been used as cosmetics. Herbal cosmetics have utilized a variety of extracted compounds, plant exudates, fixed oils, essential oils, juices, and isolated compounds from natural sources due to their potential effect on skin care. Because of their lesser side effects, availability, safety, increased biocompatibility, affordability, and historical applications, herbal cosmetics are becoming more and more favoured by the public. The consistency of herbal cosmetics inspires the investigation of more herbs, plant materials, and natural components for their potential cosmetic benefits (Dwivedi, 2021).

These are the key factors that have contributed to a recent global increase in interest and demand for traditional medicine, as traditional medicinal practices have a very long trend of maintaining health in addition to prevention and treatment of illnesses including skin issues. Many conventional medications have been originated after considering traditional wisdom (Manna and Mishra, 2018).

1.2 WORLDWIDE STATUS OF TRADITIONAL MEDICINE FOR SKIN HEALTH

According to the WHO, over 75% of the global population relies on treatments based on plants mostly for their healthcare. Herbs or plants have not only given people food and shelter, but also sustain the human race by treating various dysfunctions. Nearly 80% of population from Africa and Asia hold their reliance on conventional medicine to meet their basic medicinal requirements. Conventional drugs include conventional medicine from China, Korea, and other traditional systems, that is, Ayurveda, Siddha, Unani, traditional African medicine, European medicine, and Iran's medical system, etc. (Ansari and Inamdar, 2010).

Humans have been found to hold faith in nature throughout history to fulfil their basic needs, including medicines to treat a broad spectrum of illnesses. Particularly, plants have served as the foundation for the earliest accounts of complex conventional medical systems, dated from roughly 2600 BCE, describing the uses from Mesopotamia that included 1000 compounds extracted from plants. These contain the oils of *Cedrus* species (cedar), *Commiphora* species (myrrh), *Cupressus sempevirens* (cypress), *Glycyrrhiza glabra* (liquorice), and *Papaver somniferum* (poppy), which are still used in contemporary medicine to treat ailments like parasite infections, swelling, common colds and coughs. Egyptian medicine dates back to around 2900 BCE, in contrast, the "Ebers Papyrus" from 1500 BCE is the best known in terms of record, listing over 700 medications, most of which were derived from plants (Cragg and Newman, 2013).

Over the ages, Chinese traditional medicine has gained widespread recognition with the information found in Shennong herbal, a handbook on the utility of medicinal herbs encompassing 365 medicines (100 BCE), the Tang herbal describing 850 medicines (659 CE), and Wu Shi Er Bing Fang which is also pronounced as Wushi'er Bing Fang, translating recipes for 52 ailments and was written in ancient Chinese seal script during 1100 to 800 BCE (Adhikari and Paul, 2018). In China, Western medicine was introduced in the sixteenth century, but it could not be developed until the nineteenth century. Preceding that period the foremost form of medical care was the Traditional

Chinese Medicine (TCM) in the country (Dong, 2013). To date, TCM plays a key role in China and it is continuously developing. TCM is grounded on 5000 times of medical exercise and experiences and is very rich with documents from "clinical trials" that guarantee its usefulness and efficiency. Presently TCM is a significant part of the Chinese public health system. In recent times, TCM has gained significant approval as a reciprocal or indispensable drug in the Western world.

The most important part of TCM, the Chinese herbal drug is presently used in the healthcare of an estimated 1.5 billion people globally (Dobos et al., 2005; Qi et al., 2013). TCM with time has also become popular for the treatment of dermatological diseases. TCM is applicable in various dermatological disorders such as psoriasis, atopic dermatitis etc. (Koo and Arain, 1998).

Traditional medicine is extensively accepted in Africa. Traditional medicine also illustrates admiration for the artistic heritage. Eighty per cent of the African population relies on traditional medicine either itself or along with conventional drugs. Moreover, 80% of Ghanaians and Ethiopians depend on traditional medicine either alone or in confluence with conventional medicine for their main healthcare demands (Boakye et al., 2015). Ghana's traditional medical system is tied to the public healthcare system and thus, it is comparatively well organized. Research on *Hydnora africana*, which is used as ethno-medicine in Africa, has established the twenty antioxidant and antibacterial actions of indigenous products.

Moreover, the use of plants with traditional medicinal aptitude for the treatment of dermatological infections is very common in many rural areas. Traditional medicines are highly effective in treating dermatological problems because of their ability to stop bleeding, fast wound healing, and improve other skin conditions (Naidoo and Coopoosamy, 2011).

The Traditional Medicine of Japan is Kampo. TCM was introduced to Japan around the fifth and sixth centuries from China. From that period TCM has been substantially modified and acclimated by Japanese physicians to meet their specific requirements and with time developed into Kampo (Watanabe et al., 2001). From a recent study, it has been identified that some Japanese physicians use Kampo medicine as their preferred remedy (Watanabe et al., 2001; Yakubo et al., 2014; Mogami and Hattori, 2014). Together with chemotherapy or radiotherapy, some physicians of Japan constantly use Kampo traditional medicine for the treatment of cancer cases. This indicates how traditional medicine is well integrated with Western medicine in Japan (Yakubo et al., 2014; Yu et al., 2006). The derivatives of natural compounds from Kampo traditional medicine may lead to the development of novel drugs that may be used for cancer treatment. The natural products from Kampo traditional medicine represent a significant reservoir for tumour treatment strategies in the future (Efferth et al., 2007).

Due to the specialized geographical terrain of Russia, Russian herbal remedies have collected and incorporated traditional medicine styles that were adopted from Asia and Europe. Herbal remedy is an official and sovereign department of medicine in Russia, therefore herbal medicinal products are viewed as sanctioned therapies. A recent check shows that 14% of the Russian population constantly practise herbal remedies and 44% utilize them sometimes. Soviet Russian experimenters have concentrated substantially on the formation of adaptogens derivative of plants. The use of Russian Herbal Medicine for different medical issues is gradually heightening in all areas of medicine including dermatology. These traditional dermatologic herbal remedies are predominantly used in Russia as the rich collection of eleven different plant zones has resulted from a great range of medicinal plants. Herbal medicine requires extensive research and investigations to understand its active substance content. A large number of Russian traditional herbal medicines are used for the treatment of pyodermas, vitiligo, acne, parasitic and infectious skin diseases, hyperpigmentation, rhytides, dermatitides, psoriasis, rosacea, and hyperhidrosis (Olisova et al., 2018).

Sasang Constitutional Medicine (SCM) is a branch of Korean traditional medicine. It was incepted first in the mid-nineteenth century. SCM is theoretically analogous to individualized medicine. SCM supplies personalized and constitution-specific actions for numerous problems (WHO, 2013; Kim et al., 2013; Yoon et al., 2013; Kim and Noble, 2014). Korean traditional medicine has become gradually popular for the treatment of inflammatory skin diseases, for example, atopic dermatitis (Yun et al., 2013). Although the conventional healthcare association is relatively good in Korea, 86% of the population still has faith in SCM. Traditional medicine doctors can provide Korean SCM in both public and private hospitals. Both public and private medical insurance cover Korean SCM services. It has received numerous significant achievements enclosing constitution-diagnostic means, constitution-specific diseases, susceptibilities, and genetic research.

1.3 TRADITIONAL MEDICINE IN INDIA FULFILLING SKIN HEALTH

The Rigveda (1700–1100 BCE), Yajurveda (1400–1000 BCE), and Atharvaveda (1200 BCE) are three ancient texts that list 81 plants for therapeutic uses, 290 plants as utilities practised as traditional medicine in India (Kumar et al., 2007). Since the beginning of time, people in India have relied heavily on plant-based herbal medicines for treating and preventing various illnesses due to their intimate ties to nature, forests, and plants in their daily lives. Over 700 million people in India still mostly utilize the tenets of three traditional Indian medical systems; Ayurvedic Siddha and Unani medicines. India is recognized as a momentous hub for biodiversity, with more than 45,000 plant species, more than 6,000 of which are employed in folklore and traditional herbal medicine. Despite only making about 2.4% of the planet's land, India is home to more than a per cent of the world's diversity. The Eastern Himalayas and Western Ghats, two of the world's richest, most distinctive, and most endangered eco-regions are two of India's key ecological hotspots. India contributes 15–20% of the 20,000 medicinal plants identified by the WHO as being used throughout the world. A list of a few significant medicinal plants from India used to maintain skin health along with their bioactive components is presented in Table 1.1. Captured images of certain important Indian medicinal plants in maintenance of skin health are presented in Figure 1.1.

The ethnic groups known as "Adivasis," or native social groups are well-versed in the traditional uses of plants and other biological resources. The majority of healers and practitioners mix their own concoctions and give them to patients. The majority of tribal or ethnic populations lack written languages and scripts; therefore, knowledge of the formulations, treatments, and methods for doctoring illnesses that are part of the long-standing, indigenous tribal traditional medical practices is mostly untapped (Chakraborty and Paul, 2014).

The aforementioned medical practices are thought to have Indian origins or have arrived in India from other countries and gradually assimilated into Indian culture to create the Indian traditional medicine. India is the only country with such a wide variety of renowned traditional medical practices, such as naturopathy homoeopathy, yoga, Ayurveda, Siddha, and Unani (Kumar et al., 2007).

1.3.1 AYURVEDA

Ayurveda is one of the oldest medicinal practices that has been cultured according to ancient Indian texts Rigveda and the Atharvaveda, which were written between 2500 and 500 BCE (Mukherjee, 2001).

The quality, safety, stability, and efficacy of Ayurvedic medications are currently being assured through considerable research. Skin (*Twak*), the most important component of humans' natural beauty also gives clues to the Ayurvedic constitution. Clean and glowing skin represents true relaxation and inner happiness. Knowledge of Ayurveda includes uses of amalaki (*Emblica officinalis*), pippali (*Piper longam*) for *Vayasthapana* (anti-aging) and glowing skin, kumkum (*Crocus sativus*) for improving the complexion of skin and removing hyperpigmentation, red sandalwood (*Petrocarpus santalinus*) for antimicrobial, anti-allergic treatments. Ayurveda also uses some traditional techniques to alleviate stress or any negative thought or fear or emotion because it believes these can down-regulate skin's natural glow and shine (Deshmukh *et al.*, 20 15). Tridosha functions to properly maintain skin health in humans. *Vata* improves nutrient circulation among different layers of skin, *Pitta* balances the chemical and hormonal activities of the skin, and *Kapha* moisturizes the skin (Pawar and Sharma, 2023). Antioxidants with natural flavonoids (anthocyanins; red grapes), lycopene (tomatoes), quercetin (onions), carotenoids (carrots, turmeric, oranges), catechins (green tea) containing alpha hydroxyl acids and beta hydroxy acids cause skin elasticity improvement, added skin thickness and rise in collagen composition (Datta et al., 2011; ENVIS Centre on Floral Diversity, Govt. of India).

TABLE 1.1
Medicinal Plants Commonly Used for Maintaining Skin Health in India

Plants (Scientific Name)	Local name	Family	Types of Plants	Parts used	Main Phytochemicals	Mode of use	References
Argania spinosa	Argan	Sapotaceae	Tree	Seed	Tocopherols, N-heterocycles (1 methyl 1 pyrrole, 2,6 dimethyl pyrazine), furans, alcohols (2,3 butanediol, 1 hexanol, 2 methyl 1 propanol) acetoin.	Oil, extracted from argan seed used on skin to keep skin wet, prevent skin tanning and to help the pores open.	Gharby and Charrouf, 2022; El Monfalouti et al., 2012; Michalak et al., 2021.
Coffea arabica	Coffee	Rubiaceae	Tree	Green/ Fermented seeds	Phenolic compounds as caffeine chlorogenic acid (trans-cinnamic acid esters and quinic acid ester).	Mature green/fermented seeds are used as paste for glowing skin. Extracted seed oil is also used on skin for maintaining skin health.	ALAsmariet al., 2020
Cocos nucifera	Dab or Narkel	Arecaceae	Tree	Tender coconut water, Endosperm	Caffeic acid, ascorbic acid, catechin/ epicatechin, vitamin E cytokinin (kinetin, kinetin riboside and trans-zeatin). Fatty acids (lauric, capric, palmitic, stearic, myristic, oleic and linoleic acid).	Tender coconut water is used as skin toner. Coconut oil highly moisturizes the skin and protect the skin from different infections.	Michalak et al., 2021 Halim et al., 2023
Oenothera biennis	Evening primrose	Onagraceae	Herb	Seeds	Flavonoid glycosides (quercetin-3-O-galactoside/rhamnoside glucoside of kaempferol and myricetin), phenolic acids (coumaric, caffeic ellagic, gallic), tannins (oenothein A and B), Linoleic acid, ɤ-linolenic acid.	Primrose oil, extracted from seeds enhances skin elasticity. It moisturizes and soothes the skin and also reduces roughness, dryness, wrinkles of the skin.	Michalak et al., 2021 Simon et al., 2014 Muggli, 2005 Fecker et al., 2020
Cannabis sativa	Cannabi, Hemp	Cannabaceae	Herb	Flowers, Leaves	Delta-9-tetrahydrocannabinol, cannabidiol, α-linolenic acid Linolenic acid and Oleic acid.	Cold-pressing of de-hulled hemp seeds produce oils. Oils are used as moisturizer of skin.	Michalak et al., 2021

(*Continued*)

TABLE 1.1 (*Continued*)
Medicinal Plants Commonly Used for Maintaining Skin Health in India

Plants (Scientific Name)	Local name	Family	Types of Plants	Parts used	Main Phytochemicals	Mode of use	References
Pterocarpus santalinus	Raktochandan	Fabaceae	Tree	Heartwood	Phenols, triterpenoids, alkaloids, saponins, glycosides, sterols, tannins, flavonoids, isoflavones, savinin, calocedrin, and glucosides.	Sandalwood powder is mixed with rose water to make a slurry and applied to the face for preventing acne.	Soundararajan et al., 2016
Camellia sinensis	Cha	Theaceae	Shrub	Leaves	Flavonoids (Catechins, epicatechin etc.) caffeines, theobrownins, thearubigins, and theaflavins.	Cold tea soaked in cotton pad, are used to clear skin and to shrink pores. It also regulates sebum production and maintains skin elasticity.	Koch et al., 2019
Rosa rubiginosa or Rosa damascena	Golap	Rosaceae	Shrub	Flower (petals)	Rose oxide, 2-phenylethanol, acetic acid, octanoic acid, acetoin, β- citronellol, isoamyl-acetate, phenylacetic acid, linalool etc.	Sap is extracted from flowers (petals) and prepared rose water. It is applied on face as tonner.	Özdemir and Budak, 2022
Citrus limon	Lebu	Rutaceae	Tree	Mesocarp and endocarp (Epicarp sometimes)	Phenolic (coumarins, flavonoids, phenolic acid etc.), terpenoids (carotenoids and limonoids), pectin, ascorbic acid, tocopherols and tocotrienols, minerals (Se, Cu, Fe, Zn, Mn) Naringenin.	Peels are pasted and used on face for brighter skin. It is enriched with high Vit-C, that's why reduces skin pigmentation and tanning.	Saini et al., 2022
Citrus sinensis	Komla	Rutaceae	Tree	Mesocarp and endocarp (Epicarp sometimes)	Phenolics (flavonoids, phenolic acid), provitamin (β-carotene as carotenoids),	Dried orange peel mixed with raw turmeric is used as face pack for enhancing the brightness of the skin	Cebadera-Miranda et al., 2020
Phyllanthus emblica	Amlaki	Euphorbiaceae	Tree	Fruit	3-6-di-o-galloyl-glucose, alanine, arginine, ascorbic acid, aspartic acid, β-carotene, carbohydrates, chebulagic acid, chebulaginic acid, chebulic acid, chibulinic acid, corilagic acid, corilagin, cystine, ellagic acid, emblicanins, emblicol, ethyl gallate, flavonoids, gallic acid, gallotannins, gibberellin-a-1, kaempferol, methionine, myo-inositol, myristic acid.	Fruit extract mixed with neem inhibits pimples and moisturizes the skin. It also lightens the skin tone and protects skin from aging.	Priya and Islam, 2019

Crocus sativus	Kesar	Iridaceae	Herb	Stigma	Crocin, picrocrocin, safranal.	Soaked stigma with virgin coconut oil or olive oil and some amount of raw milk mixture is very useful to enhance the circulation of face skin for preventing fine lines and wrinkles.	Mzabriet al., 2019
Cucumis sativus	Shoshaa	Cucurbitaceae	Herb	Fruit	Tannins, cardiac glycosides, terpenoids, saponins, phytosterols, resins, alkaloids, flavonoids, glycosides, phylobatamin, and terpenes.	Cucumber fruit juice mixed with honey and few drops of lemon, make the skin clean, supple, and radiant. Cucumber fruits are also used in case of sunburn and maintain the softness of skin.	Akhtar et al., 2020
Santalum album	Shwet chandan	Santalaceae	Tree	Wood	α-Santalol, β-Santalol, nor-tricyclo-kasantalal ($C_{11}H_{16}O$), alcohols-santenol ($C_9H_{16}O$), aldehydes-nortricyclo-ekasantalene ($C_{11}H_{18}$), teresantalol ($C_{10}H_{16}O$).	Sandalwood powder, mixed with rosewater is used on face for boosting skin health and removing excess oil. Sandalwood oil mixed with hot water is useful to decrease wrinkles.	Dulalet al., 2019
Ocimum sanctum	Tulsi	Lamiaceae	Sub-shrub	Leaf	Linalool, eugenol, caryophyllene oxide, 1,8-cineole, methyl eugenol, β-bisabolene, β-caryophyllene, Iso-caryophyllene.	Tulsi leaves extract decreases wrinkles. Tulsi leaf paste is also essential for removing blemishes.	Kumar et al., 2022; Pattanayaket al., 2010

(Continued)

TABLE 1.1 *(Continued)*
Medicinal Plants Commonly Used for Maintaining Skin Health in India

Plants (Scientific Name)	Local name	Family	Types of Plants	Parts used	Main Phytochemicals	Mode of use	References
Aloe vera	Ghrito- kumara	Asphodelaceae	Succulent	Leaf	Aloe-emodin, aloectic acid, anthranol, aloin A and B, isobarbaloin, emodin, ester of cinnamic acid, ure mannan, acetylated mannan, acetylated glucomannan, glucogalactomannan, galatan, galactogalacturan, arabinogalactan, galactoglucoarabinomannan, pectic substance, xylan, cellulose, chromones, isoaloeresin-D, isoarabaichromone, neoaloesin A. Arachidonic acid, linolenic acid, triglycerides, triterpenoid, gibberellins, lignin, potassium sorbate, salicylic acid, uric acid, campesterol, cholesterol and β-sitosterol.	*Aloe vera* hydrates the skin and also reduces the skin fragility.	Raksha et al., 2014
Azadiracta indica	Neem	Meliaceae	Tree	Leaf	Azadirachtin, nimbolinin, nimbin, nimbidin, nimbidol, sodium nimbinate, gedunin, salannin, Quercetin, nimbanene, 6-desacetylnimbinene, nimbandiol, nimbolide, ascorbic acid, n-hexacosanol and amino acid, 7-desacetyl-7-benzoylazadiradione, 7-desacetyl-7-benzoylgedunin, 17-hydroxyazadiradione, ß-sitosterol, and nimbiol.	Neem leaves paste makes the skin firmer and also sooths dry skin. Additionally, neem water (leaves are boiled in water, cooled and filtered to use as neem water) is used as good toner and also has anti-aging property as it promotes collagen production.	Alzohairy, 2016

Centella asiatica L.	Thankuni	Apiaceae.	Herb	Leaf	3-glucosylquercetin, 3-glucoside- and 7-glucoside-Kaempferol, polyacetylenes I-V and nine other acetylenes; amino acids, asiatic, centic, centellic, centoic, pectic, madasiatic acids, carotene, centellose (oligosaccharide), hydrocotylin (alkaloid), lipid, protein, pectin, saponins, vallerine, isothankuniside, medecassoside, thankcuriside, brahmic (= made cassic or made gascaric acid, 2α, 3β, 6β, 23- tetrahydroxy-urs-12-en-28-oic acid), and thankunic acids; β-caryophyllene, trans-β-farnesene, germacrene D, mesoinositol, polyphenols, α-terpinene, thymol methyl ether, betulic acid, β-sitosterol, campesterol, stigmasterol, tannins, vitamins B1, B2, and C.	Thankuni hydrates the dry and sensitive skin. It also has anti-aging property as it improves collagen.	Chonsut et al., 2024; Zakaria et al., 2019
Curcuma longa L	Halud	Zingiberaceae	Herb	Rhizome	Bisacurone, α-turmerone, ß-turmerone, ar-curcumyl alcohol, curcuminoids, coronadiene, cyclocurcumin, 8,12-epoxygermacra-1(10),4,7,11-tetraen-6-one, cyclohexanecarboxylic acid methyl ester, isopulegol, 2-menthen-1-ol, menth-1-en-9-ol, octahydrocurcumin, and labda-8(17)-12-diene-15,16-dial.	Raw turmeric paste revitalizes the skin by enhancing its natural radiance. It also reduces dark spots and sun tanning.	Rajkumari and Sanatombi, 2017
Withania somnifera	Ashwa-gandha	Solanaceae	Shrub	Leaf	Withanolide iv, withanoside v, vi, withaferin A, withastromonaloide, withanolide A, withanone, withanolide B, ergostane.	It enhances skin cells regeneration and also removes dead skin cells.	Kumar et al., 2015; Dwivediet al., 2015

(Continued)

TABLE 1.1 (*Continued*)
Medicinal Plants Commonly Used for Maintaining Skin Health in India

Plants (Scientific Name)	Local name	Family	Types of Plants	Parts used	Main Phytochemicals	Mode of use	References
Hibiscus rosasinensis	Joba	Malvaceae	Shrub	Flower	Ethanimidic acid, ethyl ester, propanal, 2,3dihydroxy, propanamide, N Ethylethylenediamine, O-Methylisourea hydrogen sulfate, hexadecanoic acid, methyl ester, 7-Formylbicyclo heptanes, 2-Butanamine, (S)-, 1,3,5-Triazine-2,4,6-triamine, N-Formylβ-alanine, (Z)6, (Z)9-Pentadecadien-1-ol, Butanedial, 1-Propanol, 2-methyl, and methane carboxylic acid.	It serves similarly to an astringent by tightening the skin's open pores. This regulates the degree of sebum production in the skin, saving from the annoyance of excess oiliness.	Missoum, 2018
Calendula officinalis	Calendula	Asteraceae	Herb	Flower	Hydroxyacetic Acid Hydrazide, 3-Ethoxy-propanenitrile, 2-Methyl, Adenosin, Phthalazine-1,4-dione, Benzoicacid, 2,6-bis(trimethylsilyl) ester, etanephrinebis(trimethylsilyl) ether, 2-Chloro-1-Propanol, 2,4,6-trimethyldecane, Alpha-Cubebene, beta-copaene, beta-guaiene, Selina-3,7(11)-diene, gamma-muurolene, pentadecanoic acid, 14-methyl-methyl ester.	Calendula promotes the hydration of the skin. Essential oil extracted from flowers, has beneficial role for reducing dryness of the skin.	Al-Mussawi and Al-Hussani, 2019
Lavandula angustifolia	Lavender	Lamiaceae	Herb	Flower	Alpha-Thujene, alpha-pinene, sabinene, camphene, beta-myrcene, D-limonene, beta-phellandrene, eucalyptol (1,8-cineole), terpinen-4-ol, caryophyllene, santalene.	Lavender oil has the ability to maintain even skin tone and also reduce wrinkles.	Jianuet al., 2013

Helianthus annuus	Suryamukhi	Asteraceae	Herb	Flower	α-Pinene, camphene, 2,4-Thujadiene, β-Terpinene, L-β-Pinene, Epoxycyclooctane, 2,3-Dehydro-1,8-cineole, 1,3,8-p-Menthatriene, E,E-2,6-Dimethyl 1,3,5,7 octatetraene, 4-Isopropenyl toluene, o-Cymene, Isosylvestrene, γ-Terpinen, trans-p-Mentha-2,8-dienol, Berbenol, Campholenal, 4Isopropenyltoluene, Terpinolene, Benzyl ethyl carbinol, L-Pinocarveol, cis-Verbenol, d-Verbenol, Pinocarvone, L-Terpinen-4-ol, p-Cymen-8-ol, Myrtenal, α-Terpineol, Verbenone, cis-Carveol, L-Carveol, Carvol, Hotrienol, 3,5-Diethylphenol, trans-2-Caren-4-ol, Bornyl acetate, (−)-trans-Pinocarvyl acetate, 4-Vinylguaiacol.	Retains moisture in skin, Protects from solar UV radiation. By maintaining the collagen and elastin composition of the skin, it helps to prevent premature aging.	Liu et al., 2020
Solanum tuberosum	Alu	Solanaceae	Herb	Tuber	α-Solanine, $β_2$-Solanine, γ-Solanine, Solanidine, α-Chaconine.	Potato slices or juice reduce the dark spots, scars, blemishes and hyperpigmentation of skin.	Friedman, 2004

1.3.2 Siddha

Another old medicinal practice known as "Siddha," was mainly founded in Southern India. It was formed via the regular use of natural resources to support excellent health. According to this medical philosophy, the human body is made up of three humours, three basic components, and undesired by-products.

A harmony of humours displays a state of health, and inequalities in it result in illness or disease (Pillai, 1998). Siddha constitutes an oil massage before a bath. For this, medicated oil like gingelly oil gets absorbed into the skin and causes skin improvement. Traditional oils have many natural ingredients that increase sensation in the skin, remove dry skin and make it shine. Holistic approach of Siddha treatment is the therapeutic emesis that normalizes Azal (Pittam) causing lustre to the skin (Ministry of AYUSH, 2019) Another potential use has been found in Southern India that the leaf extract of *Acalypha indica* has antioxidant, antibacterial antimicrobial, and antifungal properties and is effective against *Candida sp, Aspergillus spp*, (Umate et al., 2020).

1.3.3 Unani

Unani physicians believe that different naturally occurring products such as plant-based chemicals and animal objects along with Dalak (Unani massages) improve skin smoothness, textures, and shine. It also reduces toxins from the skin, enhances immune power by maintaining proper pH balance, and influences long-term rejuvenation. The motto of Unani is to absorb biologically active

FIGURE 1.1 Some significant medicinal plants for maintaining skin health. (a) *Pterocarpus santalinus* (Rakto chandan), (b) *Hibiscus rosasinensis* (Joba), (c) *Citrus limon* (Lebu), (d) *Ocimum sanctum* (Tulsi), (e) *Santalum album* (Shwet chandan), (f) *Azadiracta indica* (Neem), (g) *Helianthus annuus* (Suryamukhi), (h) *Calendula officinalis* (Calendula), (i) *Aloe vera* (Ghrito kumari), (j) *Rosa rubiginosa* (Golap), (k) *Centella asiatica* L. (Thankuni), (l) *Curcuma longa* L. (Holud), (m) *Phyllanthus emblica* (Amlaki), (n) *Solanum tuberosum* (Alu), (o) *Camellia sinensis* (Cha).

ingredients into deeper skin layers and those functioning at the cellular level. Clover and fruit-based gelatin beautify the skin by clearing the blood, whereas sweet basil, chickpea, and ripe dried figs improve blood circulation in the skin. Sesame oil contains sesamin and sesamolin that moisturize the skin (Saqlain et al., 2015). Different Unani therapies maintain the level of hydration and normal temperature in the skin. Hydrating skin prevents early ageing of the skin, decreases reduction in collagen production, changes in elastin fibres, alleviates premature wrinkles, and resists skin from being flaccid (Faiz and Anas, 2019).

1.4 CONCLUSION

Indian sages used a variety of plants along with other natural ingredients throughout the Vedic era to heal people and shield them from harm. They combined them with spiritual convictions, traditional knowledge, maxims, folklore, and techniques when they realized the benefits and implemented them for subsequent generations. The knowledge of Ayurveda, Siddha, and Unani was developed by utilizing a wide range of locally accessible natural resources with therapeutic potential.

Certain indigenous plants can be taken in, and people generally consume them as green vegetables. These serve as a good diet to maintain the overall wellness of the skin. Only a few recent studies have brought to light the bioactive substances, such as vitamins, fatty acids, carotenoids, and polyphenols, which have particular emphasis on skin-related conditions. Traditional materials have a significant impact in maintaining skin characteristics, including wrinkles, dryness/hydration, elasticity, firmness, and pigmentation. Polyphenols are an additional essential chemical that controls the hydration, smoothing, softening, soothing, and other attributes of the skin by suppressing the functions of the enzymes collagenase, hyaluronidase, and elastase.

Oils are extremely essential for skin due to their ability to keep the skin moisturized while providing vital unsaturated fatty acids that preserve the texture of the skin and prevent tanning and other signs of ageing. Numerous studies have also demonstrated the antioxidant properties of traditional oils containing vitamin C and vitamin E. Not only that, vitamins C and E are rich dietary supplements and medicinal herbs that protect the skin, especially the epidermal layer, from oxidation and also shield the skin from damaging ultraviolet (UV) rays.

Indigenous medicines have a significant role in healthcare, especially in the prevention of numerous dermatological conditions such as eczema, psoriasis, skin rashes, acne, etc., as they comprise a variety of bioactive components. Moreover, it has the potential to enhance cutaneous circulation, moisturize the skin, and have antimicrobial, anti-inflammatory, anticancer, and cell-stimulating abilities in addition to lowering microbial loading. Apart from providing the beneficial medicinal products for topical application as wound-healing agents, traditional medicine also has the potential to develop novel drugs with specific properties for maintaining skin health and eventually being implemented as an alternative therapeutic strategy. In order to provide a universal healthcare system that combines the potential benefits of various complementary and traditional medicine practices with conventional medicines that will be broadly recognized and affordable, an exclusive research target must be emphasized with the aim to validate the effectiveness as well as the security of these products.

REFERENCES

Adhikari, P.P. and Paul, S.B., 2018. History of Indian traditional medicine: a medical inheritance. *History*, 11(1), p. 421.

Afaq, F. and Katiyar, S.K., 2011. Polyphenols: skin photoprotection and inhibition of photocarcinogenesis. *Mini Reviews in Medicinal Chemistry*, 11(14), pp. 1200–1215.

Afnan, Q., Kaiser, P.J., Rafiq, R.A., Nazir, L.A., Bhushan, S., Bhardwaj, S.C., Sandhir, R. and Tasduq, S.A., 2016. Glycyrrhizic acid prevents ultraviolet-B-induced photodamage: a role for mitogen-activated protein kinases, nuclear factor kappa B and mitochondrial apoptotic pathway. *Experimental Dermatology*, 25(6), pp. 440–446.

Akhtar, P., Ahmad, I., Jameela, A., Ashfaque, M. and Begum, Z., 2020. Energizing effectiveness of cucumber (Khayarain) for health. A review article. *Journal of Emerging Technologies and Innovative Research*, 7(11).

ALAsmari, K.M., Zeid, I.M.A. and Al-Attar, A.M., 2020. Medicinal properties of Arabica coffee (Coffea arabica) oil: an overview. *Advancements in Life Sciences*, 8(1), pp. 20–29.

Al-Mussawi, Z.K. and Al-Hussani, I.M., 2019. Phytochemical study of Calendula officinalis plant by used GC-MS and FTIR techniques. *Plant Arch*, 19(1), pp. 845–851.

Alves, R.R. and Rosa, I.M., 2007. Biodiversity, traditional medicine and public health: where do they meet? *Journal of Ethnobiology and Ethnomedicine*, 3, pp. 1–9.

Alzohairy, M.A., 2016. Therapeutics role of Azadirachta indica (Neem) and their active constituents in diseases prevention and treatment. *Evidence-Based Complementary and Alternative Medicine*, 2016.

Ansari, J.A. and Inamdar, N.N., 2010. The promise of traditional medicines. *IJP-International Journal of Pharmacology*, 6(6), pp. 808–812.

Boakye, M.K., Pietersen, D.W., Kotzé, A., Dalton, D.L. and Jansen, R., 2015. Knowledge and uses of African pangolins as a source of traditional medicine in Ghana. *PLoS ONE*, 10(1), p. e0117199.

Cebadera-Miranda, L., Morales, P. and Cámara, M., 2020. Bioactive compounds in oranges from the Mediterranean climate area. In *The Mediterranean Diet* (pp. 293–309). Academic Press.

Celleno, L. and Tamburi, F., 2009. Structure and function of the skin. In *Nutritional Cosmetics* (pp. 3–45). William Andrew Publishing.

Chakraborty, N.R. and Paul, A., 2014. Traditional knowledge on medicinal plants used by the tribal people of birbhum district of West Bengal in India. *International Journal of Agriculture, Environment and Biotechnology*, 7(3), p. 547.

Chonsut, P., Romyasamit, C., Konyanee, A., Niyomtham, N., Goodla, L. and Mordmuang, A., 2024. Potential Activities of *Centella asiatica* Leaf Extract against Pathogenic Bacteria-Associated Biofilms and Its Anti-Inflammatory Effects. *Advances in Pharmacological and Pharmaceutical Sciences*, 2024(1), p. 5959077.

Chopra, R.N. and Nayar, S.L., 1956. Glossary of Indian medicinal plants. *Council of Scientific and Industrial Research*, pp. 2–79.

Cragg, G.M. and Newman, D.J., 2013. Natural products: a continuing source of novel drug leads. *Biochimica et Biophysica Acta (BBA)-General Subjects*, 1830(6), pp. 3670–3695.

Datta, H.S., Mitra, S.K., Paramesh, R. and Patwardhan, B., 2011. Theories and management of aging: modern and ayurveda perspectives. *Evidence-Based Complementary and Alternative Medicine*, 2011.

Deshmukh, S., Nikhode, K., Thakre, T. and Talnikar, M., 2015. Concept of beauty through Ayurveda. *International Journal of Ayurveda and Pharma Research*, 3(9), pp. 22–25.

Dobos, G.J., Tan, L., Cohen, M.H., McIntyre, M., Bauer, R., Li, X. and Bensoussan, A., 2005. Are national quality standards for traditional Chinese herbal medicine sufficient?: Current governmental regulations for traditional Chinese herbal medicine in certain Western countries and China as the Eastern origin country. *Complementary Therapies in Medicine*, 13(3), pp. 183–190.

Dong, J., 2013. The relationship between traditional Chinese medicine and modern medicine. *Evidence-Based Complementary and Alternative Medicine*, 2013.

Dulal, S.R., Taher, M.A. and Sheikh, H., 2019. Sandalwood oil can be a miraculous tackle on skin aging, skin appearance and wrinkle skin – a review. *World Journal of Pharmaceutical and Medical Research*, 5, pp. 51–55.

Dwivedi, A.K., 2021. Skin care potential of selected medicinal plants: a review. *International Journal of Pharmaceutical Research and Applications*, 6(4).

Dwivedi, D., Thanwar, M. and Gharia, A.K., 2015. Study of phytochemical active compounds in extract of Withaniasomnifera. *RASAYAN Journal of Chemistry*, 8, pp. 522–526.

Efferth, T., Miyachi, H. and Bartsch, H., 2007. Pharmacogenomics of a traditional Japanese herbal medicine (Kampo) for cancer therapy. *Cancer Genomics & Proteomics*, 4(2), pp. 81–91.

El Monfalouti, H., Charrouf, Z., Belviso, S., Ghirardello, D., Scursatone, B., Guillaume, D., Denhez, C. and Zeppa, G., 2012. Analysis and antioxidant capacity of the phenolic compounds from argan fruit (Argania spinosa (L.) Skeels). *European Journal of Lipid Science and Technology*, 114(4), pp. 446–452.

Fabricant, D.S. and Farnsworth, N.R., 2001. The value of plants used in traditional medicine for drug discovery. *Environmental Health Perspectives*, 109(suppl 1), pp. 69–75.

Faiz, A. and Anas, M., 2019. Concept of aging of skin in Unani medicine and its management through hydration therapy: a review. *Journal of Drug Delivery and Therapeutics*, 9(3–s), pp. 825–827.

Fecker, R., Buda, V., Alexa, E., Avram, S., Pavel, I.Z., Muntean, D., Cocan, I., Watz, C., Minda, D., Dehelean, C.A. and Soica, C., 2020. Phytochemical and biological screening of Oenothera biennis L. hydroalcoholic extract. *Biomolecules*, 10(6), p. 818.

Friedman, M., 2004. Analysis of biologically active compounds in potatoes (Solanum tuberosum), tomatoes (Lycopersicon esculentum), and jimson weed (Datura stramonium) seeds. *Journal of Chromatography A*, 1054(1–2), pp. 143–155.

Gharby, S. and Charrouf, Z., 2022. Argan oil: chemical composition, extraction process, and quality control. *Frontiers in Nutrition*, 8, p. 804587.

Halim, H.H., Dek, M.S.P., Hamid, A.A., Saari, N., Lazim, M.I.M., Abas, F., Ngalim, A., Ismail, A. and Jaafar, A.H., 2023. Novel sources of bioactive compounds in coconut (Cocos nucifera L.) water from different maturity levels and varieties as potent skin anti-aging strategies and anti-fatigue agents. *Food Bioscience*, 51, p. 102326.

Im, A.R., Song, J.H., Lee, M.Y. and Chae, S., 2015. Magnolol reduces UVB-induced photodamage by regulating matrix metalloproteinase activity. *Environmental Toxicology and Pharmacology*, 39(1), pp. 417–423.

Jarić, S., Kostić, O., Mataruga, Z., Pavlović, D., Pavlović, M., Mitrović, M. and Pavlović, P., 2018. Traditional wound-healing plants used in the Balkan region (Southeast Europe). *Journal of ethnopharmacology*, 211, pp. 311–328.

Jianu, C., Pop, G., Gruia, A.T. and Horhat, F.G., 2013. Chemical composition and antimicrobial activity of essential oils of lavender (Lavandula angustifolia) and lavandin (Lavandula x intermedia) grown in Western Romania. *International Journal of Agriculture and Biology*, 15(4), pp. 772–776.

Katiyar, S.K., Matsui, M.S., Elmets, C.A. and Mukhtar, H., 1999. Polyphenolic antioxidant (-)-epigallocatechin-3-gallate from green tea reduces UVB-induced inflammatory responses and infiltration of leukocytes in human skin. *Photochemistry and Photobiology*, 69(2), pp. 148–153.

Kim, J.U., Ku, B., Kim, Y.M., Do, J.H., Jang, J.S., Jang, E., Jeon, Y.J., Kim, K.H. and Kim, J.Y., 2013. The concept of sasang health index and constitution-based health assessment: an integrative model with computerized four diagnosis methods. *Evidence-based Complementary and Alternative Medicine*, 2013.

Kim, J.Y. and Noble, D., 2014. Recent progress and prospects in Sasang constitutional medicine: a traditional type of physiome-based treatment. *Progress in Biophysics and Molecular Biology*, 116(1), pp. 76–80.

Koch, W., Zagórska, J., Marzec, Z. and Kukula-Koch, W., 2019. Applications of tea (Camellia sinensis) and its active constituents in cosmetics. *Molecules*, 24(23), p. 4277.

Koo, J. and Arain, S., 1998. Traditional Chinese medicine for the treatment of dermatologic disorders. *Archives of Dermatology*, 134(11), pp. 1388–1393.

Kumar, B., Vijayakumar, M., Govindarajan, R. and Pushpangadan, P., 2007. Ethnopharmacological approaches to wound healing – exploring medicinal plants of India. *Journal of Ethnopharmacology*, 114(2), pp. 103–113.

Kumar, R., Saha, P., Lokare, P., Datta, K., Selvakumar, P. and Chourasia, A., 2022. A systemic review of Ocimum sanctum (Tulsi): morphological characteristics, phytoconstituents and therapeutic applications. *International Journal for Research in Applied Sciences and Biotechnology*, 9(2), pp. 221–226.

Kumar, V., Dey, A., Hadimani, M.B. and Emeraldjavascript, M., 2015. Chemistry and pharmacology of Withaniasomnifera: an update. *CellMed*, 5(1), pp. 1–1.

Lai-Cheong, J.E. and McGrath, J.A., 2013. Structure and function of skin, hair and nails. *Medicine*, 41(6), pp. 317–320.

Liu, X.S., Gao, B., Li, X.L., Li, W.N., Qiao, Z.A. and Han, L., 2020. Chemical composition and antimicrobial and antioxidant activities of essential oil of sunflower (Helianthus annuus L.) receptacle. *Molecules*, 25(22), p. 5244.

Manna, S.S. and Mishra, S.P., 2018. Ethnomedicinal survey of plants used by tribal in Lalgarh forest range, WB, India. *The Journal of Phytopharmacology*, 7(2), pp. 199–202.

Michalak, M., Pierzak, M., Kręcisz, B. and Suliga, E., 2021. Bioactive compounds for skin health: a review. *Nutrients*, 13(1), p. 203.

Ministry of AYUSH, Government of India, New Delhi, 2019. *SIDDHA SYSTEM OF MEDICINE: The Science of Holistic Health*. Ministry of AYUSH, Government of India, New Delhi, 2019. www.ayush.gov.in. ISBN: "978-81-937426-3-1".

Missoum, A., 2018. An update review on Hibiscus rosa sinensis phytochemistry and medicinal uses. *Journal of Ayurvedic and Herbal Medicine*, 4(3), pp. 135–146.

Mogami, S. and Hattori, T., 2014. Beneficial effects of rikkunshito, a Japanese kampo medicine, on gastrointestinal dysfunction and anorexia in combination with Western drug: a systematic review. *Evidence-Based Complementary and Alternative Medicine*, 2014.

Muggli, R., 2005. Systemic evening primrose oil improves the biophysical skin parameters of healthy adults. *International Journal of Cosmetic Science*, 27(4), pp. 243–249.

Mukherjee, P.K., 2001. Evaluation of Indian traditional medicine. *Drug Information Journal*, 35(2), pp. 623–632.

Mzabri, I., Addi, M. and Berrichi, A., 2019. Traditional and modern uses of saffron (Crocus sativus). *Cosmetics*, 6(4), p. 63.

Naidoo, K.K. and Coopoosamy, R.M., 2011. A comparative analysis of two medicinal plants used to treat common skin conditions in South Africa. *African Journal of Pharmacy and Pharmacology*, 5(3), pp. 393–397.

Olisova, O.Y., Snarskaya, E.S., Gladko, V.V. and Burova, E.P., 2018. Russian traditional medicine in dermatology. *Clinics in Dermatology*, 36(3), pp. 325–337.

Özdemir, N. and Budak, N.H., 2022. Bioactive compounds and volatile aroma compounds in rose (Rosa damascena Mill.) vinegar during the aging period. *Food Bioscience*, 50, p. 102062.

Pattanayak, P., Behera, P., Das, D. and Panda, S.K., 2010. Ocimum sanctum Linn. A reservoir plant for therapeutic applications: an overview. *Pharmacognosy Reviews*, 4(7), p. 95.

Pawar, J.B. and Sharma, G.B., 2023. An Ayurveda approach of skin care WSR to deha prakriti (body constitution): a review study. *Journal of Ayurveda and Integrated Medical Sciences*, 8(4), pp. 111–115.

Pillai, N.K., 1998. *History of Siddha Medicine*. Chennai: Department of Indian Medicine and Homoeopathy.

Priya, F.F. and Islam, M.S., 2019. Phyllanthus emblica Linn. (Amla) – a natural gift to humans: an overview. *Journal of Diseases and Medicinal Plants*, 5, pp. 1–9.

Qi, F., Wang, Z., Cai, P., Zhao, L., Gao, J., Kokudo, N., Li, A., Han, J. and Tang, W., 2013. Traditional Chinese medicine and related active compounds: a review of their role on hepatitis B virus infection. *Drug Discoveries & Therapeutics*, 7(6), pp. 212–224.

Rajkumari, S. and Sanatombi, K., 2017. Nutritional value, phytochemical composition, and biological activities of edible Curcuma species: a review. *International Journal of Food Properties*, 20(sup3), pp. S2668–S2687.

Raksha, B., Pooja, S. and Babu, S., 2014. Bioactive compounds and medicinal properties of Aloe vera L.: an update. *Journal of Plant Sciences*, 2(3), pp. 102–107.

Saini, R.K., Ranjit, A., Sharma, K., Prasad, P., Shang, X., Gowda, K.G.M. and Keum, Y.S., 2022. Bioactive compounds of citrus fruits: a review of composition and health benefits of carotenoids, flavonoids, limonoids, and terpenes. *Antioxidants*, 11(2), p. 239.

Saqlain, M., Ali, F., Parveen, A. and Riaz, A., 2015. Cosmetology in Unani system of medicine: a review. *World Journal of Pharmaceutical Research*, 5(1), pp. 292–301.

Shi, Q., Li, L., Huo, C., Zhang, M. and Wang, Y., 2010. Study on natural medicinal chemistry and new drug development. *Zhongcaoyao= Chinese Traditional and Herbal Drugs*, 41(10), pp. 1583–1589.

Simon, D., Eng, P.A., Borelli, S., Kägi, R., Zimmermann, C., Zahner, C., Drewe, J., Hess, L., Ferrari, G., Lautenschlager, S. and Wüthrich, B., 2014. Gamma-linolenic acid levels correlate with clinical efficacy of evening primrose oil in patients with atopic dermatitis. *Advances in Therapy*, 31, pp. 180–188.

Soundararajan, V., Ravi Kumar, G., Murugesan, K. and Chandrashekar, B.S., 2016. A review on red sanders (Pterocarpus santalinus Linn.)-phytochemistry and pharmacological importance. *World Journal of Pharmacy and Pharmaceutical Sciences*, 5(6), pp. 667–689.

Umate, S.K., Marathe, V.R. and Pawar, P.V., 2020. Evaluation of Charu prepared from Acalypha indica L-an important medicinal plant of traditional Siddha system useful in treating skin diseases. *Advances in Zoology and Botany*, 8(3), pp. 116–121.

Watanabe, S., Imanishi, J., Satoh, M. and Ozasa, K., 2001. Unique place of Kampo (Japanese traditional medicine) in complementary and alternative medicine: a survey of doctors belonging to the regional medical association in Japan. *The Tohoku Journal of Experimental Medicine*, 194(1), pp. 55–63.

World Health Organization, 2000. *General Guidelines for Methodologies on Research and Evaluation of Traditional Medicine* (No. WHO/EDM/TRM/2000.1). World Health Organization.

World Health Organization, 2013. *WHO Traditional Medicine Strategy: 2014–2023*. World Health Organization.

Yakubo, S., Ito, M., Ueda, Y., Okamoto, H., Kimura, Y., Amano, Y., Togo, T., Adachi, H., Mitsuma, T. and Watanabe, K., 2014. Pattern classification in Kampo medicine. *Evidence-Based Complementary and Alternative Medicine*, 2014.

Yoon, D.W., Lee, S.K., Yi, H., Hong, J.H., Soichiro, M., Lee, S.W., Kim, J.Y. and Shin, C., 2013. Total nasal resistance among Sasang constitutional types: a population-based study in Korea. *BMC Complementary and Alternative Medicine*, 13(1), pp. 1–6.

Yu, F., Takahashi, T., Moriya, J., Kawaura, K., Yamakawa, J., Kusaka, K., Itoh, T., Morimoto, S., Yamaguchi, N. and Kanda, T., 2006. Traditional Chinese medicine and Kampo: a review from the distant past for the future. *Journal of International Medical Research*, 34(3), pp. 231–239.

Yun, Y., Lee, S., Kim, S. and Choi, I., 2013. Inpatient treatment for severe atopic dermatitis in a traditional Korean medicine hospital: introduction and retrospective chart review. *Complementary Therapies in Medicine*, 21(3), pp. 200–206.

Zakaria, F., Ibrahim, W.N.W., Ismail, I.S., Ahmad, H., Manshoor, N., Ismail, N., Zainal, Z. and Shaari, K., 2019. LCMS/MS Metabolite Profling and Analysis of Acute Toxicity Effect of the Ethanolic Extract of *Centella asiatica on Zebrafsh Model. Pertanika Journal of Science & Technology*, 27(2), pp. 985–1003.

2 Innovative Extraction Methods for Phytochemicals Used to Combat Dermatological Disorders

Avrajit Manna and Prasenjit Maity

2.1 INTRODUCTION

Phytochemicals, the naturally occurring compounds in vegetation, are acknowledged for their substantial roles in mankind specifically in the medical and pharmaceutical fields (Ntemafack et al., 2023). The modern extraction techniques of these bioactive compounds are a pivotal method in the pharmaceutical industry, affecting the effectiveness and quality of the consequential products. This chapter delves into several modern, significant extraction methods of phytochemicals, highlighting their principles, recent progressions, and their specific applications in the context of the prevention of various dermatological disorders.

Phytochemicals are nothing but plant-derived compounds that have played in medicine and other fields since mankind has looked to Mother Nature to deliver treatments for several ailments and diseases. These plant metabolites, which are secondary, can produce pharmacologic and toxicologic effects in mankind. They are very useful biologically active compounds derived from plants. These compounds can be classified as alkaloids, terpenoids, nitrogenous compounds, sulphur-comprising compounds, and compounds having phenolic groups. Phytochemicals are further divided into several subclasses (Ferrazzano et al., 2011; Sahebi et al., 2017; Ivanovíc et al., 2020). Processing and extraction of medicinal plants is done for immediate consumption as herbal medicine or research interest. Medicinal plants have significant potential because they are a prominent source of therapeutic drugs. Most of the plant phytochemicals like phenolics and flavonoids show a meaningful impact on human health and the prevention of fatal diseases like cancer (Venugopal and Liu, 2012). On the other hand, cosmetic markets are among the most vigorously growing markets on the earth. Phytochemicals are used in herbal makeup to defend the skin from ultraviolet radiation, air contaminants, as well as weather changes. The continuously progressing studies on manufacturers of cosmetics are growing attention in high quality and improved cosmetic results in skin care. Epigallocatechin gallate, Erianin, paeoniflorin, caffeine, baicalein, gallic acid, and alpinetin are some significant phytochemical compounds (Strzępek-Gomółka et al., 2021; Amberg and Fogarassy, 2019).

Hence, extraction techniques to achieve different significant phytochemicals is a topic of huge scientific interest. Various approaches are utilized for the extraction of therapeutic plants such as decoction, percolation, maceration, ultrasound, microwave-assisted extractions, soxhlet extraction, superficial extraction, etc. Also, various procedures of chromatography like TLC, paper chromatography, GLC, and HPLC are used in the separation and purification of

 DOI: 10.1201/9781003394846-2

biologically active phytochemicals (Azwanida, 2015; Ingle et al., 2017; Doughari, 2012; Pandey and Tripathi, 2014).

2.2 DIFFERENT SOURCES OF PHYTOCHEMICALS

Medicinal plants deliver advantageous tools for treating numerous diseases in many indigenous societies. The phytochemical name comes from the Greek word Phyto, which means plants. Generally, plants derive a chemical compound known as phytochemicals, which defend plant cells from viruses, contaminants, UV exposure, bacteria, and other attacks. These are often found in cereal grains, garlic, fruits, vegetables, olive leaf, onion, tea, coffee, and many other plants (Figure 2.1). According to the metabolism of plants, phytochemicals can be classified into two categories such as primary metabolites (proteins, amino acids, carbohydrates, etc.) and secondary metabolites (terpenes, alkaloids, steroids, polyphenols, etc.) (Ramawat et al., 2009). Sources of some significant phytochemicals are gradually discussed here. Alangine A can be extracted from the plant *Alangium salviifolium* whereas carotenoids, polyphenols, curcumin, and flavonoids can be obtained from *Allium sativum* which is nothing but garlic. Phytochemicals quinine and caffeine can be isolated from the bark of the *Cinchona* tree and coffee beans, respectively. Antimicrobial flavonoids and diterpenes are extracted from *Helichry sumplicatum*. Last but not least, morphine and nicotine is obtained from *Papaver somniferum* (opium poppy) and tea leaves, respectively. *A. millefolium L.* extract is the chief ingredient for the cosmetic industries.

Various plant-based compounds have been utilized recently in topical dermatological treatments, for instance, 8-methoxypsoralen, salicylic acid, black tea, etc. (Saraswathy et al., 2010; Hadjzadeh et al., 2006; Babbar, 2015; Strzępek-Gomółka et al., 2021).

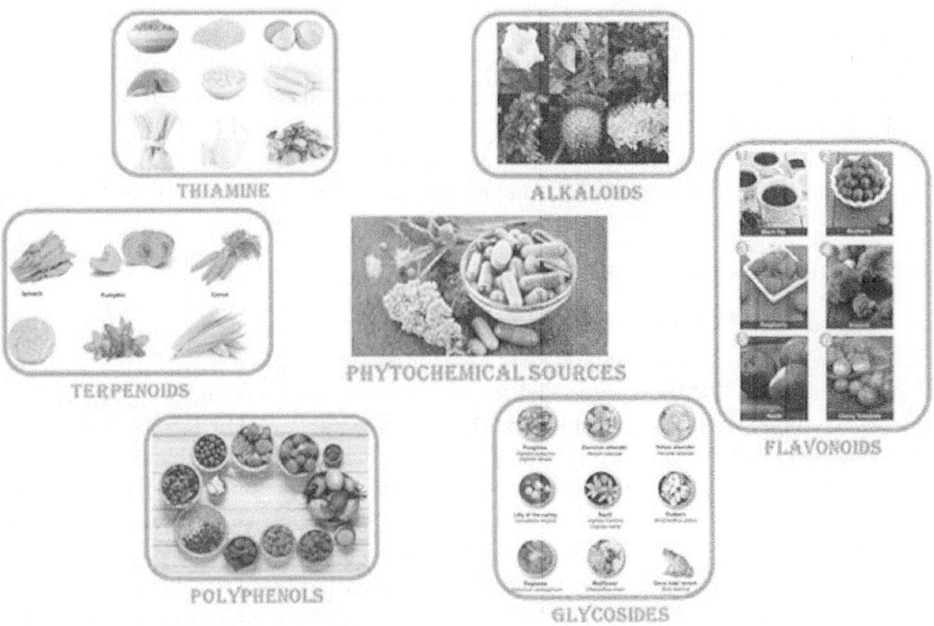

FIGURE 2.1 Different sources of plant phytochemicals.

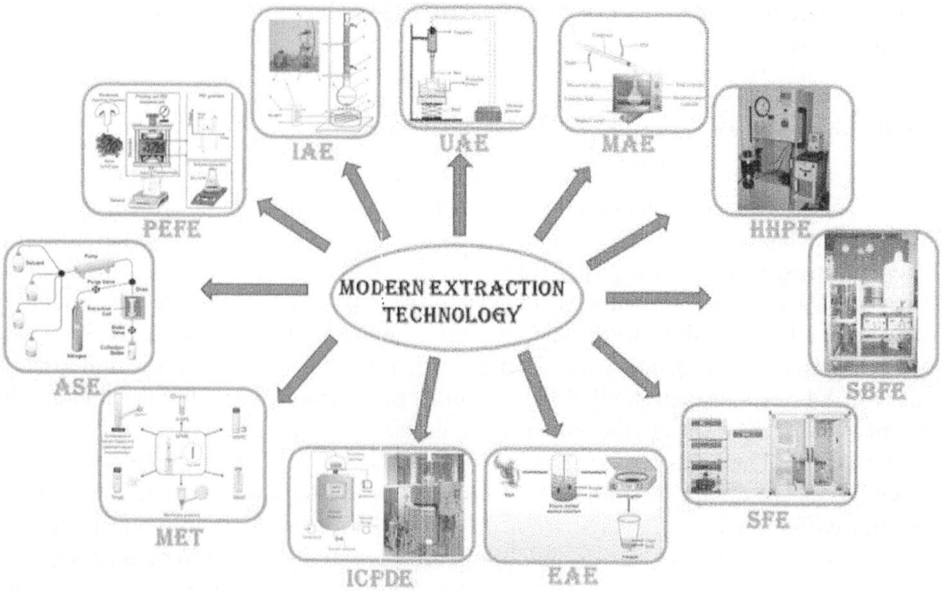

FIGURE 2.2 Modern frequently used technologies for the extraction of phytochemicals used to combat dermatological disorders.

2.3 EXTRACTION METHODS OF NATURAL BIOACTIVE PHYTOCHEMICALS

The extraction of phytochemicals was the first and primary step involved in phytochemical properties. Conventional solvent extraction is the ancient method and mainly comprises of SLE and LLE. Several other innovative extraction methods are accelerated solvent extraction (ASE), supercritical fluid extraction (SFE), ultrasound-assisted extraction (UAE), infrared-assisted extraction (IAE), enzyme-assisted extraction (EAE), high hydrostatic pressure extraction (HHPE), instant controlled pressure drop extraction (ICPDE), microwave-assisted extraction (MAE), microextraction techniques (MET), subcritical fluid extraction (SbFE), and pulsed electric field extraction (PEFE) (Figure 2.2).

2.3.1 Traditional Solvent Extraction Method

2.3.1.1 Solid-Liquid Extraction (SLE)

The SLE method is an easy and common technique for the extraction of phytochemicals from various plants. Zainol et al. (2023) reported that the phytochemical compound was extracted from the herbal plant Asystasia gangetica by using methanol as a solvent followed by incubator shaker (80 °C, 5 h), centrifugation (5800 rpm, 15 min) and freezing (4 °C). Zekovic's research group extracted phenolic compounds like ρ-coumaric caffeic and ferulic by SLE method from Salvia officinalis L (Zekovic et al., 2017). Here, sample material was isolated through various solvents like water and a water-ethanol mixture (20%, 40%, 60%, 80%, and 96% ethanol). Different phytochemical compounds like ferulic acid, rutin, and isorhamnetin were extracted from Asparagus officinalis L by SLE techniques (Fan et al., 2015). In another study, a polyphenol compound was obtained by the SLE method from fresh tea leaves (FD-QMETLJ) and different concentrations (95%, 85%, 75%, 65%, and 55% v/v) of EtOH, used as a solvent (Gadkari et al., 2014). Different phytochemicals were isolated from a dry sample of Haplophyllum myrtifolium via the SLE method, where different solvents such as petroleum ether, ethyl acetate, and methanol are used at room temperature (Zengin et al., 2014). Among them, ethyl acetate extract contains a high phenolic compound.

2.3.1.2 Liquid-Liquid Extraction (LLE)

Liquid-liquid extraction was also a general traditional isolation method. Gadkari et al. (2014) reported that catechins were extracted by the LLE technique from fresh tea leaves (QMETLJ). Here, different solvents like dichloromethane, hexane, $CHCl_3$, diethyl ether, and $CH_3COOC_2H_5$ were used in a volumetric proportion of 1:1.

2.3.1.3 Heat-Assisted Extraction (HAE)

HAE was a simple and economic isolation process normally utilized for the isolation of different biologically active substances. This method involves mixing of a solid and solvent for some time at a particular temperature (Caleja et al., 2017). Lopez Research Group (2018) reported that anthocyanin was extracted by two methods HAE and UAE from Arbutus unedo L. fruits. In the HAE method, the 600 mg sample and 20 mL ethanol were mixed followed by time and temperature (30 to 90 °C). This method is effective at 90 °C in five minutes, with 80% of acidified EtOH. Moongngarm et al. (2022) reported that phytochemicals were extracted from Stevia rebaudiana leaves by OH-AWE procedure where three applied electric power are seventy-five, one hundred and two hundred Vcm^{-1}. Another study reported that the HAE method was used for the extraction of biologically active anthocyanin dyes from F. carica (Backes et al., 2018). Here 1 g peel was added in 20 mL acidic liquid under nonstop electromagnetic stirring at temperature (20–90 °C) and time (5–68.8 min) range.

2.3.1.4 Hydrogen-Rich Water Extraction (HRW-E)

Alwazeer's research group isolated phytochemicals (phenolic compounds, flavonoids, and anthocyanins) from various plant wastes such as orange carrot, tomato peel, green apple peel, red cabbage, and lemon peel via HRW-E procedure (Alwazeer et al., 2023). HRW is synthesized by adding clean H_2 (99.99%) into pure water.

2.3.2 Innovative Extraction Methods

The huge solvent used, lower quantity of yield, long extraction period, as well as low-quality product issues hamper the extraction efficacy in the traditional solvent extraction method. So, to overcome these problems scientists have developed ecological, economical, and innovative extraction methods to effectively obtain phytochemicals (Chemat et al., 2019). Among these innovative methods are:

2.3.2.1 Ultrasound-Assisted Extraction (UAE)

UAE was extensively used to isolate different natural biologically active compounds because of its simplicity, less time required, use of less solvent than other approaches and also raise both the value and efficacy of the yields. Devi's research group isolated phytochemicals (8.24%) from bottle gourd seed (Lagenaria siceraria) by ultrasonic extraction using a 70% ultrasound amplitude at 20 kHz frequency (Devi et al., 2023). Sharma et al. (2023) reported that phytochemicals were isolated from Lagenaria siceraria peel using ultrasound by ultrasonicator at 40 to 45 °C temperature. This extraction yielded 9.85% phytochemicals. Piechowiak's research group found flavonoids and other phytochemicals from Pinus sylvestris L. by ultrasound extraction procedure using an ultrasonic homogeniser at 20 kHz frequency with a specific time period (Piechowiak et al., 2023). In another study, anthocyanin was isolated and used in the UAE method for Arbutus unedo L. fruits using an ultrasonic device (QSonica Sonicators) by mixing 1.5 g compound and 50 mL of acidic EtOH tested ranges and variables such as 100 to 500 W ultrasound power, time (3 to 45 min), and temperature (30–35 °C) (López et al., 2018). Backes Research Group described that anthocyanin was extracted from ficus peel by three separate extraction procedures (heat, ultrasound, and microwave) (Backes et al., 2018). The UAE method was proceeded by mixing of 2.5 g lyophilized sample and 50 mL acidified solvent where the ultrasound power range was 100–400 W. This was the active extraction technique of this learning yielding 3.82 mg C/g extracted residue. Total phenolic and flavonoid compounds were isolated from Inga marginata Willd fruits with UAE procedures (Flores et al., 2023).

2.3.2.2 Microwave Assisted Extraction (MAE)

MAE method was a simple, fiscal as well as fast technique for the isolation of phytochemicals (Hiranvarachat and Devahastin, 2014). This method involves microwave energy transfer to heat. These procedures have many advantages like less time process, solvent consumption, and increased productivity of the products. Backes Research Group reported that anthocyanin was isolated from Ficus L. peel by the MAE method (Backes et al., 2018). This method was executed in a Biotage Initiator Microwave by mixing 0.5 g peel samples with 10 mL acidified solvent in a closed vessel where the microwave power was set at 400 W. Phytochemical compounds were extracted by MAE method from the root of *Paramignya trimera* (Nguyen et al., 2015). Maximum phytochemical materials such as p-coumaric acid, gallic acid, syringic acid, caffeic acid, kaempferol, and rutin were produced in these methods.

Phytochemical compounds (polyphenols and fucoxanthin) were isolated from A. nodosum using the MAE method where the optimal conditions are time (3 min), pressure (10.4 bar), and solvent (46.8 % ethanol) (Cassani et al., 2024). Bioactive phytochemical compounds were extracted from Inga marginata Willd fruits using MAE methods (Flores et al., 2023). Monomeric anthocyanin was isolated from Lavandula pedunculata L. fresh plant with MAE techniques using an ordinary microwave oven (Farzaneh and Carvalho, 2017). Here the experiment was done by three different parameters such as liquid/compound ratios (20, 30, and 40 mL/g), microwave power (300, 450, 600 W) and times (60, 120, 180 s). The maximum anthocyanins were found when the liquid/compound ratio, microwave power and time were 30.3 mL/g, 464.9 W, and 114.3 s, respectively.

2.3.2.3 Accelerated Solvent Extraction (ASE)

The ASE method similarly known as the pressurised liquid extraction (PLE) technique and executed in a huge pressure and temperature for a short period (Viganó et al., 2016). Phenolic and flavonoids were extracted from three plants rosemary, oregano, and marjoram by this method extraction (Hossain et al., 2011). Garcia-Mendoza's research group described that anthocyanins, as well as phenolic samples, were extracted from juçara residues using PLE method (Garcia-Mendoza et al., 2017). PLE was carried out at 10 MPa high pressure and different temperatures (40, 60, and 80 °C) using four different solvents such as 99.5% ethanol, distilled water, acidified water and also ethanol and water acidified mixture. The highest phenolic compounds were found in ethanol and water acidified mixture at 80 °C and at 40 °C, the maximum anthocyanin was found in acidified water. In another study, PLE was executed using 15 g Theobroma cacao L. hulls with 99.8% EtOH at 70 °C temperature and 10 MPa pressure for 20 minutes (Mazzutti et al., 2018). The highest phenolic and antioxidant compounds were found in this condition. Saha's research group used ASE techniques for the carotenoid isolation from orange carrots at 40, 50, and 60 °C temperatures, 5, 10, and 15 minutes and at 1500 psi pressure (Saha et al., 2015). Three different solvent mixtures were used in this experiment such as ethanol: hexane; acetone: hexane; and ethanol: hexane: acetone. Significant amounts of carotenoids were found at optimised conditions when the temperature was at 60 °C and time 15 minutes.

2.3.2.4 High Hydrostatic Pressure Extraction (HHPE)

Polyphenols were isolated from the plants using the HHPE method. Ma's research group extracted pepper seed oil (PSO) using the HHPE method using an HHP-700 hydrostatic pressure (Ma et al., 2019). The maximum (83.0%) oil isolation efficacy was found at optimal conditions such as pressure (370 MPa), temperature (50 °C), and time (5.7 min). PSO by HHPE exhibited a significantly high amount of γ-tocopherol, and fatty acids compared to conventional methods and UAE methods. In another study, ginsenoside was extracted from ginseng roots using the HHPE method. Entire ginsenosides and ginsenoside metabolite maximum found at optimum pressure reached at 80 MPa.

2.3.2.5 Subcritical Fluid Extraction (SbFE)

SbFE, similarly known as warm water isolation and high-temperature water isolation, was an innovative environment-friendly method for the isolation of phytochemicals from different pants. In this method H_2O is used as a solvent at 100–374 °C temperatures and 10–60 bar pressure under the critical point of H_2O (Yahya et al., 2018). Luo's research group reported that polyphenolic compounds were extracted from sorghum bran using the SbFE methodology (Luo et al., 2018). The maximum polyphenolic compounds were found at the optimised conditions such as temperature (144.5 °C), period (21 min), and solid-liquid fraction (35 mL/g). Vladić et al. (2017) reported that total phenol and flavonoid compounds were isolated using a subcritical water extraction method by mixing a 5.0 g Satureja montana sample with 50 mL of water at various temperatures and isolation time with constant pressure. The whole phenols and flavonoids are obtained at effective isolation conditions such as temperature (200 °C) and isolation time (20.8 min). In another study, polyphenolic compounds were extracted from A. uva-ursi herbal powder using the SbFE process. In this process temperature, and time were independent variables, whereas the other parameters were constant. The predicted whole phenolic materials and flavonoids originated at the optimal SbFE conditions at 151.2 °C temperature, and 10 minutes extraction period. Nastić's research group extracted the highest total phenolic compound from Teucrium montanum L. by the SbFE method at 160 °C temperature and 10 bar pressure (Nastić et al., 2018).

2.3.2.6 Supercritical Fluid Extraction (SFE)

SFE is more efficient than conventional methods due to the use of supercritical solvents (Roselló-Soto et al., 2019). CO_2 was commonly used as a supercritical solvent in this technique while various supercritical solvents are present such as pentane, NO, ammonia, CHF_3, butane, C_2H_6, and H_2O, owing to its little viscosity and quite large diffusivity (Saini and Keum, 2018). Other advantages, compared to other methods, SFE have a safe, less extraction time, higher effectiveness, increased safety, and yields (Herrero et al., 2006). Alwazeer et al. (2023) reported that the different phytochemicals were extracted from many plant wastes by using a supercritical CO_2 fluid extraction procedure in a SFE machine. Garcia-Mendoza research group (2017) reported that significant anthocyanin was isolated from juçara materials using the SFE method where the carbon dioxide flow proportion is 2.08×10^{-4} kgs^{-1}. Here, an acid-ethanol and water mixture were used as solvent and pressure (20 MPa) as well as temperature (60 °C). In another SFE study, phenolic and antioxidant materials were isolated from cocoa beans (Theobroma cacao L.) using 99.9% carbon dioxide (Mazzutti et al., 2018). The SFE procedure was done in three different temperatures such as 40, 50, and 60 °C, as well as pressures (20 and 30 MPa) with continuous liquid flow rate (11 g CO_2 min^{-1}). In these experiments, maximum phenolic and antioxidant compounds were found when temperatures and pressures were 40 °C and 20 MPa, respectively. Sanjaya et al. (2014) extracted phenolic and phytochemicals compounds from a Myrmecodia pendans plant by using supercritical carbon dioxide extraction techniques at pressures (9–22.5 MPa) and temperatures (313.15–343.15 K). Catechin, gallic acid, luteolin, as well as kaempferol, were isolated in this investigation.

2.3.2.7 Enzyme Assisted Extraction (EAE)

EAE of phytochemicals from plant materials was a significant substitute for the old general techniques (Choudhari and Ananthanarayan, 2007). Enzymes are biocompatible, biodegradable Nature's sustainable catalysts to carry out the isolation, reform, or synthesis of biologically active phytochemical materials from natural sources (Puri et al., 2012). Enzymes such as α-amylase, cellulose, and pectinase have the capability to destroy or disrupt the cell structure and membranes to improve the extraction and overall yield (Patindol et al., 2007; Pinelo et al., 2006). Swer's research group isolated anthocyanins from Prunus nepalensis fruit using the EAE method under varied times and temperatures (Swer et al., 2016). The highest yield of anthocyanins was found using cellulase action at 40 °C and 180 minutes whereas 14.61% lower yield was found in the conventional process. Boulila's research group described that the enzyme action increases the extraction efficacy of important oils from Laurus L. leaves (Boulila et al., 2015). The yields of essential oil increased by 243, 227, 240.54, and 0.48% when the samples were mixed with cellulase, hemicellulase, xylanase, and the three-enzyme combination (1:1:1), respectively. Bioactive compounds like α-pinene, α-terpinyl acetate, cineole, sabinene, and β-pinene were present in all essential oils. Pectinase and cellulose enzymes were used to isolate the maximum whole carotenoid (127 mg/kg) and lycopene (89.4 mg/kg) from tomatoes (Strati et al., 2015). Researchers extracted anthocyanins from saffron tepals (Crocus sativus) using different enzymes such as cellulase, hemicellulase, and pectinase at 40 °C and different times (20, 40, 60, 120, and 180 min) and concentrations (1%, 2.5%, 5%, 7%, and 10%) (Lotfi et al., 2015). Carotenoid, chlorophyll, tocopherol, and polyphenol compounds were obtained from industrial tomato waste by EAE using cellulolytic and pectinolytic enzymes (Catalkaya and Kahveci, 2019).

2.3.2.8 Instant Controlled Pressure Drop Extraction (ICPDE)

Allaf and Vidal (1989) first introduced the ICPDE method for the isolation of several biomolecules and also improved this method. In this method, high temperature and a rapid pressure drop over a short time ensure an increase in the quality of extracts (Rombaut et al., 2014). ICPDE is a useful modern method to isolate the major biologically active materials like essential oil, polyphenols, and antioxidants (Berka-Zougali et al., 2010; Sánchez-Valdepenas et al., 2015). Its other merits are reduced isolation time and solvent consumption. Sánchez-Valdepenas et al. (2015) described the phenolic materials extracted in grape stalk powder by using the ICPDE method. Abi-Khattar's research group isolated polyphenol compounds from olive leaves using this method (Abi-Khattar et al., 2021). This extraction process increased the whole phenolic compound by nearly three times more than negative controls. Also, oleuropein increased to 600% and hydroxytyrosol increased to 238%. (Mkaouar et al., 2015) also reported polyphenol compounds such as oleuropein, luteolin-7-glucoside, and vanillic acid were extracted from O. europaea leaf by the ICPDE method.

2.3.2.9 Microextraction Techniques

Microextraction methods were innovative extraction methods that reduced sample action stages, minimum solvent, and sample use as well as less waste generation. These methods have enormous applications in different fields due to being environment friendly, avoiding hazardous reagents, energy saving, cost-effectiveness, and high extraction efficacies (Casado et al., 2020). Many microextraction methods were utilized for the isolation of biologically active phytochemical materials from various plants used in sorbent-based adsorption procedures, such as MEPS, μ-MSPD, d-MSPE, SPME, etc. Porto-Figueira et al. (2015) reported that a rapid and advanced microextraction procedure, μSPEed, was used to extract the most abundant phenolic compounds such as rutin and quercetin 3-glucoside in teas. Musteata's research group extracted non-volatile phytochemicals using solid-phase microextraction from twenty Amazonian plant species (Musteata et al., 2016). Another study reported that two flavonoid compounds (neohesperidin and naringin) were extracted from olive fruit via a miniaturized MSPD technique using an ionic solvent as elution (Xu et al., 2016).

2.3.2.10 Pulsed Electric Field Extraction (PEFE)

PEFE was a non-thermal emerging procedure mainly used to isolate valuable bioactive materials, such as carotenoids, polyphenols, and flavonoids (Chatzimitakos et al., 2023). This technology increases the extraction rate in a shorter time approach as well as minimizing solvent and energy consumption (Mohamed and Amer Eiss, 2012; Yan et al., 2017). Moradi and Rahimi (2018) reported that PEFE action (1.1 kV/cm) improved the oil isolation yield by 12.7% compared to hexane solvent isolation from sunflower seeds. Kantar et al. (2018) isolated polyphenol materials from citrus fruits via PEF treatment.

2.3.2.11 Infrared Assisted Extraction (IAE)

The IAE is a globally-friendly advanced effective process for the extraction of bioactive phytochemicals from beneficial plants (Escobedo et al., 2016). This recently developed technique has several benefits such as reduced time period and high isolation efficiencies due to maximum cellular and structural damage to plant materials. The disadvantage of this method is the specific and expensive equipment used. Duan et al. (2010) reported that rutin, gentisic acid, and quercetin bioactive constituents were extracted by this method from leaves of Lycium barbarum Linn. The IAE method reduces isolation time to six minutes compared to three hours for the HSE technique. Abi-Khattar et al. (2019) reported that intensified polyphenol compounds were extracted from olive leaves using the IAE method. This green technology increases 30% the entire phenolic materials compared to the water bath isolation method. Another study reported that polyphenol compounds were found in pomegranate peels by using this technique (Rajha et al., 2019).

2.4 CONCLUSION AND FUTURE REMARKS

Researchers are interested in the isolation of biologically active phytochemicals from different parts of plants is to harvest their valuable uses. These specialized extraction methods with innovative stationary and mobile phases are used to optimize the concentration of biologically active constituents in the extract. These modern extraction proceedings have shown much-improved retrieval with decreased time and usage of solvent. These advanced techniques will be helpful for increasing expectations among consumers for safe, environment-friendly, and as per supervisory guidelines. With continuously growing awareness and numerous encouraging phytochemical drugs and cosmetics in recent progression which are from natural resources, and with continuous progressive technical abilities associated with phytochemical study. There are some improved prospects to discover the significant biological activity of earlier unreachable natural product resources.

Furthermore, the gradual rise in reception of the chemical miscellany of phytochemicals is well-matched to deliver the fundamental frameworks for upcoming drugs and cosmetics. In future, there will be additional advancement in the utilization of innovative phytochemicals and chemical libraries founded on drug discovery related to natural products and skincare crusades.

REFERENCES

Abi-Khattar, A. M., Rajha, H. N., Abdel-Massih, R. M., et al. 2021. "Intensification of Vaporization by Decompression to the Vacuum" (IVDV), a novel technology applied as a pretreatment to improve polyphenols extraction from olive leaves. *Food Chemistry*, 342: 128236.

Abi-Khattar, A. M., Rajha, H. N., Abdel-Massih, R. M., Maroun, R. G., Louka, N., & Debs, E. 2019. Intensification of polyphenol extraction from olive leaves using IredIrrad®, an environmentally-friendly innovative technology. *Antioxidants*, 8(7): 227.

Allaf, K., & Vidal, P. 1989. Feasibility study of a new process of drying/swelling by instantaneous decompression towards vaccum of vegetables pieces in view of a rapid re-hydration. In: *Gradient Activity Plotting at Compiegne University of Technology UTC NCR/89/103*, Industrial SILVA-LAON Partner, Compiegne, France.

Alwazeer, D., Elnasanelkasim, M. A., Çiçek, S., Engin, T., Çigdem, A., & Karaogul, E. 2023. Comparative study of phytochemical extraction using hydrogen-rich water and supercritical fluid extraction methods. *Process Biochemistry*, 128: 218–226.

Amberg, N., & Fogarassy, C. 2019. Green consumer behavior in the cosmetics market. *Resources*, 8(3): 137.

Azwanida, N. N. 2015. A review on the extraction methods use in medicinal plants, principle, strength, and limitation. *Medicinal & Aromatic Plants*, 4: 196.

Babbar, N. 2015. An introduction to alkaloids and their applications in pharmaceutical chemistry. *The Pharma Innovation Journal*, 4(10): 74–75.

Backes, E., Pereira, C., Barros, L., Prieto, M., Genena, A. K., Barreiro, M. F., & Ferreira, I. C. (2018). Recovery of bioactive anthocyanin pigments from Ficus carica L. peel by heat, microwave, and ultrasound based extraction techniques. *Food Research International*, 113: 197–209.

Berka-Zougali, B., Hassani, A., Besombes, C., & Allaf, K. 2010. Extraction of essential oils from Algerian myrtle leaves using instant controlled pressure drop technology. *Journal of Chromatography A*, 1217: 6134–6142.

Boulila, A., Hassen, I., Haouari, L., et al. 2015. Enzyme-assisted extraction of bioactive compounds from bay leaves (Laurus nobilis L.). *Industrial Crops and Products*, 74: 485–493.

Caleja, C., Barros, L., Prieto, M. A., Barreiro, F. M. F., Oliveira, M. B. P., & Ferreira, I. C. F. R. 2017. Extraction of rosmarinic acid from Melissa officinalis L. by heat-, microwave- and ultrasound-assisted extraction techniques: A comparative study through response surface analysis. *Separation and Purification Technology*, 186: 297–308.

Casado, N., Morante-Zarcero, S., Pérez-Quintanilla, D., et al. 2020. Two novel strategies in food sample preparation for the analysis of dietary polyphenols: Micro-extraction techniques and new silica-based sorbent materials. *Trends in Food Science and Technology*, 98: 167–180.

Cassani, L., Silva, A., Carpena, M., et al. 2024. Phytochemical compounds with promising biological activities from Ascophyllum nodosum extracts using microwave-assisted extraction. *Food Chemistry*, 438: 138037.

Catalkaya, G., & Kahveci, D. 2019. Optimization of enzyme assisted extraction of lycopene from industrial tomato waste. *Separation and Purification Technology*, 219: 55–63.

Chatzimitakos, T., Athanasiadis, V., Kalompatsios, D., Mantiniotou, M., Bozinou, E., & Lalas, S. I. 2023. Pulsed electric field applications for the extraction of bioactive compounds from food waste and by-products: A critical review. *Biomass*, 3(4): 367–401.

Chemat, F., Abert-Vian, M., Fabiano-Tixier, A. S., Strube, J., Uhlenbrock, L., Gunjevic, V., & Cravotto, G. 2019. Green extraction of natural products. Origins, current status, and future challenges. *TrAC Trends in Analytical Chemistry*, 118: 248–263.

Choudhari, S. M., & Ananthanarayan, L. 2007. Enzyme aided extraction of lycopene from tomato tissues. *Food Chemistry*, 102: 77–81.

Devi, S. R., Kumari, T., & Deka, S. C. 2023. Extraction of dietary fiber and phytochemicals from bottle gourd seeds (Lagenaria siceraria), its physicochemical properties and application in food model. *Food Chemistry Advances*, 2: 100252.

Doughari, J. H. 2012. Extraction methods, basic structures, and mode of action as potential chemotherapeutic agents, phytochemicals – a global perspective of their role in nutrition and health. In: *A Global Perspective of Their Role in Nutrition and Health. Phytochemicals*, Venketeshwer, R., Ed. InTech. www.intechopen.com.

Duan, H. T., Chen, Y., & Chen, G. 2010. Far infrared-assisted extraction followed by capillary electrophoresis for the determination of bioactive constituents in the leaves of Lycium barbarum Linn. *Journal of Chromatography A*, 1217: 4511–4516.

Escobedo, R., Miranda, R., & Martínez, J. 2016. Infrared irradiation: Toward green chemistry, a review. *International Journal of Molecular Sciences*, 17: 453.

Fan, R., Yuan, F., Wang, N., Gao, Y., & Huang, Y. 2015. Extraction and analysis of antioxidant compounds from the residues of Asparagus officinalis L. *Journal of Food Science & Technology*, 52: 2690–2700.

Farzaneh, V., & Carvalho, I. S. 2017. Modelling of Microwave Assisted Extraction (MAE) of anthocyanins (TMA). *Journal of Applied Research on Medicinal and Aromatic Plants*, 6: 92–100.

Ferrazzano, G. F., Amato, I., Ingenito, A., Zarrelli, A., Pinto, G., & Pollio, A. 2011. Plant polyphenols and their anti-cariogenic properties: A review. *Molecules*, 16: 1486–1507.

Flores, D. C. B., Boeira, C. P., Weis, G. C. C., et al. 2023. Extraction of antioxidant and antimicrobial compounds from Inga marginata Willd bark and pulp using different extraction techniques and phytochemical characterization. *Innovative Food Science and Emerging Technologies*, 83: 103244.

Gadkari, P. V., Kadimi, U. S., & Balaraman, M. 2014. Catechin concentrates of garden tea leaves (Camellia sinensis L.): Extraction/isolation and evaluation of chemical composition. *Journal of the Science of Food and Agriculture*, 94: 2921–2928.

Garcia-Mendozaa, M. D. P., Espinosa-Pardoa, F. A., Baseggio, A. M., et al. 2017. Extraction of phenolic compounds and anthocyanins from jucara (Euterpe edulis Mart.) residues using pressurized liquids and supercritical fluids. *The Journal of Supercritical Fluids*, 119: 9–16.

Hadjzadeh, M., Tavakol Afshari, J., Ghorbani, A., & Shakeri, M. 2006. The effects of aqueous extract of Garlic (*Allium sativum* L.) on laryngeal cancer cells (Hep-2) and L929 cells in vitro. *JMP*, 2(18): 41–48.

Herrero, M., Cifuentes, A., & Ibañez, E. 2006. Sub- and supercritical fluid extraction of functional ingredients from different natural sources: Plants, food-by-products, algae and microalgae. *Food Chemistry*, 98: 136–148.

Hiranvarachat, B., & Devahastin, S. 2014. Enhancement of microwave-assisted extraction via intermittent radiation: Extraction of carotenoids from carrot peels. *Journal of Food Engineering*, 126: 17–26.

Hossain, M. B., Barry-Ryan, C., Martin-Diana, A. B., & Brunton, N. P. 2011. Optimisation of accelerated solvent extraction of antioxidant compounds from rosemary (Rosmarinus officinalis L.), marjoram (Origanum majorana L.) and oregano (Origanum vulgare L.) using response surface methodology. *Food Chemistry*, 126: 339–346.

Ingle, K. P., Deshmukh, A. G., Padole, D. A., Dudhare, M. S., Moharil, M. P., & Khelurkar, V. C. 2017. Phytochemicals: Extraction methods, identification, and detection of bioactive compounds from plant extracts. *Journal of Pharmacognosy & Phytochemistry*, 6(32): 6.

Ivanović, M., Islamčević Razboršek, M., & Kolar, M. 2020. Innovative extraction techniques using deep eutectic solvents and analytical methods for the isolation and characterization of natural bioactive compounds from plant material. *Plants*, 9: 1428.

Kantar, S. E., Boussetta, N., Lebovka, N., Foucart, F., Rajha, H. N., Maroun, R. G., . . . Vorobiev, E. 2018. Pulsed electric field treatment of citrus fruits: Improvement of juice and polyphenols extraction. *Innovative Food Science and Emerging Technologies*, 46: 153–161.

López, C. J., Caleja, C., Prieto, M., Barreiro, M. F., Barros, L., & Ferreira, I. C. 2018. Optimization and comparison of heat and ultrasound assisted extraction techniques to obtain anthocyanin compounds from Arbutus unedo L. fruits. *Food Chemistry*, 264: 81–91.

Lotfi, L., Kalbasi-Ashtari, A., Hamedi, M., & Ghorbani, F. 2015. Effects of enzymatic extraction on anthocyanins yield of saffron tepals (Crocos sativus) along with its color properties and structural stability. *Journal of Food and Drug Analysis*, 23(2): 210–218.

Luo, X., Cui, J., Zhang, H., & Duan, Y. 2018. Subcritical water extraction of polyphenolic compounds from sorghum (Sorghum bicolor L.) bran and their biological activities. *Food Chemistry*, 262: 14–20.

Ma, Y., Wu, X., Zhao, L., Wang, Y., & Liao, X. 2019. Comparison of the compounds and characteristics of pepper seed oil by pressure-assisted, ultrasound-assisted and conventional solvent extraction. *Innovative Food Science & Emerging Technologies*, 54: 78–86.

Mazzutti, S., Rodriguesa, L. G. G., Mezzomo, N., et al. 2018. Integrated green-based processes using supercritical CO2 and pressurized ethanol applied to recover antioxidant compounds from cocoa (Theobroma cacao) bean hulls. *The Journal of Supercritical Fluids*, 135: 52–59.

Mkaouar, S., Bahloul, N., Gelicus, A., Allaf, K., & Kechaou, N. 2015. Instant controlled pressure drop texturing for intensifying ethanol solvent extraction of polyphenols. *Separation and Purification Technology*, 145: 139–146.

Mohamed, M. E. A., & Amer Eiss, A. H. 2012. Pulsed electric fields for food processing technology. In: *Structure and Function of Food Engineering*, Amer Eissa, A., Ed. InTech. ISBN 978-953-51-0695-1.

Moongngarm, A., Sriharboot, N., Loypimai, P., & Moontree, T. 2022. Ohmic heating-assisted water extraction of steviol glycosides and phytochemicals from Stevia rebaudiana leaves. *LWT*, 154: 11279.

Moradi, N., & Rahimi, M. 2018. Effect of simultaneous ultrasound/pulsed electric field pretreatments on the oil extraction from sunflower seeds. *Separation Science and Technology*, 53: 2088–2099.

Musteata, F. M., Sandoval, M., Ruiz-Macedo, J. C., et al. 2016. Evaluation of in vivo solid phase microextraction for minimally invasive analysis of nonvolatile phytochemicals in Amazonian plants. *Analytica Chimica Acta*, 933: 124–133.

Nastić, N., Švarc-Gajić, J., Delerue-Matos, C., et al. 2018. Subcritical water extraction of antioxidants from mountain germander (Teucrium montanum L.). *The Journal of Supercritical Fluids*, 138: 200–206.

Nguyen, V. T., Bowyer, M. C., Vuong, Q. V., Altena, I. A. V., & Scarlett, C. J. 2015. Phytochemicals and antioxidant capacity of Xao tam phan (Paramignya trimera) root as affected by various solvents and extraction methods. *Industrial Crops and Products*, 67: 192–200.

Ntemafack, A., Ayoub, M., Hassan, Q. P., & Gandhi, S. G. 2023. A systematic review of pharmacological potential of phytochemicals from Rumex abyssinicus Jacq. *South African Journal of Botany*, 154: 11–25.

Pandey, A., & Tripathi, S. 2014. Concept of standardization, extraction, and pre-phytochemical screening strategies for herbal drug. *Journal of Pharmacognosy & Phytochemistry*, 2: 115–119.

Patindol, J., et al. 2007. Cellulase-assisted extraction of oligosaccharides from defatted rice bran. *Journal of Food Science*, 72: C517–C521.

Piechowiak, T., Matłok, N., & Balawejder, M. 2023. Extraction of phytochemicals from young shoots of Pinus sylvestris L. and analysis of their selected biological properties. *LWT*, 188: 115404.

Pinelo, M., et al. 2006. Upgrading of grape skins: Significance of plant cell-wall structural components and extraction techniques for phenol release. *Trends in Food Science and Technology*, 17: 579–590.

Porto-Figueira, P., Figueira, J. A., Pereira, J. A. M., & Câmara, J. S. 2015. A fast and innovative microextraction technique, µSPEed, followed by ultrahigh performance liquid chromatography for the analysis of phenolic compounds in teas. *Journal of Chromatography A*, 1424: 1–9.

Puri, M., Sharma, D., & Barrow, C. J. 2012. Enzyme-assisted extraction of bioactives from plants. *Trends in Biotechnology*, 30: 37–44.

Rajha, H. N., Mhanna, T., El Kantar, S., El Khoury, A., Louka, N., & Maroun, R. G. 2019. Innovative process of polyphenol recovery from pomegranate peels by combining green deep eutectic solvents and a new infrared technology. *LWT*, 111: 138–146.

Ramawat, K. G., Dass, S., & Mathur, M. 2009. The chemical diversity of bioactive molecules and Therapeutic potential of medicinal plants. In: *Herbal Drugs: Ethnomedicine to Modern Medicine*. Springer-Verlag Berlin Heidelberg.

Rombaut, N., Tixier, A. S., Bily, A., & Chemat, F. 2014. Green extraction processes of natural products as tools for biorefinery. *Biofuels, Bioproducts and Biorefining*, 8: 530–544.

Roselló-Soto, E., Barba, F. J., Lorenzo, J. M., Dominguez, R., Pateiro, M., Mañes, J., et al. 2019. Evaluating the impact of supercritical-CO2 pressure on the recovery and quality of oil from "horchata" by-products: Fatty acid profile, α-tocopherol, phenolic compounds, and lipid oxidation parameters. *Food Research International*, 120: 888–894.

Saha, S., Walia, S., Kundu, A., Sharma, K., & Paul, R. K. 2015. Optimal extraction and fingerprinting of carotenoids by accelerated solvent extraction and liquid chromatography with tandem mass spectrometry. *Food Chemistry*, 177: 369–375.

Sahebi, M., Hanafi, M. M., van Wijnen, A. J., Akmar, A. S. N., Azizi, P., Idris, A. S., Taheri, S., & Foroughi, M. 2017. Profiling secondary metabolites of plant defence mechanisms and oil palm in response to Ganoderma boninense attack. *International Biodeterioration & Biodegradation*, 122: 151–164.

Saini, R. K., & Keum, Y.-S. 2018. Carotenoid extraction methods: A review of recent developments. *Food Chemistry*, 240: 90–103.

Sánchez-Valdepeñas, V., Barrajón, E., Vegara, S., Funes, L., Martí, N., Valero, M., & Saura, D. 2015. Effect of instant controlled pressure drop (DIC) pre-treatment on conventional solvent extraction of phenolic compounds from grape stalk powder. *Industrial Crops and Products*, 76: 545–549.

Sanjaya, R. E., Tedjo, Y. Y., Kurniawan, A., et al. 2014 Investigation on supercritical CO2 extraction of phenolicphytochemicals from an epiphytic plant tuber (Myrmecodia pendans). *Journal of CO2 Utilization*, 6: 26–33.

Saraswathy, A., Meena, A. K., Shakila, R., Sunil Kumar, K. N., & Ariyanathan, S. 2010. Pharmacognostic studies on *Alangiums alvifolium* (Linn.f.) Wang. root bark. *Phcog.Net*, 2(11).

Sharma, S., Kumari, T., Choudhury, N., & Deka, S. C. 2023. Extraction of dietary fiber and encapsulated phytochemical enriched functional pasta from Bottle gourd (Lagenaria siceraria) peel waste. *Food Chemistry Advances*, 3: 100492.

Strati, I. F., Gogou, E., & Oreopoulou, V. 2015. Enzyme and high pressure assisted extraction of carotenoids from tomato waste. *Food and Bioproducts Processing*, 94: 668–674.

Strzępek-Gomółka, M., Gaweł-Bęben, K., & Kukula-Koch, W. 2021. Achillea species as sources of active phytochemicals for dermatological and cosmetic applications. *Oxidative Medicine and Cellular Longevity*, Article ID 6643827: 14.

Swer, T. L., Chauhan, K., Paul, P. K., & Mukhim, C. 2016. Evaluation of enzyme treatment conditions on extraction of anthocyanins from Prunus nepalensis L. *International Journal of Biological Macromolecules*, 92: 867–871.

Venugopal, R., & Liu, R. H. 2012. Phytochemicals in diets for breast cancer prevention: The importance of resveratrol and ursolic acid. *Food Science and Human Wellness*, 1: 1–13.

Viganó, J., Aguiar, A. C., Moraes, D. R., Jara, J. L. P., Eberlin, M. N., Cazarin, C. B. B., et al. 2016. Sequential high pressure extractions applied to recover piceatannol and scirpusin B from passion fruit bagasse. *Food Research International*, 85: 51–58.

Vladić, J., Canli, O., Pavlić, B., Zekovic, Z., et al. 2017. Optimization of Satureja montana subcritical water extraction process and chemical characterization of volatile fraction of extracts. *The Journal of Supercritical Fluids*, 120: 86–94.

Xu, J. J., Yang, R., Ye, L. H., et al. 2016. Application of ionic liquids for elution of bioactive flavonoid glycosides from lime fruit by miniaturized matrix solid-phase dispersion. *Food Chemistry*, 204: 167–175.

Yahya, N. A., Attan, N., & Wahab, R. A. 2018. An overview of cosmeceutically relevant plant extracts and strategies for extraction of plant-based bioactive compounds. *Food and Bioproducts Processing*, 112: 69–85.

Yan, L.-G., He, L., & Xi, J. 2017. High intensity pulsed electric field as an innovative technique for extraction of bioactive compounds – A review. *Critical Reviews in Food Science and Nutrition*, 57: 2877–2888.

Zainol, N., Aziz, N. H., & Baharudin, A. S. 2023. Influence of agitation and solvent percentage on the extraction of phytochemical compound from Asystasia gangetica. *Food Chemistry Advances*, 3: 100538.

Zekovic, Z., Pintac, D., Majkic, T., Vidovic, S., Mimica-Dukic, S., Teslic, N., et al. 2017. Utilization of sage by-products as raw material for antioxidants recovery – ultrasound versus microwave-assisted extraction. *Industrial Crops and Products*, 99: 49–59.

Zengin, G., Sarikurkcu, C., Aktumsek, A., Ceylan, R., & Ceylan, O. 2014. A comprehensive study on phytochemical characterization of Haplophyllum myrtifolium Boiss. Endemic to Turkey and its inhibitory potential against key enzymes involved in Alzheimer, skin diseases and type II diabetes. *Industrial Crops and Products*, 53: 244–251.

3 Phytochemicals for Photocarcinogenesis Prevention

Priyabrata Mitra and Amitava Mandal

3.1 INTRODUCTION

Phytochemicals are biologically active chemicals constructed by plants and help in the protection of cells from environmental hazards like pathogenic attack, pollution, UV exposure, etc. Phytochemicals are a wide variety of compounds that occur naturally in plants. They are also referred to as secondary plant metabolites which are also advantageous for human health. Such chemicals are thought to act as synergistic agents and allow nutrients to be used more efficiently by the body. There are some advantageous biological roles of phytochemicals due to low toxicity, easy accessibility, and biological activities of the chemicals like antioxidants, antimicrobial effects, minimization of platelet accumulation, regulation of detoxification enzymes, and incentive of the immune system, etc. They have constructive effects on human health and have significant health benefits. Such chemicals are of great interest and have a momentous antioxidant potential. There are two different subcategories of natural antioxidants, namely *in vitro* antioxidants and *in vivo* antioxidants. In accordance with epidemiological research, intake of grains, fruits, and green vegetables on a daily basis may reduce the risk of growing a number of diseases that are related to oxidative damage.

Nowadays chronic diseases (CDs) like cancers, cardiovascular diseases, diabetes, etc. are global health issues and cause death. Cancer is a disease that involves the abnormal expansion of malignant cells inside our body. The largest organ in the body, the skin, serves as an external defence against the elements. Sunlight or ultraviolet (UV-emitting) tanning devices can expose skin to UV radiation, which can lead to skin cancer. Changes in the genes that control how the body normally functions are one of many elements that contribute to the development of cancer. There is growing interest in various cancer prevention strategies as a result of the steadily rising cancer incidence in the world as well as the growing issues with drug resistance. Herbal prescriptions have drawn a lot of attention in recent years as an alternative strategy to make up for orthodox pharmacotherapy's alleged shortcomings all over the world. Recently, the use of herbal medicine (collected from plants) has increased considerably though there is no proper medical evidence to support their mechanistic pathway, therapeutic efficiency, and toxicity. According to the World Health Organization (WHO) report in developing and underdeveloped countries, almost 80% of the population relies on traditional medicine practices and uses herbal medicine for their primary health treatment (WHO, 2017; Bode and Dong, 2015; Ng et al., 2018; George et al., 2021; Kotecha et al., 2016; Filippi et al., 2017; Farzaei et al., 2016). "Ayurveda" and "Siddha" are two medicinal practices practised by Indians from time immemorial. Chinese traditional medicines also claimed the cure for cancer or related diseases from ancient times. Medicinal chemistry research is ongoing to find the modern scientific base for these practices.

3.2 WHAT IS PHOTOCARCINOGENESIS?

Skin cancer is ultimately brought on by a series of synchronal and consecutive biochemical incidents collectively known as photocarcinogenesis. The photocarcinogenesis process is a complex multistep

DOI: 10.1201/9781003394846-3

process of tumour (malignant cell) growth and development which involves three distinct stages, the initiation stage, propagation stage and metastasis stage. The carcinogenesis process is influenced by the formation of DNA photoproducts such as the mutation of proto-oncogenes and tumour suppressor genes. UV exposure makes radical species have successive results on mutation and extra-nuclear action. The phenotypic influences on DNA photoproducts may consist of immunological effects, dietary factors, and antioxidant protection. Photocarcinogen is a substance which causes cancer when an organism is exposed to it and then illuminated. When combined with exposure to light, especially UV light, many chemicals that are not carcinogenic can become photocarcinogenic. The largest body organ, the skin, which sits at the point where the body and the environment converge, is directly harmed by genotoxic and xenobiotic agents, such as UV radiation. UV-radiation-induced skin cancer is a promptly increasing problem; more than 40% of human cancers in the United States are due to UV radiation (Ng et al., 2018; Kotecha et al., 2016; Farzaei et al., 2016; Buckman et al., 1998; Dreher and Junod, 1996).

3.3 CLASSIFICATION OF PHYTOCHEMICALS

Since ancient times, people have believed that certain biologically active compounds found in plants have therapeutic benefits for treating a variety of diseases, including cancer, hepatic, digestive, and respiratory issues, as well as skin, respiratory, and urinary complications. Phytochemicals are the name for these active substances and their activity. Due to the presence of chemical constituents in the plants that are beneficial for the physiological system of the human body. The medicinal value of such compounds provides an idea for the discovery and evolution of new bio-relevant pharmaceuticals. Secondary plant metabolites also known as phytochemicals can be divided into a number of groups according to their chemical structure, biosynthetic process, and solvent solubility (Bode and Dong, 2015; Ng et al., 2018; George et al., 2021; Kotecha et al., 2016; Filippi et al., 2017; Farzaei et al., 2016).

3.3.1 Alkaloid

Naturally occurring basic organic compounds with at least one nitrogen atom are referred to as alkaloids. Alkaloids are naturally occurring toxic amine-type compounds synthesized by plants. They mainly play a role in the defence mechanism of plants to protect them against herbivores. There are some toxic effects of alkaloids including interruption in the immune system, digestive system, reproductive system, respiratory system, and central nervous system (CNS). Few alkaloids are identified in lower organisms, whereas alkaloids are typically present in higher plants, especially in dicots.

Alkaloids can be found in a particular plant organ or throughout the entire plant body. Alkaloids that are synthesized from amino acids and have one or more than one carbon ring with hetero atom nitrogen and exhibit drug activity. Many indole derivatives have antihypertensive properties, as do quinidine and spareien derivatives. Alkaloids like quinine have antimalarial properties and some have anticancer properties like vincristine, dimeric indoles, and vinblastine (Figure 3.1).

Few alkaloids contain caffeine, nicotine, quinidine, morphine, etc. exhibiting stimulant properties and are used as an analgesic and quinine is used as an anti-malarial drug. Three skeletal structures of some alkaloids are given below (Han et al., 2018; Martino et al., 2018; Vincken et al., 2007).

3.3.2 Carotenoids

Carotenoids are found in a wide range of foods such as fruits and various vegetables. A variety of antioxidant actions are exhibited by carotenoids. Intake of foods containing carotenoids reduces the risk of cardiovascular disease, night blindness, cancer, etc. Green vegetables and fruits contain high amounts of oxygenated carotenoids (lutein, zeaxanthin) and hydrocarbon carotenoids (α-carotene

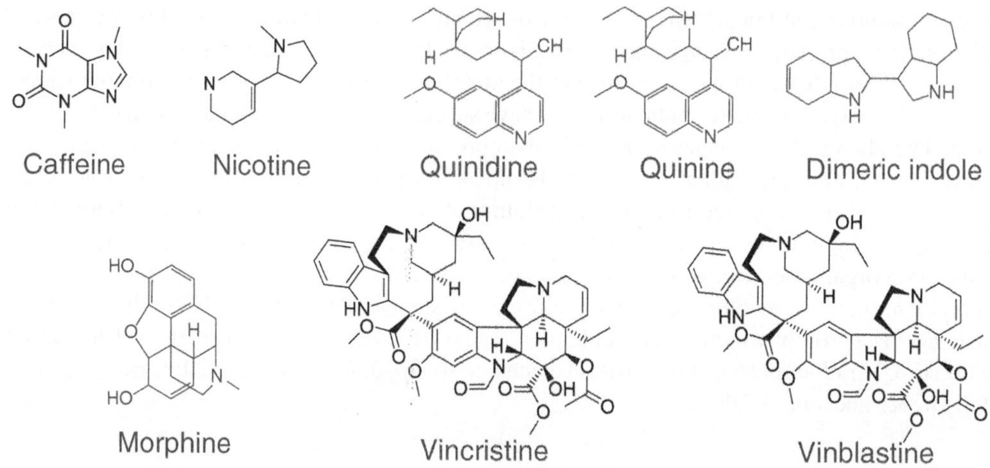

FIGURE 3.1 The chemical structure of the most important and popular alkaloids.

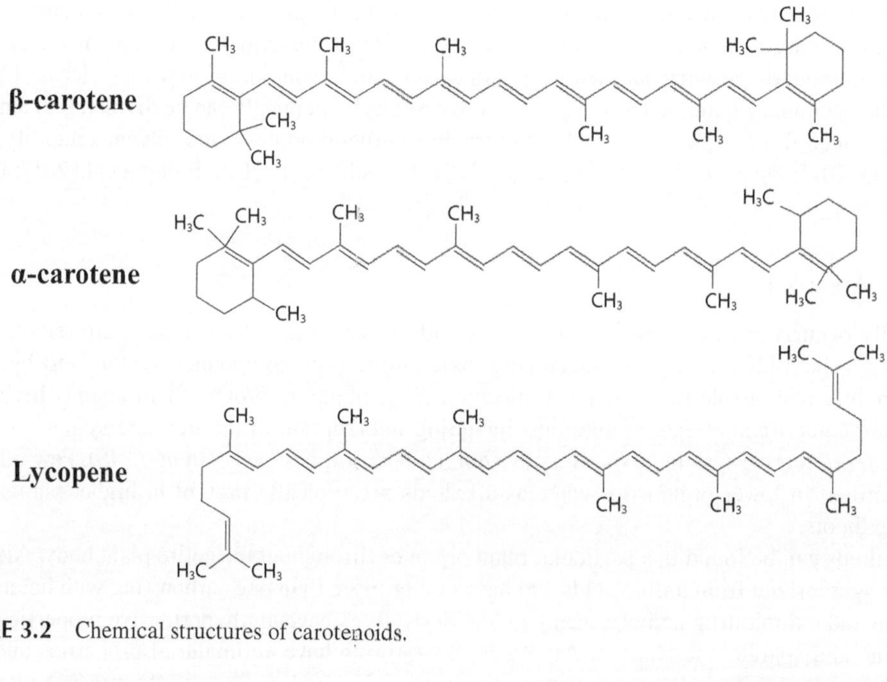

FIGURE 3.2 Chemical structures of carotenoids.

and β-carotene), yellow or orange colour vegetables (carrot) and fruits (papaya) contain a large amount of β-carotene, tomatoes, watermelon, and other red fruits are good source of lycopene (Figure 3.2) (Krinsky and Johnson, 2005; Bendich, 1988; Silva et al., 2021).

3.3.3 Phenols

Phenolic phytochemicals are the most abundant phytochemicals found in the plant kingdom. The three most significant phenolic phytochemical subgroups are flavonoids, phenolic acids, and polyphenols.

3.3.3.1 Flavonoids

The most researched and largest group of plant phenols are flavonoids. Flavonoids are a group of phenolic phytochemicals present in different parts of higher plants with a variety of applications in the medicinal field. Numerous health benefits of flavonoids include their antiviral, anticancer, antioxidant, and anti-inflammatory properties (Figure 3.3).

Flavonoids contain a benzopyrone ring with phenolic or polyphenolic groups linked at various positions of the benzopyrone ring. Flavonoids are divided into various types depending on different parameters such as composition, degree of active unsaturation, etc. (Figure 3.4). A large number of anticancer drugs are available on the market, but a very small number exhibit inhibition activity against oncogenesis, and most of them are toxic and have side effects. Most of the flavonoids behave as anticancer agents and prevent the cell growth. Colon cancer is a global problem nowadays.

Quercetin is a naturally occurring flavonoid exists in green tea, grains and different types of berries and is extensively used in colon cancer. Flavonoid-Modified drugs (FMDs) are extensively used in lung cancer (Han et al., 2018; Croft, 1998; Rice-Evans et al., 1996; McCullough et al., 2012; Sak, 2014; Rasulev et al., 2005; Castellano et al., 2013; Garcia-Lafuente et al., 2009).

3.3.3.2 Phenolic Acids

One of the major classes of phytochemicals that contain phenolic compounds is phenolic acids. Plant-based foods like fruits, vegetables, seeds, etc. contain a wide range of phenolic acids. Phenolic acids are frequently used phytochemicals that can be easily metabolized by natural microbes.

Recently, cancer is one of the major global health issues. Some particular antioxidants may be useful agents that are used to prevent cancer and radical formation inside our body. Consumption of

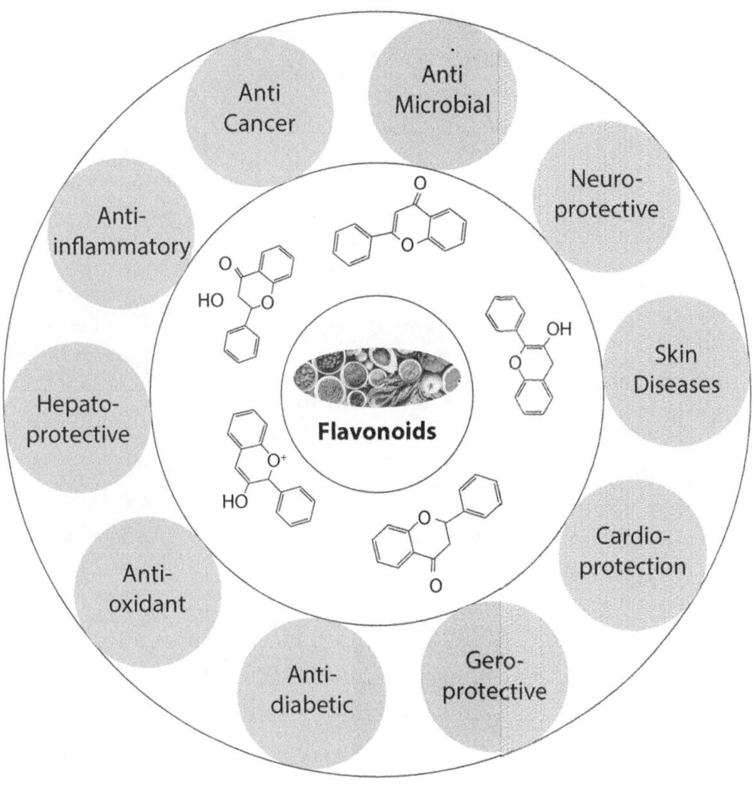

FIGURE 3.3 Some applications of flavonoids.

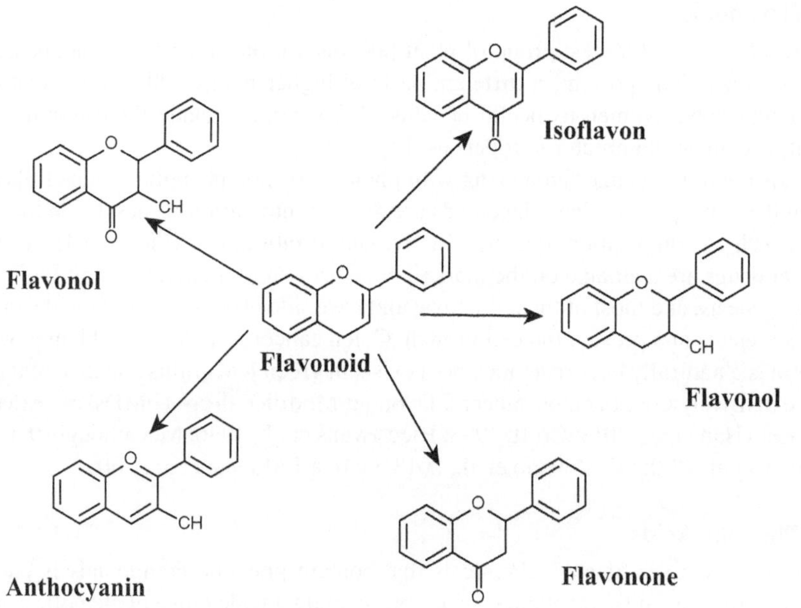

FIGURE 3.4 The main types of flavonoids and their chemical structure.

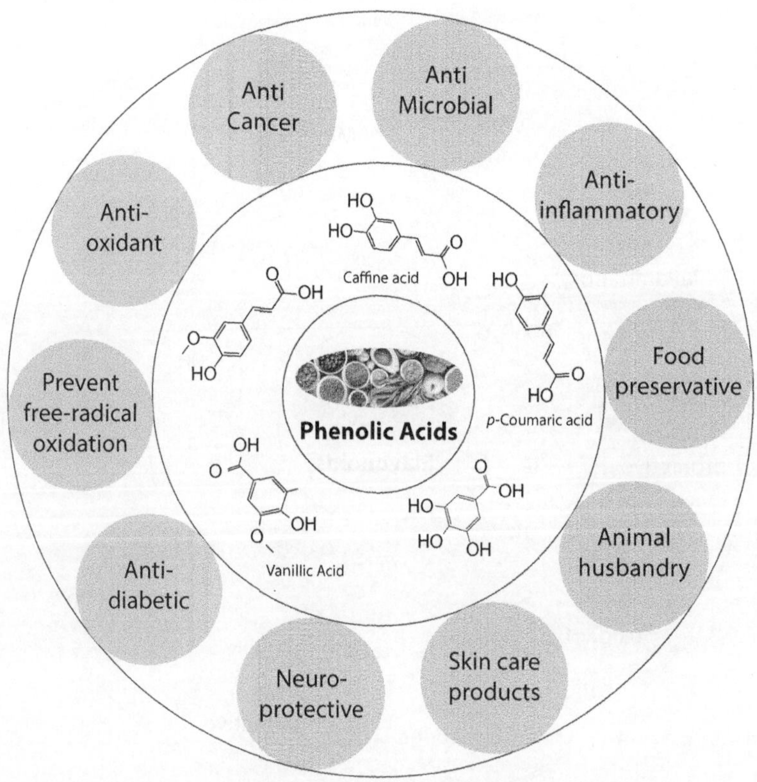

FIGURE 3.5 Some applications of phenolic acids.

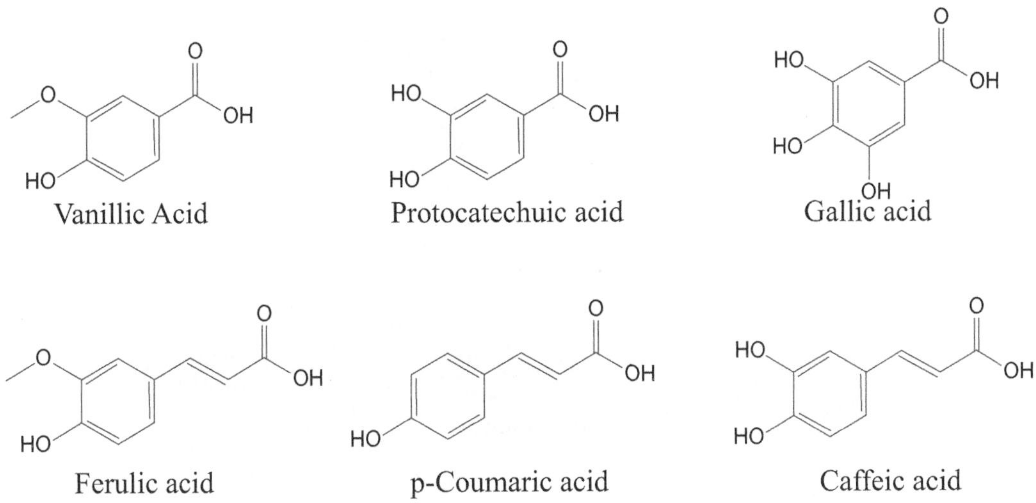

FIGURE 3.6 The chemical structures of the main phenolic acid compounds.

antioxidant-rich foods significantly reduces the risk of different types of cancers. Therapeutic treatment and prevention of the different types of cancer are greatly aided by phenolic acids and their derivatives, such as hydroxybenzoic and hydroxycinnamic acids.

Phenolic acids prevent cellular damage brought on by free-radical oxidation reactions and are easily absorbed through the walls of your intestinal tract (Figure 3.5). Coffee consumption has been linked to a lower risk of some cancers and other chronic illnesses. In addition to caffeine and other potentially advantageous compounds, it contains the antioxidants caffeic acid and chlorogenic acid (Figure 3.6) (Xie et al., 2011; Croft, 1998; Rice-Evans et al., 1996).

3.3.3.3 Polyphenols

Organic substances known as natural polyphenols contain phenol units built into their structures. Numerous experimental findings imply that polyphenols may impact antitumor growth and spread and carcinogenesis (Figure 3.7).

Oranges, apples, grapes, blueberries, spinach, and onions are some of the best sources of polyphenols. Resveratrol, quercetin, and curcumin are examples of polyphenols which are diversely distributed in nature and exhibit some protective roles inside the biological system such as antioxidative activity, anti-inflammatory activity, antiproliferative activity, anticancer activity, etc. (Figure 3.8) (Li et al., 2014; Kim et al., 2014; Vattem et al., 2005; Wang et al., 2015).

3.3.4 Tannin

Plants contain tannins, astringent-tasting polyphenols that can bind to and precipitate proteins. Tannins are mostly abundant and important biologically active molecules found in plants. The molecular weight of tannins is very high, ranging from 500 to 3000 kDa. Two different types of tannins are hydrolysable tannins and condensed tannins. Hydrolysable tannin contains a carbohydrate moiety such as D-glucose at the centre. Most of the hydroxyl groups of carbohydrate moiety situated at the centre of the hydrolysable tannins are esterified by phenolic acids.

Condensed tannins are also known as proanthocyanidins, and they are widely distributed in different parts of plants. Quality of foods is strongly affected by such types of tannins.

Tannins also have a variety of biological activities such as anticancer activity, antibacterial activity, antimicrobial activity, etc. They also act as antioxidants. Tannic acid also inhibits the proliferation of various cancer cells (Figure 3.9) (Sp, 2020; Geng, 2019).

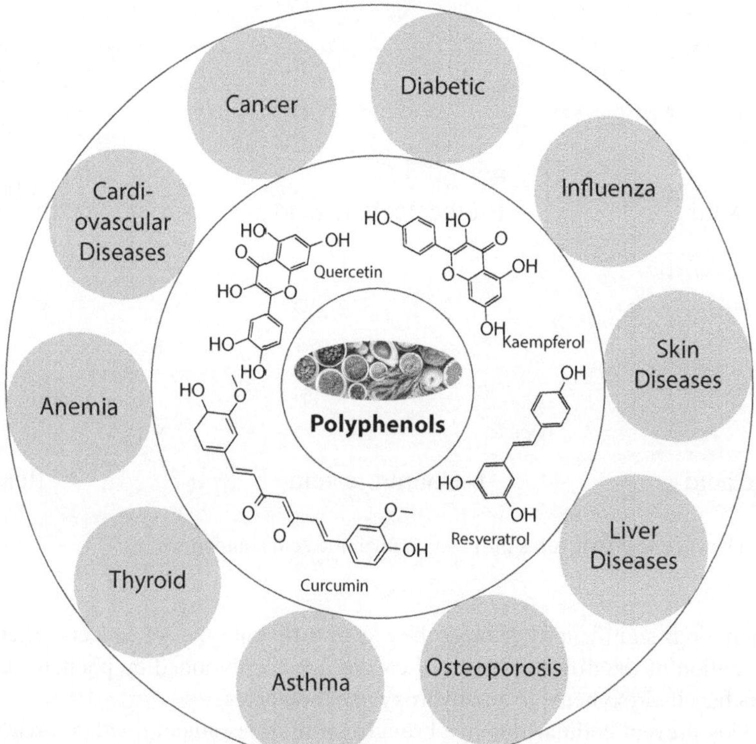

FIGURE 3.7 Some applications of polyphenolic phytochemicals.

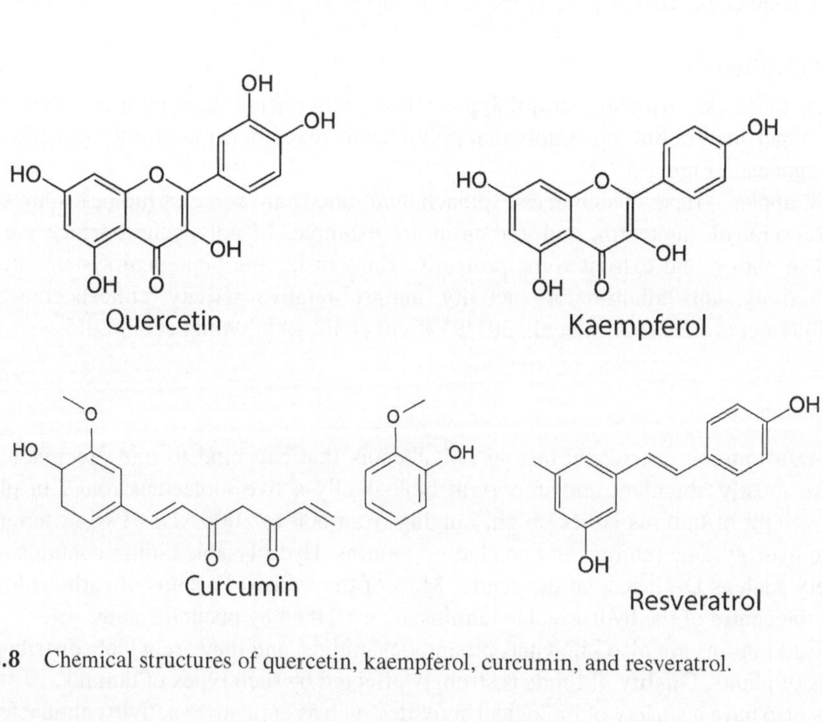

FIGURE 3.8 Chemical structures of quercetin, kaempferol, curcumin, and resveratrol.

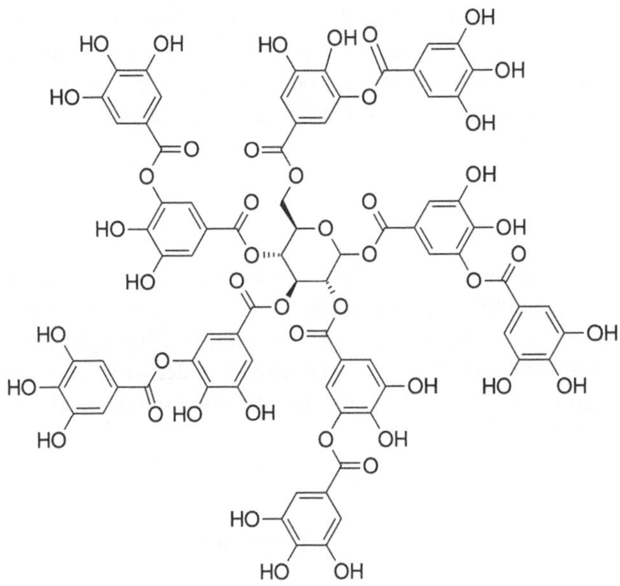

Tannic acid

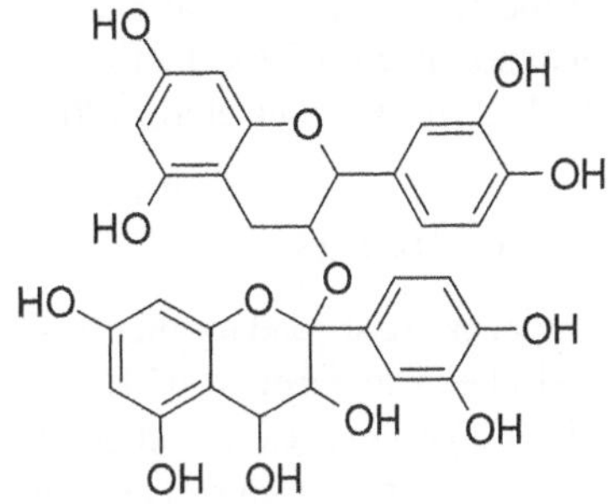

Proanthocyanidin

FIGURE 3.9 Chemical structures of tannins with anticancer effect.

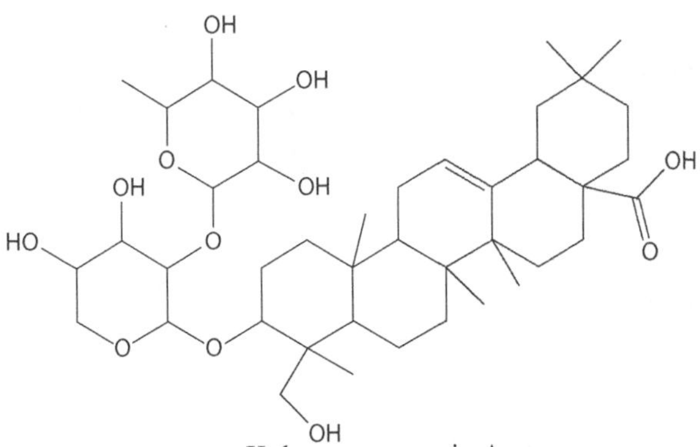

Kalopanaxsaponin A

FIGURE 3.10 Chemical structures of kalopanaxsaponin A.

3.3.5 Saponin

Saponins are naturally occurring phytochemicals with a wide range of structural and functional diversity that are present in all types of plants. Saponins are formed by linking oligosaccharide (containing 2–10 monosaccharide units) moieties with triterpenoid or steroidal aglycones (Figure 3.10). They are a group of chemically diverse compounds with complex structure. The saponins are glycosidic plant products made up of a parent compound and an erratic sugar component.

Triterpene glycosides, spirostanol glycosides, and steroidal alkaloid glycosides are the three categories under which saponins fall. Aglycone and a sugar moiety are the byproducts of the hydrolysis of the saponin molecule. They have shown effective anticancer activity in various cancer cells by inhibiting cell growth and inducing apoptosis. Generally speaking, saponins lower blood lipid levels, cancer risks, and blood glucose response. Saponins play an important role in the prevention of the progression of malignant cells by interfering with the replication of cellular DNA. Saponins enhance neurotrophic factors, restore monoaminergic tone, and promote neurogenesis (Farzaei et al., 2016; Vincken et al., 2007; Jang et al., 2014).

3.3.6 Terpenes

Terpenes are secondary metabolites that contain isoprene units. Terpenes are represented by a general chemical formula $(C_5H_8)_n$. Terpenes are mainly responsible for the way most plants smell. Many terpenoids exhibit anticancer activity by triggering various stages of cancer cell progression. Terpenes are also known as terpenoids when some organic functional groups (aldehydes, ketones, alcohols) are attached to the chemical structure of terpene. Terpenes are classified into several categories depending on the number of isoprene units [15, 33].

3.3.6.1 Hemiterpenoids

Terpenoids containing single isoprene unit ($-C_5H_8$). Isovaleric acid and prenol are examples of hemiterpenoids (Figure 3.11).

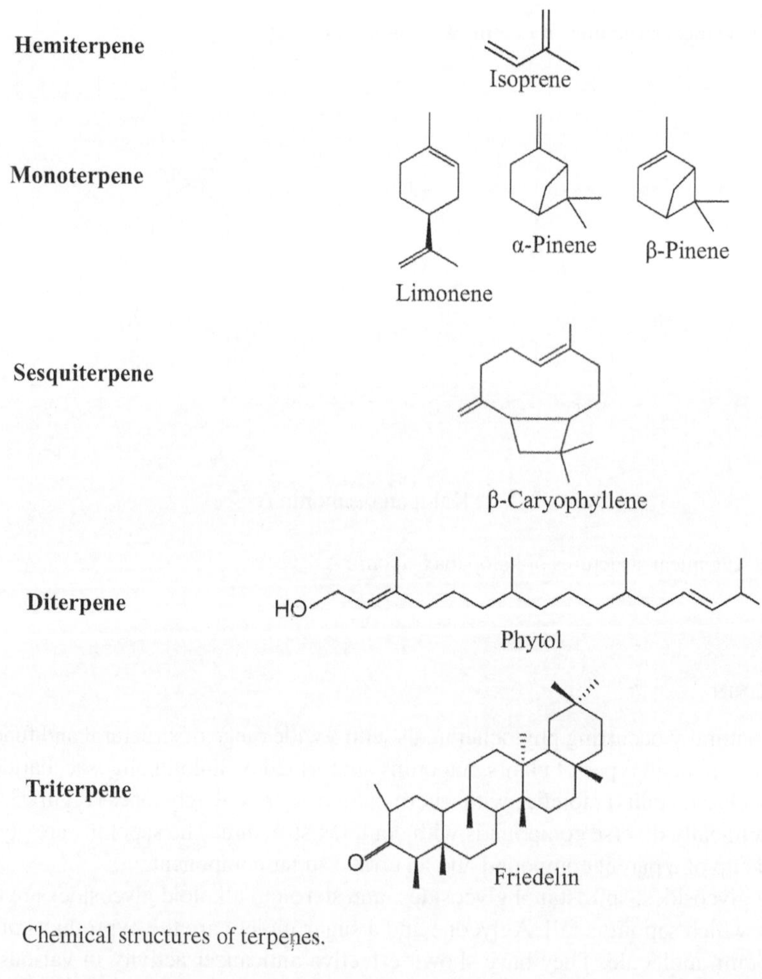

FIGURE 3.11 Chemical structures of terpenes.

3.3.6.2 Monoterpenoids

Monoterpenoids contain two isoprene units having the general formula $C_{10}H_{16}$. Different arrangements of monoterpenoids exist around us such as acyclic terpenoids, monocyclic terpenoids, and bicyclic terpenoids (Figure 3.11).

3.3.6.3 Sesquiterpenes

They contain three isoprene units ($C_{15}H_{24}$). Structures of such terpenoids are simple to complex containing monocyclic and polycyclic units. Artemisinin and bisabol are examples of sesquiterpenes which are responsible for the smell of flowers (Figure 3.11).

3.3.6.4 Diterpenes

They contain four isoprene units ($C_{20}H_{32}$). Retinol, retinal, and phytol are some examples of diterpenes (Figure 3.11).

3.3.6.5 Triterpenes

They contain six isoprene units ($C_{30}H_{48}$). Triterpenoids are widely distributed forms of terpenes containing different carbon skeletons (Figure 3.11). Examples of some triterpene skeletons are lupane, ursane, friedelane, etc. Betulinic acid is the most important anticarcinogenic triterpenoid on the lupine skeleton. Other triterpenes having anticarcinogenic activity are ursolic acid, friedelin, Asiatic acid, betulin, lupeol, etc.

3.4 MORTALITY RATE OF PHOTOCARCINOGENESIS

The largest organ in the body, the skin, serves as an external defence against the elements. Skin cancer is primarily brought on by solar ultraviolet radiation, which also causes pre-malignant actinic keratoses. Basal cell carcinomas are a specific type of skin cancer and are also responsible for solar UV radiation. Cutaneous squamous cell carcinoma is a common variety of skin cancers developed due to the abnormal growth of squamous cells promoted by UV radiation. In the United States, skin cancers are becoming a more common disease and make up more than 40% of all cancers. According to the WHO report, nearly ten million people will die from various types of cancer worldwide in 2020 [1]. Among the different types of cancer, the most common were:

- Breast cancer (2.26 million cases)
- Lung cancer (2.21 million cases)
- Colon and rectum cancer (1.93 million cases)
- Prostate gland cancer (1.41 million cases)
- Skin (non-melanoma) cancer (1.20 million cases)
- Stomach cancer (1.09 million cases)

3.5 MECHANISM OF PHOTOAGEING PROCESS AND PHOTOCARCINOGENESIS

Photoaging is a natural and biological technique that leads to the continuous degradation of the inner and outer organs and causes cancer. Cancer is a common chronic disease caused by the mutation of certain genes inside the DNA, resulting in uncontrolled and undisciplined cellular augmentation

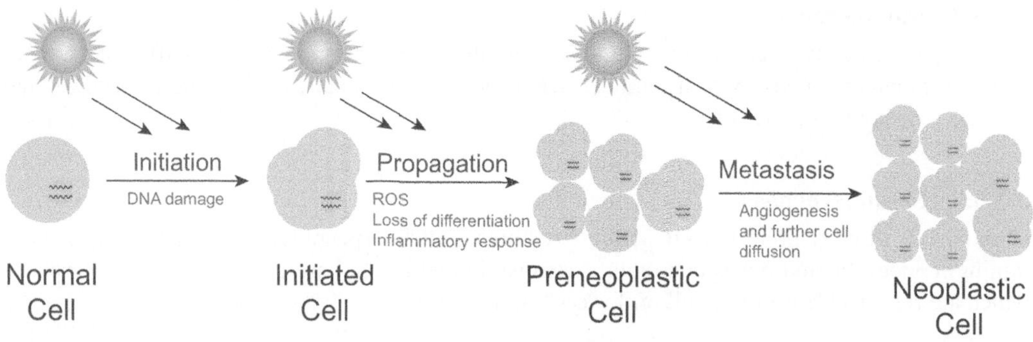

FIGURE 3.12 Diagrammatic representation of the mechanism of photocarcinogenesis.

and cell division. The multistep photocarcinogenesis follows a three-step mechanism, that is, initiation, propagation, and metastasis (Figure 3.12) (Jang et al., 2014; Yang et al., 2020; Yaar and Gilchrest, 2007).

3.5.1 Initiation Stage

DNA mutations inside keratinocytes are the most common phenomenon that occurs during the initial stage of photocarcinogenesis. UV radiation induces certain genes inside the DNA and promotes photocarcinogenesis.

Photochemical changes in DNA structures are mainly responsible for photocarcinogenesis. The structure of cellular molecules like proteins and lipids is changed in the presence of UV radiation and promotes mutation of DNA and abnormal cell division. UV-B (280 nm–315 nm) from solar radiation is mainly responsible for the mutation of DNA structures and generates some reactive oxygen species by accelerating the dimerization of thymine and pyrimidine-pyrimidone.

The dimeric products of thymine are mainly responsible for tumour when the dimer affects the tumour arrester gene p53. Skin with actinic keratosis, skin cancer, and chronic sun damage exhibits mutated p53.

3.5.2 Propagation Stage

DNA photoproducts formation like DNA repair, mutation of proto-induced oncogenes, tumour arrestor genes, etc. are promoted upon repeated UV exposure. UV rays produce active oxygen species which promotes radical species formation with subsequent effects on DNA mutation and different epigenetic functions that influence the carcinogenesis process. Ornithine decarboxylase (ODC) is an example of a rate-limiting enzyme involved in the biosynthesis of polyamine which is also related to the promotion stage.

3.5.3 Metastasis Stage

In this stage of carcinogenesis, ultraviolet radiation constructs some genetic and biochemical transformation which are responsible for invasive squamous cell carcinomas. Prostaglandins promote photocarcinogenesis which ultimately spreads the cancer cells.

Some phytochemicals traditionally used by Indians for photo-ageing are presented in Table 3.1.

TABLE 3.1
Example of Some Phytochemicals Traditionally Used by Indians for Photo-ageing

Class	Name of the Bioactive Component	Source	Function
Alkaloids	Caffeine Morphine Quinine Nicotine	Coffee beans, Tea Opium poppy Bark of cinchona Tobacco plant	Protect plants from predators and regulate their growth. They have antihypertensive effects, antimalarial activity, and anticancer action.
Flavonoids	Quercetin Malvidin Cyanidin derivatives	Fruits, Vegetables, Grains, Bark, Roots, Stems, Flowers, Tea, Wine	Exhibit anticancer, antioxidant, anti-inflammatory, and antiviral properties. Protect from UV radiation-induced damage.
Polyphenols	Resveratrol Quercetin	Oranges, Apples, Grapes, Blueberries, Spinach, Onions	Reduce skin wrinkling and skin oxidative stress. Also exhibit hypolipidemic, antioxidative, antiproliferative, and anti-inflammatory effects.
Polyphenolic acids	Hydroxycinnamic acids p-Coumaric acid Caffeic acid Gallic acid	Kiwifruit, Cocoa, Plum, Blueberry, Grape	Exhibit anticancer, antioxidant, anti-inflammatory, and antimicrobial properties.
Carotenoids	β-Carotene Lycopene Lutein	Sweet potatoes, Carrot, Lycopene, Mangoes, Papaya	Antioxidants, reduce the risk of cancer, cardiovascular disease, macular degeneration, etc.
Vitamins	Vitamin C Vitamin E	Citrus fruits, Oranges, Amla, Almond, Sunflower oil, Bell pepper, etc.	Protect our cells from damage caused by environmental toxins and the body's natural chemical (metabolic) processes.
Tannin	Tannic acid	Chestnut, Oak trees	It has an astringent effect, antiviral effect, and antibacterial effect.

3.6 CONCLUSION

Phytochemicals are naturally occurring substances widely found in different parts of plants which exhibit diverse medical applications and offer significant health benefits, including protection against environmental hazards, low toxicity, affordability, and accessibility. They have antioxidant, antimicrobial, immune system support, and anticancer properties. Consuming phytochemical-rich foods reduces the risk of such diseases. The rising incidence of cancer and drug resistance has sparked interest in cancer prevention strategies by employing phytochemicals. The present book chapter clearly elaborates on the photocarcinogenic activity of different groups of phytochemicals. We believe the present chapter would provide a short glimpse into the topic, its importance in the present context, and would be helpful to readers working in this field.

REFERENCES

Bendich, A. 1988. The safety of beta-carotene. *Nutr. Cancer*, 11, 207–214.

Bode, A.M.; Dong, Z. 2015. Toxic phytochemicals and their potential risks for human cancer. *Cancer Prev. Res.*, 8, 1–8.

Buckman, S.Y.; Gresham, A.; Hale, P.; Hruza, G.; Anast, J.; Masferrer, J.; Pentland, A.P. 1998. COX-2 expression is induced by UVB exposure in human skin: Implications for the development of skin cancer. *Carcinogenesis*, 19, 723–729.

Castellano, G.; Gonzalez-Santander, J.L.; Lara, A.; Torrens, F. 2013. Classification of flavonoid compounds by using entropy of information theory. *Phytochemistry*, 93, 182–191.

Croft, K.D. 1998. The chemistry and biological effects of flavonoids and phenolic acids. *Ann. N. Y. Acad. Sci.*, 854, 435–442.

Dreher, D.; Junod, A.F. 1996. Role of oxygen free radicals in cancer development. *Eur. J. Cancer*, 32A, 30–38.

Farzaei, M.H.; Bahramsoltani, R.; Rahimi, R. 2016. Phytochemicals as adjunctive with conventional anticancer therapies. *Curr. Pharm. Des.*, 22, 4201–4218.

Filippi, A.; Ciolac, O.A.; Ganea, C.; Mocanu, M.M. 2017. ErbB proteins as molecular target of dietary phytochemicals in malignant diseases. *J. Oncol.*, 1532534.

Garćıa-Lafuente, A.; Guillamon, E.; Villares, A.; Rostagno, M.A.; Martínez J.A. 2009. Flavonoids as anti-inflammatory agents: Implications in cancer and cardiovascular disease. *J. Inflamm. Res.*, 58(9), 537–552.

Geng, N. 2019. Tannic acid synergistically enhances the anticancer efficacy of cisplatin on liver cancer cells through mitochondria-mediated apoptosis. *Oncol. Rep.*, 42, 2108–2116.

George, B.P.; Chandran, R.; Abrahamse, H. 2021. Role of phytochemicals in cancer chemoprevention: Insights. *Antioxidants*, 10, 1455.

Han, Q.-T.; Ren, Y.; Li, G.-S.; Xiang, K.-L.; Dai, S.-J. 2018. Flavonoid alkaloids from Scutellariamoniliorrhiza with anti-inflammatory activities and inhibitory activities against aldose reductase. *Phytochemistry*, 152, 91–96.

Jang, H.J.; Han, I.H.; Kim, Y.J.; Yamabe, N.; Lee, D.; Hwang, G.S.; Oh, M.; Choi, K.C.; Kim, S.N.; Ham, J.; et al. 2014. Anticarcinogenic effects of products of heat-processed ginsenoside Re, a major constituent of ginseng berry, on human gastric cancer cells. *J. Agric. Food Chem.*, 62, 2830–2836.

Kim, H.-S.; Quon, M.J.; Kim, J.-A. 2014. New insights into the mechanisms of polyphenols beyond antioxidant properties; lessons from the green tea polyphenol, epigallocatechin 3-gallate. *Redox Biol.*, 2, 187–195.

Kotecha, R.; Takami, A.; Espinoza, J.L. 2016. Dietary phytochemicals and cancer chemoprevention: A review of the clinical evidence. *Oncotarget*, 7, 52517–52529.

Krinsky, N.I.; Johnson, E.J. 2005. Carotenoid actions and their relation to health and disease. *Mol. Asp. Med.*, 26, 459–516.

Li, A.-N.; Li, S.; Zhang, Y.-J.; Xu, X.-R.; Chen, Y.-M.; Li, H.-B. 2014. Resources and biological activities of natural polyphenols. *Nutrients*, 6(12), 6020–6047.

Martino, E.; Casamassima, G.; Castiglione, S.; Cellupica, E.; Pantalone, S.; Papagni, F.; et al. 2018. Vinca alkaloids and analogues as anti-cancer agents: Looking back, peering ahead. *Bioorg. Med. Chem. Lett.*, 28(17), 2816–2826.

McCullough, M.L.; Peterson, J.J.; Patel, R.; Jacques, P.F.; Shah, R.; Dwyer, J.T. 2012. Flavonoid intake and cardiovascular disease mortality in a prospective cohort of US adults. *Am. J. Clin. Nutr.*, 95, 454–464.

Ng, C.Y.; Yen, H.; Hsiao, H.Y.; Su, S.C. 2018. Phytochemicals in skin cancer prevention and treatment: An updated review. *Int. J. Mol. Sci.*, 19, 941.

Rasulev, B.F.; Abdullaev, N.D.; Syrov, V.N.; Leszczynski, J. 2005. A quantitative structure-activity relationship (QSAR) study of the antioxidant activity of flavonoids. *QSAR Comb. Sci.*, 24(9), 1056–1065.

Rice-Evans, C.A.; Miller, N.J.; Paganga, G. 1996. Structure-antioxidant activity relationships of flavonoids and phenolic acids. *Free Radical Biol. Med.*, 20(7), 933–956.

Sak, K. 2014. Site-specific anticancer effects of dietary flavonoid quercetin. *Nutr. Cancer*, 66, 177–193.

Silva, B.I.M.; Nascimento, E.A.; Silva, C.J.; Silva, T.G.; Aguiar, J.S. 2021. Anticancer activity of monoterpenes: A systematic review. *Mol. Biol. Rep.*, 48, 5775–5785.

Sp, N. 2020. Tannic acid inhibits non-small cell lung cancer (NSCLC) stemness by inducing G0/G1 cell cycle arrest and intrinsic apoptosis. *Anticancer Res.*, 40, 3209–3220.

Vattem, D.A.; Ghaedian, R.; Shetty, K. 2005. Enhancing health benefits of berries through phenolic antioxidant enrichment: Focus on cranberry. *Asia Pac. J. Clin. Nutr.*, 14, 120–130.

Vincken, J.P.; Heng, L.; de Groot, A.; et al. 2007. Saponins, classification and occurrence in the plant kingdom. *Phytochemistry*, 68(3), 275–297.

Wang, J.; Tang, L.; Wang, J.-S. 2015. Biomarkers of dietary polyphenols in cancer studies: Current evidence and beyond. *Oxid. Med. Cell. Longevity*, 732302.

World Health Organization. World Health Organization: Geneva, 2017. www.who.int/en/

Xie, C.H.; Kang, J.; Chen, J.R.; Nagarajan, S.; Badger, T.M.; Wu, X.L. 2011. Phenolic acids are in vivo atheroprotective compounds appearing in the serum of rats after blueberry consumption. *J. Agric. Food Chem.*, 59, 10381–10387.

Yaar, M.; Gilchrest, B.A. 2007. Photoageing: Mechanism, prevention and therapy. *Br. J. Dermatol.*, 157, 874–887.

Yang, W.; Chen, X.; Li, Y.; Guo, S.; Wang, Z.; Yu, X. 2020. Advances in pharmacological activities of terpenoids. *Nat. Prod. Commun.*, 15, 1934578X20903555.

4 Liposomal Formulations with Topical Application and In Vitro Testing Methods

Miray Karamese, Seyfure Adıguzel, Elif Ayse Kacar and Sevde Altuntas

4.1 INTRODUCTION

The skin, the largest organ of the human body, is a potential anatomical site for the application of pharmaceuticals and cosmetic products (Schlich et al., 2022). However, barrier mechanisms prevent the delivery of topically applied exogenous substances to the skin structures. The stratum corneum (SC), the outer layer of the epidermis, forms the first physical hydrophobic barrier, the tight connections of keratinocytes in the stratum granulosum in the epidermis constitute the second barrier, and the basal lamina connecting the epidermis to the dermis forms the third barrier layer (Mojumdar et al., 2017; Gorzelanny et al., 2020). The problem of limited drug penetration into the skin due to barrier mechanisms makes it important to investigate topical applications that can increase drug penetration and provide a long-term therapeutic effect. In recent years, various innovative nanosystems have been developed that show good biocompatibility, low cytotoxicity, and the ability to increase the stability and skin penetration of drugs (Krishnan and Mitragotri, 2020a; Carter et al., 2019; Casamonti et al., 2019).

Among nanomedicine strategies, liposomes have significant potential for delivering active compounds into skin structures (Chacko et al., 2020). Previous studies have revealed that liposomes penetrate the SC and beyond following topical application (Liu et al., 2021). Improved liposome formulations have facilitated the topical application of, not only synthetic drugs but also phytochemicals. These advanced systems can protect phytochemicals from physical and chemical degradation and improve their bioavailability and pharmacological activity. Liposomes are also promising carriers for topical applications of phytochemicals due to their similarities with biological membranes, their affinity with skin components, and their ability to load hydrophilic and lipophilic molecules simultaneously (Sogut et al., 2021; Allaw et al., 2022).

4.2 PROPERTIES OF LIPOSOMES AND THEIR USE IN THE TOPICAL FIELD

A drug delivery system (DDS) allows pharmacologically active compounds to reach and be effectively released in the target tissue or organ. Furthermore, DDS aims to maximize the therapeutic efficiency of medications while minimizing their negative effects (Komatani, 2016). There are several routes for medications to enter the body. These routes include oral delivery, transdermal application, inhalation, mucosal application, and intravenous injection (Stark, 2015). The transdermal drug delivery system (TDDS) is the favoured method due to its numerous benefits (Jeong et al., 2021). Since TDDS does not pass through the gastrointestinal tract, drugs are prevented from encountering a first-pass effect, preventing fluctuations in plasma levels of drugs. This can increase the therapeutic effectiveness of medications and reduce their side effects (Patel et al., 2021). The

DOI: 10.1201/9781003394846-4

skin, the largest organ of the human body, is a potential anatomical site for the application of pharmaceuticals and cosmetic products (Schlich et al., 2022).

Transdermal/topical medication delivery systems are an appealing technique because of their numerous benefits, but their application is limited by the skin's innate barrier. The skin's multilayered structure protects it against environmental threats like chemicals, heat, and poisons. Skin is typically classified into three layers: epidermis, dermis, and hypodermis. The epidermis is the skin's outermost layer that provides protection. The dermis is the layer where blood vessels and skin cells are found. The hypodermis is the basal layer of fat tissue (Jeong et al., 2021). The skin is the first organ to be targeted with topical or transdermal drugs. The stratum corneum (SC), the skin's outermost layer, functions as a barrier to drug penetration and dispersion (Bouwstra and Honeywell-Nguyen, 2002; Choi and Maibach, 2005). Due to the issue of limited drug penetration via topical administration to the skin, looking for safe drug carriers that can improve drug permeation through the skin and offer a sustained therapeutic impact is critical (Antimisiaris et al., 2021). Increasing drug permeability by lowering the SC's barrier resistance allows for better drug administration to the target tissue by methods such as SC modification, the use of lipid-based drug carriers, bypassing the SC, and electrical approaches (Williams and Barry, 1992). Liposomes, lipid-based drug carriers, permeate the SC membrane and integrate directly into the upper layer of the skin, boosting drug absorption via the skin (Choi and Maibach, 2005). Figure 4.1 illustrates that liposomes can connect with cells and deliver their substance using different major ways.

Nanotechnology, a multidisciplinary field, contributes to every discipline of science by being applied in various fields such as medicine, biology, chemistry, physics, and material science (Jaafar Hawraz Jawdat, 2023). Liposomes and nanoparticles, tiny drug carriers for therapeutic agents in the nanotechnology area, have gained prominence in the biomedical field for use in clinical trials

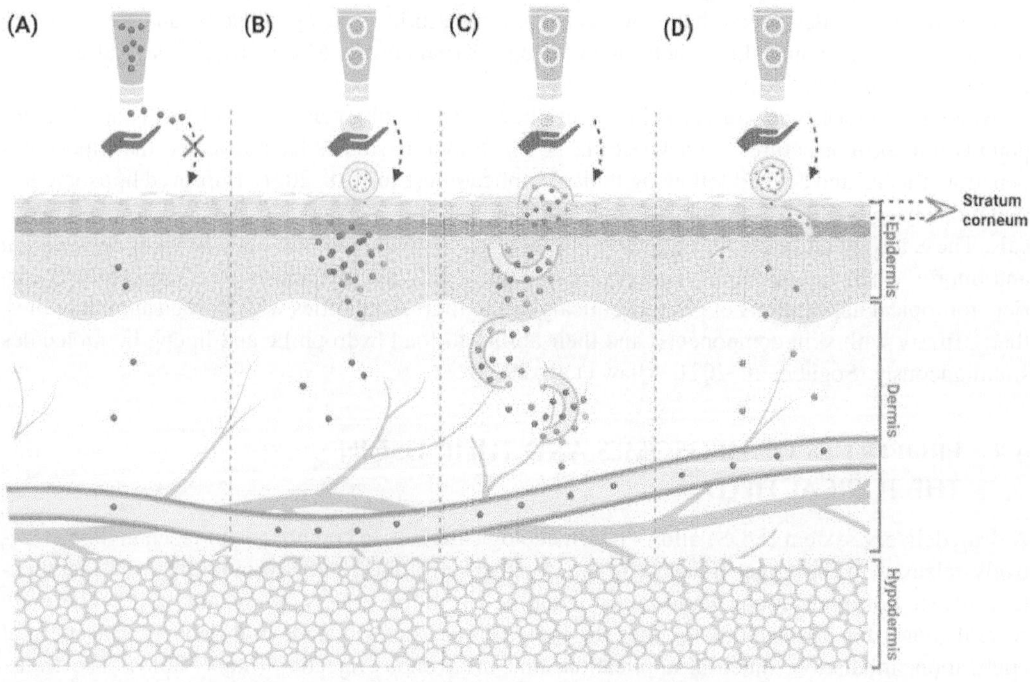

FIGURE 4.1 A summary of numerous potential processes of liposome-skin interactions. (A) Drug delivery method without liposomal transport, (B) free drugs that release from liposomes on the surface of the skin after surface interaction, (C) liposome adsorption to the stratum corneum, (D) disintegration of the liposome into its components and release of the drugs.

or on the market (Malam et al., 2009; Bozzuto and Molinari, 2015). Alec Banghamm and colleagues discovered in 1965 that phospholipid aqueous solutions spontaneously form double-layer membrane-like structures called liposomes. The phospholipid bilayer closes on itself and creates liposome structures because of the interaction of the hydrophobic chains of the phospholipid bilayer with each other and the polar groups in the phospholipid structure with water (Allen and Cullis, 2013; Malam et al., 2009; Park, 2019). The van der Waals force between the hydrophobic chains of the double-layer structure influences the architecture while the hydrogen bond between polar groups and water molecules influences the structure's stability. Furthermore, the methods utilized in liposome synthesis, the type of phospholipid used, the surface charge, structural aggregation, and size also affect liposome opsonization (Bozzuto and Molinari, 2015). While the first nanocarrier systems, liposomes, were vesicles made wholly of natural phospholipids, they are now made up of both natural and synthetic phospholipids. Drugs with varying degrees of solubility can be enclosed in different sections of liposomes. Hydrophilic drugs can be found in the liposome's aqueous core region, adjacent to the water-lipid interface, or near-polar head groups, whereas hydrophobic drugs are found in the liposome's phospholipid bilayer (Bozzuto and Molinari, 2015; Bulbake et al., 2017). Liposomes deliver genes, nucleic acids, proteins, and imaging agents, as well as medications such as chemotherapeutics, antifungals, antibiotics, and analgesics; they are utilized in diagnostics, therapies, and theranostics.

Furthermore, liposomes are also used in non-biomedical industries such as nutrition, agriculture, and cosmetics (Sercombe et al., 2015; Tenchov et al., 2021; Xing et al., 2016). The encapsulated substance is not metabolized before reaching the target tissue, accumulates in a minimal amount in healthy tissues, and does not dilute in the circulation, contributing to the pharmacokinetic properties of the substance changing and increasing the therapeutic index and bioavailability (Bozzuto and Molinari, 2015; Bulbake et al., 2017; Sercombe et al., 2015).

Figure 4.2 illustrates liposomal delivery techniques are classified into four types: conventional liposomes, ligand-targeted liposomes, sterically stabilized liposomes (PEGylation), and theranostic

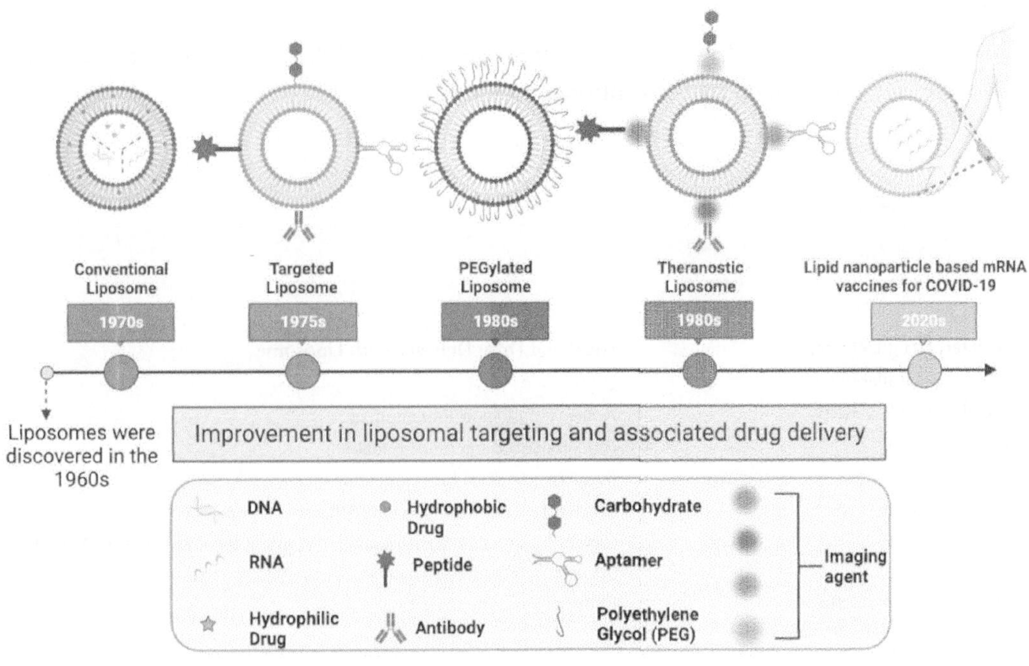

FIGURE 4.2 History and evolution of liposomes from beginning to date.

liposomes (Sercombe et al., 2015). Conventional liposomes, the first generation, are made up of only a natural or synthetic phospholipid layer and can encapsulate a variety of compounds (Nsairat et al., 2022). Cholesterol has been added to phospholipid structures to increase the fluidity of typical liposomes while also altering the stiffness and stability of the phospholipid layer (Nsairat et al., 2022). In clinical uses of traditional liposomes, research has shown that liposomal encapsulation improves the therapeutic index of medications compared to the free application of doxorubicin and amphotericin pharmaceuticals encapsulated in liposomes. Conventional liposomes are highly sensitive to rapid accumulation and elimination in the spleen and liver due to the presence of opsonization proteins such as complement proteins, laminin, fibronectin, C-reactive protein, Type I collagen, and immunoglobulins, which are responsible for the mononuclear phagocyte system's (MPS) uptake of the reticuloendothelial system (RES) in the plasma (Nag and Awasthi, 2013; Nsairat et al., 2022; Sercombe et al., 2015). To improve the circulation time and durability of traditional liposomes, polyethylene glycol (PEG), a biocompatible, hydrophilic, inert, non-toxic, and low immunogenic polymer, was attached to their surface (Nag and Awasthi, 2013; Karamese et al., 2023). The PEGylation process reduces electrostatic and hydrophobic interactions between liposomes and opsonization proteins, as well as RES targeting liposomes by producing steric hindrance with opsonization proteins. (Sercombe et al., 2015). Liposomes that can reach the target tissue without being removed through the PEGylation process have higher pharmacokinetic aspects than conventional liposomes. According to studies, the half-life of liposomes in circulation increases from two hours to 24 hours in rodents and up to 45 hours in people (Sercombe et al., 2015). Many liposomal products are available on the market, with some in clinical development (Sercombe et al., 2015). The first nano formulation medicine was liposomal encapsulated, PEGylated doxorubicin (Doxil®), which was approved by the FDA in 1995 for the treatment of chemotherapy-resistant Kaposi sarcoma and epithelial ovarian cancer in AIDS patients (Bozzuto and Molinari, 2015). To improve the functionality of liposomes, the surface has been enriched with diverse ligands over time (Trucillo et al.,

TABLE 4.1

Comparison of localized drug delivery systems and localized drug delivery systems with liposomes for drugs taken by conventional methods

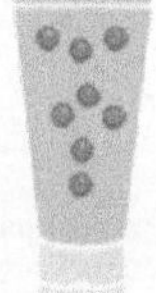

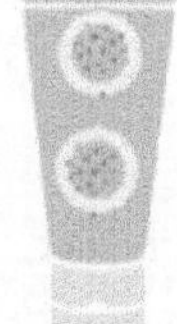

Localized Drug Delivery	**Localized Drug Delivery with Liposome**
Increased bioavailability	Highly safe
Reduction of side toxicity	Non-toxic and non-immunogenic
Off-target effects	Biocompatible and biodegradable
Reduced cost	Reducing the need for frequent dosing
Insufficient effect	Protect the drug from potentially harsh environments
Higher "effective" doses	Capability to retain high amounts of drugs at localized sites, and to release the drug in a controlled manner
Minimizes untoward effects associated with systemic delivery	Reduced systemic toxicities

Source: From Antimisiaris et al., 2021; Wen et al., 2015.

2020). The interaction of ligands with receptors or cell adhesion molecules expressed in the target cell or organ *in vitro* is like a key-lock mechanism. Ligands on the liposome surface include primarily antibodies, peptides, carbohydrates, and aptamers. As a result, ligand-decorated liposomes allow the encapsulated substance to reach the target region at a higher rate (Sercombe et al., 2015). Furthermore, theranostic liposomes allow for high-contrast ultrasound imaging of tumour areas in cancer detection and treatment, which increases the intracellular distribution of anticancer medicines and genes. (Yoon et al., 2014). In 2020, liposomes developed against SARS-CoV-2 infection contained the incorporation of mRNA encoding the spike protein into liposomes and were effective among COVID-19 vaccines (Chatzikleanthous et al., 2021; Huang et al., 2021).

4.3 IN VITRO TESTING OF LIPOSOMAL FORMULATIONS

Comprehensive physicochemical characterization tests are required to determine the effects of liposomes on the biological system. Thus, the first step is taken in the comparison of biological and toxicological tests (Costa and Fadeel, 2016). Because the composition, size, and surface charge of liposomes can affect their activities and, therefore, their cellular fate, leading to different biological responses. (Pourtalebi Jahromi et al., 2023). For a comprehensive characterization of nanomaterials, Sayes and Warheit (2009) proposed a three-stage model. They emphasized that the primary characterization should be carried out in the natural state of the nanomaterial, which is in the dry state. On the other hand, secondary characterization should be conducted in the solution or suspension state of the nanomaterial. Thus, they explained that the effects of materials with determined physicochemical properties on biological systems should be demonstrated with *in vitro* and *in vivo* models, which are tertiary characterizations (Soares et al., 2018; Sayes and Warheit, 2009).

The following sections are intended to provide an overview of the physicochemical characterization methods and biopharmaceutical methodologies most commonly used in the *in vitro* evaluation of topical liposomal formulations.

4.3.1 Physical and Chemical Characterization of Liposomes

The determination of the physical and chemical properties of liposomes and the in-depth characterization of their possible interactions with drugs constitute a preliminary step in revealing the biological effects of the formulation. To verify and test that the formulation has the desired properties, techniques such as dynamic light scattering, infrared spectroscopy, differential scanning calorimetry, and electron microscopy are used to provide information about the size, shape, composition, charge, and surface properties of liposomes. In the following section, the most frequently used methods in the physical and chemical characterization of liposomes are mentioned. These methods are introduced in Figure 4.3.

4.3.1.1 Dynamic Light Scattering (DLS)

The average size, size distribution, and charge of liposomes are important in *in vitro* and preclinical evaluations because these properties directly affect the biodistribution of liposomes and their interaction with the skin (Gioria et al., 2018). Dynamic light scattering (DLS) is the most frequently chosen technique for size determination due to its affordability and ease of use (Maguire et al., 2018). In DLS experiments, a monochromatic laser enters a cuvette made of plastic or glass containing a homogeneous and transparent suspension of the sample. The Brownian motion of the suspended nanoparticles in the suspension causes the laser light to scatter at different intensities. The charge on the surface of liposomes also affects their mobility and changes the scattering of light. Therefore, DLS is commonly used to estimate zeta potential (Giordani et al., 2023). In a study in which curcumin-loaded liposomes with high encapsulation efficiency, good stability, and outstanding skin permeability were developed for wound healing, it was determined that the particle size increased

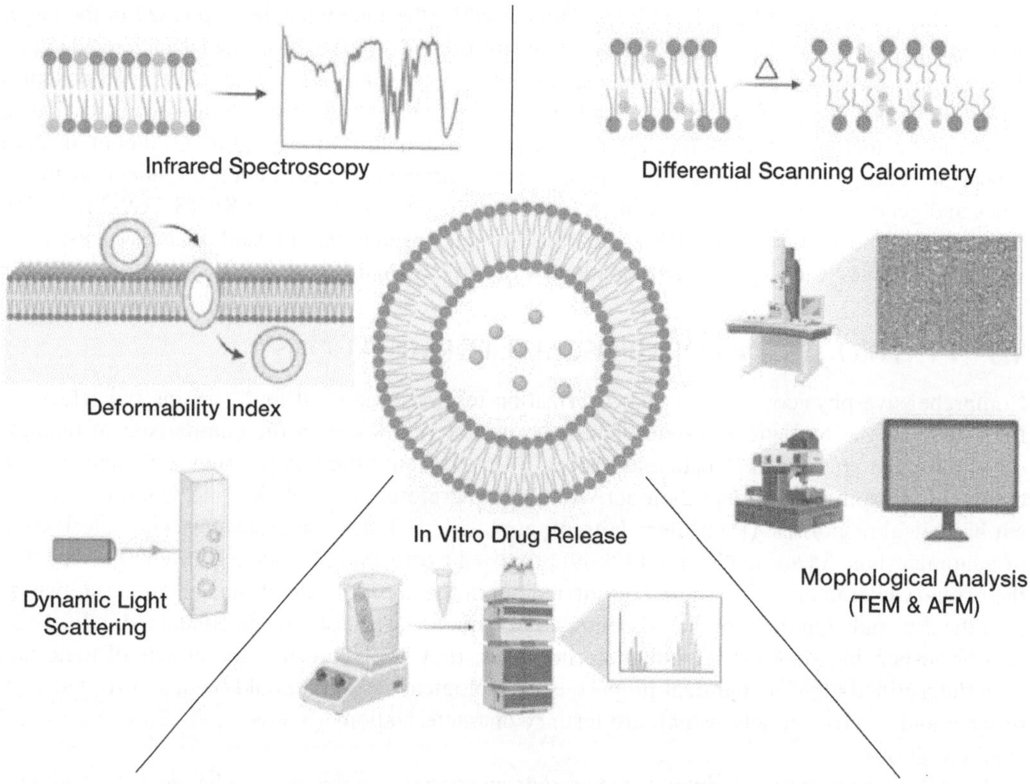

FIGURE 4.3 *In vitro* physical and chemical characterization methods of liposomes.

with higher lipid concentration. This has been shown to cause changes in release and penetration abilities (Choudhary et al., 2019).

4.3.1.2 Morphological Analysis

Transmission electron microscopy (TEM) is the most commonly used imaging technique to visualize the structure and surface modifications of lipid vesicles (Guillot et al., 2023). In an exemplary study, the morphological feature of the curcumin-containing liposome (RC-Lip) that mimics the red blood cell membrane (RBCM) was examined by TEM, and the fusion of the RBC membrane with the liposomal membrane was determined by this method (Tang et al., 2023).

Atomic force microscopy (AFM) is an alternative method for visualizing nanoparticles and detecting their surface properties, thanks to its high resolution on the order of a nanometer (Guillot et al., 2023). It is used as a complement to TEM in imaging the size and surface properties of liposomes (Kapoor et al., 2018).

4.3.1.3 Infrared Spectroscopy

Infrared (IR) spectroscopy is a well-established analytical technique used to characterize the structure of materials at the molecular scale (Baudot et al., 2010). Fourier transform infrared spectroscopy (FTIR), a type of IR, is used to analyze the structure of the lipid membrane and provides a kind of "fingerprint" of the sample being examined (Large et al., 2021). FTIR is used to detect the lipid composition of the liposome membrane, interactions between components, and also their impact on drug encapsulation and stability (Briuglia et al., 2015).

For example, in a study conducted to investigate the mechanisms of increased penetration of Licochalcone A (LA)-loaded skin keratin liposomes (LALs) in the skin and B16F10 mouse melanoma cells, two significant peaks representing binding in SC lipids were observed to migrate after LAL treatment. As a result of FTIR analysis, it was revealed that LALs can disrupt the lipid order in the SC structure to increase LA penetration (Wu et al., 2022).

4.3.1.4 Deformability Index

In topical applications, the flexibility of liposomes is a very important parameter in determining their penetration properties (Tardi et al., 2001). Liposome flexibility is associated with its ability to promote the accumulation and penetration of active substances into the skin (Bangham et al., 1965). Flexible liposomes can squeeze through skin pores and penetrate layers of the skin with a tendency to undergo deformation (Sharma and Verma, 2017; Zhang, 2017).

The deformability index of the liposome is measured by extrusion and microfluidic system methods, which involve passing liposomes through a porous polycarbonate membrane under a known pressure of MPa (Andra et al., 2022). The basic principle of the method is that flexible liposomes pass through the membrane while preserving their morphology because they can be deformed while non-flexible liposomes are subject to fragmentation that blocks the small pores of the membrane when forced to pass through the membrane (Cevc et al., 2002). The deformability index has been associated with the ability of liposomes to penetrate the skin and affect the accumulation of active molecules in the skin. In a study in which DPPC liposomes loaded with different amounts of curcumin (CURC) were produced, it was shown that DPPC led to an increase in the fluidity and deformability index of liposomes, thus leading to increased epidermis accumulation of curcumin (Campani et al., 2020).

4.3.1.5 Differential Scanning Calorimetry (DSC)

Differential scanning calorimetry (DSC) is a thermo-analytical technique used to detect various thermal phenomena of materials (Gu and Burgess, 2014). In DSC experiments, the difference in heat flow to a sample and a reference at a given temperature program is monitored over time or temperature (Odlyha, 2003). DSC examines the thermal transitions of phospholipids to characterize the thermotropic preference of liposomes and is of great importance in the development of drug carrier nanoliposome formulations based on their thermal behaviour (Scholtysek et al., 2012). At the same time, monitoring thermal transitions is also used to observe the changes in the skin as a result of liposome application. The decrease in the melting temperature and increased enthalpy of the skin after liposome application reflects the changing keratin structure. In a study conducted with keratin liposomes, revealed that skin keratin changes with the changes in the thermodynamic properties of the skin after application, and drug permeability increases thanks to the detection of this thermal change (Wu et al., 2022).

4.3.1.6 In Vitro Drug Release

Liposomes must deliver active ingredients to the target site at the desired time and amount under specified conditions. Therefore, drug release kinetics must be determined accurately to develop and test the effectiveness of liposomes as an optimal drug carrier in biological systems (Kozik et al., 2023). In evaluating the *in vitro* release of a drug from liposomes, it is necessary to mimic the *in vivo* release conditions. Since small changes in the physiochemical properties of liposomes can have a huge impact on the release behaviour of the drug, testing their release behaviour *in vivo* and preclinically is of great importance (Solomon et al., 2017; Adiguzel et al., 2024).

The most commonly used methods to measure drug release profiles are high-performance liquid chromatography (HPLC) and spectrophotometry methods. The amount and kinetics of drugs collected from dialysis are calculated by drug release under acidic or neutral conditions that mimic the conditions of the target region (Solomon et al., 2017). For instance, a good sustained pharmacokinetic release behaviour of shikonin, used in the treatment of infected skin wounds, was demonstrated by HPLC when loaded into liposomes. It has been demonstrated that the successful increase in antimicrobial activity with topical application of shikonin-loaded liposomes occurs through continuous release to the target area (Shu et al., 2022).

4.3.1.7 In Vitro Membrane Permeation Studies

Depending on the targeted area and purpose, topically applied active substances may be intended to have a local effect within the skin or be used for systemic absorption. The target area may also be subcutaneous areas. In such a case, the active ingredient is expected to pass through all skin layers and show high penetration and retention in the target area (Oh et al., 2019). Hence, evaluating the skin penetration of drug-carrier liposomes is an important step in the development of an effective formulation. *In vitro*, skin penetration studies of topical drug delivery systems are usually performed using a diffusion cell with donor and acceptor compartments separated by a membrane (Ng et al., 2012). Excised human skin is considered the standard for *in vitro* penetration studies (Flaten et al., 2015). However, the use of human skin is subject to limitations due to ethical concerns and physiological differences between donor samples. (Neupane et al., 2020). For these reasons, some researchers prefer to obtain experimental data using skin from animals such as pigs, rabbits, and rodents (Supe and Takudage, 2021; Ioele et al., 2015; Phatale et al., 2022).

Most skin penetration studies are performed using the vertical Franz diffusion cell system, which is extremely simple to use (Supe and Takudage, 2021). This system consists of two parts: transmitter and receiver. The tested formulation is applied through the recipient chamber, and the recipient site is filled with a suitable liquid to mimic the physiological environment. The skin membrane is placed between the two compartments and the rate of penetration of the drug through the skin is obtained by measuring the amount of drug recovered into the receptor compartment (Figure 4.4A). The penetration of the drug into the skin layers is determined by extraction (Sinico et al., 2005). For example, in research where allantoin, used in the treatment of skin ulcers, was incorporated into argan oil-enriched liposomes to improve its local accumulation in deeper skin layers, newborn pig skin, and Franz diffusion cells were used for *in vitro* penetration studies. It was found as a result of penetration studies that loaded liposomes containing argan oil supported the accumulation of allantoin in the dermis and its penetration through the skin (Manca et al., 2016).

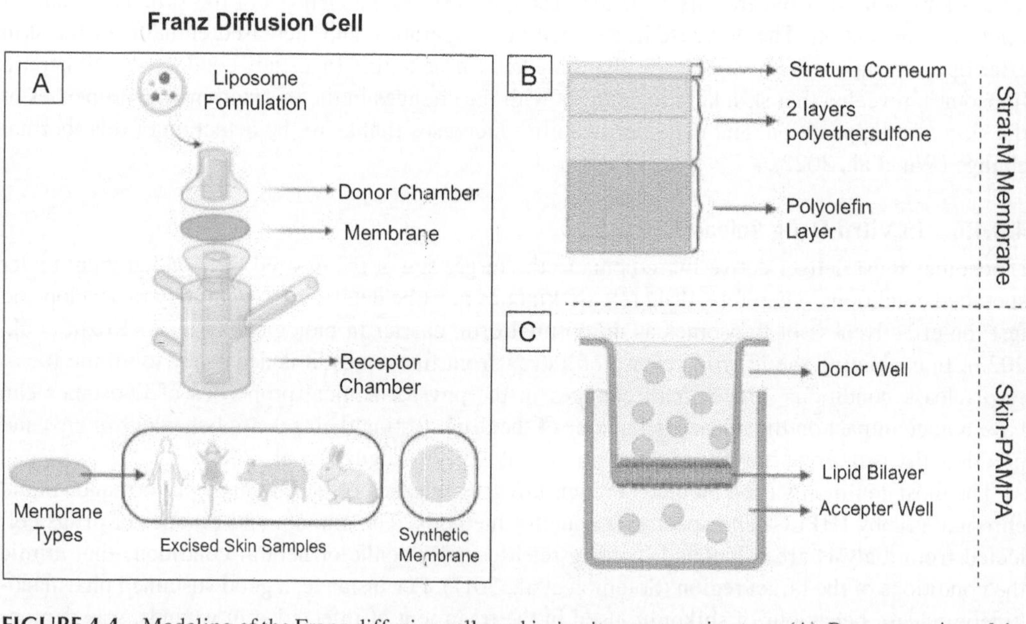

FIGURE 4.4 Modeling of the Franz diffusion cell used in *in vitro* permeation tests (A), Representation of the layers of the Strat-M membrane (B), illustration of the Skin-PAMPA permeation test (C).

Skin models used in *in vitro* skin permeability experiments have some disadvantages, such as low storage stability and inter-sample skin variations (Schlich et al., 2022). For such reasons, *in vitro* models based on artificial mimetic membranes have been developed that can offer an alternative. The most studied artificial membranes are lipid-based and non-lipid-based models. Among non-lipid-based models, microporous membranes based on cellulose acetate, polydimethylsiloxane, and polyethersulfone are frequently used to determine the permeability of drugs across membranes (Chaturvedi and Garg, 2021). To obtain more information about the stability and thermodynamic activity of Tretinoin, which plays a role in epithelial cell growth, in liposomal formulations, Tretinoin diffusion was studied using lipophilic membranes (Sinico et al., 2005). However, due to the low success of these synthetic membranes in testing hydrophilic compounds (Miki et al., 2015) such as Strat-M®, which is designed to mimic human skin new artificial membranes have been developed. Strat-M® consists of an upper layer that mimics the lipid composition of the human stratum corneum and the epidermis, consisting of polyether sulfonates that exhibit a similar morphology to the dermis (Figure 4.4B) (Rasmussen et al., 2010). Recent studies suggest the use of Strat-M® in the evaluation of topical nanoformulations. A recent study aimed to develop a topical formulation to actively transduce basal cell carcinoma tumour cells. In this study, membrane permeability test analysis of 5FU-loaded liposomes modified with aptamer was performed using the Strat-M® artificial membrane, which contains the target markers of the aptamer on its surface (Cadinoiu et al., 2021). One instance of model membranes based on lipids is the Parallel Artificial Membrane Permeability Test (PAMPA).

PAMPA consists of a hydrophobic filter coated with phosphatidylcholines, a donor compartment containing the drug formulation, and an acceptor compartment containing buffer liquid (Figure 4.4C) Skin-PAMPA models were developed by adding synthetic ceramides, cholesterol, and stearic acid to the filter coating that simulates the SC (Chi et al., 2019). The skin-PAMPA test was applied to evaluate *in vitro* the permeability of liposome formulations produced to improve the topical delivery of resveratrol. As a result of the Skin-PAMPA test, it has been proven that liposome formulations increase the cutaneous permeability of resveratrol (Casamonti et al., 2019).

Cell-based *in vitro* skin models can provide useful information about the interaction of liposomes with living cells in a 3D tissue and are therefore among the systems used to evaluate the permeation of drugs. In some skin penetration studies performed using the *in vitro* skin model, there is no need to use devices such as Frenz Cells (van Ravenzwaay and Leibold, 2004). In particular, the EpiSkin model has been used to test the absorption of vitamins C and E into the skin (Meena and Mohandass, 2019). Despite these advantages, there are suggestions that *in vitro* cell-based models have not yet developed sufficiently to be used as the sole tool for testing the skin penetration of liposomes. In addition, there is a thought that studies using cell-based 3D models should be supported by penetration studies using human or porcine skin explants (Schlich et al., 2022).

4.3.2 In Vitro Cell Studies

Product safety tests and risk/benefit assessments are performed to evaluate the effectiveness of topical products and to identify outcomes that may cause adverse skin reactions. Instead of previous studies focusing on animals, 2D and 3D cell-based skin models are preferred today (Ponec, 2002). The majority of cell-based analyses by researchers are performed in two-dimensional monolayer cell cultures cultured on flat and rigid substrates (Rasmussen et al., 2010; Lukowski et al., 2017). Nonetheless, as assessments relying on 2D cell culture models fall short in accurately representing physiological processes like the intracellular transport of nanocarriers, 3D cell culture and *in vivo* methodologies are employed (Rasmussen et al., 2010). As a result, the standard procedure for drug discovery and screening of product formulations begins with 2D cell culture-based testing, followed by 3D cell-based models and animal model testing (Edmondson et al., 2014).

4.3.2.1 Three-Dimensional Skin Models

Cell-based 3D skin models are obtained by culturing cells in layers consisting of one or more skin cell types under *in vitro* conditions to mimic the histological organization and barrier function of the skin (Randall et al., 2018). In an effort to reduce dependence on animal models for drug and chemical testing research, bioengineered human skin equivalents (HSE) have been created as cell-based skin models. Due to the presence of live and metabolically active cells, these models are ideal for performing tests on skin irritation, corrosion, toxicity, etc. HSEs are divided into two parts: reconstructed human epidermis (RHE) and full thickness (FT) (containing both epidermal and dermal layers) (Phatale et al., 2022). While RHE is commercially available under brands such as EpiDerm™, Episkin, and SkinEthic, brands such as EpiDermFT®, Vitrolife-Skin™, and GraftSkin® are available for FT (Ponec, 2002). In particular, the use of RHE systems in skin irritation (test no 439) and skin corrosion (test no 431) tests has been accepted by the OECD (www.oecd.org/.) From this perspective, HSE models, which can be designed according to the needs of the experiments, are an extraordinary tool for the pharmacological and toxicological evaluation of topical liposome formulations on skin cells (Meena and Mohandass, 2019). For instance, the HSE model was used to demonstrate that liposomes loaded with antioxidants can reduce mustard gas analogue-induced cell damage (Schlich et al., 2022). In a different study, the irritation potential on healthy cells of the 5-fluorouracil liposomal formulation, which was developed for the topical treatment of basal cell carcinoma, was determined using RHE (Cadinoiu et al., 2021).

4.3.3 In Vitro Toxicity Evaluation

Since some drug carriers can penetrate the deep dermal layer and reach the systemic circulation, the toxicology of their interactions with the skin needs to be investigated for their possible risks (Filon et al., 2015). *In vitro* toxicity assays are used to investigate the overall cytotoxicity or genotoxicity of formulations. *In vitro*, assays allow simpler, faster, and more cost-effective assessment of toxicity (Kroll et al., 2009).

4.3.3.1 In Vitro Cytotoxicity Analyses

The first step in evaluating the cytotoxicity of nanodrug carriers is through *in vitro* analysis of cell lines (Soares et al., 2018). Cell viability is the most commonly used method for cytotoxicity analysis of lipid vesicles. Different *in vitro* cytotoxicity assays are available for their analysis (Guillot et al., 2023). The most commonly used ones are the evaluation of mitochondrial activity (Perez et al., 2016) apoptotic-necrotic cell staining, and detection of Caspase-3 (Kumar et al., 2017).

4.3.3.1.1 Detection of Mitochondrial Activity

Colourimetric MTT assay is a widely used method for the evaluation of metabolically active cells based on the reduction of the yellow tetrazolium dye 3-(4,5-dimethylthiazol-2-yl)-2,5-diphenyltetrazolium bromide (MTT) to purple water-insoluble formazan is a cell viability assay (Mosmann, 1983).

4.3.3.1.2 Annexin V/Propidium Iodide Staining for Apoptotic and Necrotic Cells

One of the main methods observed in the *in vitro* evaluation of nanoparticle toxicity. Fluorescence-labeled Annexin V (VAC alpha), which binds strongly to phosphatidylserine in a calcium-dependent manner, is used to detect apoptotic cells (Trotter et al., 1995). Phosphatidylserine, retained on the extracellular side of the plasma membrane, crosses the membrane with the onset of apoptosis, and Annexin V uses it to detect apoptotic cells. It enables the detection of necrotic cells by staining with propidium iodide (3,8-diamino-5-[3-(diethylmethylammonio)propyl]-6-phenylphenanthridinium diiodide) (Aubry et al., 1999).

4.3.3.1.3 Detection of the Apoptosis Marker Caspase-3

Once Caspase-3 is activated, cell death is inevitable, so detection of Caspase-3 is one of the most widely used apoptosis assays (Waterhouse and Green, 1999). Activated Caspase-3 is assayed by measuring the cleavage of Caspase-3 substrates (Srinivasula et al., 2001).

4.3.3.1.4 Genotoxicity Assay

The genotoxicity effect of any ingredient in pharmaceutical formulations is a change in the structure, sequence, and segregation of DNA molecules and needs to be detected at an early stage. Currently, in addition to 2D cell lines, various 3D skin models such as EpisSkinTM, SkinEthic RHE, and EpiDerm are routinely used in various laboratories in the genotoxicological evaluation of topical formulations (Meena and Mohandass, 2019).

Comet assay (single-cell gel electrophoresis) is the standard protocol for determining the genotoxicity of any compound. Nanocarriers can cause DNA damage either directly by interacting with DNA or indirectly by inducing inflammation-causing reactive oxygen species (Kohl et al., 2020). The comet assay is used to measure these DNA strand breaks in eukaryotic cells. Cells embedded in agarose on a microscope slide are lysed in a solution containing detergent and high salt to remove histones from the DNA. During electrophoresis, DNA breaks release the supercoiling, resulting in comet-like appearances. The density of the comet tail reflects the number of DNA breaks (Magdolenova et al., 2014). In a study in which 5-fluorouracil and resveratrol were co-encapsulated into liposomes, Comet assay, and Caspase-3 assay were used to examine the synergistic antitumor effect of this formulation on skin cancer cells. The detection of genomic DNA breaks and caspase-3 activation tests showed a high correlation with each other and were included in this study as complementary to each other (Cosco et al., 2015). Additionally, another study demonstrated the protective effect of a new azelaic acid liposomal formulation against hydrogen peroxide-induced cellular DNA damage using the comet assay (Pasca et al., 2022).

4.3.3.2 Skin Irritation and Skin Corrosion Test

Skin irritation and allergic reactions may occur on the skin exposed to some nanocarriers and the high concentration of the drug released (Ema et al., 2013; Lotem et al., 2000). Traditionally, assessment of skin irritation and skin corrosion potential has been made by the Draize test, in which test compounds are applied to the skin of albino rabbits (Veloso et al., 2022). Instead of this method, researchers prefer reconstructed human epidermis (RHE) tissue models, which are accepted by the Food and Drug Administration (FDA) and European Medicines Agency (EMA), in the evaluation of skin irritation. Unlike traditional monolayer cell culture systems, RHE tissue models mimic skin physiology and thus allow the application of topical formulations (Oecd, 2023). Assessment of skin irritation potential is based on the principle of detecting cell damage by penetration of the tested substances into the RHE tissue model and subsequent activation of the inflammatory cascade. In this method, cell viability is measured using MTT, and test compounds are classified as irritating or corrosive if cell viability is less than or equal to 50% after exposure (Veloso et al., 2022). A study found that although the 5-FU loaded liposome had good compatibility with topical formulations, it had irritating potential when used with SkinEthic™ RHE tissues (Cadinoiu et al., 2021).

4.3.3.2.1 Evaluation of Inflammatory Response

Liposomes are considered safe ingredients, but chemical modifications and other additional ingredients in the formulation can lead to increased toxicity of liposomes and the emergence of immune responses (Inglut et al., 2020). In particular, its interaction with proteins and vesicular systems triggers the production of cytokines, an innate immune response (Sklenarova et al., 2023; van Och et al., 2005).

Enzyme-linked immunosorbent assay (ELISA) is a frequently used method to detect inflammatory marker production *in vitro*. It enables the detection of inflammatory markers in cell culture supernatants through antibodies and enzymatic reactions. The most commonly tested inflammatory markers are the chemokine Interleukin-8 (IL-8), Interleukin-6 (IL-6), and Tumour Necrosis Factor-Alpha (TNF-α) (Lequin, 2005). ELISA method is also used to test the treatment effectiveness of liposomal formulations developed for wound healing and treatment of skin diseases. In a recent study, curcumin-containing liposomes (RC-Lip) that mimic the red blood cell membrane, developed for the treatment of infected diabetic wounds, were produced and the response of treatment with RC-Lip to the inflammatory response was measured by ELISA method. As a result of the test, it was shown that the level of interleukin-1β decreased, and the results showed that the healing of the infected wound was achieved by RC-Lip's modulation of the inflammatory response (Tang et al., 2023). In another study, the decrease in the levels of interleukins, which are seen at high levels in psoriasis, after treatment with Ibrutinib and curcumin-loaded liposomes was detected by ELISA. It was concluded through this method that there is a synergistic combination of the two components in healing psoriasis inflammation (Jain et al., 2022).

4.3.4 Topical Applications of Liposomal Formulations

Liposomes have been used for anti-fungal treatment (Garg et al., 2020; Asadi et al., 2021), acne treatment (Rahman et al., 2016; Eroğlu et al., 2020), diabetic ulcer treatment (Fukui et al., 2017; Pandey et al., 2023) skin cancer (Krishnan and Mitragotri, 2020b), and also for cosmetic purposes such as antiaging, body moisturizer, skin repair, anti-cellulite, sunscreen creams, and/or gels (Barge et al., 2023; Tenchov et al., 2021). For the first time, vesicles' efficacy for skin delivery was documented in 1980, indicating that liposomal formulations may improve the topical release of medications (Mezei and Gulasekharam, 1980). The global liposome cosmetics market was estimated to be worth 2861.58 million USD in 2020 and is projected to increase at a compound annual growth rate (CAGR) of 7.67% from 2020 to 2027, according to the liposome cosmetic market size and share study 2021 (Barge et al., 2023). Christian Dior's anti-ageing gel "Capture" (Mueller-Goymann and Frank, 1986) was the first liposomal product in the cosmetics field (Zhou et al., 2021). Since then, many liposomal products have been manufactured as reviewed elsewhere (Antimisiaris et al., 2021; Barge et al., 2023; Pourtalebi Jahromi et al., 2023; Zhou et al., 2021). For example, Lancome noisome cream by L'Oreal is used as a moisturizing and skin-lightening agent, just as Citrolumine 8 cream by Lipoid Kosmetik. On the other hand, Liposome Proteos oil-free formula by Martiderm and Azelac Ru liposome cream and serum by Sesderma are applied for acne treatment and skin lightening (Barge et al., 2023).

In a recent study, a transdermal liposomal chrysomycin A was fabricated, and has shown that it crossed the skin barrier. Thus, it is concluded that it can be used to treat skin-related diseases such as skin cancer. Also, it displays antibacterial activity against methicillin-resistant *Staphylococcus aureus* (Cai et al., 2023). In another study, an antifungal liposomal formulation which is called PQA-Az-13 was designed. This compound showed antifungal activity against *Candida auris.* It has been reported that it is non-toxic to human dermal fibroblasts. Furthermore, this compound was used to treat fungal biofilms on porcine skin. Compared to the untreated control group, PQA-Az-13 liposomes reduced 83% of *C. auris* biofilm (Jaromin et al., 2023). Xia et al. fabricated a drug delivery system by using tetramethylpyrazine-loaded liposomes in sodium alginate and chitosan hydrogel for people suffering from atopic eczema. This hydrogel system was demonstrated to lessen atopic eczema symptoms in mice models compared to the tetramethylpyrazine group alone (Xia et al., 2024).

In a study, carboxymethyl, chitosan and hyaluronic acid hydrogel with liposomes were generated to enhance skin retention and support anti-ageing efficacy. Additionally, this hydrogel with liposomes improves collagen density and melanin expression (Wu et al., 2023). Eroglu et al. produced tetracycline HCl and tretinoin hydrogels containing liposomes to treat acne. They observed no toxicity in mice fibroblast cells. Furthermore, the hydrogels exhibited antibacterial and comedolytic activity (Eroğlu et al., 2020).

In a study, rifampicin was used in chronic wounds loaded with liposomes and encapsulated in hydroxypropyl methylcellulose gel. As a result, it showed stability over six months and the intended drug release was achieved. According to the *in vitro* cytotoxicity assay, cell viability remained at around 95% compared to the untreated control group (Jahromi et al., 2023). In another study, liposomes containing naringenin, an antioxidant, anti-inflammatory agent extracted from grapefruit, embedded in hydroxypropyl methylcellulose and hydroxyethyl cellulose gels to treat skin cancer. Human dermal fibroblast and keratinocyte cells were treated with this hydrogel. It was determined that the release of naringenin was impeded by the hydrogels as opposed to the naringenin solution. It is concluded that this controlled drug delivery system can be a good candidate for treating skin cancer (Marwah et al., 2022; Nasr et al., 2019). Yu et al. produced liposomal stromal cell-derived factor-1 alpha (SDF-1α) in a hydrogel platform and observed its effects on diabetic mice. As a result, this platform promotes angiogenesis and epidermal improvement by reducing the open wound area after a month (Yu et al., 2021). Nasr et al. used a natural biomaterial known as ferrous chlorophyllin and chitosan nanoparticles coated by liposomes to treat skin cancer. Additionally, they compared the nanoparticles on both basic cell culture and spheroid model. As a result, it showed no toxicity and bioavailability. Thus, it can be utilized as a topical drug delivery system (Nasr et al., 2019). In another study, the anti-ageing effects of collagen liposomes were studied *in vitro*. First, human keratinocytes were cultured 3D and applied 30% ethanol to cause damage. Then, the damaged and healthy cells were treated with collagen liposomes. Later, collagen and keratin gene expression levels were compared by qRT-PCR. Thus, it has been proven that collagen liposomes diffused into the skin and enhanced skin-related genes (Lee et al., 2023). In a recent article, self-adaptable liposomes with cefpodoxime proxetil, an antibiotic, were generated to overcome skin infections. Thereby, cefpodoxime proxetil is delivered in a regulated way in the infected area. Rats were used as a model organism and it was observed that even ten times higher concentrations of the drug were not lethal (Maheshwari et al., 2024). Leon et al. produced a topical anaesthetic gel by using lidocaine-containing liposomes in rats. The aim was to increase lidocaine penetration through skin which is a challenge. As a result, lidocaine containing liposomes can be efficient even at lower concentrations and also stay in the skin longer (Leon et al., 2024). In another study, liposomal lidocaine was used to reduce pain during botulinum toxin A therapy. It was observed by the patients that less pain was felt when liposomal lidocaine was used compared to basic analgesic creams. (Marani et al., 2024).

Recent findings suggest that liposomes containing ethanol and propylene glycol can encapsulate nicotine. Moreover, it was reported that the liposomes did not cause any damage to the skin. Hence, they are thought to be a safe way for nicotine to penetrate the skin (Hui Wang, 2024). Fang et al. fabricated lycopene (a natural compound derived from tomatoes) containing liposomes. Later they were coated with tobramycin, an antibiotic. Then these modified liposomes were embedded into methacrylated hyaluronic acid hydrogel. Rats with diabetic ulcer wounds are treated with this hydrogel system. It was observed that the hydrogel helped to close the wound area in two weeks (Fang et al., 2024). An important feature sought in the topical application of liposomes for the skin is bioavailability which means a drug achieves its target organ/site at a particular extent and rate (Antimisiaris et al., 2021). Liposomes are also preferred because they are safe, biodegradable, and biocompatible. As mentioned in this section, dermal liposomes are used for various purposes such as increasing penetration of some drugs, being effective even at low concentrations and controlling drug delivery (Table 4.1).

4.3.5 Conclusion

Liposomes are commonly investigated topically as drug delivery systems on the skin due to their ideal properties, which include being biocompatible, similar to the structure of a cell membrane, increasing the permeability of the skin's innate stratum corneum, providing higher bioavailability of the drug, protecting the encapsulated components from the harmful effects of the environment, and reducing dosing frequency. These advantages of liposomes have enabled the development of liposomal formulations of phytochemicals, thus facilitating the topical application of phytochemicals and increasing their bioavailability.

However, despite dozens of clinical and preclinical studies, few topical liposomal formulations have been licensed in both the United States and the European Union. One of the underlying factors is the lack of standardized *in vitro* methods. Authorities such as the FDA (Center for Drug Evaluation and Research and U.S. Food and Drug Administration, 2018) and EMA (European Medicines Agency, 2013) emphasize the need for a complete and detailed characterization of liposomal systems (Schlich et al., 2022). This issue is particularly emphasized because small changes in the purity of lipids in the formulation can greatly affect the biopharmaceutical performance of liposomal systems. Hence, all parameters affecting the physicochemical quality of liposomes and their behaviour in physiological environments should be analyzed in depth. If all these obstacles are overcome, liposomes will continue to be promising carriers for topical applications of phytochemicals due to their similarity to biological membranes, affinity with skin components, and ability to simultaneously load hydrophilic and lipophilic molecules.

REFERENCES

Adiguzel, S., Karamese, M., Kugu, S., Kacar, E. A., Esen, M. F., Erdogan, H., . . . & Altuntas, S. 2024. "Doxorubicin-loaded liposome-like particles embedded in chitosan/hyaluronic acid-based hydrogels as a controlled drug release model for local treatment of glioblastoma". *International Journal of Biological Macromolecules*, 278: 135054.

Allaw, Mohamad, Maria Letizia Manca, Ines Castangia, and Maria Manconi. 2022. "From Plants to Phospholipid Vesicles: A Comprehensive Review on the Incorporation of Phytochemicals into Phospholipid Vesicles Designed for Skin Applications with Special Focus on Scalability and In Vitro and In Vivo Efficacy." *Journal of Drug Delivery Science and Technology*. Editions de Sante. https://doi.org/10.1016/j.jddst.2021.103049.

Allen, Theresa M., and Pieter R. Cullis. 2013. "Liposomal Drug Delivery Systems: From Concept to Clinical Applications." *Advanced Drug Delivery Reviews*. https://doi.org/10.1016/j.addr.2012.09.037.

Andra, Veera Venkata Satya Naga Lakshmi, S. V.N. Pammi, Lakshmi Venkata Krishna Priya Bhatraju, and Lakshmi Kalyani Ruddaraju. 2022. "A Comprehensive Review on Novel Liposomal Methodologies, Commercial Formulations, Clinical Trials and Patents." *BioNanoScience*. Springer. https://doi.org/10.1007/s12668-022-00941-x.

Antimisiaris, S. G., A. Marazioti, M. Kannavou, E. Natsaridis, F. Gkartziou, G. Kogkos, and S. Mourtas. 2021. "Overcoming Barriers by Local Drug Delivery with Liposomes." *Advanced Drug Delivery Reviews*. Elsevier B.V. https://doi.org/10.1016/j.addr.2021.01.019.

Asadi, Peyman, Ahmad Mehravaran, Nahid Soltanloo, Mahdi Abastabar, and Javad Akhtari. 2021. "Nanoliposome-Loaded Antifungal Drugs for Dermal Administration: A Review." *Current Medical Mycology*. Mazandaran University of Medical Sciences. https://doi.org/10.18502/CMM.7.1.6247.

Aubry, J. P., A. Blaecke, S. Lecoanet-Henchoz, P. Jeannin, N. Herbault, G. Caron, et al. 1999. "Annexin V Used for Measuring Apoptosis in the Early Events of Cellular Cytotoxicity." *Cytometry* 37: 197–204. https://doi.org/10.1002/(SICI)1097-0320(19991101)37:3<197::AID-CYTO6>3.0.CO;2-L.

Bangham, A. D., M. M. Standish, and G. Weissmann. 1965. "The Action of Steroids and Streptolysin S on the Permeability of Phospholipid Structures to Cations." *Journal of Molecular Biology* 13 (1): 253–259. https://doi.org/10.1016/S0022-2836(65)80094-8.

Barge, Kajal, Shivali Tank, Bhagyashri Parab, Mangal Nagarsenker, and Supriya Shidhaye. 2023. "Liposomal Cosmeceuticals as Skin Protectives and Curatives." *Indian Journal of Pharmaceutical Education and Research* 58 (1): 34–44. https://doi.org/10.5530/ijper.58.1.4.

Baudot, Charles, Cher Ming Tan, and Jeng Chien Kong. 2010. "FTIR Spectroscopy as a Tool for Nano-Material Characterization." *Infrared Physics and Technology* 53 (6): 434–438. https://doi.org/10.1016/j.infrared.2010.09.002.

Bouwstra, J. A., and P. L. Honeywell-Nguyen. 2002. "S Kin Structure and Mode of Action of Vesicles." *Advanced Drug Delivery Reviews* 54. www.elsevier.com/locate/drugdeliv.

Bozzuto, Giuseppina, and Agnese Molinari. 2015. "Liposomes as Nanomedical Devices." *International Journal of Nanomedicine*. Dove Medical Press Ltd. https://doi.org/10.2147/IJN.S68861.

Briuglia, Maria Lucia, Chiara Rotella, Amber McFarlane, and Dimitrios A. Lamprou. 2015. "Influence of Cholesterol on Liposome Stability and on In Vitro Drug Release." *Drug Delivery and Translational Research* 5 (3): 231–242. https://doi.org/10.1007/s13346-015-0220-8.

Bulbake, Upendra, Sindhu Doppalapudi, Nagavendra Kommineni, and Wahid Khan. 2017. "Liposomal Formulations in Clinical Use: An Updated Review." *Pharmaceutics*. MDPI AG. https://doi.org/10.3390/pharmaceutics9020012.

Cadinoiu, Anca N., Delia M. Rata, Leonard I. Atanase, Cosmin T. Mihai, Simona E. Bacaita, and Marcel Popa. 2021. "Formulations Based on Drug Loaded Aptamer-Conjugated Liposomes as a Viable Strategy for the Topical Treatment of Basal Cell Carcinoma – In Vitro Tests." *Pharmaceutics* 13 (6). https://doi.org/10.3390/pharmaceutics13060866.

Cai, Yue, Yuteng Chu, Yubei Gong, Yulu Hong, Fuhang Song, Hong Wang, Huawei Zhang, and Xuanrong Sun. 2023. "Enhanced Transdermal Peptide-Modified Flexible Liposomes for Efficient Percutaneous Delivery of Chrysomycin A to Treat Subcutaneous Melanoma and Intradermal MRSA Infection." *Advanced Healthcare Materials* 12 (26). https://doi.org/10.1002/adhm.202300881.

Campani, Virginia, Lorena Scotti, Teresa Silvestri, Marco Biondi, and Giuseppe De Rosa. 2020. "Skin Permeation and Thermodynamic Features of Curcumin-Loaded Liposomes." *Journal of Materials Science: Materials in Medicine* 31 (2). https://doi.org/10.1007/s10856-019-6351-6.

Carter, Prerana, Balaji Narasimhan, and Qun Wang. 2019. "Biocompatible Nanoparticles and Vesicular Systems in Transdermal Drug Delivery for Various Skin Diseases." *International Journal of Pharmaceutics* 555 (January): 49–62. https://doi.org/10.1016/j.ijpharm.2018.11.032.

Casamonti, Marta, Vieri Piazzini, Anna Rita Bilia, and Maria Bilia Bergonzi. 2019. "Evaluation of Skin Permeability of Resveratrol Loaded Liposomes and Nanostructured Lipid Carriers Using a Skin Mimic Artificial Membrane (Skin-PAMPA)." *Drug Delivery Letters* 9 (2): 134–145.

Center for Drug Evaluation and Research and U.S. Food and Drug Administration. 2018. *Liposome Drug Products: Chemistry, Manufacturing, and Controls; Human Pharmacokinetics and Bioavailability, and Labeling Documentation*. FDA.

Cevc, Gregor, Andreas Schätzlein, and Holger Richardsen. 2002. "Ultradeformable Lipid Vesicles Can Penetrate the Skin and Other Semi-Permeable Barriers Unfragmented. Evidence from Double Label CLSM Experiments and Direct Size Measurements." www.bba-direct.com.

Chacko, Indhu A., Vivek M. Ghate, Leonna Dsouza, and Shaila A. Lewis. 2020. "Lipid Vesicles: A Versatile Drug Delivery Platform for Dermal and Transdermal Applications." *Colloids and Surfaces B: Biointerfaces*. Elsevier B.V. https://doi.org/10.1016/j.colsurfb.2020.111262.

Chaturvedi, Shashank, and Anuj Garg. 2021. "An Insight of Techniques for the Assessment of Permeation Flux across the Skin for Optimization of Topical and Transdermal Drug Delivery Systems: 'Modelling the Topical and Transdermal Drug Delivery Systems'." *Journal of Drug Delivery Science and Technology*. Editions de Sante. https://doi.org/10.1016/j.jddst.2021.102355.

Chatzikleanthous, Despo, Derek T. O'Hagan, and Roberto Adamo. 2021. "Lipid-Based Nanoparticles for Delivery of Vaccine Adjuvants and Antigens: Toward Multicomponent Vaccines." *Molecular Pharmaceutics*. American Chemical Society. https://doi.org/10.1021/acs.molpharmaceut.1c00447.

Chi, Cheng Ting, Ming Han Lee, Ching Feng Weng, and Max K. Leong. 2019. "In Silico Prediction of PAMPA Effective Permeability Using a Two-QSAR Approach." *International Journal of Molecular Sciences* 20 (13). https://doi.org/10.3390/ijms20133170.

Choi, M. J., and H. I. Maibach. 2005. "Liposomes and Niosomes as Topical Drug Delivery Systems." *Skin Pharmacology and Physiology*. https://doi.org/10.1159/000086666.

Choudhary, Vikram, Hosakote Shivakumar, and Himanshu Ojha. 2019. "Curcumin-Loaded Liposomes for Wound Healing: Preparation, Optimization, In Vivo Skin Permeation and Bioevaluation." *Journal of Drug Delivery Science and Technology* 49 (February): 683–691. https://doi.org/10.1016/j.jddst.2018.12.008.

Cosco, Donato, Donatella Paolino, Jessica Maiuolo, Luisa Di Marzio, Maria Carafa, Cinzia A. Ventura, and Massimo Fresta. 2015. "Ultradeformable Liposomes as Multidrug Carrier of Resveratrol and 5-Fluorouracil for Their Topical Delivery." *International Journal of Pharmaceutics* 489 (1–2): 1–10. https://doi.org/10.1016/j.ijpharm.2015.04.056.

Costa, Pedro M., and Bengt Fadeel. 2016. "Emerging Systems Biology Approaches in Nanotoxicology: Towards a Mechanism-Based Understanding of Nanomaterial Hazard and Risk." *Toxicology and Applied Pharmacology* 299 (May): 101–111. https://doi.org/10.1016/j.taap.2015.12.014.

Edmondson, Rasheena, Jessica Jenkins Broglie, Audrey F. Adcock, and Liju Yang. 2014. "Three-Dimensional Cell Culture Systems and Their Applications in Drug Discovery and Cell-Based Biosensors." *Assay and Drug Development Technologies*. Mary Ann Liebert Inc. https://doi.org/10.1089/adt.2014.573.

Ema, Makoto, Akitaka Matsuda, Norihiro Kobayashi, Masato Naya, and Junko Nakanishi. 2013. "Dermal and Ocular Irritation and Skin Sensitization Studies of Fullerene C60 Nanoparticles." *Cutaneous and Ocular Toxicology* 32 (2): 128–134. https://doi.org/10.3109/15569527.2012.727937.

Eroğlu, İpek, Minela Aslan, Ümran Yaman, Merve Gultekinoglu, Semih Çalamak, Didem Kart, and Kezban Ulubayram. 2020. "Liposome-Based Combination Therapy for Acne Treatment." *Journal of Liposome Research* 30 (3): 263–273. https://doi.org/10.1080/08982104.2019.1630646.

European Medicines Agency. 2013. "Data Requirements for Intravenous Liposomal Products Developed with Reference to an Innovator Liposomal Product." https://www.ema.europa.eu/en/data-requirements-intravenous-liposomal-products-developed-reference-innovator-liposomal-product-0 (Accessed October 31, 2024).

Fang, Yifei, Tianqi Nie, Guangze Li, Liying Wang, Jianhang Du, and Jun Wu. 2024. "Multifunctional Antibiotic Hydrogel Doped with Antioxidative Lycopene-Based Liposome for Accelerative Diabetic Wound Healing." *Chemical Engineering Journal* 480 (January). https://doi.org/10.1016/j.cej.2023.147930.

Filon, Larese, Francesca, Marcella Mauro, Gianpiero Adami, Massimo Bovenzi, and Matteo Crosera. 2015. "Nanoparticles Skin Absorption: New Aspects for a Safety Profile Evaluation." *Regulatory Toxicology and Pharmacology*. Academic Press Inc. https://doi.org/10.1016/j.yrtph.2015.05.005.

Flaten, Gøril Eide, Zora Palac, André Engesland, Jelena Filipović-Grčić, Željka Vanić, and Nataša Škalko-Basnet. 2015. "In Vitro Skin Models as a Tool in Optimization of Drug Formulation." *European Journal of Pharmaceutical Sciences*. Elsevier B.V. https://doi.org/10.1016/j.ejps.2015.02.018.

Fukui, Tsuyoshi, Akira T. Kawaguchi, Susumu Takekoshi, Muneo Miyasaka, Hideaki Sumiyoshi, and Rica Tanaka. 2017. "Liposome-Encapsulated Hemoglobin Accelerates Skin Wound Healing in Diabetic DB/DB Mice." *Artificial Organs* 41 (4): 319–326. https://doi.org/10.1111/aor.12864.

Garg, Abhinava, Ganti S. Sharma, Amit K. Goyal, Goutam Ghosh, Sudam Chandra Si, and Goutam Rath. 2020. "Recent Advances in Topical Carriers of Anti-Fungal Agents." *Heliyon*. Elsevier Ltd. https://doi.org/10.1016/j.heliyon.2020.e04663.

Giordani, Stefano, Valentina Marassi, Andrea Zattoni, Barbara Roda, and Pierluigi Reschiglian. 2023. "Liposomes Characterization for Market Approval as Pharmaceutical Products: Analytical Methods, Guidelines and Standardized Protocols." *Journal of Pharmaceutical and Biomedical Analysis*. Elsevier B.V. https://doi.org/10.1016/j.jpba.2023.115751.

Gioria, Sabrina, Fanny Caputo, Patricia Urbán, Ciarán Manus Maguire, Susanne Bremer-Hoffmann, Adriele Prina-Mello, Luigi Calzolai, and Dora Mehn. 2018. "Are Existing Standard Methods Suitable for the Evaluation of Nanomedicines: Some Case Studies." *Nanomedicine*. Future Medicine Ltd. https://doi.org/10.2217/nnm-2017-0338.

Gorzelanny, Christian, Christian Mess, Stefan W. Schneider, Volker Huck, and Johanna M. Brandner. 2020. "Skin Barriers in Dermal Drug Delivery: Which Barriers Have to Be Overcome and How Can We Measure Them?" *Pharmaceutics*. MDPI AG. https://doi.org/10.3390/pharmaceutics12070684.

Gu, Bing, and Diane J. Burgess. 2014. "Polymeric Materials in Drug Delivery." In *Natural and Synthetic Biomedical Polymers*, 333–349. Elsevier Inc. https://doi.org/10.1016/B978-0-12-396983-5.00021-1.

Guillot, Antonio José, Miquel Martínez-Navarrete, Teresa M. Garrigues, and Ana Melero. 2023. "Skin Drug Delivery Using Lipid Vesicles: A Starting Guideline for Their Development." *Journal of Controlled Release*. Elsevier B.V. https://doi.org/10.1016/j.jconrel.2023.02.006.

Huang, Hai, Caili Zhang, Shuping Yang, Wen Xiao, Qian Zheng, and Xiangrong Song. 2021. "The Investigation of MRNA Vaccines Formulated in Liposomes Administrated in Multiple Routes against SARS-CoV-2." *Journal of Controlled Release* 335 (July): 449–456. https://doi.org/10.1016/j.jconrel.2021.05.024.

Inglut, Collin T., Aaron J. Sorrin, Thilinie Kuruppu, Shruti Vig, Julia Cicalo, Haroon Ahmad, and Huang Chiao Huang. 2020. "Immunological and Toxicological Considerations for the Design of Liposomes." *Nanomaterials*. MDPI AG. https://doi.org/10.3390/nano10020190.

Ioele, Giuseppina, Lorena Tavano, Michele De Luca, Gaetano Ragno, Nevio Picci, and Rita Muzzalupo. 2015. "Photostability and Ex-Vivo Permeation Studies on Diclofenac in Topical Niosomal Formulations." *International Journal of Pharmaceutics* 494 (1): 490–497. https://doi.org/10.1016/j.ijpharm.2015.08.053.

Jaafar, Hawraz Jawdat. 2023. "The Importance of Nanopharmaceuticals and Nanomedicines Applications and Properties: A Current Review." *Technium BioChemMed: Journal of Multidisciplinary Research, Biology, Chemistry and Medicine* 5.

Jahromi, Pourtalebi, Leila, Markus Rothammer, and Gregor Fuhrmann. 2023. "Polysaccharide Hydrogel Platforms as Suitable Carriers of Liposomes and Extracellular Vesicles for Dermal Applications." *Advanced Drug Delivery Reviews*. Elsevier B.V. https://doi.org/10.1016/j.addr.2023.115028.

Jain, Harsha, Devabattula Geetanjali, Harshita Dalvi, Aditi Bhat, Chandraiah Godugu, and Saurabh Srivastava. 2022. "Liposome Mediated Topical Delivery of Ibrutinib and Curcumin as a Synergistic Approach to Combat Imiquimod Induced Psoriasis." *Journal of Drug Delivery Science and Technology* 68 (February). https://doi.org/10.1016/j.jddst.2022.103103.

Jaromin, Anna, Robert Zarnowski, Adam Markowski, Agnieszka Zagórska, Chad J. Johnson, Haniyeh Etezadi, Shinji Kihara, et al. 2023. "Liposomal Formulation of a New Antifungal Hybrid Compound Provides Protection against Candida Auris in the Ex Vivo Skin Colonization Model." *Antimicrobial Agents and Chemotherapy* (January). https://doi.org/10.1128/aac.00955-23.

Jeong, Woo Yeup, Mina Kwon, Hye Eun Choi, and Ki Su Kim. 2021. "Recent Advances in Transdermal Drug Delivery Systems: A Review." *Biomaterials Research*. BioMed Central Ltd. https://doi.org/10.1186/s40824-021-00226-6.

Kapoor, Mudra Saurabh, Anisha D'Souza, Noorjahan Aibani, Swathi Sivasankaran Nair, Puja Sandbhor, Durga kumari, and Rinti Banerjee. 2018. "Stable Liposome in Cosmetic Platforms for Transdermal Folic Acid Delivery for Fortification and Treatment of Micronutrient Deficiencies." *Scientific Reports* 8 (1). https://doi.org/10.1038/s41598-018-34205-0.

Karamese, Miray, Necati Atalay, Senanur Kugu, Ozge Dinc, Ahmet Katı, Merve Bacanlı, Hakan Erdogan, and Sevde Altuntas. 2023. "FOLFIRINOX-Loaded Immunoliposome-like Particles for Localized Pancreatic Cancer Treatment." *Journal of Drug Delivery Science and Technology* 86 (September). https://doi.org/10.1016/j.jddst.2023.104595.

Kohl, Yvonne, Elise Rundén-Pran, Espen Mariussen, Michelle Hesler, Naouale El Yamani, Eleonora Marta Longhin, and Maria Dusinska. 2020. "Genotoxicity of Nanomaterials: Advanced In Vitro Models and High Throughput Methods for Human Hazard Assessment – A Review." *Nanomaterials*. MDPI AG. https://doi.org/10.3390/nano10101911.

Komatani, Takeshi S. 2016. "Patenting Drug Delivery Technology: Tips for Formulating Practical and Effective Patent Portfolio." *Pharmaceutical Patent Analyst* 5 (2): 97–100. https://doi.org/10.4155/ppa.15.42.

Kozik, Violetta, Danuta Pentak, Marlena Paździor, Andrzej Zięba, and Andrzej Bąk. 2023. "From Design to Study of Liposome-Driven Drug Release Part 1: Impact of Temperature and PH on Environment." *International Journal of Molecular Sciences* 24 (14). https://doi.org/10.3390/ijms241411686.

Krishnan, Vinu, and Samir Mitragotri. 2020a. "Nanoparticles for Topical Drug Delivery: Potential for Skin Cancer Treatment." *Advanced Drug Delivery Reviews*. Elsevier B.V. https://doi.org/10.1016/j.addr.2020.05.011.

Krishnan, Vinu, and Samir Mitragotri. 2022b. "Design and Development of Topical Liposomal Formulations in a Regulatory Perspective." *Drug Delivery and Translational Research* 12 (8): 1811–1828. https://doi.org/10.1007/s13346-021-01089-z.

Kroll, Alexandra, Mike H. Pillukat, Daniela Hahn, and Jürgen Schnekenburger. 2009. "Current In Vitro Methods in Nanoparticle Risk Assessment: Limitations and Challenges." *European Journal of Pharmaceutics and Biopharmaceutics*. https://doi.org/10.1016/j.ejpb.2008.08.009.

Kumar, Vinay, Neha Sharma, and S. S. Maitra. 2017. "In Vitro and In Vivo Toxicity Assessment of Nanoparticles." *International Nano Letters* 7 (4): 243–256. https://doi.org/10.1007/s40089-017-0221-3.

Large, Danielle E., Rudolf G. Abdelmessih, Elizabeth A. Fink, and Debra T. Auguste. 2021. "Liposome Composition in Drug Delivery Design, Synthesis, Characterization, and Clinical Application." *Advanced Drug Delivery Reviews*. Elsevier B.V. https://doi.org/10.1016/j.addr.2021.113851.

Lee, Mi So, Hoai Thuong Duc Bui, Su Ji Kim, Jun Bae Lee, and Hyuk Sang Yoo. 2023. "Liposome-Assisted Penetration and Antiaging Effects of Collagen in a 3D Skin Model." *Journal of Cosmetic Dermatology* (January). https://doi.org/10.1111/jocd.15912.

Leon, Maria Magdalena, Alexandra Maştaleru, Andra Oancea, Teodora Alexa-Stratulat, Cătălina Anişoara Peptu, Bogdan Ionel Tamba, Valeria Harabagiu, Cristina Grosu, Anisia Iuliana Alexa, and Elena Cojocaru. 2024. "Lidocaine–Liposomes – A Promising Frontier for Transdermal Pain Management." *Journal of Clinical Medicine* 13 (1). https://doi.org/10.3390/jcm13010271.

Lequin, Rudolf M. 2005. "Enzyme Immunoassay (EIA)/Enzyme-Linked Immunosorbent Assay (ELISA)." *Clinical Chemistry* 51 (12): 2415–2418. https://doi.org/10.1373/clinchem.2005.051532.

Liu, Junye, Anjie Zheng, Baowei Peng, Yuhong Xu, and Ning Zhang. 2021. "Size-Dependent Absorption through Stratum Corneum by Drug-Loaded Liposomes." *Pharmaceutical Research* 38: 1429–1437. https://doi.org/10.1007/s11095-021-03079-9/Published.

Lotem, Michal, Ayala Hubert, Olga Lyass, Michael A. Goldenhersh, Arieh Ingber, Tamar Peretz, and Alberto Gabizon. 2000. "Skin Toxic Effects of Polyethylene Glycol-Coated Liposomal Doxorubicin." *Archives of Dermatology* 136. http://archderm.jamanetwork.com/.

Lukowski, Jessica K., Eric M. Weaver, and Amanda B. Hummon. 2017. "Analyzing Liposomal Drug Delivery Systems in Three-Dimensional Cell Culture Models Using MALDI Imaging Mass Spectrometry." *Analytical Chemistry* 89 (16): 8453–8458. https://doi.org/10.1021/acs.analchem.7b02006.

Magdolenova, Zuzana, Andrew Collins, Ashutosh Kumar, Alok Dhawan, Vicki Stone, and Maria Dusinska. 2014. "Mechanisms of Genotoxicity. A Review of In Vitro and In Vivo Studies with Engineered Nanoparticles." *Nanotoxicology*. https://doi.org/10.3109/17435390.2013.773464.

Maguire, Ciarán Manus, Matthias Rösslein, Peter Wick, and Adriele Prina-Mello. 2018. "Characterisation of Particles in Solution–a Perspective on Light Scattering and Comparative Technologies." *Science and Technology of Advanced Materials*. Taylor and Francis Ltd. https://doi.org/10.1080/14686996.2018.1517587.

Maheshwari, Rahul, Piyush Ghode, and Mayank Sharma. 2024. "Lab on Chip Based Self-Adjustable Liposomes for Rapid Wound Healing: An in Depth In Vitro, In Vivo and Higher Dose Toxicity Investigation." *Biomaterials Advances* 158 (April): 213777. https://doi.org/10.1016/j.bioadv.2024.213777.

Malam, Yogeshkumar, Marilena Loizidou, and Alexander M. Seifalian. 2009. "Liposomes and Nanoparticles: Nanosized Vehicles for Drug Delivery in Cancer." *Trends in Pharmacological Sciences*. https://doi.org/10.1016/j.tips.2009.08.004.

Manca, Maria Letizia, Pietro Matricardi, Claudia Cencetti, Josè Esteban Peris, Virginia Melis, Claudia Carbone, Elvira Escribano, Marco Zaru, Anna Maria Fadda, and Maria Manconi. 2016. "Combination of Argan Oil and Phospholipids for the Development of an Effective Liposome-like Formulation Able to Improve Skin Hydration and Allantoin Dermal Delivery." *International Journal of Pharmaceutics* 505 (1–2): 204–211. https://doi.org/10.1016/j.ijpharm.2016.04.008.

Marani, Andrea, Helena Gioacchini, Matteo Paolinelli, Ivan Bobyr, Emanuela Martina, Giulia Radi, Federico Diotalallevi, and Anna Campanati. 2024. "Pain Control during the Treatment of Primary Palmar Hyperhidrosis with Botulinum Toxin A by a Topical Application of Liposomal Lidocaine: Clinical Effectiveness." *Toxins* 16 (1): 28. https://doi.org/10.3390/toxins16010028.

Marwah, Mandeep, Raj K. S. Badhan, and Deborah Lowry. 2022. "Development of a Novel Polymer-Based Carrier for Deformable Liposomes for the Controlled Dermal Delivery of Naringenin." *Journal of Liposome Research* 32 (2): 181–194. https://doi.org/10.1080/08982104.2021.1956529.

Meena, Surya Nandan, and Chellandi Mohandass. 2019. "Application of 3D Cell Culture Techniques in Cosmeceutical Research." In *Advances in Biological Science Research: A Practical Approach*, 469–484. Elsevier. https://doi.org/10.1016/B978-0-12-817497-5.00029-X.

Mezei, Michael, and Vijeyalakshmi Gulasekharam. 1980. "Liposomes-a Selective Drug Delivery System for the Topical Route of Administration I. Lotion Dosage Form." *Life Sciences* 26.

Miki, Ryotaro, Yasuna Ichitsuka, Takumi Yamada, Soichiro Kimura, Yuya Egawa, Toshinobu Seki, Kazuhiko Juni, Hideo Ueda, and Yasunori Morimoto. 2015. "Development of a Membrane Impregnated with a Poly(Dimethylsiloxane)/Poly(Ethylene Glycol) Copolymer for a High-Throughput Screening of the Permeability of Drugs, Cosmetics, and Other Chemicals Across the Human Skin." *European Journal of Pharmaceutical Sciences* 66 (January): 41–49. https://doi.org/10.1016/j.ejps.2014.09.024.

Mojumdar, Enamul Haque, Quoc Dat Pham, Daniel Topgaard, and Emma Sparr. 2017. "Skin Hydration: Interplay between Molecular Dynamics, Structure and Water Uptake in the Stratum Corneum." *Scientific Reports* 7 (1). https://doi.org/10.1038/s41598-017-15921-5.

Mosmann, Tim. 1983. "Rapid Colorimetric Assay for Cellular Growth and Survival: Application to Proliferation and Cytotoxicity Assays." *Journal of ImmunologicalMethods* 65.

Mueller-Goymann, C. C., and S. G. Frank. 1986. "Interaction of Lidocaine and Lidocaine-HCl with the Liquid Crystal Structure of Topical Preparations." *International Journal of Pharmaceutics* 29 (2–3): 147–159.

Nag, Okhil K., and Vibhudutta Awasthi. 2013. "Surface Engineering of Liposomes for Stealth Behavior." *Pharmaceutics*. https://doi.org/10.3390/pharmaceutics5040542.

Nasr, Soad, Mai Rady, Iman Gomaa, Tatiana Syrovet, Thomas Simmet, W. Fayad, and Mahmoud Abdel-Kader. 2019. "Ethosomes and Lipid-Coated Chitosan Nanocarriers for Skin Delivery of a Chlorophyll Derivative: A Potential Treatment of Squamous Cell Carcinoma by Photodynamic Therapy." *International Journal of Pharmaceutics* 568 (September). https://doi.org/10.1016/j.ijpharm.2019.118528.

Neupane, Rabin, Sai H. S. Boddu, Jwala Renukuntla, R. Jayachandra Babu, and Amit K. Tiwari. 2020. "Alternatives to Biological Skin in Permeation Studies: Current Trends and Possibilities." *Pharmaceutics*. MDPI AG. https://doi.org/10.3390/pharmaceutics12020152.

Ng, Shiow Fern, Jennifer J. Rouse, Francis D. Sanderson, and Gillian M. Eccleston. 2012. "The Relevance of Polymeric Synthetic Membranes in Topical Formulation Assessment and Drug Diffusion Study." *Archives of Pharmacal Research*. https://doi.org/10.1007/s12272-012-0401-7.

Nsairat, Hamdi, Dima Khater, Usama Sayed, Fadwa Odeh, Abeer Al Bawab, and Walhan Alshaer. 2022. "Liposomes: Structure, Composition, Types, and Clinical Applications." *Heliyon*. Elsevier Ltd. https://doi.org/10.1016/j.heliyon.2022.e09394.

Odlyha, Marianne. 2003. "Chapter 2 the Applications of Thermoanalytical Techniques to the Preservation of Art and Archaeological Objects." In *Handbook of Thermal Analysis and Calorimetry*. Vol. 2. https://doi.org/10.1016/S1573-4374(03)80006-7.

Oecd, Ocde. 2023. "Section 4 Health Effects OECD Guideline for Testing of Chemicals." www.oecd.org/termsandconditions/.

Oh, Luke, Sojeong Yi, Da Zhang, Soo Hyeon Shin, and Edward Bashaw. 2019. "In Vitro Skin Permeation Methodology for Over-The-Counter Topical Dermatologic Products." *Therapeutic Innovation and Regulatory Science*. SAGE Publications Inc. https://doi.org/10.1177/2168479019875338.

Pandey, Supriya, Arshiya Shamim, Mohammad Shaif, and Poonam Kushwaha. 2023. "Development and Evaluation of Resveratrol-Loaded Liposomes in Hydrogel-Based Wound Dressing for Diabetic Foot Ulcer." *Naunyn-Schmiedeberg's Archives of Pharmacology* 396 (8): 1811–1825. https://doi.org/10.1007/s00210-023-02441-5.

Park, Kinam. 2019. "Probing the Mechanism of Drug Release from Liposomes." *Journal of Controlled Release*. Elsevier B.V. https://doi.org/10.1016/j.jconrel.2019.01.003.

Pasca, Paula Melania, Florina Miere, Angela Antonescu, Luminita Fritea, Florin Banica, Simona Ioana Vicas, Vasile Laslo, Dana Carmen Zaha, and Simona Cavalu. 2022. "Novel Liposomal Formulation with Azelaic Acid: Preparation, Characterization, and Evaluation of Biological Properties." *Applied Sciences (Switzerland)* 12 (24). https://doi.org/10.3390/app122413039.

Patel, Ravı V., Gazala Y. Ansari, Suraj R. Chauhan, and Jitendra O. Bhangale. 2021. "Review: Transdermal Patch." *International Journal of Biology, Pharmacy and Allied Sciences* 10 (12 (Special Issue)). https://doi.org/10.31032/ijbpas/2021/10.12.1010.

Perez, Ana Paula, Maria Julia Altube, Priscila Schilrreff, Gustavo Apezteguia, Fabiana Santana Celes, Susana Zacchino, Camila Indiani de Oliveira, Eder Lilia Romero, and Maria Jose Morilla. 2016. "Topical Amphotericin B in Ultradeformable Liposomes: Formulation, Skin Penetration Study, Antifungal and Antileishmanial Activity in Vitro." *Colloids and Surfaces B: Biointerfaces* 139 (March): 190–198. https://doi.org/10.1016/j.colsurfb.2015.12.003.

Phatale, Vivek, Klaudi K. Vaiphei, Shikha Jha, Dnyaneshwar Patil, Mukta Agrawal, and Amit Alexander. 2022. "Overcoming Skin Barriers through Advanced Transdermal Drug Delivery Approaches." *Journal of Controlled Release*. Elsevier B.V. https://doi.org/10.1016/j.jconrel.2022.09.025.

Ponec, Maria. 2002. "S Kin Constructs for Replacement of Skin Tissues for In Vitro Testing." *Advanced Drug Delivery Reviews* 54. www.elsevier.com/locate/drugdeliv.

Rahman, Salwa Abdel, Nevine Shawky Abdelmalak, Alia Badawi, Tahany Elbayoumy, Nermeen Sabry, and Amany El Ramly. 2016. "Tretinoin-Loaded Liposomal Formulations: From Lab to Comparative Clinical Study in Acne Patients." *Drug Delivery* 23 (4): 1184–1193. https://doi.org/10.3109/10717544.2015.1041578.

Randall, Matthew J., Astrid Jüngel, Markus Rimann, and Karin Wuertz-Kozak. 2018. "Advances in the Biofabrication of 3D Skin in Vitro: Healthy and Pathological Models." *Frontiers in Bioengineering and Biotechnology*. Frontiers Media S.A. https://doi.org/10.3389/fbioe.2018.00154.

Rasmussen, Cathy, Ken Gratz, Frank Liebel, Michael Southall, Michelle Garay, Surjya Bhattacharyya, Nick Simon, et al. 2010. "The StrataTest® Human Skin Model, a Consistent In Vitro Alternative for Toxicological Testing." *Toxicology In Vitro* 24 (7): 2021–2029. https://doi.org/10.1016/j.tiv.2010.07.027.

Sayes, Christie M., and David B. Warheit. 2009. "Characterization of Nanomaterials for Toxicity Assessment." *WIREs Nanomedicine and Nanobiotechnology* 1: 660–670. https://doi.org/10.1002/wnan.058.

Schlich, Michele, Umberto M. Musazzi, Virginia Campani, Marco Biondi, Silvia Franzé, Francesco Lai, Giuseppe De Rosa, Chiara Sinico, and Francesco Cilurzo. 2022. "Design and Development of Topical Liposomal Formulations in a Regulatory Perspective." *Drug Delivery and Translational Research* 12 (8): 1811–1828. https://doi.org/10.1007/s13346-021-01089-z.

Scholtysek, Peggy, Anja Achilles, Claudia Viktoria Hoffmann, Bob Dan Lechner, Annette Meister, Carsten Tschierske, Kay Saalwächter, Katarina Edwards, and Alfred Blume. 2012. "A T-Shaped Amphiphilic Molecule Forms Closed Vesicles in Water and Bicelles in Mixtures with a Membrane Lipid." *Journal of Physical Chemistry B* 116 (16): 4871–4878. https://doi.org/10.1021/jp207996r.

Sercombe, Lisa, Tejaswi Veerati, Fatemeh Moheimani, Sherry Y. Wu, Anil K. Sood, and Susan Hua. 2015. "Advances and Challenges of Liposome Assisted Drug Delivery." *Frontiers in Pharmacology*. Frontiers Media S.A. https://doi.org/10.3389/fphar.2015.00286.

Sharma, N., and S. Verma. 2017. "Current and Future Prospective of Liposomes as Drug Delivery Vehicles for the Effective Treatment of Cancer. *International Journal of Green Pharmacy* 11(3): 8377. https://doi.org/10.22377/IJGP.V11I03.1145.

Shu, Gang, Dan Xu, Wei Zhang, Xiaoling Zhao, Haohuan Li, Funeng Xu, Lizi Yin, et al. 2022. "Preparation of Shikonin Liposome and Evaluation of Its In Vitro Antibacterial and In Vivo Infected Wound Healing Activity." *Phytomedicine* 99 (May). https://doi.org/10.1016/j.phymed.2022.154035.

Sinico, Chiara, Maria Manconi, Marcello Peppi, Francesco Lai, Donatella Valenti, and Anna Maria Fadda. 2005. "Liposomes as Carriers for Dermal Delivery of Tretinoin: In Vitro Evaluation of Drug Permeation and Vesicle-Skin Interaction." *Journal of Controlled Release* 103 (1): 123–136. https://doi.org/10.1016/j.jconrel.2004.11.020.

Sklenarova, Renata, Mohamad Allaw, Matteo Perra, Ines Castangia, Jana Frankova, Josè Luis Pedraz, Maria Letizia Manca, and Maria Manconi. 2023. "Co-Delivering of Oleuropein and Lentisk Oil in Phospholipid Vesicles as an Effective Approach to Modulate Oxidative Stress, Cytokine Secretion and Promote Skin Regeneration." *European Journal of Pharmaceutics and Biopharmaceutics* 185 (April): 126–136. https://doi.org/10.1016/j.ejpb.2023.02.018.

Soares, Sara, João Sousa, Alberto Pais, and Carla Vitorino. 2018. "Nanomedicine: Principles, Properties, and Regulatory Issues." *Frontiers in Chemistry*. Frontiers Media S.A. https://doi.org/10.3389/fchem.2018.00360.

Sogut, Oguz, Umran Aydemir Sezer, and Serdar Sezer. 2021. "Liposomal Delivery Systems for Herbal Extracts." *Journal of Drug Delivery Science and Technology*. Editions de Sante. https://doi.org/10.1016/j.jddst.2020.102147.

Solomon, Deepak, Nilesh Gupta, Nihal S. Mulla, Snehal Shukla, Yadir A. Guerrero, and Vivek Gupta. 2017. "Role of In Vitro Release Methods in Liposomal Formulation Development: Challenges and Regulatory Perspective." *AAPS Journal*. Springer New York LLC. https://doi.org/10.1208/s12248-017-0142-0.

Srinivasula, S., A. Saleh, M. Ahmad, T. Fernandes-Alnemri, and E. Alnemri. 2001. "Isolation and Assay of Caspases." *Methods in Cell Biology* 66: 1–27. https://doi.org/10.1016/S0091-679X(01)66002-3.

Stark, M. M. 2015. "Substance Misuse: Methods of Administration." In *Encyclopedia of Forensic and Legal Medicine: Second Edition*, 406–410. Elsevier Inc. https://doi.org/10.1016/B978-0-12-800034-2.00352-9.

Supe, Shibani, and Pooja Takudage. 2021. "Methods for Evaluating Penetration of Drug into the Skin: A Review." *Skin Research and Technology*. John Wiley and Sons Inc. https://doi.org/10.1111/srt.12968.

Tang, Qinghan, Mei Dong, Zeyu Xu, Nannan Xue, Ruihan Jiang, Xuchao Wei, Jingyue Gu, et al. 2023. "Red Blood Cell-Mimicking Liposomes Loading Curcumin Promote Diabetic Wound Healing." *Journal of Controlled Release* 361 (September): 871–884. https://doi.org/10.1016/j.jconrel.2023.07.049.

Tardi, C., M. Drechsler, K. H. Bauer, and M. Brandl. 2001. "Steam Sterilisation of Vesicular Phospholipid Gels." *International Journal of Pharmaceutics* 217. www.elsevier.com/locate/ijpharm.

Tenchov, Rumiana, Robert Bird, Allison E. Curtze, and Qiongqiong Zhou. 2021. "Lipid Nanoparticles from Liposomes to MRNA Vaccine Delivery, a Landscape of Research Diversity and Advancement." *ACS Nano*. American Chemical Society. https://doi.org/10.1021/acsnano.1c04996.

Trotter, Patrick J., Margaret A. Orchard, and John H. Walker. 1995. "Ca2+ Concentration during Binding Determines the Manner in Which Annexin V Binds to Membranes." *Biochemical Journal* 308.

Trucillo, Paolo, Roberta Campardelli, and Ernesto Reverchon. 2020. "Liposomes: From Bangham to Supercritical Fluids." *Processes*. MDPI AG. https://doi.org/10.3390/pr8091022.

van Och, François M. M., Henk Van Loveren, Josien C. Van Wolfswinkel, Anja J. C. Machielsen, and Rob J. Vandebriel. 2005. "Assessment of Potency of Allergenic Activity of Low Molecular Weight Compounds Based on IL-1α and IL-18 Production by a Murine and Human Keratinocyte Cell Line." *Toxicology* 210 (2–3): 95–109. https://doi.org/10.1016/j.tox.2005.01.004.

van Ravenzwaay, Bennard, and E. Leibold. 2004. "A Comparison Between In Vitro Rat and Human and In Vivo Rat Skin Absorption Studies." *Human and Experimental Toxicology* 23 (9): 421–430. https://doi.org/10.1191/0960327104ht471oa.

Veloso, Cláudia, Romeu A. Videira, Paula B. Andrade, Catarina Cardoso, and Carla Vitorino. 2022. "Topical Fixed-Dose Combinations: Current In Vitro Methodologies for Pre-Clinical Development." *International Journal of Pharmaceutics*. Elsevier B.V. https://doi.org/10.1016/j.ijpharm.2022.121621.

Wang, Hui, Qing Shao, Yan Zhang, Jing Ding, Miaomiao Yang, and Liu Yang. 2024. "Preparation and Evaluation of Liposomes Containing Ethanol and Propylene Glycol as Carriers for Nicotine." *Current Drug Delivery* 21 (2): 249–260.

Waterhouse, Nigel J., and Douglas R. Green. 1999. "Mitochondria and Apoptosis: HQ or High-Security Prison?" *Journal of Clinical Immunology* 19.

Wen, Hong, Huijeong Jung, and Xuhong Li. 2015. "Drug Delivery Approaches in Addressing Clinical Pharmacology-Related Issues: Opportunities and Challenges." *AAPS Journal* 17 (6): 1327–1340. https://doi.org/10.1208/s12248-015-9814-9.

Williams, A. C., and B. W. Barry. 1992. "Skin Absorption Enhancers." *Critical Reviews™ in Therapeutic Drug Carrier Systems*, 3–4.

Wu, Sijie, Gaodan Liu, Ping Shao, Xingyu Lin, Jiahao Yu, Hanchi Chen, Huiliang Li, and Simin Feng. 2023. "Transdermal Sustained Release Properties and Anti-Photoaging Efficacy of Liposome-Thermosensitive Hydrogel System." *Advanced Healthcare Materials* (January). https://doi.org/10.1002/adhm.202301933.

Wu, Wenfeng, Zhuxian Wang, Yufan Wu, Huiyi Wu, Tingting Chen, Yaqi Xue, Yuan Wang, et al. 2022. "Mechanisms of Penetration Enhancement and Transport Utilizing Skin Keratine Liposomes for the Topical Delivery of Licochalcone A." *Molecules* 27 (8). https://doi.org/10.3390/molecules27082504.

Xia, Ying, Keang Cao, Ruoyang Jia, Xue Chen, Yang Wu, Yu Wang, Zhiqing Cheng, Hongmei Xia, Yinxiang Xu, and Zili Xie. 2024. "Tetramethylpyrazine-Loaded Liposomes Surrounded by Hydrogel Based on Sodium Alginate and Chitosan as a Multifunctional Drug Delivery System for Treatment of Atopic Dermatitis." *European Journal of Pharmaceutical Sciences* 193 (February). https://doi.org/10.1016/j.ejps.2023.106680.

Xing, Hang, Kevin Hwang, and Yi Lu. 2016. "Recent Developments of Liposomes as Nanocarriers for Theranostic Applications." *Theranostics*. Ivyspring International Publisher. https://doi.org/10.7150/thno.15464.

Yoon, Young I. L., Yong Su Kwon, Hee Sang Cho, Sun Hee Heo, Kyeong Soon Park, Sang Gyu Park, Soo Hong Lee, et al. 2014. "Ultrasound-Mediated Gene and Drug Delivery Using a Microbubble-Liposome Particle System." *Theranostics* 4 (11): 1133–1144. https://doi.org/10.7150/thno.9945.

Yu, Justine R., Pranav Varrey, Barry J. Liang, Huang Chiao Huang, and John P. Fisher. 2021. "Liposomal SDF-1 Alpha Delivery in Nanocomposite Hydrogels Promotes Macrophage Phenotype Changes and Skin Tissue Regeneration." *ACS Biomaterials Science and Engineering* 7 (11): 5230–5241. https://doi.org/10.1021/acsbiomaterials.1c01140.

Zhang, Hongwei. 2017. "Thin-Film Hydration Followed by Extrusion Method for Liposome Preparation." In *Methods in Molecular Biology*, 1522, 17–22. Humana Press Inc. https://doi.org/10.1007/978-1-4939-6591-5_2.

Zhou, Hong, Dan Luo, Dan Chen, Xi Tan, Xichen Bai, Zhi Liu, Xiangliang Yang, and Wei Liu. 2021. "Current Advances of Nanocarrier Technology-Based Active Cosmetic Ingredients for Beauty Applications." *Clinical, Cosmetic and Investigational Dermatology*. Dove Medical Press Ltd. https://doi.org/10.2147/CCID.S313429.

5 Mechanism Involved in Wound Healing

A Serological and Pathological Aspect

Jannatun Nesa and Abdul Sadat

5.1 INTRODUCTION

In the daily health care system, wounds are a serious problem, with associated complications that lead to the detrimental results of mortality (Alonso et al., 1996; Natarajan et al., 2000). To counteract this detrimental challenge, much endeavour has been focused to understand the physiology of wound healing with an emphasis on the development of new therapeutic techniques to cure patients suffering from acute and chronic wounds (Szycher and Lee, 1992; Robson et al., 2001).

The widespread occurrence and persistence of wounds, particularly among the elderly, has emerged as a critical global socio-economic challenge that demands urgent attention. Moreover, long-lasting, acute and chronic wounds in association with other diseases cause severe damage to the cutaneous covering along with the vascular system (Velnar et al., 2009). The occurrence of these wounds augments with increasing age. For instance, it has been recorded that out of 0.1 million 120 people suffering from chronic wounds falling in the age range of 45 to 65 years. While, these numbers rise up to 800 people for the age range greater than 75 years (Natarajan et al., 2000; Robson et al., 2001). Moreover, failure to manage the complications associated with acute wounds resulted in the development of chronic wounds, which became a more serious problem for the healthcare system (Natarajan et al., 2000).

5.2 WOUNDS AND THEIR TYPES

A wound is any disruption or damage to the normal anatomical arrangement and function (Velnar et al., 2009). A wound can range from a simple rupture in the epithelial layer of the body membrane, or it may be deeper, penetrating into the subcutaneous tissue with indentation to other architectural components, that is to say, muscles, tendons, vessels, parenchymal organs, nerves, and even deep into the bone (Alonso et al., 1996).

As time is the key factor for the treatment of injury and wound healing, based on the time-frame wounds are classified as acute wounds and chronic wounds (Lazarus et al., 1994; Bischoff et al., 1999).

5.2.1 Acute Wound

Wounds that heal on their own following an orderly and timely healing mechanism, which culminate with the restoration of normal anatomical architecture and functions, are categorized as acute wounds. Acute wounds take five to ten days of repair time or sometime within 30 days. This wound generally occurs due to any surgical operation or traumatic tissue loss (Velnar et al., 2009). For instance, the surgical removal of a tumour (soft tissue) located in the skin and, as a consequence, a large wound

DOI: 10.1201/9781003394846-5

developed in the underlying parenchymal layer, which cannot be repaired, is attributable to the large defect in the affected tissue. Acute wounds are also frequent due to traumatic injury. It may result in only damage to soft tissue or occasionally might be allied with fractures in bone. These collective injuries have been categorized by the globally accepted AO Foundation (Arbeitsgemeinschaft fuer Osteosynthesefragen/Association for the Study of Internal Fixation) (AO Foundation).

5.2.2 Chronic Wounds

Wounds that cannot be healed in an orderly and timely manner through normal progression of healing are termed as chronic wounds (Velnar et al., 2009). In this case, the wound repair process is imperfect and distressed by a variety of factors that prolong one or more stages in the wound healing process. For instance, these affecting factors are infection, excess levels of inflammatory cytokines, tissue hypoxia, exudate and necrosis (Vanwijck, 2001). Prolonged inflammatory stage in the wound leads to the development of a reaction cascade in the tissue, which results in the non-healing of the wound. Due to this uncoordinated progression of the healing process, the outcome is the poor functional and anatomical aspects of the affected tissue and the wound relapse again and again (Degreef, 1998). Chronic wounds generally occur for various reasons, such as pressure, naturopathic, cardio-vascular deficiency, vasculitis burns, etc. (Degreef, 1998; Komarcevic, 2000).

5.2.3 Complicated Wounds

This is a special type of wound that develops in combination with tissue damage and infection (Bischoff et al., 1999). This wound resulted from post-infection/traumatic conditions, or prolonged tissue destruction due to chemo/radiotherapy during tumour treatment. Such wounds are infected irrespective of their causative agents, locations, sizes and management techniques. Nevertheless, the development of infection is dependent on the type, number and virulence of the invading microbial pathogens, along with the blood supply to the infected area and the patient's immune resistance. Such infections are characterized by well-documented signs and symptoms such as redness, heat, pain, oedema and loss or limited function in the affected area (Attinger et al., 2006).

5.3 THE PROCESS OF WOUND HEALING

Our skin is the outermost protective layer that provides a physiological barrier for the whole organism from an external harsh environment (Gushiken et al., 2021). In addition to this protective action, the skin also plays a crucial role in homeostatic functions, ranging from receiving external stimuli to the regulation of thermo-stability. Critically, our skin provides the first line of defence to protect from desiccation and thermal, photic, mechanical and chemical damage to internal organs (Takeo et al., 2015). This physiological barrier adapted to a highly complicated immune response that confront the entry of hazardous agents and microbes while allowing movements of the symbiotic microbes (Naik et al., 2015). The skin has also developed well-organized and quick mechanisms to close its breaks in the barrier system, which is termed a wound-healing process. Wound healing is categorized into four phases: Coagulation and haemostatic phase, inflammatory phase, proliferative phase and dermal remodelling phase (Broughton et al., 2006) that resulted into structural and functional restoration of the damaged tissue (Figure 5.1). Present section illustrates all the stages of wound healing process in detail.

5.3.1 Coagulation and Hemostatic Phase

Immediately following injury, a rapid vaso-constriction occurs in close vicinity of the tissue damage and blood coagulation prevents excessive blood loss due to vascular damage (Velnar et al., 2009).

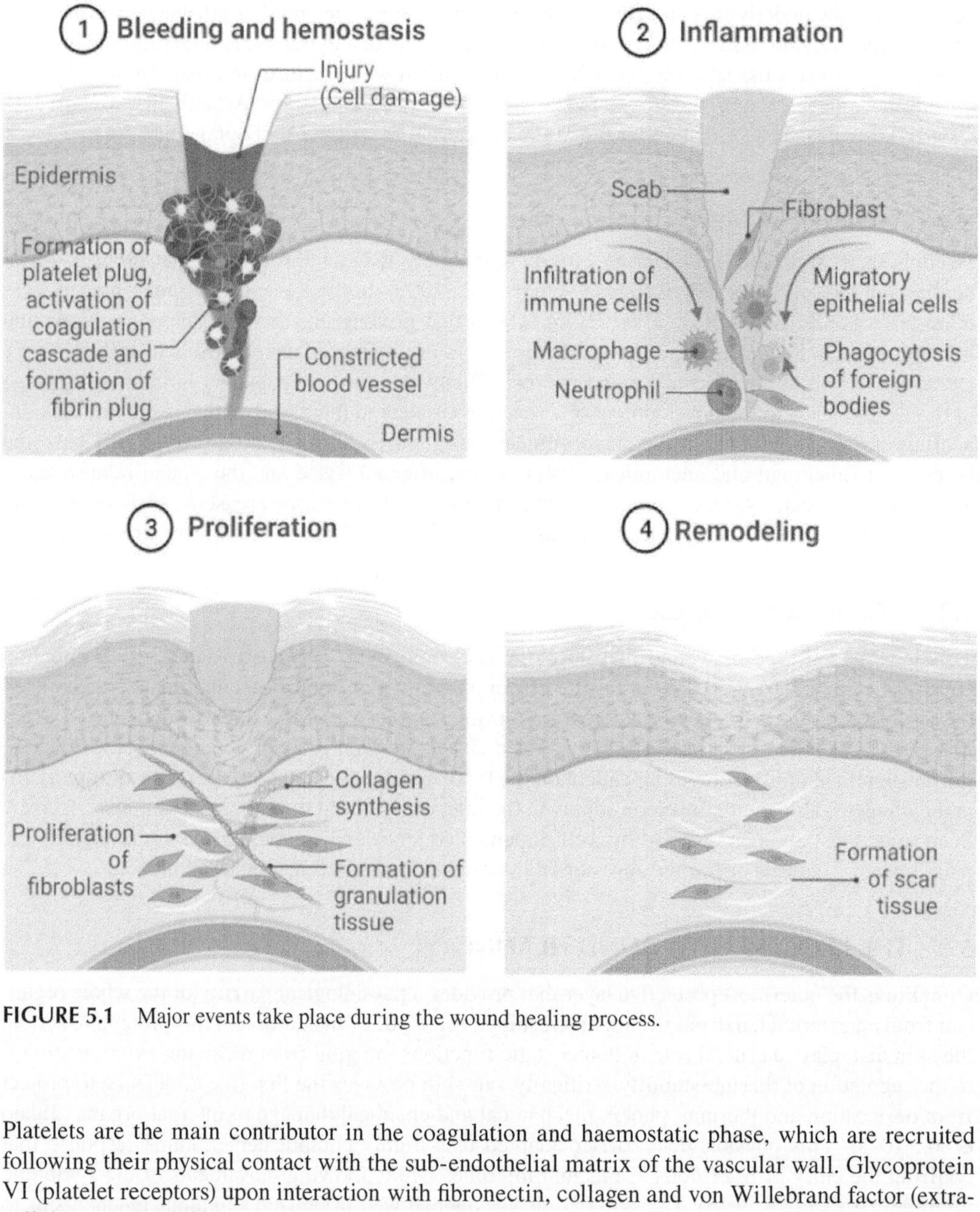

FIGURE 5.1 Major events take place during the wound healing process.

Platelets are the main contributor in the coagulation and haemostatic phase, which are recruited following their physical contact with the sub-endothelial matrix of the vascular wall. Glycoprotein VI (platelet receptors) upon interaction with fibronectin, collagen and von Willebrand factor (extracellular matrix (ECM) proteins) promotes adhesion to the vascular wall. Following this adherence, thrombin triggers activation of the platelets that involves its conformational change, and discharge of dense granules enclosing bioactive molecules which strengthen the coagulation process (Golebiewska and Poole, 2015).

The coagulation process culminates as the insoluble clot is formed in association with fibrin, vitronectin, fibronectin and thrombo-spondin at the site of injury (Zaidi and Green, 2019), which plugs the rupture, occurs in the blood vessel and prevents blood loss. In addition to this, the blood clot shielding the entry point that prevents microbial invasion provides a harbouring site for effecter cells (white blood cells) and effecter molecules (growth factors and cytokines). Those are key elements for the modulation of serological behaviour of the injured cells and tissue during early repair (Van der Veer et al., 2011).

The wound healing process is triggered by tissue damage and bleeding, followed by *hemostasis*, where a platelet plug and a prelude fibrin plug are formed at the injury site to prevent blood loss. The *inflammation* phase prevents further infection and eliminates tissue debris with the help of infiltrating immune cells, that is, macrophages and neutrophils.

The *proliferative* phase ensures the closing of the wound gap by migratory keratinocytes; the fibrin plug is replaced by proliferating fibroblasts and granulation tissue. In the final *remodelling* phase, fibroblasts and myofibroblast-guided remodelling of the deposited matrix are conducted to form scar tissue.

Platelets play a key role in the engagement of immune effecter cells to the site wound, either by entrapping leukocytes in the clot plug, or platelets exuding secretomes through de-granulation that act as chemokine attractants (Golebiewska and Poole, 2015). Indeed, the secretome also possesses growth factors, which activate resident epidermal cells, keratinocytes and fibroblasts that assist in tissue reformation (Scully et al., 2020). As the platelets play an active role in the suppression of bacterial population during the early wound repair phase, they are capable of producing antimicrobial peptides under the regulation of toll-like receptors (TLRs) (Tang et al., 2002; Shiraki et al., 2004; Cognasse et al., 2005). The coagulation process is turned off, while adequate clot has been formed at the injury site, avoiding unnecessary thrombosis. This suppression of the coagulation process is under the control of a cascade of reactions mediated by protein C that degrades coagulation factors V and VII, and prostacyclin and antithrombin III that inhibits thrombin formation (Mann, 2003). Simultaneously, the damaged wall of blood the vessel is restored by endothelial cells and smooth muscle cells, which are amplified in response to PDGF (platelet-derived growth factor) secreted by the platelets (Kingsley et al., 2002). Progenitors of endothelium also help in the restoration as they have greater proliferative capacity in comparison to mature cells of vaso-endothelium (Rennert et al., 2012).

5.3.2 Inflammatory Phase

Both the cellular and humoral inflammatory responses are triggered at the late coagulation to protect the injury site from invading microbes. The onset of the inflammatory phase is marked by the release of DAMPs (damage-associated molecular patterns) by necrotic cells and damaged tissue and bacteria-secreted PAMPs (pathogen-associated molecular patterns). These two components eventually bind with macrophages, Langerhans cells, mast cells and T cells through binding pattern recognition receptors, which switch on the inflammatory cascade pathways (Chen and DiPietro, 2017). Moreover, chemokines and cytokines (Table 5.1) act as pro-inflammatory molecules that pull leucocytes towards the wound site (Martin and Leibovich, 2005). In addition to this, chemokines and cytokines cause vasodilatation and stimulate bio-synthesis of adhesion molecules, that is, selectins, which assist the adhesion of monocytes and neutrophils with the damaged endothelial cells (Vestweber, 2015).

In early inflammatory response neutrophils recruited at the injury site by the chemo-attractants such as bacterial endotoxins (lipopolysaccharide, LPS), tumour necrosis factor-alpha (TNF-α) and interleukin 1 (Kolaczkowska and Kubes, 2013). In response to the pro-inflammatory signal cascade, neutrophils discharge cytokines that activate the inflammatory signal cascade through NF-κB (Kolaczkowska and Kubes, 2013). Neutrophils eradicate pathogens and damaged tissue by phagocytic action and also produce proteolytic enzymes, antimicrobial peptides, reactive oxygen species (ROS) and eicosanoids that play an important role in wound healing (Segel et al., 2011).

The inflammation is a complex response, which is under the control of different intrinsic and extrinsic factors. Excessive inflammation leads to severe tissue damage and delays in wound healing (Boniakowski et al., 2017). On the other hand, insufficient recruitment of immune cell at the injury site suppresses timely healing of the wound (Lin et al., 2011). Therefore, orderly recruitment and responses by immune cells are very critical to maintaining the wound-healing process. However, after the cure of infection, the number of recruited neutrophils reduces from the wound within a

TABLE 5.1
Major Serological Components Involved in Wound Healing

Serological components	Origin	Biological activity
	Growth factors	
TGF β, β1, β2, β3	Platelets, Fibroblasts, Macrophages	• Chemotaxis of fibroblasts and activates deposition of ECM through decreasing collagen and TIMP synthesis and increasing MMP synthesis • Diminishes scar formation by rendering collagen and fibronectin synthesis
PDGF-AA, BB, VEGF	Platelets, Fibroblasts, Macrophages, Keratinocytes	• Activates cells of the immune system, fibroblasts and deposition of ECM through decreasing collagen and TIMP synthesis and increasing MMP synthesis • Angiogenesis
Fibroblast growth factors (FGF) and keratinocyte growth factors (KGF)	Fibroblasts, Macrophages, Endothelial cells	• Activation of endothelial cells • Angiogenesis • Proliferation and migration of keratinocytes • Deposition of ECM
	Pro-inflammatory cytokines	
TNF-α	Macrophages	• Migration of poly-morphonuclear neutrophils (PMN) and cytotoxicity • Hinder collagen fibre synthesis
IFN-γ	Macrophages, T-Lymphocytes	• Activation of PMN and Macrophages • Check synthesis of collagen fibre • Trigger collagenase activity to check cross-link formation among collagen fibres
IL-1	Macrophages, Keratinocytes	• Trigger chemotaxis of fibroblasts and keratinocytes • Check collagen synthesis
IL-2	T-Lymphocytes	• Enhances fibroblast infiltration at the injury site
IL-6	Macrophages	• Proliferation of fibroblast • Synthesis of hepatic-acute phase protein
IL-8	Macrophages	• Chemotaxis of PMN and macrophages • Maturation of keratinocytes
	Anti-inflammatory cytokines	
IL-4	Basophils, T-Lymphocytes, Mast cells	• Inhibits TNF-α, IL-1 and IL-6 • Triggers proliferation of fibroblast and collagen synthesis
IL-8	Macrophages, T-Lymphocytes, Keratinocytes	• Inhibits macrophages, PMN, TNF-α, IL-1 and IL-6

few days (Kim et al., 2008). Removal of the recruited neutrophils from the injury site takes place by several means, that is, efferocytosis by macrophage cells (Martin and Leibovich, 2005) and intact neutrophils return to the blood circulation by reverse migration through the trans-endothelial layer (Buckley et al., 2006; Yoo and Huttenlocher, 2011; Wang et al., 2017). Following the recruitment of neutrophils, a wave of monocytes migrating to the wound takes place. After this recruitment, circulating monocytes transformed into macrophages that eventually engaged in the damage tissue repair with plasticity and versatility. Macrophages engulf pathogens and damage cellular debris through conserved membrane receptors, while a response to cytokines exhibits altered morphological and behavioural characteristics (Mantovani et al., 2005).

Wound-healing macrophages are categorized into two main sets such as pro-inflammatory (activated) and anti-inflammatory (alternatively activated) based on their phenotypes under the stimulation of injured tissue and signals (Mosser and Edwards, 2008; Snyder et al., 2016). Classically pro-inflammatory macrophages are activated by pro-inflammatory stimuli, that is, interferon-gamma (IFN-γ) and bacterial LPS, which promote the development of inflammation through the action of ROS, growth factors (e.g. PDGF and vascular endothelial growth factor, VEGF) and inflammatory cytokines (e.g. TNF-α, IL-1 and IL-6) (Table 5.1). These macrophages remove apoptotic neutrophils through phagocytic action and recruit themselves as the main effecter cell for the inflammatory phase (Scully et al., 2020).

Later, inflammation is marked by a transition to the anti-inflammatory stage through alternative activation of the lately recruited monocytes or switching on anti-inflammatory function by the macrophages in the site of inflammation. This total game change of phenotypic switch is stimulated by the local environment, that is, change in cytokine population, efferocytosis, transcription factors, miRNAs, and alteration of receptors during the pro-inflammatory phase and anti-inflammatory phase (Khallou-Laschet et al., 2010; Nelson et al., 2011; Das et al., 2014; Das et al., 2015; Lin et al., 2016). Moreover, pro-inflammatory macrophages release arginase and cytokine IL-4, IL-10 and IL-13 (Table 5.1) that actively assist wound healing (Barrientos et al., 2008; Campbell et al., 2013), whereas anti-inflammatory macrophages secrete numerous growth factors (Table 5.1) that ensure the restoration of epithelial cells and fibrous tissue (Delavary et al., 2011) and development of angiogenesis (Jetten et al., 2014). Therefore, the combined function of the recruited macrophages is to remove tissue derrises and pro-inflammatory cells (neutrophils), kill invading pathogens, as well as stimulate wound healing to restore normal architecture and physiology of the injured tissue.

The dominance of the recruited neutrophils and macrophages in wound healing eventually shadows the importance of other blood cells for the injury response. Recent investigations revealed that T cells are critical for both early responses to the injury site as well as exert anti-inflammatory action later (Nosbaum et al., 2016). The role of T cells is strongly supported by the different clinical assays, that is, reduction in the resident T cells leads to the delayed healing process in experimental mice (aged and hyperglycemic), whereas administration of epidermal (resident) T cells resulted in restoration of wound healing (Liu et al., 2016; Keyes et al., 2016). Further studies revealed that the removal of circulating T cells hinders the tissue refurbishment process in the experimental mice suggesting the role of T cells in anti-inflammatory regulations (Nosbaum et al., 2016). Recently mast cells are also unmasked as an important regulatory component of wound healing, as they secrete histamine that helps to recruit the neutrophils during the early inflammatory phase (Weller et al., 2006).

5.3.3 Proliferation Phase

When the current injury has resolved, haemostasis is completed and the immune reaction is effectively established, the injured tissue shifts toward its repair (Velnar et al., 2009). Proliferation is largely involved in tissue repair, which starts after two days of wound formation and lasts for about two weeks.

This proliferative phase of wound repair is characterized by activation of the macrophages, fibroblasts, keratinocytes, and endothelial cells (Table 5.1) that leads to closure of the wound, deposition of matrix and angiogenesis. After 12 hours of wounding, due to mechanical tension and fluctuation in the electrical gradients at the injury site, recruitment of cytokines and growth factors at the injury site, and exposure of the injured tissue to hydrogen peroxide and pathogens resulted in activation of keratinocytes (Shaw and Martin, 2016). The activated keratinocytes further develop migratory phenotype through epithelial-mesenchymal transition (EMT) (Li et al., 2007). The acquisition of top-to-bottom polarity enables keratinocytes to invade the wound in a lateral direction that causes reformation of the epidermal layer (re-epithelialization) (Wager and Leavesley, 2015). Studies reported that alteration in cell-adhesion property of the keratinocytes is achieved via PCKα and Eph-mediated changes in desmosome and adherens junctions (Thomason et al., 2012; Nunan et al., 2015), which allow the rearrangement of the keratinocytes in response to epithelial sheet migration (Shaw and Martin, 2016). Moreover, keratinocytes present in the neo-epidermis exude matrix metalloproteinases (MMPs, MMP-1 and MMP-9) that broaden their migratory path, whereas new ECM proteins lay down to reform the basement membrane for the epidermis (Rousselle et al., 2019). Following this specific stem cells of hair follicles (Krt15+ve and Krt19+ve) are activated to proliferate with newly formed epidermal cells to address the cellular need for resurfacing the wound (Ito et al., 2005; Ito et al., 2007; Driskell et al., 2015; Garcin et al., 2016). Keratinocytes settle in the wound bed along with necrotic tissue and tissue debris via the interactions between ECM proteins and their integrin receptors (Santoro and Gaudino, 2005). Migration of keratinocytes ceases when they come in contact with opposite ends, which results in the formation of a new epithelial layer and keratinocytes settle down to its underlying matrix. Following this, keratinocytes undergo terminal differentiation and a fully reconstructed basement membrane to stratify and renew the epidermis. (Baum and Arpey, 2005).

Fibroblasts are mainly engaged in the replacement of fibrin-rich matrix by more extensive granulated tissue mass. Both residential and derived (mesenchymally) fibroblasts act in response to different signalling molecules (Table 5.1) such as endothelial cells, platelets, macrophages, PDGF and transforming growth factor (TGF-β). In response to these signals, fibroblasts either settle down with ECM proteins or transform into myofibroblasts that cause contraction of wound (Li et al., 2007).

In fact, this role of fibroblasts is described in a simple way, as they show functional diversity that repairs dermal tissue in a diverse mechanism. An experimental study revealed two distinct lineages for fibroblast origin, where lower lineage aids in ECM deposition, while upper lineage fibroblasts contribute to re-epithelialization (Driskell et al., 2013). Furthermore, current studies have confronted the conventional understanding of the origin of wound fibroblasts and reported that two-thirds of wound fibroblasts in granulation tissue are derived from macrophages present in the injury site (Sinha et al., 2018). Fibroblasts eventually produce MMPs that degrade the temporary matrix which is replaced by granulation tissue that contains large amounts of proteoglycans, immature collagens and fibronectin (Xue and Jackson, 2015). This newly formed granulation tissue aids in migration and differentiation of damaged cells, formation of mature ECM and angiogenesis.

New blood vessels are created during the process of angiogenesis to meet the metabolic demands of the highly proliferative healing tissue. The process of angiogenesis is stimulated by hypoxia under the control of different factors like hypoxia-inducible factors (HIFs) and cyclooxygenase 2, which leads to the production and secretion of VEGF (Huang et al., 2005). Consequently, microvascular endothelial cells start proliferating and drift into the wound bed, where they build new tiny blood vessels that fuse with each other to form continuous tubular networks in the area of the wound (Honnegowda et al., 2015). During this process, VEGF checks apoptosis of the endothelial cells through the upregulation of BCL-2 protein (Cai et al., 2003), whereas the fibrin matrix aids the phenotypic changes of endothelial cells that promote their migration to the wound bed (Kalebic et al., 1983). Furthermore, macrophages have an important contribution to angiogenesis. These myeloid cells secrete MMPs (proteases) and chemotactic factors (e.g. TGF-β, TNF-α and VEGF) that degrade the fibrin network and aid the migration of endothelial cells respectively (Du Cheyne, 2020). In addition to this, macrophages contribute to vasculature remodelling by joining vessel tips, degrading redundant vessels and checking unnecessary vessel formation (Fantin et al., 2010; Stefater et al., 2011; Poché et al., 2015).

Skin possesses an extensive network of nervous systems that control both the sensory and locomotory actions of human beings. Therefore, regeneration of nerve fibre is crucial in the wound healing process.

Unfortunately, there is very scanty information regarding this process. An experimental study suggested that during the repair process, regenerating neurons exude substance P (neuropeptide) exerts important processes like cellular proliferation and regeneration of the vascular system at the injury site (Jung et al., 2016; Um et al., 2016). Furthermore, it was revealed that its absence in diabetic mice model resulted in delayed wound healing, whereas topical application of substance P restores normal healing process along with regeneration of the nerves (Leal et al., 2015; Um et al., 2017; Zhu et al., 2016).

5.3.4 Matrix Remodelling Phase

As the concluding phase of wound repair, the matrix remodelling phase involves the formation of a new epithelial layer and scar tissue. This phase starts with the synthesis of ECM in the proliferating phase with the developing granulation tissue and may last for several years which concludes with the formation of collagen-rich scar (Type I) (Velnar et al., 2009).

Fibroblasts play a key role in matrix remodelling, which replace fibrin clots with fibronectin, hyaluronan and proteoglycans, and later form the ultimate collagen fibrils (Darby et al., 2014). Proteoglycans are major components that aid in cellular migration and are responsible for the development of mature collagen fibrils which are cross-linked in nature (Schultz and Wysocki, 2009). There is a clear contrast in collagen composition in wound-repairing granulation tissue (Type I collagen:Type III collagen: 10%:30%) and uninjured skin tissue (Type I collagen:Type III collagen:: 80%:10%) (Witte and Barbul, 1997). Along with the improvement of the wound repair process Type I collagen replaces Type III collagen, which enhances tensile strength in the forming scar (Diegelmann and Evans, 2004). However, the scar ECM never gains its completely identical structural design and integrity as when it was uninjured skin. Collagen fibrils in unwounded skin showed basket weave-like orientation, while in scar tissue, it largely adopts a parallel bundle-like orientation. As a result of this architectural alteration, post-injury scar tissue gains up to 80% strength of pre-injury condition (Witte and Barbul, 1997; Young and McNaught, 2011).

These orderly modulations in the matrix involve fine equilibrium between synthesis and degradation of collagen fibre, which is under the control of MMPs. These enzymes are secreted by keratinocytes, fibroblasts and anti-inflammatory macrophages, which cleave the newly formed helical collagens during the repair process (Darby et al., 2014). Elastin is another key component of ECM that maintains the elasticity skin. Surprisingly, damaged ECM releases elastokines (elastin fragments) that perform as a signalling molecule to trigger the synthesis of native elastin from its precursor molecule tropoelastin (Duca et al., 2004). In the early phase, native elastin shows irregular arrangement and fully oriented mature elastin appears in the scar tissue after several months of injury (Almine et al., 2012; Amadeu et al., 2004).

TGF-β expression and mechanical tension in the scar tissue stimulate differentiation of myofibroblast, which contains augmented alpha-smooth muscle actin (α-SMA) that causes focal adhesions and strong contractile forces on the myofibroblast (Hinz et al., 2001). Further, pseudopodial extensions aid in binding cytoplasmic actin with the fibronectin of the matrix scaffold, which facilitates myofibril (Li et al., 2007). Myofibroblasts join with each other through desmosomes and then bind with the matrix fibrils, hence this firm network holds the entire ECM together (Stadelmann et al., 1998). The response for wound healing subsides when the endothelial cells, macrophages and fibroblasts either endure apoptosis or exit from the injured area to leave the scar (Larouche et al., 2018).

5.4 CONCLUSION

The healthcare system is facing terrific challenges due to the occurrence and prevalence of wounds in everyday life. Severe wound formation causes tissue damage and impinges on the local environment, which alters both structure and function of the injured area. The immune response triggered by this injury is established through diverse physiological processes to cure the wound, which includes four uninterrupted phases, that is, coagulation and hemostatic phase, inflammatory phase, proliferation phase and matrix remodelling phase. A deep understanding of the mechanism of the wound healing process will help medical practitioners conduct proper clinical management that may have a positive impact on the course of wound healing and also trim down the potential complications which are associated with the injury.

REFERENCES

Almine, J. F., Wise, S. G. & Weiss, A. S. 2012. Elastin signaling in wound repair. *Birth Defects Research Part C: Embryo Today: Reviews* 96(3): 248–257.

Alonso, J. E., Lee, J., Burgess, A. R. & Browner, B. D. 1996. The management of complex orthopedic injuries. *Surgical Clinics of North America* 76(4): 879–903.

Amadeu, T. P., Braune, A. S., Porto, L. C., Desmoulière, A. & Costa, A. M. 2004. Fibrillin-1 and elastin are differentially expressed in hypertrophic scars and keloids. *Wound Repair and Regeneration* 12(2): 169–174.

AO Foundation (Arbeitsgemeinschaft fuer Osteosynthesefragen/Association for the Study of Internal Fixation). Available at: www.aofoundation.org/wps/portal/.

Attinger, C. E., Janis, J. E., Steinberg, J., Schwartz, J., Al-Attar, A. & Couch, K. 2006. Clinical approach to wounds: Debridement and wound bed preparation including the use of dressings and wound-healing adjuvants. *Plastic and Reconstructive Surgery* 117(7S): 72S-109S.

Barrientos, S., Stojadinovic, O., Golinko, M. S., Brem, H. & Tomic-Canic, M. 2008. Growth factors and cytokines in wound healing. *Wound Repair and Regeneration* 16(5): 585–601.

Baum, C. L. & Arpey, C. J. 2005. Normal cutaneous wound healing: Clinical correlation with cellular and molecular events. *Dermatologic Surgery* 31(6): 674–686.

Bischoff, M., Kinzl, L. & Schmelz, A. 1999. The complicated wound. *Der Unfallchirurg* 102(10): 797–804 [in German].

Boniakowski, A. E., Kimball, A. S., Jacobs, B. N., Kunkel, S. L. & Gallagher, K. A. 2017. Macrophage-mediated inflammation in normal and diabetic wound healing. *The Journal of Immunology* 199(1): 17–24.

Buckley, C. D., Ross, E. A., McGettrick, H. M., Osborne, C. E., Haworth, O., Schmutz, C. & Rainger, G. E. 2006. Identification of a phenotypically and functionally distinct population of long-lived neutrophils in a model of reverse endothelial migration. *Journal of Leukocyte Biology* 79(2): 303–311.

Cai, J., Ahmad, S., Jiang, W. G., Huang, J., Kontos, C. D., Boulton, M. & Ahmed, A. 2003. Activation of vascular endothelial growth factor receptor-1 sustains angiogenesis and Bcl-2 expression via the phosphatidylinositol 3-kinase pathway in endothelial cells. *Diabetes* 52(12): 2959–2968.

Campbell, L., Saville, C. R., Murray, P. J., Cruickshank, S. M. & Hardman, M. J. 2013. Local arginase 1 activity is required for cutaneous wound healing. *Journal of Investigative Dermatology* 133(10): 2461–2470.

Chen, L. & DiPietro, L. A. 2017. Toll-like receptor functions in acute wounds. *Advances in Wound Care* 6(10): 344–355.

Cognasse, F., Hamzeh, H., Chavarin, P., Acquart, S., Genin, C. & Garraud, O. 2005. Evidence of Toll-like receptor molecules on human platelets. *Immunology and Cell Biology* 83(2): 196–198.

Darby, I. A., Laverdet, B., Bonté, F. & Desmoulière, A. 2014. Fibroblasts and myofibroblasts in wound healing. *Clinical, Cosmetic and Investigational Dermatology*: 301–311.

Das, A., Ganesh, K., Khanna, S., Sen, C. K. & Roy, S. 2014. Engulfment of apoptotic cells by macrophages: A role of microRNA-21 in the resolution of wound inflammation. *The Journal of Immunology* 192(3): 1120–1129.

Das, A., Sinha, M., Datta, S., Abas, M., Chaffee, S., Sen, C. K. & Roy, S. 2015. Monocyte and macrophage plasticity in tissue repair and regeneration. *The American Journal of Pathology* 185(10): 2596–2606.

Degreef, H. J. 1998. How to heal a wound fast. *Dermatologic Clinics* 16(2): 365–375.

Delavary, B. M., van der Veer, W. M., van Egmond, M., Niessen, F. B. & Beelen, R. H. 2011. Macrophages in skin injury and repair. *Immunobiology* 216(7): 753–762.

Diegelmann, R. F. & Evans, M. C. 2004. Wound healing: An overview of acute, fibrotic and delayed healing. *Frontiers in Bioscience* 9(1): 283–289.

Driskell, I., Oeztuerk-Winder, F., Humphreys, P. & Frye, M. 2015. Genetically induced cell death in bulge stem cells reveals their redundancy for hair and epidermal regeneration. *Stem Cells* 33(3): 988–998.

Driskell, R. R., Lichtenberger, B. M., Hoste, E., Kretzschmar, K., Simons, B. D., Charalambous, M. & Watt, F. M. 2013. Distinct fibroblast lineages determine dermal architecture in skin development and repair. *Nature* 504(7479): 277–281.

Duca, L., Floquet, N., Alix, A. J., Haye, B. & Debelle, L. 2004. Elastin as a matrikine. *Critical Reviews in Oncology/Hematology* 49(3): 235–244.

Du Cheyne, C., Tay, H. & De Spiegelaere, W. 2020. The complex TIE between macrophages and angiogenesis. *Anatomia, Histologia, Embryologia* 49(5): 585–596.

Fantin, A., Vieira, J. M., Gestri, G., Denti, L., Schwarz, Q., Prykhozhij, S. & Ruhrberg, C. 2010. Tissue macrophages act as cellular chaperones for vascular anastomosis downstream of VEGF-mediated endothelial tip cell induction. Blood. *The Journal of the American Society of Hematology* 116(5): 829–840.

Garcin, C. L., Ansell, D. M., Headon, D. J., Paus, R. & Hardman, M. J. 2016. Hair follicle bulge stem cells appear dispensable for the acute phase of wound re-epithelialization. *Stem Cells* 34(5): 1377–1385.

George Broughton, I. I., Janis, J. E. & Attinger, C. E. 2006. Wound healing: An overview. *Plastic and Reconstructive Surgery* 117(7S): 1e-S.

Golebiewska, E. M. & Poole, A. W. 2015. Platelet secretion: From haemostasis to wound healing and beyond. *Blood Reviews* 29(3): 153–162.

Gushiken, L. F. S., Beserra, F. P., Bastos, J. K., Jackson, C. J. & Pellizzon, C. H. 2021. Cutaneous wound healing: An update from physiopathology to current therapies. *Life* 11(7): 665.

Hinz, B., Mastrangelo, D., Iselin, C. E., Chaponnier, C. & Gabbiani, G. 2001. Mechanical tension controls granulation tissue contractile activity and myofibroblast differentiation. *The American Journal of Pathology* 159(3): 1009–1020.

Honnegowda, T. M., Kumar, P., Udupa, E. G. P., Kumar, S., Kumar, U. & Rao, P. 2015. Role of angiogenesis and angiogenic factors in acute and chronic wound healing. *Plastic and Aesthetic Research* 2: 243–249.

Huang, S. P., Wu, M. S., Shun, C. T., Wang, H. P., Hsieh, C. Y., Kuo, M. L. & Lin, J. T. 2005. Cyclooxygenase-2 increases hypoxia-inducible factor-1 and vascular endothelial growth factor to promote angiogenesis in gastric carcinoma. *Journal of Biomedical Science* 12: 229–241.

Ito, M., Liu, Y., Yang, Z., Nguyen, J., Liang, F., Morris, R. J. & Cotsarelis, G. 2005. Stem cells in the hair follicle bulge contribute to wound repair but not to homeostasis of the epidermis. *Nature Medicine* 11(12): 1351–1354.

Ito, M., Yang, Z., Andl, T., Cui, C., Kim, N., Millar, S. E. & Cotsarelis, G. 2007. Wnt-dependent de novo hair follicle regeneration in adult mouse skin after wounding. *Nature* 447(7142): 316–320.

Jetten, N., Roumans, N., Gijbels, M. J., Romano, A., Post, M. J., de Winther, M. P. & Xanthoulea, S. 2014. Wound administration of M2-polarized macrophages does not improve murine cutaneous healing responses. *PLoS ONE* 9(7): e102994.

Jung, N., Yu, J., Um, J., Dubon, M. J. & Park, K. S. 2016. Substance P modulates properties of normal and diabetic dermal fibroblasts. *Tissue Engineering and Regenerative Medicine* 13: 155–161.

Kalebic, T., Garbisa, S., Glaser, B. & Liotta, L. A. 1983. Basement membrane collagen: Degradation by migrating endothelial cells. *Science* 221(4607): 281–283.

Keyes, B. E., Liu, S., Asare, A., Naik, S., Levorse, J., Polak, L. & Fuchs, E. 2016. Impaired epidermal to dendritic T cell signaling slows wound repair in aged skin. *Cell* 167(5): 1323–1338.

Khallou-Laschet, J., Varthaman, A., Fornasa, G., Compain, C., Gaston, A. T., Clement, M. & Caligiuri, G. 2010. Macrophage plasticity in experimental atherosclerosis. *PLoS ONE* 5(1): e8852.

Kim, M. H., Liu, W., Borjesson, D. L., Curry, F. R. E., Miller, L. S., Cheung, A. L. & Simon, S. I. 2008. Dynamics of neutrophil infiltration during cutaneous wound healing and infection using fluorescence imaging. *Journal of Investigative Dermatology* 128(7): 1812–1820.

Kingsley, K., Huff, J. L., Rust, W. L., Carroll, K., Martinez, A. M., Fitchmun, M. & Plopper, G. E. 2002. ERK1/2 mediates PDGF-BB stimulated vascular smooth muscle cell proliferation and migration on laminin-5. *Biochemical and Biophysical Research Communications* 293(3): 1000–1006.

Kolaczkowska, E. & Kubes, P. 2013. Neutrophil recruitment and function in health and inflammation. *Nature Reviews Immunology* 13(3): 159–175.

Komarcević, A. 2000. The modern approach to wound treatment. *Medicinski Pregled* 53(7–8): 363–368. [in Croatian].

Larouche, J., Sheoran, S., Maruyama, K. & Martino, M. M. 2018. Immune regulation of skin wound healing: Mechanisms and novel therapeutic targets. *Advances in Wound Care* 7(7): 209–231.

Lazarus, G. S., Cooper, D. M., Knighton, D. R., Margolis, D. J., Percoraro, R. E., Rodeheaver, G. & Robson, M. C. 1994. Definitions and guidelines for assessment of wounds and evaluation of healing. *Wound Repair and Regeneration* 2(3): 165–170.

Leal, E. C., Carvalho, E., Tellechea, A., Kafanas, A., Tecilazich, F., Kearney, C. & Veves, A. 2015. Substance P promotes wound healing in diabetes by modulating inflammation and macrophage phenotype. *The American Journal of Pathology* 185(6): 1638–1648.

Li, J., Chen, J. & Kirsner, R. 2007. Pathophysiology of acute wound healing. *Clinics in Dermatology* 25(1): 9–18.

Lin, Q., Fang, D., Fang, J., Ren, X., Yang, X., Wen, F. & Su, S. B. 2011. Impaired wound healing with defective expression of chemokines and recruitment of myeloid cells in TLR3-deficient mice. *The Journal of Immunology* 186(6): 3710–3717.

Lin, Y. W., Lee, B., Liu, P. S. & Wei, L. N. 2016. Receptor-interacting protein 140 orchestrates the dynamics of macrophage M1/M2 polarization. *Journal of Innate Immunity* 8(1): 97–107.

Liu, Z., Xu, Y., Chen, L., Xie, J., Tang, J., Zhao, J. & Liu, X. 2016. Dendritic epidermal T cells facilitate wound healing in diabetic mice. *American Journal of Translational Research* 8(5): 2375.

Mann, K. G. 2003. Factor VII-activating protease: Coagulation, fibrinolysis, and atherothrombosis?. *Circulation* 107(5): 654–655.

Mantovani, A., Sica, A. & Locati, M. 2005. Macrophage polarization comes of age. *Immunity* 23(4): 344–346.

Martin, P. & Leibovich, S. J. 2005. Inflammatory cells during wound repair: The good, the bad and the ugly. *Trends in Cell Biology* 15(11): 599–607.

Mosser, D. M. & Edwards, J. P. 2008. Exploring the full spectrum of macrophage activation. *Nature Reviews Immunology* 8(12): 958–969.

Naik, S., Bouladoux, N., Linehan, J. L., Han, S. J., Harrison, O. J., Wilhelm, C. & Belkaid, Y. 2015. Commensal–dendritic-cell interaction specifies a unique protective skin immune signature. *Nature* 520(7545): 104–108.

Natarajan, S., Williamson, D., Stiltz, A. J. & Harding, K. 2000. Advances in wound care and healing technology. *American Journal of Clinical Dermatology* 1: 269–275.

Nelson, S. M., Lei, X. & Prabhu, K. S. 2011. Selenium levels affect the IL-4–induced expression of alternative activation markers in murine macrophages. *The Journal of Nutrition* 141(9): 1754–1761.

Nosbaum, A., Prevel, N., Truong, H. A., Mehta, P., Ettinger, M., Scharschmidt, T. C. & Rosenblum, M. D. 2016. Cutting edge: regulatory T cells facilitate cutaneous wound healing. *The Journal of Immunology* 196(5): 2010–2014.

Nunan, R., Campbell, J., Mori, R., Pitulescu, M. E., Jiang, W. G., Harding, K. G. & Martin, P. 2015. Ephrin-Bs drive junctional downregulation and actin stress fiber disassembly to enable wound re-epithelialization. *Cell Reports* 13(7): 1380–1395.

Poché, R. A., Hsu, C. W., McElwee, M. L., Burns, A. R. & Dickinson, M. E. 2015. Macrophages engulf endothelial cell membrane particles preceding pupillary membrane capillary regression. *Developmental Biology* 403(1): 30–42.

Rennert, R. C., Sorkin, M., Garg, R. K. & Gurtner, G. C. 2012. Stem cell recruitment after injury: Lessons for regenerative medicine. *Regenerative Medicine* 7(6): 833–850.

Robson, M. C., Phillips, T. J., Falanga, V., Odenheimer, D. J., Parish, L. C., Jensen, J. L. & Steed, D. L. 2001. Randomized trial of topically applied repifermin (recombinant human keratinocyte growth factor-2) to accelerate wound healing in venous ulcers. *Wound Repair and Regeneration* 9(5): 347–352.

Rousselle, P., Braye, F. & Dayan, G. 2019. Re-epithelialization of adult skin wounds: Cellular mechanisms and therapeutic strategies. *Advanced Drug Delivery Reviews* 146: 344–365.

Santoro, M. M. & Gaudino, G. 2005. Cellular and molecular facets of keratinocyte reepithelization during wound healing. *Experimental Cell Research* 304(1): 274–286.

Schultz, G. S. & Wysocki, A. 2009. Interactions between extracellular matrix and growth factors in wound healing. *Wound Repair and Regeneration* 17(2): 153–162.

Scully, D., Sfyri, P., Wilkinson, H. N., Acebes-Huerta, A., Verpoorten, S., Muñoz-Turrillas, M. C. & Matsakas, A. 2020. Optimising platelet secretomes to deliver robust tissue-specific regeneration. *Journal of Tissue Engineering and Regenerative Medicine* 14(1): 82–98.

Segel, G. B., Halterman, M. W. & Lichtman, M. A. 2011. The paradox of the neutrophil's role in tissue injury. *Journal of Leukocyte Biology* 89(3): 359–372.

Shaw, T. J. & Martin, P. 2016. Wound repair: A showcase for cell plasticity and migration. *Current Opinion in Cell Biology* 42: 29–37.

Shiraki, R., Inoue, N., Kawasaki, S., Takei, A., Kadotani, M., Ohnishi, Y. & Yokoyama, M. 2004. Expression of toll-like receptors on human platelets. *Thrombosis Research* 113(6): 379–385.

Sinha, M., Sen, C. K., Singh, K., Das, A., Ghatak, S., Rhea, B. & Roy, S. 2018. Direct conversion of injury-site myeloid cells to fibroblast-like cells of granulation tissue. *Nature Communications* 9(1): 936.

Snyder, R. J., Lantis, J., Kirsner, R. S., Shah, V., Molyneaux, M. & Carter, M. J. 2016. Macrophages: A review of their role in wound healing and their therapeutic use. *Wound Repair and Regeneration* 24(4): 613–629.

Stadelmann, W. K., Digenis, A. G. & Tobin, G. R. 1998. Physiology and healing dynamics of chronic cutaneous wounds. *The American Journal of Surgery* 176(2): 26S–38S.

Stefater III, J. A., Lewkowich, I., Rao, S., Mariggi, G., Carpenter, A. C., Burr, A. R. & Lang, R. A. 2011. Regulation of angiogenesis by a non-canonical Wnt–Flt1 pathway in myeloid cells. *Nature* 474(7352): 511–515.

Szycher, M. & Lee, S. J. 1992. Modern wound dressings: A systematic approach to wound healing. *Journal of Biomaterials Applications* 7(2): 142–213.

Takeo, M., Lee, W. & Ito, M. 2015. Wound healing and skin regeneration. *Cold Spring Harbor Perspectives in Medicine* 5(1): a023267.

Tang, Y. Q., Yeaman, M. R. & Selsted, M. E. 2002. Antimicrobial peptides from human platelets. *Infection and Immunity* 70(12): 6524–6533.

Thomason, H. A., Cooper, N. H., Ansell, D. M., Chiu, M., Merrit, A. J., Hardman, M. J. & Garrod, D. R. 2012. Direct evidence that PKCα positively regulates wound re-epithelialization: Correlation with changes in desmosomal adhesiveness. *The Journal of Pathology* 227(3): 346–356.

Um, J., Jung, N., Chin, S., Cho, Y., Choi, S. & Park, K. S. 2016. Substance P enhances EPC mobilization for accelerated wound healing. *Wound Repair and Regeneration* 24(2): 402–410.

Um, J., Yu, J. & Park, K. S. 2017. Substance P accelerates wound healing in type 2 diabetic mice through endothelial progenitor cell mobilization and Yes-associated protein activation. *Molecular Medicine Reports* 15(5): 3035–3040.

Van der Veer, W. M., Van Egmond, M., Niessen, F. B. & Beelen, R. H. 2011. Macrophages in skin injury and repair. *Immunobiology* 216(7): 753–762.

Vanwijck, R. 2001. Surgical biology of wound healing. *Bulletin et memoires del'Academie Royale de Medecine de Belgique* 156(3–4): 175–184.

Velnar, T., Bailey, T. & Smrkolj, V. 2009. The wound healing process: An overview of the cellular and molecular mechanisms. *Journal of International Medical Research* 37(5): 1528–1542.

Vestweber, D. 2015. How leukocytes cross the vascular endothelium. *Nature Reviews Immunology* 15(11): 692–704.

Wager, L. J. & Leavesley, D. I. 2015. MicroRNA regulation of epithelial-to-mesenchymal transition during re-epithelialisation: assessing an open wound. *Wound Practice & Research: Journal of the Australian Wound Management Association* 23(3): 132–142.

Wang, J., Hossain, M., Thanabalasuriar, A., Gunzer, M., Meininger, C. & Kubes, P. 2017. Visualizing the function and fate of neutrophils in sterile injury and repair. *Science* 358(6359): 111–116.

Weller, K., Foitzik, K., Paus, R., Syska, W., Maurer, M., Weller, K. & Maurer, M. 2006. Mast cells are required for normal healing of skin wounds in mice. *The FASEB Journal* 20(13): 2366–2368.

Witte, M. B. & Barbul, A. 1997. General principles of wound healing. *Surgical Clinics of North America* 77(3): 509–528.

Xue, M. & Jackson, C. J. 2015. Extracellular matrix reorganization during wound healing and its impact on abnormal scarring. *Advances in Wound Care* 4(3): 119–136.

Yoo, S. K. & Huttenlocher, A. 2011. Spatiotemporal photolabeling of neutrophil trafficking during inflammation in live zebrafish. *Journal of Leukocyte Biology* 89(5): 661–667.

Young, A. & McNaught, C. E. 2011. The physiology of wound healing. *Surgery (Oxford)* 29(10): 475–479.

Zaidi, A. & Green, L. 2019. Physiology of haemostasis. *Anaesthesia & Intensive Care Medicine* 20(3): 152–158.

Zhu, F. B., Fang, X. J., Liu, D. W., Shao, Y., Zhang, H. Y., Peng, Y. & Liu, D. M. 2016. Substance P combined with epidermal stem cells promotes wound healing and nerve regeneration in diabetes mellitus. *Neural Regeneration Research* 11(3): 493.

6 The Skin Microbiome and Its Implications for Maintaining Skin Health

Bilge Olceroglu, Gamze Balci and Ahmet Kati

6.1 SKIN MICROBIOME: ITS COMPLEX STRUCTURE AND COMPONENTS

Skin is the human body's largest organ and serves several important functions. A complex organ is a protective barrier between the internal organs and the external environment. The skin constitutes an ecosystem encompassing 1.8 m^2, featuring diverse habitats with intricate folds, invaginations, and specialized niches that harbour a myriad of microorganisms, including bacteria, fungi, viruses, micro-eukaryotes (mites), archaea, and phages. The microbiome refers to the collective genomes of all the microorganisms living in a particular environment. Notably, a symbiotic relationship between these microorganisms and the skin contributes to maintaining a balanced and harmonious state. Symbiotic microorganisms inhabit various skin niches, protecting against potential invasion by pathogenic entities. Imbalances in this symbiotic equilibrium may precipitate skin disorders or infections.

The stratum corneum, the outermost layer of the epidermis, comprises terminally differentiated, enucleated keratinocytes known as squames. These squames, characterized by keratin fibrils and crosslinked, cornified envelopes embedded in lipid bilayers, serve as integral components, constituting the structural foundation of the epidermis. Disruptions in the intricate interplay of these elements can have consequential implications for skin health (1).

The skin microbiome is influenced by various factors such as age, sex, health, and hygiene habits. The skin's surface is slightly acidic and dry, with a lower body temperature. Skin thickness, fold depth, and hair follicle density also affect the microbiome. Water content also plays a role in the composition (2). The skin is composed of two layers: the epidermis and the dermis. The epidermis comprises differentiated keratinocytes, while the top layer, the stratum corneum, comprises terminally differentiated, enucleated keratinocytes. Body sites provide diverse microenvironments based on ultraviolet light exposure, pH, temperature, moisture, sebum content, and topography. These sites are sebaceous or oily, moist, and dry. Appendages like sweat glands, hair follicles, and sebaceous glands influence these environments. Sweat glands are more abundant in moist sites and are crucial for thermoregulation and antimicrobial properties. Sebaceous glands, connected to hair follicles, secrete lipid-rich sebum, a hydrophobic coating that lubricates and protects hair and skin (3). *Propionibacterium acnes* and *Staphylococcus spp.* are the most common bacteria in oily sites, *Corynebacterium spp.* and *Staphylococcus spp.* in moist sites, and there are many members of the phyla proteobacteria and flavobacteriales in dry sites. Notably, Thaumarchaeota species constitute approximately 88% of observed operational taxonomic units (OTUs) within the archaea domain, with Euryarchaeota comprising the remainder. Dust mites, predominantly on facial skin, around sebaceous glands, and hair follicles, are considered commensals in 23–100% of healthy individuals. Demodex mites, associated with conditions such as rosacea and chronic blepharitis, also inhabit the skin (4).

DOI: 10.1201/9781003394846-6

The abundance of microorganisms on the skin varies across regions, ranging from 10^3 to 10^4 CFU/cm^2 on the skin surface to over 10^6 CFU/cm^2 in humid areas like the groin, armpits, and nostrils. The scalp harbours approximately 10^5 CFU/cm^2, while the upper back, chest, and arms range from 10^4 to 10^6 CFU/cm^2. Hair and beards, with their grooved cuticle surface, accumulate a significant number of microorganisms, including Gram-Positive and Gram-Negative bacteria, such as *Staphylococcus aureus* and *Staphylococcus epidermidis* (2).

The human skin microbiome predominantly consists of Actinobacteria (52%), Firmicutes (24%), Proteobacteria (16%), and Bacteroidetes (6%). Notably, coagulase-negative staphylococci, cutibacterium acnes, corynebacterium, micrococcus, streptococcus, and acinetobacter are dominant, constituting 45 to 80% of the entire skin microbiome (5). Archaea, including Thaumarchaeota, comprise approximately 4% of the microbiome genes. Fungi, mainly Malassezia, are also present, with the highest diversity observed on the feet. While viruses, particularly Papillomaviridae, Polyomaviridae, and Circoviridae families, are identified, data on their presence remain limited due to the abundance of bacteria (6).

Traditionally, skin microbiome analysis relied on microbiological techniques, but modern methods such as 16S rRNA sequencing and DNA barcoding enhance accuracy, allowing detection of non-culturable or low-quantity bacteria. About 80% of skin microbiome species are culturable, but advanced techniques reveal previously unidentified microorganisms in both surface and deeper layers of human skin (4).

6.2 EVOLUTIONARY PERSPECTIVE OF THE SKIN MICROBIOME AND ITS EFFECTS ON HUMAN HEALTH

Human skin, our body's physical barrier and biggest organ, supports a complex ecosystem that includes millions of bacteria, fungi, viruses, and minute mites. The resulting "skin microbiome," generated by the convergence of host and microbe genomes, is termed a "hologenome" and dictates the organism's general health and functioning. While the skin is normally inhabited by helpful microorganisms that help sustain metabolic processes and serve as the first line of defence on our body's outer surface, they and the host coexist in a delicate equilibrium.

Microbes and their genetic products in the skin's deeper layers directly impact overall skin health by interacting with the host immune system. Furthermore, accumulating data reveals that bacteria may switch between different barrier tissues. External and internal differences in skin structure and local cell types contribute to alterations in normal and dysbiotic microbial skin populations. As the body's exterior barrier, the skin is continually subjected to environmental factors such as water, sun, pollution exposure, temperature and pH variations, and behavioural impacts caused by hygiene or cosmetic habits. These ever-changing circumstances test the skin's adaptive capability, reducing its ability to keep stable and healthy microbial communities on the epidermal surface. Furthermore, differentiating between microbiome members (who play skin-specific and critical functional functions) and those just present on the skin temporarily owing to environmental influences is essential.

The skin microbiome faces more challenges than most other human microbiomes due to its ecological habitat, which naturally provides lower nutrient availability and a harsh and dry environment that limits the diversity of organisms compared to other human microbiome ecosystems (e.g., the gut). Despite these considerations, human skin is home to various microorganisms, including bacteria, fungi, and viruses, which form complex and dynamic community connections. These resident bacteria are well-equipped to adapt to the skin and frequently settle into favoured niches.

The intricacy of human skin and its potential to promote microbial niche specialization can be explained by two major coupled traits. First, the skin has a three-dimensional structure resembling a geographical area. The typical person's skin has a surface area of 25 m^2, allowing for countless interactions between host and microorganism. Additionally, skin thickness varies by bodily area. While bacteria in the intestines are frequently protected from direct contact with the body by an antibiotic mucus layer, skin cells, including immune system cells, come into indirect touch with microbes, particularly those in the deep dermal layers. Second, the skin has a complex topography of hair follicles and different glands under the surface, including eccrine, apocrine, and sebaceous glands. Unlike the villus epithelium in the human gut, the skin has roughly five million hair follicles (pores) and sweat ducts, with their concave structure and depth contributing greatly to the skin's complexity and diversity. The densities of these microorganisms divide the skin into three types: Sebaceous/oily (forehead, scalp, chest, and back), moist/humid (around the nose and mouth, armpits, elbow crease, abdomen, lower buttocks, back of the knees, and feet), and dry (forearm, back of elbow, buttock crease, and front of the knee/leg). However, more refined scale changes in human skin over time and owing to personal variances must be considered when determining the influence of micro-scale variety on microbial community membership patterns.

Since the emergence of genetic research tools for identifying microorganisms, our understanding of microbial life on the skin has greatly improved. As genomic technology advanced, so did the analytical approaches used to produce richer microbiological information (7).

6.2.1 Healthy Skin Microbiome and Types of Microorganisms

Bacteria, fungi, viruses, and mites are among the many microorganisms living on the human skin. These bacteria are crucial in maintaining skin homeostasis and health by consuming skin resources, as indicated in Figure 6.1. However, commensal bacteria can become harmful in some situations, contributing to inflammatory skin disorders. Skin microenvironments are often classified into three types: sebaceous/oily, moist/humid, and dry, with commensal organisms commonly dominating due to the density of hair follicles and sweat glands. Current findings indicate that sebaceous areas, particularly those in follicular microenvironments, have the lowest species diversity, owing to the dominance of *Cutibacterium*species, notably *C. acnes* (8). In wet locations, *Staphylococcus* and *Corynebacterium* species are most common on the epidermal surface. In contrast, dry regions have a mixed population of *Cutibacterium*, *Staphylococcus*, *Corynebacterium*, and *Streptococcus* species, as well as more transitory bacteria (9) *Staphylococcus* commensal species, including *S. epidermidis*, *S. capitis*, *S. hominis*, *S. lugdengensis*, *S. haemolyticus*, and *S. warneri*, are common across all skin locations due to their variety and facultative anaerobic capabilities (3).

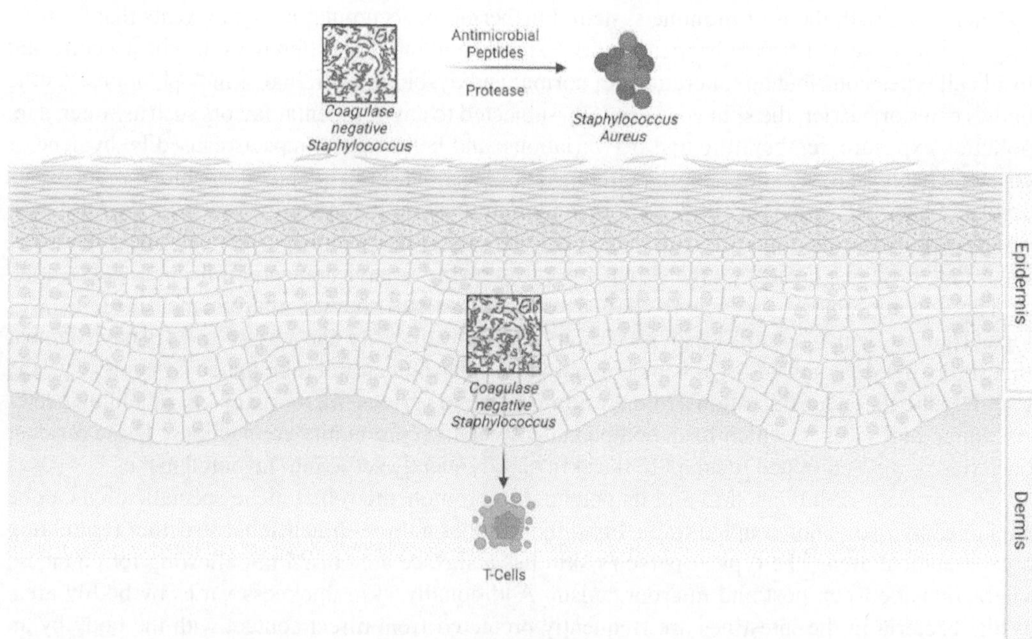

FIGURE 6.1 Microbial and immune barrier of skin microbiome.

Human skin also contains a fungal microbiome (mycobiome), which has just recently been studied using DNA-based approaches (9). *Malassezia* species, notably *M. restricta*, *M. globosa*, and *M. sympodialis*, are the most frequent fungal germs on human skin, especially in oily areas (3, 9). *Candida* species, such as *C. albicans* and *C. auris,* are common opportunistic fungal diseases that invade the skin. The human viral microbiome (virome) on the skin is less understood than the mycobiome. Bacteriophages (viruses that infect bacteria) are important components of the skin microecosystem, with discovered phages targeting *Cutibacterium*and *Staphylococcus* being conserved across people. However, the interactions and dynamics between commensal bacteria and bacteriophages remain poorly understood.

Human papillomaviruses and Merkel cell polyomaviruses, two eukaryotic viruses linked to skin cancer, have been found in human skin and are often individual-specific rather than site-specific. When shotgun metagenomics is used, viral genomes that lack taxonomic marker genes, such as 16S for bacteria or ITS1 for fungi, are readily overwhelmed by prokaryotic or eukaryotic genomes, particularly in populations that are orders of magnitude bigger. In addition, the skin contains minute arthropods such as Demodex mites, which are sometimes missed. These mites favour oily skin regions where sebum is the predominant food source, such as the face and scalp, and have been related to imbalances associated with disorders such as rosacea. While less well known, archaea have been found as part of the skin microbiome. Although there is no broad agreement on the number and significance of archaea in the skin, 16S gene signatures have been identified for the phyla Thermoproteota (previously Thaumarchaeota), Methanobacteriota, and Halobacteriota. Thaumarchaeota that oxidize ammonia is thought to help decrease the pH of the skin, so strengthening the barrier against foreign and harmful microbes. However, recent research analyzing archaeal communities on the skin found that archaea make up less than 1% of mammalian skin samples and have a small and ephemeral presence mostly influenced by environmental factors. Surfaces in touch with the skin, such as computer keyboards, phones, and door handles, have been shown to contain archaea, demonstrating that our constructed surroundings influence skin microbial ecosystems.

6.3 THE RELATIONSHIP BETWEEN SKIN MICROBIOME AND IMMUNOLOGICAL ASPECTS

6.3.1 Immune System

Immunity is characterized as resistance to illnesses, specifically infectious diseases. The immune system is a collection of cells, tissues, and chemicals that fight infections. The immune response refers to the simultaneous and coordinated reaction of these cells and chemicals to microorganisms that cause illnesses (10).

The immune system is classified into two parts based on the speed and specificity of the reaction. Neutrophils, monocytes, macrophages, complement, cytokines, and acute-phase proteins are all examples of innate immunity, also known as natural immunity. The other form of immunity is adaptive immunity, which is a unique property of the immune system in higher animals. This response involves antigen-specific responses via T and B cells (11).

The relationship between the skin microbiota and the immune system is critical for general health. The immune system, including innate and adaptive immunity, is triggered in response to microbial threats provided by the broad array of microorganisms on the skin. Understanding this dynamic interaction is critical for understanding how the skin microbiome affects the immune system's ability to protect the host against infections and preserve overall health.

6.3.2 Skin Microbiome

The skin's primary job is to provide a physical, chemical, and immunological protective barrier between the body and the outside world. The typical surface of the skin is acidic, extremely salty, dry, and aerobic, made up of a multi-layered keratinized epithelium undergoing terminal differentiation to create a strong structure. However, follicle-sebum units are anaerobic and contain more lipids (12).

The body is tightly linked to a diverse microbial ecosystem, which includes bacteria, fungi, and viruses that live on body surfaces. The microbiome is made up of microorganisms and their surroundings. The commensal microbiota actively maintains skin barrier functions by influencing fundamental physiological processes on the skin's surface (13). The microbiome contributes to skin barrier function via physical, chemical, microbial, and innate and adaptive immunological processes. These reciprocal partnerships are the outcome of a well-maintained and finely balanced microbiota. However, prolonged exposure to numerous external and internal elements can influence this balanced system, resulting in pathological difficulties and inflammatory skin illnesses such as infections, allergy diseases, and autoimmune disorders (14).

6.3.3 Composition of the Microbiome

Our skin is second only to the intestines in bacterial density, with roughly 10^4 to $10^\#$ bacteria per square centimetre and over 200 species (2). According to research, the skin microbiota plays an important role in maintaining homeostasis. These tasks include nutrition provision, vitamin and amino acid synthesis, pathogenic growth inhibition, immune system preparation for distinguishing between commensals and pathogens, and epidermal differentiation regulation (15).

While locally stable microorganisms comprise a large fraction of the microbiome, a lesser percentage are transitory microbes that might opportunistically settle when the skin is weakened. The skin has microbial communities that live in spatially different zones defined by cutaneous topography. Although the microbial makeup of a location is often retained, it can be changed by individual factors, including age, ethnicity, genetics, environment, and skincare. Skin infections can affect the microbiota, causing symptoms to vary by location. This emphasizes the importance of investigating microbial colonization in ecological niches, which can provide substantial insights into various skin disorders (16).

The different biogeography of the skin has distinct implications on microbial composition. While the intestines provide abundant micro and macronutrients for good bacteria, the skin has limited resources. The skin microbiota has adapted to the chemical environment of the stratum corneum, perspiration, and sebaceous glands. As a result, it is not unexpected that bacteria may thrive in various skin locations. UV exposure, temperature, humidity, sebum content, oxygen availability, and pH levels differ throughout these locales (1). The skin microbiome offers substantial protection at higher taxonomic levels, and although bacterial communities are classified by skin type (sebaceous, dry, and wet), fungi are only classified by body location.

6.3.4 Skin Microbiome and Barrier

The skin microbiota and skin barrier interact in a symbiotic fashion, affecting one another via physical, chemical, and immunological processes. Furthermore, the microbiome interacts directly with harmful bacteria on the skin's surface (Figure 6.1).

6.3.4.1 Physical Barrier

The skin's physical barrier is the initial defence against external attacks. This barrier, comprised of multiple thin layers of epidermal keratinocytes, creates the stratum corneum through closely regulated terminal differentiation impacted by the microbiota. Epidermal keratinocytes maintain physical continuity by forming tight junctions and adherens junctions, creating a practically impenetrable barrier against harmful germs. Tight junction proteins, such as zona occludens, may promote keratinocyte proliferation and differentiation during wound healing, restoring the barrier against microbes (17). Another function of the microbiota is to release components that aid in lipid structure.

6.3.4.2 Chemical Barrier

The epidermis and microbiota release a variety of lipids and acids that control the skin's chemical barrier. While microbial makeup is comparable in most parts of the skin, there are differences between sebaceous, moist, and dry areas (13). Both *C. acnes* and *Corynebacterium,* abundant in sebaceous regions, release lipases that hydrolyze triglycerides in sebum into free fatty acids (18, 19). Free fatty acids keep the pH low, inhibiting the development of harmful microbes. Furthermore, they suppress bacteria and boost skin immunity by promoting the production of human β-defensin 2 (hBD-2), a common antimicrobial peptide (20). The acidity of the stratum corneum is essential for both building a permeability barrier and providing antibacterial protection. An acidic skin surface provides an antagonistic chemical environment for harmful bacterial populations. Furthermore, certain lipids and free fatty acids directly inhibit harmful bacteria. Overall, the microbiome's composition and function support the chemical barriers established by skin lipids and fatty acids.

6.3.4.3 Innate Immune Barrier

When the skin barrier is disrupted, the skin microbiota triggers natural immune responses, resulting in a symbiotic relationship with healthy skin. *S. epidermidis* regulates perforin-2, a novel antimicrobial peptide that can activate γδ T cells and control the innate immune system (21). Furthermore, a particular glycan produced in *S. epidermidis* interacts with C-type lectin on human immune cells, which is required for homeostatic T-cell activation. *Candida albicans*, which causes mucocutaneous and systemic infections, protects against skin or systemic infections by activating Thl or Th17 immune responses (22). The microbiome affects the synthesis of natural substances such as IL-lα, complement C5a receptor components, and AMPs from keratinocytes and sebocytes, which improve the immune system through multiple methods (14). The commensal microbiota triggers the production of β-defensins in the skin, which have bactericidal effects against *Escherichia coli* and *S. aureus* (23).

6.3.4.4 Adaptive Immune Barrier

The link between the adaptive immune system and microorganisms is critical in developmental processes.

The immune system relies heavily on regulatory T cells (T regs) and natural killer/innate-like cells.

Experiments in a mouse model reveal that exposure to cutaneous commensal *S. epidermidis* causes a rapid buildup of neonatal T regs in the skin (24). This accumulation occurs simultaneously with hair follicle development via the Ccl20-Ccr6 pathway. MHC-Ib molecules also restricted MAIT cells, an evolutionarily primitive T cell type that recognizes microbiome antigens and metabolites. MAIT cells make up a large lymphocyte population on the skin. While microbial interactions with dermal T regs play an important role in gut microbiota colonization, T regs influence skin responses to symbionts beyond maturity. These findings highlight the importance of microorganisms in developing and activating immune cells (14).

6.3.4.5 Microbial Barrier

Microbe-microbe interactions form a barrier to invasion, colonization, and infection, allowing microorganisms to live and thrive in their habitats. The regulatory interactions between coagulase-negative *Staphylococcus* (CoNS) species and *S. aureus* are especially well recognized. CoNS species, or common microorganisms in the skin microbiome, such as *S. hominis* and S. *lugdunensis,* generate antimicrobial peptides (AMPs) that inhibit *S. aureus*. Some strains of *S. hominis* release antibiotics that inhibit *S. aureus* colonization in people with atopic dermatitis. *S. lugdunensis*generates lugdunin, a powerful *anti-S. aureus* AMP that promotes neutrophil recruitment. Furthermore, CoNS species generate compounds such as phenol-soluble modulins and cutimycin, which inhibit *S. aureus* growth and increase antimicrobial action. The quorum sensing system, a microbe-microbial communication system found in all *staphylococci*, is critical. CoNS species can disrupt *S. aureus'* quorum sensing system by suppressing agr activity, resulting in reduced toxin synthesis, tissue damage, and inflammation. *S. epidermidis* produces protease, particularly serine protease, which inhibits S. aureus biofilm formation, underscoring the importance of competitive interactions among microorganisms in maintaining skin microbiome equilibrium (14).

6.4 INFLAMMATORY SKIN DISORDERS AND THE MICROBIOME

6.4.1 Atopic Dermatitis (AD)

Atopic dermatitis is a chronic skin condition characterized by dry, scaly patches and intense itching. Scratching can lead to red, swollen spots, thickened skin, and open cuts on the skin surface. The microbial composition of skin with AD lesions differs from healthy skin without lesions. In individuals with AD, an increase in the abundance of *S. aureus* and a decrease in diversity are observed. These reductions, particularly in the genera streptococcus, corynebacterium, cutibacterium, and the proteobacteria phylum, can become more pronounced during acute flare-ups of AD. S. aureus can exacerbate AD by overcoming the symbiotic microbiome (25). This bacterium disrupts the host's immune response, damages the skin barrier, and affects adaptive immunity. This leads to inflammation due to alpha-toxin production and limits the accumulation of *S. aureus*-specific T regs. In contrast, CoNS (coagulase-negative staphylococci) are generally predominant on lesion-free skin, and these bacteria have shown a better ability to kill *S. aureus. S. aureus* strains on AD-lesioned skin can produce proteases and different surface proteins that affect the skin barrier, disrupting adaptive immune responses. Additionally, changes in the gut microbiome are associated with epithelial barrier dysfunction in AD patients, potentially worsening skin barrier functions by triggering Th2 inflammation. Consequently, alterations in the skin microbiome are believed to significantly impact AD symptoms and severity significantly. In this context, microbiome-supported treatments such as probiotics and prebiotics are suggested to have significant potential in managing AD [13]. The potential of these treatments to correct the skin microbiome and improve AD symptoms is considered a focal point for future research.

6.4.2 Psoriasis

Psoriasis is a chronic skin disorder characterized by red, scaly patches covered with psoriasis-like white scales on the skin, resulting from the body's immune response against its tissues. The relationship between psoriasis and the microbiome is complex. The microbiome, which plays a significant role in Th17 immune responses, may influence the pathogenesis of psoriasis (26). However, previous studies have shown inconsistencies in microbiome data. Conflicting trends in Firmicutes, Actinobacteria, and Proteobacteria have been reported in the microbiomes of individuals with psoriasis. Although abnormal colonization of *S. aureus* and a decrease in *S. epidermidis* abundance have been observed in psoriasis lesions, findings regarding microbial diversity are uncertain. Mouse models have supported the hypothesis that *S. aureus* can increase Th17 inflammation.

Intestinal dysbiosis is also associated with psoriasis. Intestinal factors can increase Th17 inflammation, trigger skin inflammation, and decrease T reg levels (14).

Phenol may affect abnormal skin differentiation as a result of intestinal changes. Exposure to *Candida albicans* may increase psoriasis-like skin inflammation. These findings contribute to understanding the interactions between psoriasis and the microbiome.

6.4.3 Acne

C. acnes, a bacterium in normal skin flora, contributes to acne pathogenesis (27). Microbiome diversity and imbalance between *C. acnes* phylotypes are associated with the severity of acne. *C. acnes* strains can modulate CD4+ T cell responses and activate Th17 cells. The metabolite succinic acid produced by *S. epidermidis* inhibits the growth of *C. acnes*, controlling colonization and inhibiting inflammatory responses. The gut microbiome composition changes have also been observed in acne patients, affecting skin barrier function (14). The potential of probiotics and microbiome cosmetics is being investigated in acne treatment and management. Some studies have demonstrated the anti-inflammatory and antibacterial effects of *S. thermophilus* applied to the skin. Moisturizers containing *E. faecalis* have been shown to reduce the severity of mild to moderate acne in affected individuals.

6.4.4 Lifetime Cycle of Skin Microbiota

Understanding skin physiology and microbial interactions provides insights into understanding homeostatic mechanisms and associate disturbances in normal skin functions with diseases. While the complexity and variability of skin microbiota make it challenging to determine its role in specific physiological processes, studies on how skin microbiota changes throughout life indicate an association between microbial composition and skin diseases and ageing. Microbial exposure begins in the womb, rapidly expanding upon contact with the atmospheric environment at birth. The mode of delivery influences microbiome profiles, with microbiomes of babies born via C-section mimicking the mother's skin and those born through vaginal delivery more closely resembling the vagina. Initial skin microbial colonization is critical for developing baby immunity, and disruptions in microbial transfer associated with the mode of delivery may increase the risk of Type 1 diabetes. Skin microbiota differentiates from newborns to adults, influenced by age, gender, and environmental factors. Changes in skin microbiota, particularly influenced by sex hormones, become more pronounced during adolescence. While skin microbiota in adults remains relatively stable, regional differences are observed based on the physical characteristics and topography of the skin. Consequently, changes in skin microbiota throughout life can impact skin health, and understanding these changes may facilitate the development of microbiome-based treatments and preventive measures (15).

6.4.5 Wound Pathology, Aging, and Infection

The breach of the skin barrier can significantly delay skin repair in ageing and related pathologies. This condition increases the risk of infection and may lead to the development of chronic, non-healing wounds. Chronic wounds pose a socioeconomic burden, emphasizing the importance of understanding factors contributing to delayed skin healing associated with ageing. Brief, temporary ageing is necessary for proper skin healing, but poorly healing wounds exhibit elevated levels of infection and inflammation (15). Ageing skin is more susceptible to infections due to an aged immune system. Changes in microbial composition after injury can increase the risk of infection. In elderly diabetic patients, low levels of antimicrobial peptides (AMPs) and an abundance of virulent S. aureus strains can increase the risk of infection. Oxidative stress has been observed to promote microbial dysbiosis in ageing and diabetes. Chronic wound infection contributes to poor healing, accompanied by factors such as antimicrobial resistance (AMR) and bacterial biofilms. Vascular insufficiency is a significant factor associated with infection in chronic wounds, reducing immune cell entry and diminishing systemic antibiotics' effectiveness (15).

Microbial dysbiosis can contribute to compromised interactions with commensal bacteria, and although various reports suggest an association between microbial composition and chronic wounds, studies demonstrating mechanistic functional connections are limited.

Non-healing wounds due to compromised skin barriers may harbour an environment with increased opportunistic pathogens and decreased commensal bacteria.

This underscores the importance of profiling at the strain level and highlights the need for further research to understand microbial drivers.

6.5 THE EFFECT OF EXTERNAL ENVIRONMENTAL FACTORS ON SKIN MICROBIOME

6.5.1 Understanding the Skin Microbiome

There has been growing interest in understanding the skin microbiome and its relationship with external environmental factors in recent years. These external factors, such as exposure to pollutants, UV radiation, temperature changes, and humidity levels, can significantly impact the composition and diversity of the skin microbiome (28). Studies have shown that environmental factors can influence the structure and functioning of the skin microbiota (29, 30).

For example, research has found that individuals living in urban environments with higher levels of air pollution tend to have a less diverse skin microbiome than those living in rural areas with cleaner air. Furthermore, ultraviolet radiation, particularly UVB radiation, has been shown to influence the skin microbiome (31). Exposure to UVB radiation can change the abundance and diversity of certain bacterial species on the skin. Personal care products, such as soaps, cleansers, and cosmetics, can also affect the skin microbiome (4). These products can alter the pH and moisture levels of the skin, which in turn can impact the growth and survival of different microbial species. Furthermore, lifestyle factors such as diet, exercise, and stress can also influence the skin microbiome.

Overall, it is clear that external environmental factors play a significant role in shaping the composition and functioning of the skin microbiome. Understanding how external environmental factors impact the skin microbiome is crucial for maintaining skin health and preventing skin diseases (32). The skin microbiome is highly influenced by external environmental factors such as exposure to pollutants, UV radiation, temperature changes, humidity levels, personal care products, and lifestyle factors like diet and stress (4).

These factors can alter the composition and diversity of the skin microbiome, leading to potential imbalances and skin conditions. For example, exposure to air pollution can disrupt the natural balance of the skin microbiome, leading to an overgrowth of certain bacteria and an increase in inflammatory skin conditions. Additionally, UV radiation can affect the skin microbiome by altering the abundance and diversity of microbial species. These changes can have implications for skin health and contribute to the development of skin diseases. In summary, external environmental factors such as air pollution, UV radiation, personal care products, and lifestyle factors can significantly impact the composition and diversity of the skin microbiome, potentially leading to imbalances and skin conditions. In conclusion, external environmental factors significantly impact the composition and diversity of the skin microbiome. These factors include exposure to pollutants, UV radiation, temperature changes, humidity levels, personal care products, and lifestyle factors such as diet and stress.

6.5.2 Influence of Climate on Skin Bacteria

Climate can significantly influence the composition and diversity of skin bacteria. Different regions experience varying levels of humidity, temperature, and sunlight exposure, which can directly impact the growth and survival of different microbial species on the skin (33, 34).

For example, humid environments provide favourable conditions for the growth of bacteria that thrive in moisture, while dry environments may support the growth of more resilient and drought-tolerant bacteria. Furthermore, temperature changes can affect the metabolic activity of skin bacteria, potentially altering their composition and functions (35).

Exposure to sunlight, specifically UV radiation, can also directly impact the skin microbiome. UV radiation can alter the abundance and diversity of microbial species on the skin, potentially leading to imbalances in the microbiome. Additionally, extreme weather events associated with climate change can introduce new and unfamiliar dermatologic conditions to a given location. For example, wildfires can release particulate matter and pollutants into the air, settling on the skin and disrupting the microbiome. In summary, climate factors such as humidity, temperature, sunlight exposure, and extreme weather events can greatly influence the composition and diversity of the skin microbiome and contribute to the development of skin conditions.

6.5.3 Role of UV Exposure in Shaping Skin Microbiome

UV exposure is an important external environmental factor impacting the skin microbiome. Excessive and unprotected exposure to UV radiation can disrupt the balance of the skin microbiome (36, 37).

This can lead to a decrease in microbial diversity and an increase in potentially harmful microorganisms. Additionally, UV exposure has been associated with the development of different types of skin cancers. Chronic cumulative exposure to UV radiation can result in the formation of actinic keratosis, which is a common precursor to basal cell carcinoma and squamous cell carcinoma. In summary, external environmental factors such as hygiene habits, lifestyle choices, UV exposure, and nutrition play a significant role in shaping the composition and balance of the skin microbiome it is essential to consider and address these factors to maintain a healthy and thriving skin microbiome (4, 32, 38).

Additional source: Exposure to certain environmental factors, such as UV radiation, can harm the skin microbiome. UV radiation is known to have damaging effects on the skin, including altering the composition and balance of the skin microbiome. Environmental factors such as UV radiation, hygiene habits, lifestyle choices, and nutrition can influence the skin microbiome. These external factors can disrupt the natural balance of the skin microbiome and affect its overall health. Exposure to UV radiation is a significant external environmental factor that can disrupt the balance of the skin microbiome. Additionally, UV radiation can induce DNA damage in skin cells, leading to mutations and potential changes in the skin microbiome. Furthermore, UV radiation has been linked to producing reactive oxygen species, which can harm the skin microbiome (4).

6.5.4 Effects of Temperature and Humidity on Skin Flora

Temperature and humidity are two important factors that can influence the composition and activity of skin flora. When the skin temperature is lower than the inner body temperature, it can create a favourable environment for certain microorganisms to thrive. On the other hand, higher temperatures can inhibit the growth of certain microorganisms (4, 39).

Humidity also plays a role in the skin microbiome, as higher humidity levels can lead to increased moisture on the skin, creating a more suitable environment for certain microorganisms to grow. As a result, temperature and humidity can impact the balance and diversity of the skin microbiome. Furthermore, the physical and chemical properties of the skin itself, such as its acidity, dryness, and constant shedding of keratinized cells, also influence the dominance and proportions of specific microorganisms in the skin microbiome. Exposure to these factors can disrupt the balance of the skin microbiome and potentially lead to changes in its composition and overall health.

6.5.5 Impact of Pollution on Dermal Microorganisms

Air pollution, including small particulate matter, can directly impact the skin microbiome. Exposure to environmental pollutants, such as small particulate matter and air contaminants, has been shown to disrupt the composition and diversity of dermal microorganisms (40, 41).

These pollutants can penetrate the skin barrier and interact with the microbiome, leading to imbalances and alterations in the microbial community. The resulting disruption of the skin microbiome can contribute to the developing and worsening of various skin conditions, such as inflammation, irritation, dryness, itching skin, an increased risk of inflammatory skin diseases, premature ageing, and even skin cancer. Understanding the impact of pollution on dermal microorganisms is crucial for developing strategies to protect the skin microbiome and prevent skin conditions (31).

Small particulate matter and air contaminants can penetrate the skin barrier and disrupt the natural balance of the microbiome, leading to imbalances that contribute to various skin issues. It is essential to consider protective measures such as using skincare products with antioxidant and anti-inflammatory properties to mitigate the effects of pollution on the skin microbiome. Incorporating gentle cleansing routines and maintaining a healthy lifestyle can help support a balanced and diverse skin microbiome. Research into the specific mechanisms through which pollution affects the skin microbiome will be instrumental in developing targeted skincare solutions to combat its detrimental effects (42, 43).

By addressing these concerns, a better understanding of the impact of pollution on dermal microorganisms can lead to more effective skincare practices and ultimately contribute to improved skin health and overall well-being. The impact of pollution on the skin microbiome has significant implications for skin health and the development of skin conditions.

Maintaining a healthy skin microbiome is essential for overall skin health and well-being. Given the significant impact of external environmental factors on the composition and diversity of the skin microbiome, implementing strategies to counteract these influences can help preserve a balanced and thriving microbial community on the skin.

6.5.6 Topical Products and Skin Microbiome

The use of topical products, such as cosmetics and skincare treatments, can also have a significant impact on the skin microbiome. Certain ingredients found in these products, such as antimicrobial preservatives, can disrupt the natural balance of the microbiome and decrease its diversity (44). Additionally, excessive use of cosmetics can reduce the number and variety of microbes that colonize the skin (4).

This can lead to dysbiosis in the skin microbiome, where harmful microorganisms can thrive and beneficial ones are suppressed, potentially leading to skin problems like inflammation, irritation, and even infections. Furthermore, the application and composition of skin care products can vary among different ethnicities, which may contribute to variations in the skin microbiome between populations.

Therefore, it is essential to carefully select and use skincare products that are compatible with the skin's natural microbiome and promote a healthy balance of microorganisms (45). In addition to topical products, external environmental factors such as hygiene habits, host lifestyle and environment, and even changes in the living environment can also affect the skin microbiome.

Adopting gentle cleansing routines is another vital aspect of maintaining a healthy skin microbiome. Harsh soaps and cleansers can disrupt the skin's natural pH and moisture levels, potentially leading to imbalances in microbial populations. Therefore, choosing mild and gentle cleansers is essential for preserving the diversity and equilibrium of the skin microbiome.

6.5.7 Clothing and Skin Microbiome

Clothing can also play a role in the composition of the skin microbiome. The contact between the skin and clothing can lead to the transmission of microorganisms, contributing to the formation of a textile microbiome. The presence of microbes in clothing is a common occurrence; in most cases, it does not cause any harm. However, there are instances where harmful microorganisms present in clothing can lead to infections and pose health risks, especially in healthcare settings. Healthcare-associated infections, caused by the transmission of harmful microorganisms from infected patients, healthcare workers, or environmental sources to others, are a major concern for human health (46). These infections can be exacerbated by the presence of harmful microorganisms on clothing, highlighting the importance of maintaining proper hygiene practices and implementing effective disinfection protocols in healthcare settings.

External environmental factors like living and working environments can also impact the skin microbiome (47). For example, a study conducted in Finland showed that the urban and rural environments significantly influenced the skin microbiome of children aged one to four years (48). Living in urban areas exposed children to different environmental factors than those living in rural areas, resulting in variations in their skin microbiome composition.

6.5.8 Water Quality and Skin Health

Water quality is another external environmental factor impacting skin health and the microbiome. Poor water quality, such as high levels of contaminants or chemicals, can disrupt the balance of the skin microbiome and lead to various skin issues (49). For example, water with high turbidity and the presence of suspended materials in water can introduce pollutants and potentially harmful microorganisms to the skin surface, leading to skin irritation or infections. Water quality, including chlorine content and hardness, can affect the skin microbiome. Imbalances in water composition may disrupt the microbial communities, leading to skin-related issues (45).

In conclusion, external environmental factors such as water quality, hygiene practices, living and working environments, and exposure to certain cosmetics or clothing can significantly impact the composition and diversity of the skin microbiome. Chemical parameters of dissolved oxygen, chemical oxygen demand, and biological oxygen demand that exceed environmental quality standards can increase the risk of skin disorders complaints.

6.5.9 Diet and Skin Microbiome

Diet is another external environmental factor that can influence the skin microbiome. Certain dietary components, such as prebiotics and probiotics, can positively influence the composition and diversity of the skin microbiome (50). For example, consuming foods rich in prebiotics, such as fruits, vegetables, and whole grains, can nourish beneficial bacteria on the skin and promote a healthy microbiome (51).

Additionally, a diet high in processed foods and sugars can negatively impact the skin microbiome by promoting the growth of harmful bacteria. Exposure to allergenic foods or food sensitivities can affect the diversity and composition of the skin microbiome, potentially leading to skin issues like eczema or acne. Exposure to allergenic foods or food sensitivities can affect the diversity and composition of the skin microbiome, potentially leading to skin issues like eczema or acne. Furthermore, inadequate nutrient intake can compromise the skin barrier's function, making it more susceptible to damage and infection.

6.5.10 Stress and Skin Microbiome

As an external factor, stress can also significantly impact the skin microbiome. Chronic stress can disrupt the balance of beneficial bacteria on the skin, leading to an overgrowth of harmful bacteria and a compromised skin barrier. This can increase the risk of skin conditions such as acne, eczema, and psoriasis. Stress can also affect the immune system, which is crucial in maintaining a healthy skin microbiome (52). Overall, external environmental factors such as hygiene practices, diet, living and working environments, exposure to certain cosmetics or clothing, and stress can significantly impact the composition and diversity of the skin microbiome (4).

Stress can disrupt the balance of beneficial bacteria on the skin, leading to an overgrowth of harmful bacteria and a compromised skin barrier, thereby increasing the risk of skin disorders.

6.5.11 Microbial Transfer and Social Interactions

Human-to-human microbial transfer is a dynamic and intricate process that plays a crucial role in shaping the diversity and composition of the skin microbiome (53, 54). Social interactions, particularly physical contact like hugging and kissing, create opportunities for exchanging microorganisms between individuals. The skin, the body's largest organ and a primary interface with the external environment is a complex ecosystem that harbours a diverse array of microbes. The transfer of these microbes during close interactions contributes to the interplay of different bacterial species, fungi, and other microorganisms on the skin's surface.

Direct physical contact impacts the skin microbiome, and sharing personal items also plays a significant role in microbial exchange. Personal belongings such as towels, clothing, and makeup are potential vectors for the transmission of microorganisms. When individuals share these items, they introduce a variety of microbes to their skin that may not have been present otherwise. This exchange can lead to alterations in the microbial composition of the skin, influencing its overall balance and potentially affecting skin health.

Furthermore, the nature of microbial transfer during social interactions can vary based on the intimacy and frequency of contact between individuals (55, 56).

For instance, close family members or romantic partners who engage in regular physical contact may experience a more extensive exchange of skin microbes compared to acquaintances with limited interactions. The dynamics of microbial transfer during social interactions are complex and influenced by factors such as the duration and intensity of contact and individual differences in skin health and microbiome resilience (57).

Understanding the nuances of microbial transfer during social interactions sheds light on the intricate ecology of the skin microbiome and has implications for skin health and disease. The delicate balance of microorganisms on the skin is essential for maintaining a protective barrier against harmful pathogens and supporting overall skin function.

As research in this field progresses, it may offer insights into how targeted interventions, such as personalized skincare or hygiene practices, can be developed to promote a healthy and diverse skin microbiome.

6.5.12 Strategies for Protecting Skin Microbiome from Environmental Damage

It is important to incorporate certain strategies into skincare routines to protect the skin microbiome from environmental damage. These may include:

- Avoid overusing harsh cleansers or antibacterial soaps, as they can strip the skin of beneficial bacteria and disrupt the microbiome (58).
- Regularly cleansing the skin using gentle, pH-balanced cleansers to remove dirt and excess oil without disrupting the natural microbial balance (59).
- Limit exposure to excessive UV radiation by wearing protective clothing and using sunscreen.
- Maintain proper hydration levels by drinking enough water and using hydrating skincare products.
- Avoid excessive use of cosmetics, as they can reduce the number and variety of microbes that colonize the skin.

- Practice good hygiene habits, such as washing hands regularly and using clean towels and linens, to minimize the transfer of harmful microorganisms onto the skin.
- Avoid smoking and limit alcohol consumption, as these habits can negatively impact the skin microbiome. External environmental factors can have a significant impact on the skin microbiome.
- UV radiation, for example, can lead to oxidative stress and DNA damage in the skin microbiome, disrupting its composition (4).
- Temperature and humidity can create an environment that either promotes or inhibits the growth of certain microorganisms on the skin.
- Hygiene habits and lifestyle choices, such as the use of certain skincare products or overuse of cosmetics, can alter the skin microbiome composition.
- The skin's physical and chemical properties, such as its acidity, moisture levels, and the presence of glands, can influence the diversity and balance of the skin microbiome (32).
- It is important to protect the skin microbiome from environmental damage to maintain skin health and prevent dermatologic disorders.
- To protect the skin microbiome from environmental damage, incorporating strategies such as regular gentle cleansing, avoiding harsh cleansers and antibacterial soaps, limiting UV radiation exposure, maintaining proper hydration, practising good hygiene habits, and avoiding excessive use of cosmetics can be beneficial (4).
- In the context of the COVID-19 pandemic, skin disinfection, especially hand sanitation, is crucial for preventing the spread of the virus (32).
- However, excessive or improper use of disinfectants can hurt the skin microbiome, leading to dysbiosis and potential skin in (4).
- The application of disinfectants should consider environmental factors such as temperature, humidity, and exposure time to ensure effective disinfection while minimizing disruption to the skin microbiome.
- Cosmetics, such as shampoos or creams, can also impact the skin microbiome and may cause infections if used inappropriately or by individuals with compromised immune systems.
- Overall, external environmental factors can significantly influence the skin microbiome, and it is important to take measures to protect and maintain its balance for optimal skin health.

6.5.13 Future Directions in Skin Microbiome Research

Future directions in skin microbiome research may include:

- Examine the long-term effects of skincare products and cosmetics on the skin microbiome, including the potential for microbial resistance or dysbiosis (60).
- Exploration of the role of the skin microbiome in various dermatologic disorders and skin conditions, such as acne, psoriasis, and atopic dermatitis, to better understand the underlying mechanisms and develop targeted therapies (32).
- Development of strategies to modulate the skin microbiome for therapeutic purposes, such as personalized probiotics or prebiotics tailored to individual skin types.
- Investigation into the impact of lifestyle factors, including diet, stress, and physical activity, on the diversity and composition of the skin microbiome.
- Evaluation of the role of genetics and ethnicity in shaping the skin microbiome and how it may contribute to variations in skin health and disease susceptibility.
- Assessment of the potential impact of external factors, such as urbanization or changes in living environment, on the skin microbiome and skin health in humans.
- Exploration of the potential use of the skin microbiome as a biomarker for overall health and disease and a target for therapeutic interventions (4).

- Analysis of the interactions between the skin microbiome and the immune system, including how changes in the microbiome composition may influence immune responses and inflammation (55).
- Investigate the potential role of the skin microbiome in wound healing and tissue regeneration and how it may be harnessed for therapeutic purposes.
- Development of novel technologies and techniques, such as advanced sequencing methods or *in vitro* skin models, to improve our understanding of the skin microbiome and its interactions with the host (4, 55).
- Collaboration between researchers, clinicians, and industry professionals to translate findings into practical applications for skincare products and treatments that promote a healthy skin microbiome (4).
- Overall, the effects of external environmental factors on the skin microbiome are complex and multi-faceted. They involve not only the direct impact of these factors on the microbial composition and diversity of the skin but also their potential influence on the host immune system and inflammatory responses. In conclusion, studying and understanding these effects is crucial for developing strategies to maintain a healthy skin microbiome and prevent or treat various skin conditions. In conclusion, external environmental factors such as hygiene habits, host lifestyle and environment, skin structure, cosmetics ingredients, skin disinfection practices, and even changes in living environment can have a significant impact on the diversity, composition, and overall health of the skin microbiome (4).

6.6 CONCLUSION

In conclusion, the skin microbiome is influenced by various external environmental factors, including water quality, diet, stress, microbial transfer through social interactions, and hygiene practices. These factors play a significant role in shaping the composition and diversity of the skin microbiome, ultimately impacting skin health and the development of skin disorders. Proper maintenance of hygiene practices, adherence to a balanced and nutritious diet, and managing stress levels are essential for promoting a healthy skin microbiome. Additionally, understanding the impact of external environmental factors on the skin microbiome can guide the development of effective skincare practices and interventions to support overall skin health. The complex interplay between these environmental factors and the skin microbiome underscores the importance of ongoing research and education in this field for promoting skin wellness and addressing skin-related issues.

6.7 MOLECULAR EFFECTS OF PROBIOTICS AND PREBIOTICS ON SKIN HEALTH

The introduction of probiotics and prebiotics as potential treatments for various health conditions, including skin health, has gained significant attention in recent years. Probiotics are live microorganisms that provide health benefits to the host when consumed in adequate amounts. Prebiotics, however, are non-digestible compounds that selectively stimulate the growth and activity of beneficial bacteria in the gut. These compounds can exert their effects on the skin through various mechanisms, including modulating immune responses, inhibiting the growth of pathogenic bacteria, and promoting skin barrier function and repair (4).

The molecular effects of probiotics and prebiotics on skin health extend beyond the surface, delving into intricate cellular and biochemical processes. Probiotics, primarily consisting of beneficial bacteria like Lactobacillus and Bifidobacterium species, influence the skin by modulating immune responses.

These microorganisms interact with immune cells in the gut, leading to systemic effects that can positively impact skin health. The communication between the gut and skin, often referred to as the gut-skin axis, highlights the interconnectedness of these two organ systems. Probiotics can enhance immune tolerance, reduce inflammation, and contribute to a balanced immune environment, all of which play pivotal roles in maintaining healthy skin. Additionally, prebiotics, acting as fuel for beneficial bacteria, contribute to this symbiotic relationship by fostering the growth of probiotics in the gut. As a result, the systemic effects of prebiotics indirectly influence skin health, promoting a harmonious balance of the skin microbiome and supporting the skin's natural defences.

The intricate interplay between probiotics, prebiotics, and the immune system exemplifies the multifaceted molecular mechanisms underlying the potential benefits of these bioactive compounds for skin health.

6.7.1 Probiotics and Skin Microbiome

Probiotics, live beneficial bacteria, can positively modulate the composition of the skin microbiome. By competing with pathogenic microorganisms for resources and modulating local immune responses, probiotics maintain a diverse and harmonious microbial community on the skin. Probiotics have been shown to have various beneficial effects on skin health. These include reducing inflammation, improving skin hydration, enhancing wound healing, and promoting the production of antimicrobial peptides.

6.7.1.1 Inflammation Reduction

Probiotics are pivotal in mitigating skin inflammation, a common factor in conditions such as acne, eczema, and psoriasis (61). By regulating immune responses, probiotics help dampen excessive inflammation by modulating cytokine production and promoting anti-inflammatory signals (61).

The ability of certain probiotic strains to downregulate pro-inflammatory pathways contributes to a calmer skin environment, potentially alleviating symptoms associated with inflammatory skin disorders.

6.7.1.2 Skin Hydration Improvement

Beyond their impact on the microbiome, probiotics have been associated with improvements in skin hydration (62, 63). Maintaining a balanced and diverse skin microbiome with probiotics reinforces the skin barrier. This strengthened barrier function helps reduce water loss from the skin, leading to enhanced hydration. As a result, individuals incorporating probiotics into their skincare routine may experience benefits such as smoother and more supple skin, addressing concerns related to dryness and dehydration (64).

6.7.1.3 Wound Healing Enhancement

Probiotics exhibit promising capabilities in accelerating the wound-healing process (65).

Their influence extends to acute and chronic wounds, with studies suggesting that probiotics can enhance tissue regeneration, collagen deposition, and overall wound closure. The interaction between probiotics and the immune system plays a crucial role in this process, as these microorganisms contribute to a supportive environment that fosters efficient healing. Incorporating probiotics into wound care strategies presents a novel avenue for promoting optimal recovery and minimizing scarring to the overall resilience of the skin barrier (66).

6.7.1.4 Antimicrobial Peptide Production

Probiotics contribute to the skin's defence mechanisms by promoting the production of antimicrobial peptides (AMPs) (67). These peptides play a crucial role in the innate immune response, acting as natural antibiotics against harmful microorganisms. Probiotics stimulate the skin cells to produce AMPs, boosting the local defence against pathogens (68).

This antimicrobial activity helps maintain a healthy microbial balance on the skin and provides an added layer of protection against infections, contributing to skin health and preventing various skin conditions.

In summary, the multifaceted effects of probiotics on the skin extend from their influence on the microbiome composition to their direct impact on various aspects of skin health. Incorporating probiotics into skincare routines and exploring their potential in therapeutic applications represents a promising avenue for promoting overall skin well-being.

6.7.2 Immunomodulatory Effects

Probiotics regulate cytokines and other immune mediators by influencing the skin's immune response. These immunomodulatory effects help mitigate inflammation, enhance the skin's ability to respond to environmental stressors, and promote overall skin health.

6.7.2.1 Modulation of Cytokines

One of the key mechanisms through which probiotics exert their immunomodulatory effects is by influencing the production and activity of cytokines (69). Cytokines are signalling molecules that play a crucial role in regulating immune responses. Probiotics have been shown to modulate the expression of pro-inflammatory and anti-inflammatory cytokines, creating a balanced immune environment on the skin. Fine-tuning cytokine profiles helps mitigate excessive inflammation, a common factor in various skin conditions, and supports maintaining a healthy and responsive immune system.

6.7.2.2 Enhanced Response to Environmental Stressors

The skin constantly encounters environmental stressors such as UV radiation, pollutants, and microbial challenges. Probiotics contribute to the skin's resilience by enhancing its ability to respond to these stressors. Through their immunomodulatory effects, probiotics prime the skin's immune cells to mount effective responses against external threat (38, 45, 69).

This heightened responsiveness not only aids in the immediate defence against potential irritants but also contributes to long-term skin health by promoting adaptive immune processes that improve the skin's overall resilience to environmental challenges.

6.7.2.3 Impact on Skin Microenvironment

Beyond the direct modulation of immune mediators, probiotics influence the broader microenvironment of the skin. By fostering a diverse and balanced skin microbiome, probiotics indirectly contribute to immunomodulation (70).

The intricate interplay between probiotics, the skin microbiome, and the immune system exemplifies a holistic approach to skin health. A harmonious microbial community on the skin, supported by probiotics, creates an environment that encourages appropriate immune responses and helps prevent dysregulation that may lead to skin disorders (53, 71, 72).

6.7.2.4 Potential for Therapeutic Applications

The immunomodulatory effects of probiotics hold promising implications for therapeutic applications in dermatology (61). Researchers are exploring using probiotics as adjunct treatments for various skin conditions, including inflammatory disorders and chronic wounds (73, 74).

The ability of probiotics to finely regulate immune responses opens avenues for targeted interventions, potentially offering a natural and personalized approach to managing skin health. As our understanding of the immunomodulatory effects of probiotics continues to grow, so does the potential for innovative and tailored skincare solutions that leverage the intricate interactions between probiotics and the skin's immune system.

In conclusion, probiotics have significantly affected skin health by modulating the immune response, impacting the skin microenvironment, and showing potential for therapeutic applications. Further research is needed to fully understand the molecular mechanisms underlying these effects and optimize the use of probiotics in skincare. Additionally, investigations into the specific strains and dosages of probiotics that yield the most benefits for different skin conditions are necessary.

6.7.3 Barrier Function Enhancement

Probiotics play a role in fortifying the skin barrier. Through the stimulation of ceramide production and reinforcement of tight junctions between skin cells, probiotics contribute to a more resilient skin barrier, reducing the risk of transepidermal water loss and external irritant penetration.

6.7.3.1 Stimulation of Ceramide Production

Ceramides are essential lipids that are crucial to the skin's barrier. Probiotics have been found to play a pivotal role in stimulating the production of ceramides within the skin (75). Ceramides contribute to the maintenance of the skin barrier by forming a protective layer that helps retain moisture and prevents the entry of harmful microorganisms. Probiotic-induced stimulation of ceramide production supports the structural integrity of the skin barrier, providing a foundation for improved hydration and overall skin health (76).

6.7.3.2 Reinforcement of Tight Junctions

Probiotics actively contribute to the reinforcement of tight junctions between skin cells. Tight junctions act as molecular barriers, regulating the passage of substances through the intercellular spaces (77). By promoting the formation and integrity of tight junctions, probiotics enhance the cohesion between skin cells, minimizing the chances of external irritants penetrating the skin. This reinforcement not only fortifies the physical barrier but also contributes to the overall strength and resilience of the skin, reducing susceptibility to environmental stressors and potential irritants (78).

6.7.3.3 Reduction of Transepidermal Water Loss (TEWL)

One of the primary functions of the skin barrier is to prevent excessive water loss, known as transepidermal water loss (TEWL) (75). Probiotics actively reduce TEWL by enhancing the skin's ability to retain moisture. This is achieved through a combination of mechanisms, including reinforcing the lipid barrier (comprising ceramides, fatty acids, and cholesterol) and promoting a balanced skin microbiome.

The reduction in TEWL improves skin hydration and helps maintain a supple and healthy skin texture (64, 79, 80).

6.7.3.4 Implications for Sensitive Skin and Dermatological Conditions

The fortification of the skin barrier by probiotics holds particular significance for individuals with sensitive skin and various dermatological conditions (81). Conditions such as eczema and rosacea are often characterized by a compromised skin barrier, leading to heightened sensitivity and susceptibility to irritants. Probiotics, through their ability to enhance barrier function, offer potential relief for individuals with such conditions (82).

Incorporating probiotics into skincare routines may represent a holistic and natural approach to managing sensitive skin, providing a protective shield against environmental aggressors, and promoting a more resilient skin barrier.

In conclusion, the role of probiotics in enhancing the skin barrier is multifaceted, involving the stimulation of ceramide production, reinforcement of tight junctions, and reduction of transepidermal water loss. This fortification has implications for overall skin health and individuals with sensitive skin and dermatological conditions, offering a promising avenue for developing targeted and effective skincare interventions.

6.7.4 Prebiotics and Microbial Nourishment

Prebiotics, non-digestible compounds that promote the growth of beneficial microorganisms, serve as essential nutrients for probiotics on the skin. By fostering the proliferation of beneficial bacteria, prebiotics contribute to the overall health of the skin microbiome.

6.7.4.1 Selective Nourishment for Beneficial Bacteria

Prebiotics act as selective nourishment for beneficial bacteria, providing a substrate that supports the growth and activity of specific microbial strains. These compounds, often in the form of non-digestible fibres such as inulin and oligosaccharides, are not broken down by human enzymes during digestion (83).

Instead, they reach the colon intact, becoming an energy source for beneficial bacteria. This selective nourishment creates an environment where probiotics can thrive, fostering a diverse and resilient skin microbiome (84).

6.7.4.2 Maintenance of Microbial Diversity

Promoting microbial diversity is a key aspect of prebiotic action on the skin. A rich and diverse skin microbiome is associated with overall skin health, as different microbial species contribute to various functions, including immune regulation and the production of antimicrobial substances (84).

Prebiotics, by providing a nourishing environment for various beneficial bacteria, help maintain this diversity. This not only ensures the stability of the skin microbiome but also contributes to the prevention of dysbiosis, a condition associated with imbalances in microbial communities and various skin disorders (81).

6.7.4.3 Strengthening the Skin Barrier

Prebiotics play a crucial role in reinforcing the skin barrier by supporting the growth of beneficial bacteria that actively contribute to skin health. As these bacteria flourish, they interact with skin cells and stimulate the production of essential components, such as ceramides and antimicrobial peptides (64, 69, 85).

The enhanced production of these key elements strengthens the skin barrier, reducing the risk of transepidermal water loss and providing an added defence against external pathogens. The synergy between prebiotics, probiotics, and the skin barrier exemplifies the interconnected and symbiotic relationship that contributes to overall skin well-being.

6.7.4.4 Implications for Skin Conditions and Sensitivity

Prebiotics in skincare hold particular promise for individuals with various skin conditions, including those with sensitive skin. By fostering the growth of beneficial bacteria, prebiotics help create a supportive environment for skin prone to reactivity (86, 87).

This is especially relevant for conditions such as eczema and rosacea, where a disrupted skin barrier and dysregulated immune responses contribute to symptoms. The nourishment provided by prebiotics offers a natural and proactive approach to managing skin sensitivity, potentially reducing the severity of symptoms and improving the skin's overall resilience (87).

In summary, prebiotics play a vital role in microbial nourishment by providing selective support for beneficial bacteria, maintaining microbial diversity, strengthening the skin barrier, and holding potential benefits for individuals with various skin conditions. Incorporating prebiotics into skincare formulations represents a proactive and holistic strategy to promote skin health by fostering a balanced and thriving skin microbiome.

6.7.5 Modulation of Inflammatory Pathways

Prebiotics exert anti-inflammatory effects by modulating various pathways associated with inflammation. Prebiotics contribute to a balanced inflammatory milieu within the skin by downregulating pro-inflammatory cytokines and promoting anti-inflammatory responses.

6.7.5.1 Downregulation of Pro-Inflammatory Cytokines

Prebiotics play a crucial role in modulating inflammatory pathways within the skin by downregulating the production of pro-inflammatory cytokines (88, 89).

These signalling molecules are central to the inflammatory response and are often implicated in skin conditions characterized by chronic inflammation. Prebiotics act as regulators, influencing the expression of pro-inflammatory cytokines such as interleukin-1 (IL-1), interleukin-6 (IL-6), and tumour necrosis factor-alpha (TNF-α) (90, 91).

By reducing the levels of these cytokines, prebiotics contribute to a less inflammatory microenvironment in the skin, offering potential relief for individuals with inflammatory skin conditions.

6.7.5.2 Promotion of Anti-Inflammatory Responses

Beyond simply inhibiting pro-inflammatory pathways, prebiotics actively promote anti-inflammatory responses within the skin (73).

This involves enhancing molecules and processes that counteract inflammation, fostering a more balanced immune environment. Prebiotics have been associated with the upregulation of anti-inflammatory cytokines, such as interleukin-10 (IL-10), which are crucial in resolving inflammation and preventing its chronic persistence (92, 93). This dual action – dampening pro-inflammatory signals while promoting anti-inflammatory responses – contributes to the overall modulation of inflammatory pathways and supports the maintenance of skin health.

6.7.5.3 Immunomodulation through Gut-Skin Axis

The immunomodulatory effects of prebiotics extend beyond the skin itself, emphasizing the interconnectedness of the gut-skin axis. Prebiotics can influence gut microbiota composition when consumed or applied topically, subsequently impacting systemic immune responses (51, 94).

This bidirectional communication between the gut and skin underscores the holistic approach of prebiotics in modulating inflammatory pathways. The positive alterations in systemic immune function contribute to the regulation of skin inflammation, exemplifying the intricate interplay between gut health and skin well-being (95, 96).

6.7.5.4 Implications for Inflammatory Skin Conditions

Prebiotics' modulation of inflammatory pathways directly affects individuals with inflammatory skin conditions, including acne, psoriasis, and eczema (63, 88).

These conditions are often characterized by dysregulated inflammatory responses, leading to persistent symptoms. By targeting the root cause of inflammation, prebiotics present a promising avenue for managing and potentially preventing flare-ups in these skin disorders (97). Incorporating prebiotics into skincare regimens or dietary practices may offer individuals with inflammatory skin conditions a natural and complementary approach to alleviate symptoms and promote long-term skin health.

In conclusion, the anti-inflammatory effects of prebiotics involve the downregulation of pro-inflammatory cytokines, the promotion of anti-inflammatory responses, and the modulation of inflammatory pathways through the gut-skin axis. These actions underscore the potential of prebiotics as valuable components in skincare strategies aimed at managing inflammatory skin conditions and promoting a balanced and resilient skin environment.

6.7.6 Collagen Synthesis and Tissue Repair

Probiotics and prebiotics have promoted collagen synthesis and accelerated tissue repair processes.

These effects contribute to maintaining skin elasticity, firmness, and overall skin rejuvenation.

6.7.6.1 Stimulation of Collagen Synthesis

Collagen is a fundamental protein that provides structural support to the skin, contributing to its firmness and elasticity (98). Probiotics and prebiotics play a noteworthy role in promoting collagen synthesis, a vital process for maintaining skin health. Certain probiotic strains have been shown to enhance fibroblast activity, the cells responsible for collagen production (69, 99).

Additionally, prebiotics contribute to this process by providing a nourishing environment for fibroblasts. The collaborative action of probiotics and prebiotics stimulates the production of collagen fibres, which form the scaffolding that gives the skin its strength and resilience.

6.7.6.2 Acceleration of Tissue Repair

Probiotics and prebiotics have demonstrated the ability to accelerate tissue repair processes, influencing various stages of wound healing. These include the inflammatory phase, during which immune cells respond to injury; the proliferative phase, involving the formation of new tissue; and the remodelling phase, where collagen is rearranged and strengthened. Probiotics contribute to tissue repair by modulating immune responses and promoting an environment conducive to efficient healing (86, 100).

Prebiotics, by fostering the growth of beneficial bacteria, indirectly support tissue repair by maintaining a balanced and healthy skin microbiome. The combined action of probiotics and prebiotics creates a favourable milieu for enhanced tissue regeneration and efficient recovery.

6.7.6.3 Maintenance of Skin Elasticity

Skin elasticity is closely tied to the presence and organization of collagen fibres. By promoting collagen synthesis, probiotics, and prebiotics contribute to maintaining skin elasticity (100).

Collagen allows the skin to stretch and recoil, ensuring it remains supple and responsive to movement. Skin elasticity is preserved as the collagen network is reinforced and replenished, reducing the likelihood of sagging, and forming fine lines and wrinkles. This aspect is particularly significant in ageing, where a decline in collagen production can contribute to visible signs of skin ageing.

6.7.6.4 Rejuvenation and Anti-Ageing Effects

Promoting collagen synthesis and tissue repair by probiotics and prebiotics has notable implications for skin rejuvenation and anti-ageing effects. These interventions can restore youthful characteristics, such as smooth texture and resilience. The skin's overall appearance can be rejuvenated as collagen production is stimulated and tissue repair processes are optimized (101–103).

Additionally, probiotics' anti-inflammatory and immune-modulating effects may help counteract factors contributing to premature ageing, providing a holistic approach to maintaining a youthful and vibrant complexion.

6.7.6.5 Holistic Approach to Skin Health

Incorporating probiotics and prebiotics into skincare regimens represents a holistic approach to skin health, addressing not only the surface-level concerns but also the underlying processes that contribute to the overall well-being of the skin. Promoting collagen synthesis and tissue repair exemplifies the multifaceted benefits of these bioactive compounds (64, 66, 104).

By supporting the skin's structural integrity and accelerating repair processes, probiotics and prebiotics offer a comprehensive strategy for individuals seeking to enhance and maintain the health and appearance of their skin in a natural and harmonious manner.

In conclusion, the synergistic effects of probiotics and prebiotics on skin health are profound and far-reaching.

By promoting collagen synthesis and accelerating tissue repair processes, these bioactive compounds contribute to maintaining skin elasticity, firmness, and overall rejuvenation. Stimulating collagen synthesis by probiotics and prebiotics is pivotal in preserving the skin's structural integrity, ensuring its strength and resilience.

Moreover, the acceleration of tissue repair processes of these compounds creates a conducive environment for efficient wound healing and maintenance of skin health. Furthermore, the anti-aging effects and rejuvenation potential offered by probiotics and prebiotics are significant. These compounds can help restore youthful characteristics and enhance the skin's overall appearance by optimizing collagen production and promoting tissue repair. Additionally, the holistic approach to skin health demonstrated by incorporating probiotics and prebiotics into skincare regimens addresses surface-level concerns and underlying processes, offering a comprehensive strategy for individuals seeking to enhance and maintain the health and appearance of their skin naturally and harmoniously.

Overall, the benefits of probiotics and prebiotics for skin health present a promising avenue for achieving skincare goals and long-term well-being.

In conclusion, integrating probiotics and prebiotics into skincare regimens offers a holistic approach to skin health.

6.7.7 Oxidative Stress Reduction

Probiotics and prebiotics exhibit antioxidant properties, scavenging free radicals and reducing oxidative stress within the skin. Protecting skin cells from oxidative damage contributes to a healthier, more youthful skin appearance.

6.7.7.1 Antioxidant Properties of Probiotics

Probiotics showcase notable antioxidant properties, crucial in reducing oxidative stress within the skin (103, 105).

Oxidative stress results from an imbalance between the production of free radicals and the body's ability to neutralize them. Free radicals, generated through various environmental factors such as UV radiation and pollution, can cause cellular damage and contribute to premature ageing. Certain probiotic strains, through the production of antioxidant enzymes and metabolites, act as scavengers for free radicals (106).

This antioxidant activity helps mitigate oxidative stress, providing a defence mechanism against the harmful effects of environmental aggressors on the skin.

6.7.7.2 Prebiotics as Nourishment for Antioxidant-Producing Microbes

As promoters of a healthy skin microbiome, prebiotics indirectly contribute to oxidative stress reduction by fostering the growth of antioxidant-producing microbes. The symbiotic relationship between prebiotics and beneficial bacteria creates an environment where these microbes can thrive and actively produce antioxidants (100, 107).

These antioxidants, such as certain peptides and organic acids, are vital in neutralizing free radicals and preventing oxidative damage to skin cells. The nourishment provided by prebiotics creates a supportive ecosystem that enhances the overall antioxidant defence mechanisms within the skin.

6.7.7.3 Defence against Premature Aging

The reduction of oxidative stress by probiotics and prebiotics holds significant implications for the prevention of premature ageing. If left unchecked, free radicals can accelerate the breakdown of collagen and elastin fibres, contributing to the formation of wrinkles and fine lines. The antioxidant properties of probiotics and prebiotics, by neutralizing free radicals, act as a defence against these signs of ageing (100, 108).

By preserving the skin's structural integrity and preventing oxidative damage, probiotics and prebiotics offer a proactive and natural approach to maintaining a youthful and healthy skin appearance.

6.7.7.4 Overall Contribution to Skin Health

The ability of probiotics and prebiotics to reduce oxidative stress contributes to the overall health and resilience of the skin.

Beyond addressing cosmetic concerns, this reduction in oxidative stress has broader implications for skin conditions associated with inflammation and environmental damage (109, 110).

By protecting skin cells from oxidative damage, probiotics, and prebiotics contribute to a more robust and resilient skin barrier.

This, in turn, supports the skin's ability to defend against external irritants, maintain hydration, and promote an optimal microenvironment for overall skin health.

In summary, the antioxidant properties of probiotics and prebiotics represent a valuable dimension in promoting skin health. From directly scavenging free radicals to indirectly supporting the growth of antioxidant-producing microbes, these bioactive compounds offer a comprehensive approach to reducing oxidative stress.

This multifaceted action contributes to a more youthful skin appearance and supports the skin's overall well-being by enhancing its defence mechanisms against environmental challenges.

6.7.8 Future Directions and Implications

Understanding the molecular mechanisms underlying the effects of probiotics and prebiotics on skin health opens avenues for targeted skincare interventions. Further research into specific strains, formulations, and personalized approaches will likely contribute to developing novel strategies for promoting optimal skin health through microbiome modulation.

6.7.8.1 Tailoring Skincare with Strain-Specific Approaches

As we delve deeper into the intricate molecular interactions between probiotics, prebiotics, and the skin, identifying specific microbial strains that yield distinct benefits becomes paramount (111, 112).

Future research could focus on understanding the nuanced effects of different probiotic strains on skin health. Tailoring skincare formulations incorporating specific strains with defined functionalities could lead to more targeted and effective interventions. This strain-specific approach may address various skin concerns, allowing for a more personalized and precise application of probiotics to the skin (113, 114).

6.7.8.2 Advancements in Formulations and Delivery Systems

Innovations in skincare formulations and delivery systems present exciting possibilities for enhancing the efficacy of probiotics and prebiotics. Encapsulating these bioactive compounds in advanced delivery systems, such as liposomes or microencapsulation, can protect them from environmental degradation and ensure their targeted release on the skin (114, 115).

Moreover, advancements in formulation techniques may optimize the stability and viability of probiotics in skincare products, ensuring that they reach the skin in a viable and active state (116, 117).

These developments may pave the way for more sophisticated and reliable probiotic-based skincare products.

6.7.8.3 Personalized Skincare through Microbiome Profiling

The burgeoning field of microbiome profiling holds immense potential for personalized skincare. Analyzing an individual's unique skin microbiome composition can provide insights into specific microbial imbalances or deficiencies (118–120).

Future directions may involve integrating microbiome data with probiotic and prebiotic interventions, allowing for highly personalized skincare regimens. Tailoring products based on an individual's microbiome profile could optimize the effectiveness of probiotics and prebiotics, offering a truly personalized approach to skincare that considers the unique needs of each person's skin microbiome (121).

6.7.8.4 Therapeutic Applications for Skin Disorders

Exploring probiotics and prebiotics in therapeutic applications for various skin disorders is an area ripe for future investigation (62, 83).

Understanding how these bioactive compounds modulate specific molecular pathways associated with skin conditions could lead to the development of targeted interventions. From inflammatory skin disorders to conditions characterized by impaired barrier function, probiotics, and prebiotics may offer complementary or alternative strategies for managing these conditions. The potential for therapeutic applications could extend beyond cosmetic benefits, providing relief for individuals with chronic skin conditions through microbiome-targeted interventions (111, 122).

In conclusion, the future directions and implications of research on probiotics and prebiotics in skincare are vast and promising. From tailoring skincare with strain-specific approaches to advancing formulations, exploring personalized skincare through microbiome profiling, and uncovering therapeutic applications for skin disorders, the evolving landscape of microbiome research offers many possibilities. As scientific understanding deepens, integrating these insights into skincare practices may revolutionize how we approach skin health, moving towards more targeted, personalized, and effective interventions.

REFERENCES

1. Grice EA, Segre JA. The skin microbiome. *Nature reviews microbiology*. 2011;9(4):244–253.
2. Cundell AM. Microbial ecology of the human skin. *Microb Ecol*. 2018;76(1):113–20.
3. Byrd AL, Belkaid Y, Segre JA. The human skin microbiome. *Nat Rev Microbiol*. 2018;16(3):143–55.
4. Skowron K, Bauza-Kaszewska J, Kraszewska Z, Wiktorczyk-Kapischke N, Grudlewska-Buda K, Kwiecińska-Piróg J, et al. Human skin microbiome: Impact of intrinsic and extrinsic factors on skin microbiota. *Microorganisms*. 2021;9(3):543.
5. Grice EA, Kong HH, Conlan S, Deming CB, Davis J, Young AC, et al. Topographical and temporal diversity of the human skin microbiome. *Science (1979)*. 2009;324(5931):1190–2.
6. Buerger S. *The Skin and Oral Microbiome*. Içinde: Skin Microbiome Handbook. Wiley; 2020. pp. 45–57.
7. Chen Y, Knight R, Gallo RL. Evolving approaches to profiling the microbiome in skin disease. *Front Immunol*. 2023;14:1151527.
8. Scholz CFP, Kilian M. The natural history of cutaneous propionibacteria, and reclassification of selected species within the genus propionibacterium to the proposed novel genera Acidipropionibacterium gen. nov., Cutibacterium gen. nov. and Pseudopropionibacterium gen. nov. *Int J Syst Evol Microbiol*. 2016;66(11):4422–32.
9. Nguyen UT, Kalan LR. Forgotten fungi: The importance of the skin mycobiome. *Curr Opin Microbiol*. 2022;70:102235.
10. Denman AM. Cellular and molecular immunology. *Postgrad Med J*. 1992;68(798):305.
11. Parkin J, Cohen B. An overview of the immune system. *Lancet*. 2001;357(9270):1777–89.
12. Belkaid Y, Segre JA. Dialogue between skin microbiota and immunity. *Science (1979)*. 2014;346(6212):954–9.
13. Dréno B, Araviiskaia E, Berardesca E, Gontijo G, Sanchez Viera M, Xiang LF, et al. Microbiome in healthy skin, update for dermatologists. *J Eur Acad Dermatol Venereol*. 2016;30(12):2038–47.
14. Lee HJ, Kim M. Skin barrier function and the microbiome. *Int J Mol Sci*. 2022;23(21):13071.
15. Smythe P, Wilkinson HN. The skin microbiome: Current landscape and future opportunities. *Int J Mol Sci*. 2023;24(4):3950.
16. Costello EK, Lauber CL, Hamady M, Fierer N, Gordon JI, Knight R. Bacterial community variation in human body habitats across space and time. *Science (1979)*. 2009;326(5960):1694–7.
17. Volksdorf T, Heilmann J, Eming SA, Schawjinski K, Zorn-Kruppa M, Ueck C, et al. Tight junction proteins claudin-1 and occludin are important for cutaneous wound healing. *Am J Pathol*. 2017;187(6):1301–12.
18. Bomar L, Brugger SD, Yost BH, Davies SS, Lemon KP. Corynebacterium accolens releases antipneumococcal free fatty acids from human nostril and skin surface triacylglycerols. *mBio*. 2016;7(1).
19. Ridaura VK, Bouladoux N, Claesen J, Chen YE, Byrd AL, Constantinides MG, et al. Contextual control of skin immunity and inflammation by *Corynebacterium*. *J Exp Med*. 2018;215(3):785–99.

20. Nakatsuji T, Kao MC, Zhang L, Zouboulis CC, Gallo RL, Huang CM. Sebum free fatty acids enhance the innate immune defense of human sebocytes by upregulating β-defensin-2 expression. *J Invest Dermatol.* 2010;130(4):985–94.
21. Pastar I, O'Neill K, Padula L, Head CR, Burgess JL, Chen V, et al. Staphylococcus epidermidis boosts innate immune response by activation of gamma delta T cells and induction of perforin-2 in human skin. *Front Immunol.* 2020;11.
22. Kashem SW, Igyártó BZ, Gerami-Nejad M, Kumamoto Y, Mohammed J, Jarrett E, et al. Candida albicans morphology and dendritic cell subsets determine T helper cell differentiation. *Immunity.* 2015;42(2):356–66.
23. Simanski M, Erkens A, Rademacher F, Harder J. Staphylococcus epidermidis-induced interleukin-1 beta and human beta-defensin-2 expression in human keratinocytes is regulated by the host molecule A20 (TNFAIP3). *Acta Derm Venereol.* 2019;99(2):181–7.
24. Scharschmidt TC, Vasquez KS, Truong HA, Gearty SV, Pauli ML, Nosbaum A, et al. A wave of regulatory T cells into neonatal skin mediates tolerance to commensal microbes. *Immunity.* 2015;43(5):1011–21.
25. Edslev S, Agner T, Andersen P. Skin microbiome in atopic dermatitis. *Acta Derm Venereol.* 2020;100(12):adv00164.
26. Chen L, Li J, Zhu W, Kuang Y, Liu T, Zhang W, et al. Skin and gut microbiome in Psoriasis: Gaining insight into the pathophysiology of it and finding novel therapeutic strategies. *Front Microbiol.* 2020;11.
27. Craft N, Li H. Response to the commentaries on the paper: Propionibacterium acnes strain populations in the human skin microbiome associated with acne. *J Invest Dermatol.* 2013;133(9):2295–7.
28. Wang L, Xu YN, Chu CC, Jing Z, Chen Y, Zhang J, et al. Facial skin microbiota-mediated host response to pollution stress revealed by microbiome networks of individual. *mSystems.* 2021;6(4).
29. Leung MHY, Tong X, Bastien P, Guinot F, Tenenhaus A, Appenzeller BMR, et al. Changes of the human skin microbiota upon chronic exposure to polycyclic aromatic hydrocarbon pollutants. *Microbiome.* 2020;8(1):100.
30. Wu Y, Wang Z, Zhang Y, Ruan L, Li A, Liu X. Microbiome in healthy women between two districts with different air quality index. *Front Microbiol.* 2020;11.
31. Moskovicz V, Gross A, Mizrahi B. Extrinsic factors shaping the skin microbiome. *Microorganisms.* 2020;8(7):1023.
32. Callewaert C, Ravard Helffer K, Lebaron P. Skin microbiome and its interplay with the environment. *Am J Clin Dermatol.* 2020;21(S1):4–11.
33. Zeng B, Zhao J, Guo W, Zhang S, Hua Y, Tang J, et al. High-altitude living shapes the skin microbiome in humans and pigs. *Front Microbiol.* 2017;8.
34. Isler MF, Coates SJ, Boos MD. Climate change, the cutaneous microbiome and skin disease: Implications for a warming world. *Int J Dermatol.* 2023;62(3):337–45.
35. Patra V, Wagner K, Arulampalam V, Wolf P. Skin microbiome modulates the effect of ultraviolet radiation on cellular response and immune function. *iScience.* 2019;15:211–22.
36. Patra V, Bordag N, Clement Y, Köfeler H, Nicolas JF, Vocanson M, et al. Ultraviolet exposure regulates skin metabolome based on the microbiome. *Sci Rep.* 2023;13(1):7207.
37. Rai S, Rai G, Kumar A. Eco-evolutionary impact of ultraviolet radiation (UVR) exposure on microorganisms, with a special focus on our skin microbiome. *Microbiol Res.* 2022;260:127044.
38. Patra V, Gallais Sérézal I, Wolf P. Potential of skin microbiome, pro- and/or pre-biotics to affect local cutaneous responses to UV exposure. *Nutrients.* 2020;12(6):1795.
39. Conteville LC, Vicente ACP. Skin exposure to sunlight: A factor modulating the human gut microbiome composition. *Gut Microbes.* 2020;11(5):1135–8.
40. Araviiskaia E, Berardesca E, Bieber T, Gontijo G, Sanchez Viera M, Marrot L, et al. The impact of airborne pollution on skin. *J Eur Acad Dermatol Venereol.* 2019;33(8):1496–505.
41. Janvier X, Alexandre S, Boukerb AM, Souak D, Maillot O, Barreau M, et al. Deleterious effects of an air pollutant (NO2) on a selection of commensal skin bacterial strains, potential contributor to dysbiosis? *Front Microbiol.* 2020;11.
42. Dijkhoff IM, Drasler B, Karakocak BB, Petri-Fink A, Valacchi G, Eeman M, et al. Impact of airborne particulate matter on skin: A systematic review from epidemiology to in vitro studies. *Part Fibre Toxicol.* 2020;17(1):35.
43. Ghofranian A, Maibach HI. *Effects of Air Pollution on Skin.* Içinde: Cosmetic Science and Technology. Elsevier; 2017. pp. 757–66.
44. Sfriso R, Claypool J. Microbial reference frames reveal distinct shifts in the skin microbiota after cleansing. *Microorganisms.* 2020;8(11):1634.

45. Prescott SL, Larcombe DL, Logan AC, West C, Burks W, Caraballo L, et al. The skin microbiome: Impact of modern environments on skin ecology, barrier integrity, and systemic immune programming. *World Allergy Organ J.* 2017;10:29.
46. Sanders D, Grunden A, Dunn RR. A review of clothing microbiology: The history of clothing and the role of microbes in textiles. *Biol Lett.* 2021;17(1):rsbl.2020.0700.
47. Grönroos M, Parajuli A, Laitinen OH, Roslund MI, Vari HK, Hyöty H, et al. Short-term direct contact with soil and plant materials leads to an immediate increase in diversity of skin microbiota. *Microbiologyopen.* 2019;8(3).
48. Lehtimäki J, Karkman A, Laatikainen T, Paalanen L, von Hertzen L, Haahtela T, et al. Patterns in the skin microbiota differ in children and teenagers between rural and urban environments. *Sci Rep.* 2017;7(1):45651.
49. Nielsen MC, Jiang SC. Alterations of the human skin microbiome after ocean water exposure. *Mar Pollut Bull.* 2019;145:595–603.
50. Ellis SR, Nguyen M, Vaughn AR, Notay M, Burney WA, Sandhu S, et al. The skin and gut microbiome and its role in common dermatologic conditions. *Microorganisms.* 2019;7(11):550.
51. Salem I, Ramser A, Isham N, Ghannoum MA. The gut microbiome as a major regulator of the gut-skin axis. *Front Microbiol.* 2018;9.
52. De Pessemier B, Grine L, Debaere M, Maes A, Paetzold B, Callewaert C. Gut–skin axis: Current knowledge of the interrelationship between microbial dysbiosis and skin conditions. *Microorganisms.* 2021;9(2):353.
53. Chen YE, Tsao H. The skin microbiome: Current perspectives and future challenges. *J Am Acad Dermatol.* 2013;69(1):143–155.e3.
54. Chen YE, Fischbach MA, Belkaid Y. Skin microbiota–host interactions. *Nature.* 2018;553(7689): 427–36.
55. Dimitriu PA, Iker B, Malik K, Leung H, Mohn WW, Hillebrand GG. New insights into the intrinsic and extrinsic factors that shape the human skin microbiome. *mBio.* 2019;10(4).
56. Ross AA, Doxey AC, Neufeld JD. The skin microbiome of cohabiting couples. *mSystems.* 2017;2(4).
57. Leung MHY, Tong X, Wilkins D, Cheung HHL, Lee PKH. Individual and household attributes influence the dynamics of the personal skin microbiota and its association network. *Microbiome.* 2018;6(1):26.
58. Sfriso R, Egert M, Gempeler M, Voegeli R, Campiche R. Revealing the secret life of skin – with the microbiome you never walk alone. *Int J Cosmet Sci.* 2020;42(2):116–26.
59. Diaz D, Pract CMD. The Effect of Cleansers on the Skin Microbiome. *Practical Dermatology* 2020;4:62–5.
60. Zeeuwen LJM. Skin Microbiota in Health and Disease: From Sequencing to Biology. *Journal of Dermatology*, 2020;47(10):1110–1118. Blackwell Publishing Ltd. https://doi.org/10.1111/1346-8138.15536.
61. Alesa D, Alshamrani H, Alzahrani Y, Alamssi D, Alzahrani N, Almohammadi M. The role of gut microbiome in the pathogenesis of psoriasis and the therapeutic effects of probiotics. *J Family Med Prim Care.* 2019;8(11):3496.
62. Bustamante M, Oomah BD, Oliveira WP, Burgos-Díaz C, Rubilar M, Shene C. Probiotics and prebiotics potential for the care of skin, female urogenital tract, and respiratory tract. *Folia Microbiol (Praha).* 2020;65(2):245–64.
63. Maguire M, Maguire G. The role of microbiota, and probiotics and prebiotics in skin health. *Arch Dermatol Res.* 2017;309(6):411–21.
64. Michalak M, Pierzak M, Kręcisz B, Suliga E. Bioactive compounds for skin health: A review. *Nutrients.* 2021;13(1):203.
65. Knackstedt R, Knackstedt T, Gatherwright J. The role of topical probiotics on wound healing: A review of animal and human studies. *Int Wound J.* 2020;17(6):1687–94.
66. Al-Ghazzewi FH, Tester RF. Impact of prebiotics and probiotics on skin health. *Benef Microbes.* 2014;5(2):99–107.
67. Pfalzgraff A, Brandenburg K, Weindl G. Antimicrobial peptides and their therapeutic potential for bacterial skin infections and wounds. *Front Pharmacol.* 2018;9.
68. Huttner KM, Bevins CL. Antimicrobial peptides as mediators of epithelial host defense. *Pediatr Res.* 1999;45(6):785–94.
69. Erdman SE, Poutahidis T. Probiotic 'glow of health': It's more than skin deep. *Benef Microbes.* 2014;5(2):109–19.
70. Sanford JA, Gallo RL. Functions of the skin microbiota in health and disease. *Semin Immunol.* 2013;25(5):370–7.

71. Mahmud MdR, Akter S, Tamanna SK, Mazumder L, Esti IZ, Banerjee S, et al. Impact of gut microbiome on skin health: Gut-skin axis observed through the lenses of therapeutics and skin diseases. *Gut Microbes*. 2022;14(1).
72. Ratanapokasatit Y, Laisuan W, Rattananukrom T, Petchlorlian A, Thaipisuttikul I, Sompornrattanaphan M. How microbiomes affect skin aging: The updated evidence and current perspectives. *Life*. 2022;12(7):936.
73. Mohd Zaid NA, Sekar M, Bonam SR, Gan SH, Lum PT, Begum MY, et al. Promising natural products in new drug design, development, and therapy for skin disorders: An overview of scientific evidence and understanding their mechanism of action. *Drug Des Devel Ther*. 2022;16:23–66.
74. Richmond JM, Harris JE. Immunology and skin in health and disease. *Cold Spring Harb Perspect Med*. 2014;4(12):a015339.
75. Seo SA, Park HJ, Han MG, Lee R, Kim JS, Park JH, et al. Fermented colostrum whey upregulates aquaporin-3 expression in, and proliferation of, keratinocytes via p38/c-jun N-terminal kinase activation. *Food Sci Anim Resour*. 2021;41(5):749–62.
76. Cui L, Jia Y, Cheng Z, Gao Y, Zhang G, Li J, et al. Advancements in the maintenance of skin barrier/skin lipid composition and the involvement of metabolic enzymes. *J Cosmet Dermatol*. 2016;15(4):549–58.
77. Yokouchi M, Kubo A. Maintenance of tight junction barrier integrity in cell turnover and skin diseases. *Exp Dermatol*. 2018;27(8):876–83.
78. Feingold KR. The outer frontier: The importance of lipid metabolism in the skin. *J Lipid Res*. 2009;50:S417–22.
79. Mohd Ariffin NH, Hasham R. Assessment of non-invasive techniques and herbal-based products on dermatological physiology and intercellular lipid properties. *Heliyon*. 2020;6(5):e03955.
80. Kanwar A. Skin barrier function. *Indian J Med Res*. 2018;147(1):117.
81. Asif A, Afzaal M, Shahid H, Saeed F, Ahmed A, Shah YA, et al. Probing the functional and therapeutic properties of postbiotics in relation to their industrial application. *Food Sci Nutr*. 2023;11(8):4472–84.
82. Dąbrowska AK, Spano F, Derler S, Adlhart C, Spencer ND, Rossi RM. The relationship between skin function, barrier properties, and body-dependent factors. *Skin Res Technol*. 2018;24(2):165–74.
83. Lew LC, Liong MT. Bioactives from probiotics for dermal health: Functions and benefits. *J Appl Microbiol*. 2013;114(5):1241–53.
84. Zhang X, Feng T, Tuncil YE. Editorial: Gut microbiota modulation by dietary fiber on human health: Processes and mechanisms. *Front Microbiol*. 2023;14.
85. Kim G, Kim M, Kim M, Park C, Yoon Y, Lim DH, et al. Spermidine-induced recovery of human dermal structure and barrier function by skin microbiome. *Commun Biol*. 2021;4(1):231.
86. Notay M, Foolad N, Vaughn AR, Sivamani RK. Probiotics, prebiotics, and synbiotics for the treatment and prevention of adult dermatological diseases. *Am J Clin Dermatol*. 2017;18(6):721–32.
87. Szántó M, Dózsa A, Antal D, Szabó K, Kemény L, Bai P. Targeting the gut-skin axis—Probiotics as new tools for skin disorder management? *Exp Dermatol*. 2019;28(11):1210–18.
88. Clark A, Haas K, Sivamani R. Edible plants and their influence on the gut microbiome and acne. *Int J Mol Sci*. 2017;18(5):1070.
89. Sturm C, Wagner AE. Brassica-derived plant bioactives as modulators of chemopreventive and inflammatory signaling pathways. *Int J Mol Sci*. 2017;18(9):1890.
90. Sears B, Ricordi C. Anti-inflammatory nutrition as a pharmacological approach to treat obesity. *J Obes*. 2011;2011:1–14.
91. Statovci D, Aguilera M, MacSharry J, Melgar S. The impact of western diet and nutrients on the microbiota and immune response at mucosal interfaces. *Front Immunol*. 2017;8.
92. Bennett JM, Reeves G, Billman GE, Sturmberg JP. Inflammation–nature's way to efficiently respond to all types of challenges: Implications for understanding and managing "the epidemic" of chronic diseases. *Front Med (Lausanne)*. 2018;5.
93. Fajstova A, Galanova N, Coufal S, Malkova J, Kostovcik M, Cermakova M, et al. Diet rich in simple sugars promotes pro-inflammatory response via gut microbiota alteration and TLR4 signaling. *Cells*. 2020;9(12):2701.
94. Arck P, Handjiski B, Hagen E, Pincus M, Bruenahl C, Bienenstock J, et al. Is there a 'gut–brain–skin axis'? *Exp Dermatol*. 2010;19(5):401–5.
95. Yahfoufi N, Alsadi N, Jambi M, Matar C. The immunomodulatory and anti-inflammatory role of polyphenols. *Nutrients*. 2018;10(11):1618.

96. Yoo J, Groer M, Dutra S, Sarkar A, McSkimming D. Gut microbiota and immune system interactions. *Microorganisms*. 2020;8(10):1587.
97. Luu M, Monning H, Visekruna A. Exploring the molecular mechanisms underlying the protective effects of microbial SCFAs on intestinal tolerance and food allergy. *Front Immunol*. 2020;11.
98. Barbieri JS, Wanat K, Seykora J. *Skin: Basic Structure and Function*. Içinde: Pathobiology of Human Disease. Elsevier; 2014. pp. 1134–44.
99. Song H, Zhang S, Zhang L, Li B. Effect of orally administered collagen peptides from bovine bone on skin aging in chronologically aged mice. *Nutrients*. 2017;9(11):1209.
100. Szyszkowska B, Łepecka-Klusek C, Kozłowicz K, Jazienicka I, Krasowska D. The influence of selected ingredients of dietary supplements on skin condition. *Adv Dermatol Allergol*. 2014;3:174–81.
101. Cho S. The role of functional foods in cutaneous anti-aging. *J Lifestyle Med*. 2014;4(1):8–16.
102. Zhang X, Feng T, Tuncil YE. Editorial: Gut microbiota modulation by dietary fiber on human health: Processes and mechanisms. *Front Microbiol*. 2023;14.
103. Genovese L, Sibilla S. *Innovative Nutraceutical Approaches to Counteract the Signs of Aging*. Içinde: Textbook of Aging Skin. Springer Berlin Heidelberg; 2016. pp. 1–25.
104. Kober MM, Bowe WP. The effect of probiotics on immune regulation, acne, and photoaging. *Int J Womens Dermatol*. 2015;1(2):85–9.
105. Ferrari CKB, Torres EAFS. Biochemical pharmacology of functional foods and prevention of chronic diseases of aging. *Biomed Pharmacother*. 2003;57(5–6):251–60.
106. Shahein MR, Atwaa ESH, Babalghith AO, Alrashdi BM, Radwan HA, Umair M, et al. Impact of incorporating the aqueous extract of hawthorn (C. oxyanatha) leaves on yogurt properties and its therapeutic effects against oxidative stress induced by carbon tetrachloride in rats. *Fermentation*. 2022;8(5):200.
107. Lü J, Lin PH, Yao Q, Chen C. Chemical and molecular mechanisms of antioxidants: Experimental approaches and model systems. *J Cell Mol Med*. 2010;14(4):840–60.
108. Puertollano MA, Puertollano E, Alvarez de Cienfuegos G, de Pablo MA. Dietary antioxidants: Immunity and host defense. *Curr Top Med Chem*. 2011;11(14):1752–66.
109. Cao C, Xiao Z, Wu Y, Ge C. Diet and skin aging—from the perspective of food nutrition. *Nutrients*. 2020;12(3):870.
110. Evans JA, Johnson EJ. The role of phytonutrients in skin health. *Nutrients*. 2010;2(8):903–28.
111. Lolou V, Panayiotidis MI. Functional role of probiotics and prebiotics on skin health and disease. *Fermentation*. 2019;5(2):41.
112. Puebla-Barragan S, Reid G. Probiotics in cosmetic and personal care products: Trends and challenges. *Molecules*. 2021;26(5):1249.
113. Huang MCJ, Tang J. Probiotics in personal care products. *Microbiol Discov*. 2015;3(1):5.
114. Peyrat LA, Tsafantakis N, Georgousaki K, Ouazzani J, Genilloud O, Trougakos IP, et al. Terrestrial microorganisms: Cell factories of bioactive molecules with skin protecting applications. *Molecules*. 2019;24(9):1836.
115. Baral KC, Bajracharya R, Lee SH, Han HK. Advancements in the pharmaceutical applications of probiotics: Dosage forms and formulation technology. *Int J Nanomed*. 2021;16:7535–56.
116. Jacob S, VanDaele MA, Brown JN. Treatment of *Demodex* -associated inflammatory skin conditions: A systematic review. *Dermatol Ther*. 2019;32(6).
117. Anal AK, Singh H. Recent advances in microencapsulation of probiotics for industrial applications and targeted delivery. *Trends Food Sci Technol*. 2007;18(5):240–51.
118. Jiang B, Jia Y, He C. Promoting new concepts of skincare via skinomics and systems biology—from traditional skincare and efficacy-based skincare to precision skincare. *J Cosmet Dermatol*. 2018;17(6):968–76.
119. Gueniche A, Perin O, Bouslimani A, Landemaine L, Misra N, Cupferman S, et al. Advances in microbiome-derived solutions and methodologies are founding a new era in skin health and care. *Pathogens*. 2022;11(2):121.
120. Beri K. Skin microbiome & host immunity: Applications in regenerative cosmetics & transdermal drug delivery. *Future Sci OA*. 2018;4(6).
121. Saito Y, Mihara T, Maruyama K, Saito J, Ikeda M, Tomonaga A, et al. Effects of intake of *Lactobacillus* casei subsp. *casei* 327 on skin conditions: A randomized, double-blind, placebo-controlled, parallel-group study in women. *Biosci Microbiota Food Health*. 2017;36(3):111–20.
122. França K. Topical probiotics in dermatological therapy and skincare: A concise review. *Dermatol Ther (Heidelb)*. 2021;11(1):71–7.

7 Phytochemicals and Phytoextracts Used in Skin Care Formulations

Wajid Hussain, Kishwar Sultana, Ishrat Perveen, Huan Wang, Saba Sabahat, Jawad Hussain, Muhammad Wajid Ullah and Shenqi Wang

7.1 INTRODUCTION

The skin, the largest organ in the body, protects the body from external influences by maintaining fluid equilibrium, regulating body temperature, and defending against harmful substances and microorganisms. The stratum corneum, the outermost layer of the epidermis, allows certain substances to pass through while blocking others, thereby preventing dehydration. Degradation of this barrier can cause water loss and decrease skin hydration, emphasizing the importance of skin protection. Cosmetics, often referred to as cosmeceuticals, contain active chemicals with medical qualities that offer positive topical effects and defend against deteriorating skin disorders (Sharafan et al., 2023). Cosmetics is a rapidly expanding segment within the natural personal care business, providing essential nutrients for healthy skin, enhancing aesthetic appeal, reducing wrinkles, and improving complexion, texture, and radiance (Goyal and Jerold, 2021). Natural ingredients are now more prevalent in modern formulations, influenced by their efficacy in *in vitro* and *in vivo* investigations and dermatological foundations (Alzamel, 2022).

7.2 BENEFITS OF PLANT EXTRACTS

Cosmetics use bioactive phytochemicals from various plants to enhance skin physiology and provide essential nutrients for optimal skin health.

7.2.1 Antioxidant Activity

Free radicals, chemical species with unpaired electrons, are formed by oxidative cellular metabolism. Exposure to pollution and UV rays leads to the generation of oxidants, which trigger the release of harmful enzymes, such as matrix metalloproteases (MMPs). Oxidative stress is a major factor in skin ageing and dermatological disorders. Continuous exposure to UV radiation results in skin damage, lipid peroxides, reactive oxygen species (ROS), and enzyme activity. These changes contribute to the development of several skin disorders. Antioxidants that naturally control the production of harmful free radicals protect organisms by impeding oxidative chain reactions. Exogenous reactive oxygen species disrupt skin biomolecules, leading to skin diseases (Chaudhary et al., 2023).

The antioxidant capacity of phenolic compounds depends on their unique chemical structures, with the number of hydroxyl groups being the primary determinant of their antioxidant activity. The importance of phenolic antioxidants has increased in recent years, owing to their ability to remove

 DOI: 10.1201/9781003394846-7

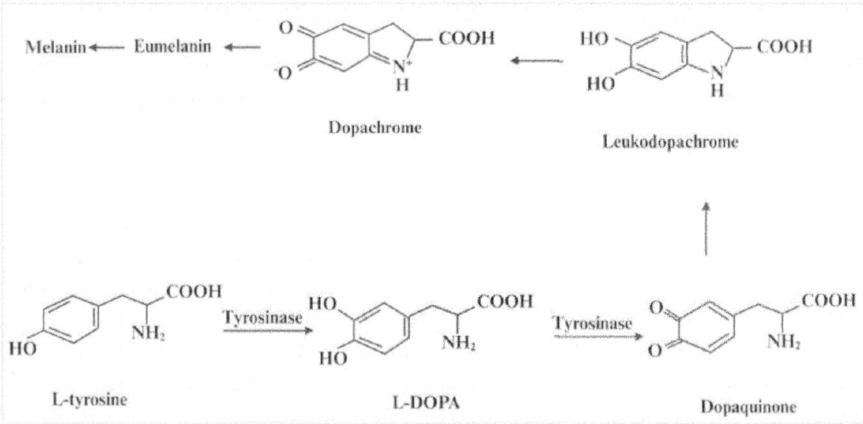

FIGURE 7.1 Mechanism indicating the melanin formation.

free radicals. Plants high in phenols can also help mitigate the harmful impact of UV radiation on the skin (del Socorro Sánchez Correa et al., 2023).

7.2.2 Tyrosinase Inhibition and Its Effects

Melanin, a pigment in humans, determines skin, hair, and eye colour. Melanocytes in the dermis produce and release melanin via melanogenesis, a physiological process. Figure 7.1 illustrates how the polyphenol oxidase tyrosinase catalyzes two separate reactions: the conversion of L-tyrosine to L-dihydroxyphenylalanine (L-DOPA) and the conversion of L-DOPA to dopaquinone. The third reaction involves the transformation of dopaquinone into leukodopachrome via a non-enzyme-catalyzed mechanism. This molecule undergoes rapid, non-enzyme-catalyzed oxidation to produce dopachrome (fourth reaction). Dopachrome undergoes chemical – and enzyme-driven reactions to transform into melanin, demonstrating that individual steps can be inhibited if any are obstructed. Thymol, despite its ability to inhibit dopachrome formation, does not qualify as a tyrosinase inhibitor (Hassan et al., 2023).

Melanin protects the skin from UV damage by blocking reactive oxygen species (ROS) and absorbing UV energy. The overproduction of tyrosinase leads to hyperpigmentation, causing skin conditions such as senile lentigo, melasma, and freckles. Studies have been conducted to identify substances that inhibit melanin release and tyrosinase catalytic activities, aiming to understand the underlying causes of these skin conditions (Upadhyay et al., 2022).

7.2.3 Antimicrobial Activity

The pharmaceutical and cosmetic industries are recognizing the need for natural antimicrobials in topical products owing to increasing public demand and bacterial resistance to conventional antibiotics (Mohanty et al., 2023). Phenolic compounds produced by plants are natural antimicrobials that can induce alterations in membrane permeability, cell death, and the penetration of bacterial cells. These compounds can extend the shelf life of products by impeding the growth of harmful bacteria, making them a valuable alternative to synthetic antimicrobials in the topical product industry. Pequi, a Brazilian fruit tree, has been found to possess antibacterial properties in its hydroethanolic extract, which has been found to be effective against various bacteria (Karnwal et al., 2023).

7.3 SKIN CARE FORMULATIONS AND PLANT EXTRACTS

The sections will discuss common plant species (Figure 7.4) with beneficial antioxidant, tyrosinase inhibitory, and antibacterial properties, which make them ideal for skin care product development.

7.3.1 Cannabis Sativa

Cannabis, a plant rich in linoleic acid, has been used to treat eczema, psoriasis, and other inflammatory conditions. It was banned in the 19th century due to its mind-altering properties. Cannabis sativa L, an annual *Cannabaceae* plant, has undergone extensive breeding and selection, resulting in numerous diverse forms. Cannabis has various uses, including in fibres, food, oil, medicine, textiles, and recreational activities. It has attracted tremendous interest in industrial hemp, marijuana, and medicinal cannabis, which are distinct plant species with unique compositions and applications. Industrial hemp has a trans-D9-tetrahydrocannabinol (THC) level of less than 0.3%, while marijuana can contain up to 30% THC and is a controlled substance (Latif et al., 2023). Hemp extracts are non-psychoactive CBD with significant medicinal potential, making it a popular choice for cannabis-derived skincare products. CBD oil from the leaves and hemp seed oil from the seeds have numerous therapeutic properties, including analgesic, antihistaminic, anti-aggregating, and anticancer, antioxidant, and antibacterial effects (Legare et al., 2022).

7.3.2 Polygonum Amplexicaule

The Polygonaceae family consists of 1,200 species and 48 genera, making it the most important plant family for therapeutic use. The Polygonaceae family comprises a multitude of medicinal plants that exhibit diverse biological functions and contain intriguing phytochemical constituents (Macêdo et al., 2021). Extensive phytochemical investigations of this plant family have verified the presence of biologically active plant compounds, including essential oils, flavonoids, tannins, unsaturated sterols, triterpenoids, alkaloids, and saponins (specifically steroidal sapogenins). Polygonum species have been used globally as both edible and therapeutic botanicals. Pakistan harbours more than 20 of the 60 species of Polygonum L. found globally (Khaliq et al., 2023). The Polygonum genus comprises a group of plants known for their medicinal and therapeutic benefits. Tea from this genus is used to treat cardiovascular issues, dysentery, menstruation, fever, ulcers, and leucorrhoea. The root sap of P. amplexicaule and D. were also used for eye injuries. Phytochemical analysis of this species assessed nutrient levels, providing valuable insights for future drug development and the evaluation of treatment efficacy.

7.3.3 Melaleuca Alternifolia (Tea Tree)

Tea tree oil (TTO) is a blend of alcohols and monoterpenes, with a minimum of 30% terpinen-4-ol and 15% 1,8-cineole. It has potent antibacterial and anti-inflammatory properties but may trigger allergic reactions. TTO is available for six basic chemotypes, including four 1,8-cineole chemotypes. TTO is a potent antioxidant substitute that is effective for treating acne vulgaris, facial acne vulgaris, and seborrheic dermatitis (Bekhof et al., 2023; Masyita et al., 2022). It is safe during pregnancy and effective in treating seborrheic dermatitis owing to its antifungal properties. The active component, terpinen-4-ol, prevents melanoma cells from proliferating, migrating, and infiltrating healthy tissues, both drug-sensitive and drug-resistant. Reducing the 1,8-cineole concentration can reduce its negative impact on TTO (Nascimento et al., 2023).

7.3.4 Aloe Vera

Aloe, a plant belonging to 550 species, is cultivated globally. The plant latex, a yellow sap, contains anthraquinones and glycosides, while its rind produces proteins and carbohydrates. Aloe vera gel (98% water) is composed of various components, including phenolic compounds, lipids, proteins, minerals, sugars, and polysaccharides. It contains 200 active ingredients including vitamins, enzymes, minerals, carbohydrates, saponins, lignin, amino acids, and salicylic acids. The topical use of aloe gel can cure acne and retain moisture in arid environments, preventing dryness and flakiness of the skin and scalp. Harvesting the plants takes six to eight weeks (Mitra et al., 2023).

7.3.5 Castanea Sativa

Chestnut trees, especially those in Portugal's Trás-os-Montes region, are rich in phenolic compounds and C. sativa flowers have antioxidant properties (Lema et al., 2023). The aqueous extract of C. sativa leaves exhibited antibacterial properties. A topical formulation without surfactants was prepared from the C. sativa leaf extract. The product has a moisturizing effect lasting at least four hours and maintains the pH and DPPH-scavenging activity after six months of storage (Hernández et al., 2023).

7.3.6 Prunus Dulcis

Almonds, also known as Prunus dulcis, are a plant in the family *Rosaceae.* They have brown skin and an edible seed enclosed by an outer hull and an inner shell. Almonds have healing properties and can be used to treat conditions, such as eczema and acne. Hulls, peels, and shells contain high levels of phenolic compounds. The seed coat is often separated from the kernel during industrial processes. Recent studies have suggested the use of by-products of almond extraction and processing as antioxidant chemicals. Almond kernel extract, which is rich in antioxidants, contains phytosterols, flavonoids, phenolics, fatty acids, and lipid-soluble vitamins, making it a valuable natural resource. Almond green husks, with their excellent antioxidant properties and low EC50 values, have the potential for use in novel antioxidant formulations, particularly in inhibiting lipid peroxidation (Fabroni et al., 2023).

7.3.7 Juglans Regia L

The walnut tree, scientifically known as Juglans regia L., yields a lucrative harvest that is consumed globally. Green walnuts, along with dried fruits (nuts), are used in the cosmetic and medicinal sectors. This includes husks (epicarps), shells, kernels, bark, and leaves. The antibacterial and astringent qualities of walnut leaves have made them popular remedies in folk medicine for the treatment of ulcers, hyperhidrosis, and skin inflammation. Certain polyphenols have also been shown to possess scavenging capabilities. The leaf extract of J. regia, diluted in a mixture of ethanol and water, had significant effectiveness against pro-oxidant species (Ara et al., 2023).

7.3.8 Olea Europaea

The Mediterranean Basin produces 98% of the global olive supply, making it a significant agricultural crop. Tyrosol and hydroxytyrosol, which are found in olive trees, fruits, oil, and mill waste, are phenolic compounds with antibacterial, anticarcinogenic, anti-inflammatory, and antioxidant properties. Hydroxytyrosol, a versatile food additive and antioxidant, also has significant potential as a cosmetic ingredient. The cosmetic industry is considering incorporating olive extracts into cosmetic goods because of their advantageous features. Olive oil extraction produces moist solid waste with high concentrations of tyrosol and hydroxytyrosol, which can be easily dissolved in water and used as an ingredient in cosmetic production (Marx, 2023). Analytical methodologies are required to monitor commercial techniques and evaluate the quality of cosmetic products. Olive leaves are popular for their health benefits, and the presence of secoiridoids and flavonoids significantly influences the antioxidant effectiveness of polar extracts (Cappelli et al., 2023).

7.3.9 Helichrysum Stoechas (L.) Moench

The genus Asteraceae, specifically Helichrysum, encompasses approximately 500 species that are distributed throughout. This genus is commonly acknowledged for its diverse therapeutic and medicinal attributes, including antioxidant, anti-allergic, antibacterial, anti-inflammatory, cold treatment, cough relief, and wound healing properties (Akinfenwa et al., 2022). It has an intricate chemical composition and encompasses a diverse range of chemical categories such as flavonoids, chalcones, phloroglucinol derivatives, essential oils, pyrones, and diterpenes. Moreover, its antioxidant properties make it beneficial for cosmetic production (Akinyede et al., 2021).

7.3.10 Quercus Robur

The indigenous oak tree Quercus robur, found in northern Africa, western Asia, and Europe, produces a variety of polyphenols. Its bark has shown antimicrobial properties against singlet oxygen, hydroxyl radicals, and superoxide anions as well as free radical scavenging. Two previously unknown and unclassified flavanols, hyperoside, ellagic acid, and rutin, in addition to phenolic chemicals, have been detected in oak leaf extracts. The leaf extract showed significant iron chelation, free radical scavenging, and UVB absorption properties, making it suitable for topical use as an antioxidant. Its skin tolerance was assessed through a patch test, showing good resistance after a single application under occlusion, highlighting its potential as a topical antioxidant against UV-induced photodamage (Sirgedaitė-Šėžienė et al., 2023).

7.3.11 Glycyrrhiza Glabra

Glycyrrhiza glabra, also known as liquorice or sweet wood, is a perennial herb native to China, Turkey, Syria, Iran, and Europe. Its primary taproot is vivid yellow with a velvety texture. Liquorice is a widely used medication for treating skin conditions, such as dermatitis, eczema, pruritus, and cysts. It has antibacterial, antiseptic, and anti-inflammatory properties, making it a popular choice for skincare. Liquorice root extract, rich in glabridin, effectively inhibits tyrosinase activity in melanoma cells and prevents hydroquinone oxidative degradation over a three-month period, outperforming other commercially available antioxidants (Zulfugarova et al., 2023).

7.3.12 Vitis Vinifera

This species is also termed wine grape or European grape and is a type of grape that contains phenols, which are the third most prevalent constituents. The pulp, seeds, and skin of grapes contain a mixture of phenolic compounds, with the seeds and skin having 60–70% and 28–35% phenolics, respectively. Vitis vinifera grape seed extract is known for its anti-caries, anti-dandruff, antifungal, antibacterial, antioxidant, flavouring, light stabilizer, and sunscreen properties.

7.3.13 Crataegus Monogyna Jacq (Hawthorn)

Hawthorn inflorescences have been used as traditional herbal remedies in Asia, America, Europe, and North Africa. The fruit, with white or rose blossoms and scarlet growth, offers lasting cardiovascular benefits without negative effects. Cosmetics and dermatology use creams and lotions containing glycolic extracts from hawthorn flowers to enhance skin firmness, mitigate signs of ageing, and prevent wrinkle development, primarily focusing on skin health. Studies have shown no negative consequences for hawthorn fruit.

7.3.14 Pinus Pinaster

The bark of *Pinus pinaster* contains numerous beneficial compounds, including polyphenols, which are produced during plant growth in response to stress factors. Pycnogenol®, a standardized extract from the bark of *Pinus pinaster* A., is commonly used as a dietary supplement in accordance with the US Pharmacopoeia regulations. Pycnogenol® is a 70% standardized extract containing bioactive compounds such as oligomers, procyanidins, taxifolin, phenolic acids, and cinnamic and benzoic acid derivatives. It has been proven to have beneficial effects on skin cells, particularly in treating chronic venous insufficiency and skin irritation (Cai et al., 2023).

7.4 ANTI-AGEING PHYTOCHEMICALS AND COSMETICS

The skin, consisting of subcutaneous, epidermal, and dermal layers, acts as a protective barrier, maintaining thermoregulation, fluid homeostasis, and defending against pathogens and ultraviolet radiation. Ageing begins at birth and is evident in skin changes over time. Advancements in science and technology have enabled invasive procedures, such as laser skin rejuvenation and plastic surgery, demonstrating the importance of maintaining skin health and avoiding harmful effects. Non-invasive therapies, such as plant extracts and phytochemicals, have gained popularity in the last decade because of their potential to slow down ageing. The current trend in anti-ageing skincare products involves the development of new plant extracts and botanical compounds with traditional medicinal properties. The herbal cosmetic industry frequently uses extracts of aloe vera, turmeric, amla, gotu kola, and Triphala to combat skin ageing. Skin ageing is influenced by intrinsic and extrinsic factors, with 90% of the skin changes attributed to extrinsic causes (Zhang and Kitts, 2021). Research on skin ageing aims to understand the molecular mechanisms driving ageing and develop strategies to stop these mechanisms. The skin ageing process is influenced by various cellular damage, including reactive oxygen species accumulation, photoaging, low-quality proinflammatory phenotype, decline in age-related autophagy, skin dryness, and decreased collagen and elastin production. Nutritional benefits may also be essential for reversing intrinsic and medical ageing (Biskanaki et al., 2023). This study aimed to explore the effectiveness of plant-derived bioactive substances in reducing extrinsic skin ageing in human beings, as shown in Figure 7.2.

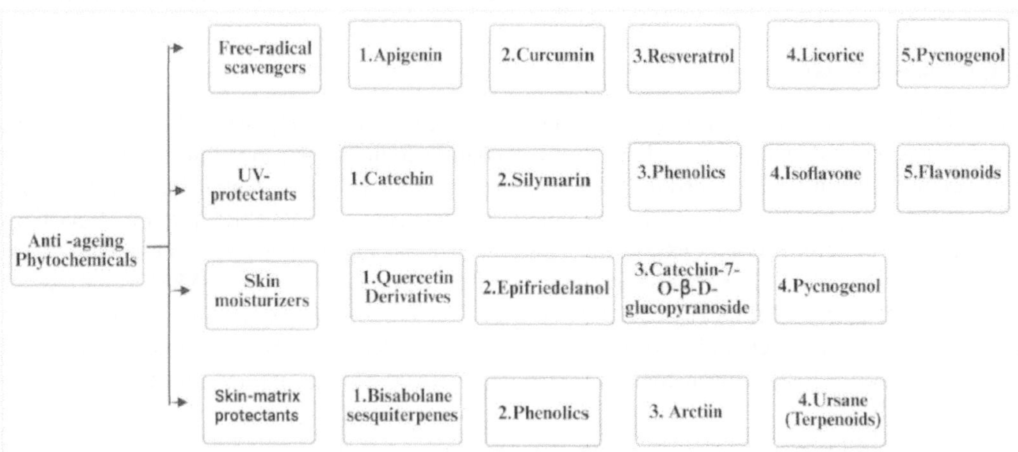

FIGURE 7.2 Broad categories of anti-ageing phytochemicals.

7.4.1 Skin Matrix Protectants

The subcutaneous tissue supports the outer layers and extracellular matrix of the integumentary system, whereas the dermis, primarily composed of collagen, provides skin integrity and increases elasticity through elastin fibres. The extracellular matrix (ECM) undergoes constant degradation and repair, with more pronounced deterioration with age, resulting in a gradual decrease in skin firmness and elasticity. The pro-inflammatory cytokine TNF, which is responsible for matrix damage, is produced when ROS and UV radiation are exposed to excess ROS and UV radiation. MMP-1 is further induced by Interleukin-6 (IL-6) following UV radiation, which eventually results in matrix breakdown. The rhizome contains bisabolane sesquiterpenes that reduce the synthesis which MMP-1 triggered by UVB exposure. UV light increases the activation of MMPs 1, 3, and 9, which break down collagen and elastin, leading to skin laxity and crease. Arctiin, found in Arctium lappa and greater burdock fruit extracts, enhances visible signs of ageing skin by inhibiting cytokine activation and promoting collagen production. Centella asiatica contains terpenoids, which enhance collagen I synthesis. Exposure to sunlight can increase the expression of matrix metalloproteinase-1 (MMP-1), which breaks down collagen and contributes to wrinkle formation. The photoactivity of M. oleifera leaves reduced MMP-1 levels, while dandelion water-based extracts protected human dermal fibroblasts (HDFs) from UV radiation by suppressing MMP activity and reducing oxidative stress (Naikoo et al., 2023). The phytochemicals which enhance collagen production and preserve the skin matrix by inhibiting collagenase and elastase are listed in Table 7.1.

7.4.2 UV Protectants

UV radiation, categorized into three bands, UVC (200–290 nm), UVA (320–400 nm), and UVB (290–320 nm), can accelerate skin cancer and photoaging. UVC radiation is the strongest and most harmful, but it can be obstructed by the ozone layer. UVB radiation causes erythema and sunburn, weakening the immune system and increasing the risk of skin cancer. It exacerbates photodamage by inducing MMPs production and stimulating the mitogen-activated protein kinase pathway. Porphyra vietnamensis contains Porphyra-334 red alga, which absorbs UV radiation and is known to have a delicious taste. Punica granatum and coffee arabica extracts are effective in preventing skin photoageing caused by UVB radiation. Punica granatum contains catechin, a bioactive compound, while coffee arabica extract suppresses MMPs, enhances pro-collagen synthesis, and reduces MPK pathway activity (Rodrigues et al., 2023). Highly unsaturated phytochemicals, such as flavonoids, polyphenols, and carotenoids listed in Table 7.1, can be utilized as UVB absorbers.

7.4.3 Skin Moisturizers

Old skin has a low concentration of hyaluronic acid, an essential component of the extracellular matrix, and glycosaminoglycan, a linear polysaccharide found in the junction and periphery of collagen and elastin fibres (Saewan, 2022). Hyaluronic acid (HA) is essential for skin hydration and anti-aging. Pycnogenol, derived from pine bark, enhances skin moisturization and elasticity by increasing the production of extracellular matrix components such as HA and collagen. These benefits are likely due to HA's water-attracting and retaining properties (Galvez-Martin et al., 2023). The lyophilized extracts from Helichrysum stoechas, including microspheres, have been used as a moisturizer, showing potential for cosmetics and dietary supplements (Dănăilă-Guidea et al., 2022).

These compounds have been found to slow down tissue ageing and age-related illnesses, as shown in Table 7.1.

7.4.4 Free Radical Scavengers

Free radicals are chemical compounds with unpaired electrons created through oxidative cellular metabolism. Exposure to pollutants and UV radiation releases harmful enzymes such as matrix

TABLE 7.1

Plant-Produced Anti-Skin Ageing Phytochemicals in Cosmetics Exhibit Antioxidant (Radical Scavenging), Skin Miniaturization, Skin-Matrix Protectant, and UV-Protective Properties

Phytochemical properties	Antioxidant Properties (Radical Scavengers)		Skin Matrix Protectant		Skin Moisturization Properties		UV-Protectant Properties	
	Plant Species	Chemical Constituents	Plant Species	Chemical Constituents	Plant Species	Chemical Constituents	Plants	Chemical Constituents
	Acanthopanaxkoreanum	Acanth panic acid, Rutin, Saponins	*Aloe vera (L.) Burm.f*	Anthraquinone glycoside	*Aloe Vere (L.) Burm. f*	Polysaccharides	*Camellia sinesis* (L.) Kuntze	Polyphenols
	Aesculashippocestenum L.	Aescin, Coumarin-a-esculin, Flavonoids	*Angelica arch angelica* L.	Furanocoumarins	*Alpinia Katsumadae* Hayata	Chalcone, Flavone	*Emblica officinalis* Gaertn	Polyphenols
	Brassica campestris L.	Hydroxycinnamic acid derivatives	*Arctiumlappa* Willd.	Arctiin	*Astragalus membranaceus* (Fisch) Bunge	Flavonoids, Polysaccharides	*Glycine max* (L.) Merr	Isoflavones
	Capsicum annuum L.	Carotenoids	*Areca catechu* L.	Phenolics	*Bowswelliaserrata Roxb. Ex Colebr*	Boswellic acid, Amyrins	*Ixoraparviflora* Lam.	Polyphenols
	Cisalpiniapulcherrima (L) Sw.	Catechin, Gallic acid, Ellagic acid	*Berberisaristata* DC.	Berberine	*Calendula officinalis* L.	Terpenes, Carotenes	*Labisiapumila* (Blume) Fern-Vill	Flavonoids
	Citrus grandis (L.) Osbeck	Coumarin phenolic acid, Cinnamic acid	*Camellia japonica* L.	Phenolics, Flavonoid glycoside	*Cinnamomum cassia* (L.) J. Presl	Phenols, Isoeugenol, Cinnamaldehyde	*Polypodium leucotomos* Poir	Phenolic acids, Caffeic acid, Ferulic acid
	Citrus Limonum Risso	Pinene, Camphene, Limonene	*Centeliaasiatica* (L.) Urb	Asiaticoside Asiatic acid	*Cucumissativus* L.	Polysaccharides	*Punicagranatum* L.	Catechin
	Coffeaarabica L.	Chlorogenic acid, Pro anthocyanidins, Ferulic acid	*Cinnamomumzeylanicum* Blume	Cinamaldehyde	*Curcuma longa* L.	Curcuminoids	*Silybummarianum* Gaertn	Flavonoid, Silymarin
	Curcuma longa L.	Curcumin	*Cipadessabaccifera* (Roth) Miq	Triterpenoid	*Dryopteriscassirrhizoma*	Dryopteroside, catechin –6 – glucopyranoside	*Vacciniumuliginosum* L.	Anthocyanidins

(*Continued*)

TABLE 7.1 ***(Continued)***

Plant-Produced Anti-Skin Ageing Phytochemicals in Cosmetics Exhibit Antioxidant (Radical Scavenging), Skin Miniaturization, Skin-Matrix Protectant, and UV-Protective Properties

Phytochemical properties	Antioxidant Properties (Radical Scavengers)		Skin Matrix Protectant		Skin Moisturization Properties		UV-Protectant Properties	
	Plant Species	**Chemical Constituents**	**Plant Species**	**Chemical Constituents**	**Plant Species**	**Chemical Constituents**	**Plants**	**Chemical Constituents**
	Glycyrrhizaglabra L.	Phenolics, Glycyrrhizin	*Curculigoorchioides Gaertn.*	Curculigosides	*Garcinia indica* (Thouars) Choisy	Gercinol, Isoeugenol, Cinnamaldehyde		
	Malus domestica Borkh.	Polyphenols	*Curcuma longa* L.	Curcumin	*Terminalia chebula* Retz.	Tannins		
	Pine species	Procyanidins, Pycnogenol	*Curcuma xanthorrhiza* Roxb.	Xanthorrhizol	*Terminalia arjuna* (Roxb. ex-DC.) Wight and Am	Catechins, Ellagic acid, Gallic acid		
	Populusnigra L.	Phenolics	*Dioscoreaviliosa* L.	Diosgenin				
	Rosmarinus Officinalis L.	Phenolic diterpenes	*Emblica officinalis* Gaertn	Gallic acid, Ellagic acid				
	Salvia Officinalis L.	Phenolic acids	*Haberlearhodopensis* Friv.	Myconoside				
	Vitisvinifera L.	Polyphenols, Resveratrol	*Humuluslupulus* L.	Flavonoids				
			Panax Ginseng C.A. Mey.	Ginsenosides				
			Polypodiumleucotomos Poir	Phenolic acids				
			Woodfordiafruticosa (L.) Kurz	Flavonoids				
			(Roscoe) *Zingiberofficinale*	p-hydroxy benzaidehyde, 6-gingerol				

metalloproteases (MMPs), which can harm the body. Plant extracts contain phytochemicals with antioxidant properties such as flavonoids and phenolic substances, which can help eliminate free radicals. The hydroxyl activity of the aromatic ring traps free radicals despite their potential for hydrogen and electron transport. Phenolic compounds, such as proanthocyanidins, quinic acid, ferulic acid, and chlorogenic acid, can scavenge free radicals (Sun and Shahrajabian, 2023). Apigenin enhances UV radiation absorption, whereas pine bark extracts contain proanthocyanidins with antioxidant properties. Acanthopanax koreanum, a plant with anti-free radical properties, contains phenylpropanoids in its leaf extracts (Rui and Liu, 2020).

The effects of UVR (UVA+UVB) on the epidermal keratinocytes were investigated. ROS are formed upon exposure to ultraviolet radiation (UVR). Elevated levels of reactive oxygen species (ROS) result in an imbalance between pro-oxidants and antioxidants, thereby causing oxidative stress and impairing the integrity of DNA, proteins, and lipids. Elevations in reactive oxygen species (ROS) levels disrupt the equilibrium between pro- and antioxidants, resulting in oxidative stress and the degradation of proteins and lipids. Moreover, ultraviolet B (UVB) radiation is recognized as a catalyst for oxidative stress that leads to DNA damage. ROS stimulates the activation of transcriptional regulators, such as JNK, NF-kB, and Nrf2. These transcription factors specifically interact with the antioxidant response element (ARE), NF-kB, and AP1, which have different DNA sequences. The downstream targets of these transcription factors include Phase II antioxidants, genes that enhance cell proliferation, and the synthesis of pro-inflammatory mediators, such as COX-2, prostaglandin E2, and interleukins. Oedema and erythema arise from inflammation, which concurrently enhances the generation of reactive oxygen species (ROS). The presence of reactive oxygen species (ROS) can lead to alterations in proteins and lipids, resulting in abnormal cellular signalling that may contribute to the development of cancer. Matrix metalloproteinases (MMPs), which are indicators of skin ageing, are synthesized and secreted in response to oxidative stress, which in turn causes collagen degradation. Figure 7.3 illustrates the process of producing natural antioxidants, exposure to solar UV radiation, and the potential generation of phytoproducts to mitigate their harmful effects.

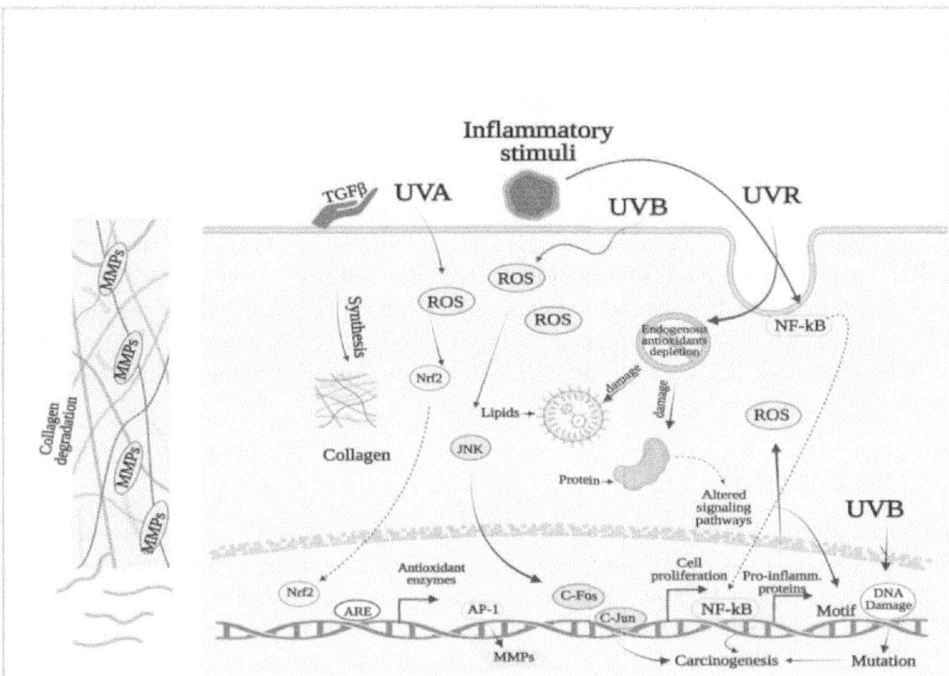

FIGURE 7.3 Exposure of skin exposure to solar UV radiation (UVA+UVB) and the potential production of natural phytoproducts to mitigate its deleterious effects.

FIGURE 7.4 Optical images of different medicinal plants mainly used for skincare formulations.

7.5 OTHER PROTECTANTS OF AGEING

Cellular senescence is an important factor in tissue aging, age-related illnesses, tissue regeneration, and cancer. Polygonum aviculare compounds such as compound 9 (quercetin-3-O-D-glucuronide) have been found to decrease cellular senescence in human primary cells, specifically human dermal fibroblasts (HDFs) and human umbilical vein endothelial cells (HUVECs). Adriamycin, used to treat HDFs and HUVECs, has also been used to treat these cells.

Thus, compound 9 could potentially aid in the development of cosmetics and nutritional supplements for age-related disorders. Ulmus davidiana's root bark contains potent chemicals that can halt cellular senescence in HUVECs and HDFs. Alkannins produced by the Arnebia nobilis R. root system have anti-ageing effects.

7.6 CONCLUSIONS

Plant-derived phytochemicals have been found to have skin-enhancing properties, including UV protection, moisturization, matrix preservation, and antioxidant activity. In recent years, several *in vitro* studies have explored the biological effects of these phytochemicals. However, further clinical trials are needed to determine their concentration, production process, safety, and anti-ageing effects. The use of natural products, particularly in cosmetics, is increasing owing to their antioxidative properties, antibacterial effects, and potential to alleviate and prevent skin conditions. Research on manufacturing formulations that incorporate plant extracts for skincare products is limited. Scientific research is needed to address consumer concerns and ensure the safe, effective, and cost-effective use of botanical extracts in cosmetic technology. Herbal products offer a gentler effect and lower risk than synthetic chemicals. The disposal of plant-derived byproducts presents challenges, but some can serve as beneficial compounds for topical use. By using botanical extracts or industrial plant processing byproducts, companies can create organic topical antioxidants, lightening agents, and preservatives, making them cost-effective alternatives to manufactured products.

ACKNOWLEDGEMENTS

This work was supported by the National Key Research and Development Program of China (Grant#: 2017YFC1104402) and Metasequoia Teacher Research Start-up Fund, Nanjing Forestry University, Nanjing 210037, PR China (163105998).

REFERENCES

Akinfenwa, A. O., Sagbo, I. J., Makhaba, M., Mabusela, W. T., & Hussein, A. A., 2022. Helichrysum Genus and Compound Activities in the Management of Diabetes Mellitus. *Plants*, *11*(10), 1386. https://doi.org/10.3390/plants11101386

Akinyede, K. A., Cupido, C. N., Hughes, G. D., Oguntibeju, O. O., & Ekpo, O. E., 2021. Medicinal Properties and In Vitro Biological Activities of Selected Helichrysum Species from South Africa: A Review. *Plants*, *10*(8), 1566. https://doi.org/10.3390/plants10081566

Alzamel, N. M., 2022. Bioactive Compounds in some Medicinal Plants from Different Habitats in KSA. *Pakistan Journal of Medical and Health Sciences*, *16*(2), 1085–1091. https://doi.org/10.53350/pjmhs221621085

Ara, T., Shafi, S., Ghazwani, M., Mir, J. I., Shah, A. H., Qadri, R. A., Hakami, A. R., Khalid, M., Hani, U., & Wahab, S., 2023. In Vitro Potent Anticancer, Antifungal, and Antioxidant Efficacy of Walnut (Juglansregia L.) Genotypes. *Agronomy*, *13*(5), 1232. https://doi.org/10.3390/agronomy13051232

Bekhof, A. M. W., van Hunsel, F. P. A. M., van de Koppel, S., & Woerdenbag, H. J., 2023. Safety Assessment and Adverse Drug Reaction Reporting of Tea Tree Oil (*Melaleucaaetheroleum*). *Phytotherapy Research*, *37*(4), 1309–1318. https://doi.org/10.1002/ptr.7687

Biskanaki, F., Kefala, V., Lazaris, A. C., & Rallis, E., 2023. Aging and the Impact of Solar Ultraviolet Radiation on the Expression of Type I and Type VI Collagen. *Cosmetics*, *10*(2), 48. https://doi.org/10.3390/cosmetics10020048

Cai, C., Zeng, B., Lin, L., Zheng, M., Burki, C., Grether-Beck, S., & Krutmann, J., 2023. An Oral French Maritime Pine Bark Extract Improves Hair Density in Menopausal Women: A Randomized, Placebo-Controlled, Double Blind Intervention Study. *Health Science Reports*, *6*(1). https://doi.org/10.1002/hsr2.1045

Cappelli, A., Lupori, L., & Cini, E., 2023. Should Extra Virgin Olive Oil Production Change the Approach? A Systematic Review of Challenges and Opportunities to Increase Sustainability, Productivity, and Product Quality. *Journal of Agricultural Engineering*, *54*(1). https://doi.org/10.4081/jae.2023.1479

Chaudhary, P., Janmeda, P., Docea, A. O., Yeskaliyeva, B., AbdullRazis, A. F., Modu, B., Calina, D., & Sharifi-Rad, J., 2023. Oxidative Stress, Free Radicals and Antioxidants: Potential Crosstalk in the Pathophysiology of Human Diseases. *Frontiers in Chemistry*, *11*. https://doi.org/10.3389/fchem.2023.1158198

Dănăilă-Guidea, S. M., Eremia, M. C., Dinu, L. D., & Miu, D.-M., 2022. Helichrysumarenarium: From Cultivation to Application. *Applied Sciences*, *12*(20), 10241. https://doi.org/10.3390/app122010241

del Socorro Sánchez Correa, M., el RocíoReyero Saavedra, M., Antonio Estrella Parra, E., Nolasco Ontiveros, E., del Carmen Benítez Flores, J., Gerardo Ortiz Montiel, J., Eduardo Campos Contreras, J., LópezUrrutia, E., Guillermo Ávila Acevedo, J., Edith Jiménez Nopala, G., & Montserrat Espinosa González, A., 2023. Ultraviolet Radiation and Its Effects on Plants. In *Abiotic Stress in Plants – Adaptations to Climate Change* [Working Title]. IntechOpen. https://doi.org/10.5772/intechopen.109474

Fabroni, S., Trovato, A., Ballistreri, G., Tortorelli, S. A., Foti, P., Romeo, F. V., & Rapisarda, P., 2023. Almond [Prunusdulcis (Mill.) DA Webb] Processing Residual Hull as a New Source of Bioactive Compounds: Phytochemical Composition, Radical Scavenging and Antimicrobial Activities of Extracts from Italian Cultivars ('Tuono', 'Pizzuta', 'Romana'). *Molecules*, *28*(2), 605. https://doi.org/10.3390/molecules28020605

Galvez-Martin, P., Soto-Fernandez, C., Romero-Rueda, J., Cabañas, J., Torrent, A., Castells, G., & Martinez-Puig, D., 2023. A Novel Hyaluronic Acid Matrix Ingredient with Regenerative, Anti-Aging and Antioxidant Capacity. *International Journal of Molecular Sciences*, *24*(5), 4774. https://doi.org/10.3390/ijms24054774

Goyal, N., & Jerold, F., 2021. Biocosmetics: Technological Advances and Future Outlook. *Environmental Science and Pollution Research*, *30*(10), 25148–25169. https://doi.org/10.1007/s11356-021-17567-3

Hassan, M., Shahzadi, S., & Kloczkowski, A., 2023. Tyrosinase Inhibitors Naturally Present in Plants and Synthetic Modifications of These Natural Products as Anti-Melanogenic Agents: A Review. *Molecules (Basel, Switzerland)*, *28*(1). https://doi.org/10.3390/molecules28010378

Hernández, D., Montalvo, A., Pérez, I., Charnay, C., Sánchez-Espejo, R., Cerezo, P., Viseras, C., Riela, S., Cinà, G., & Rivera, A., 2023. Antioxidant Efficacy and "In Vivo" Safety of a Bentonite/Vitamin C Hybrid. *Pharmaceutics*, *15*(4), 1171. https://doi.org/10.3390/pharmaceutics15041171

Karnwal, A., Shrivastava, S., Al-Tawaha, A. R. M. S., Kumar, G., Singh, R., Kumar, A., Mohan, A., Yogita, & Malik, T., 2023. Microbial Biosurfactantas an Alternate to Chemical Surfactants for Application in Cosmetics Industries in Personal and Skin Care Products: A Critical Review. *BioMed Research International*, *2023*, 1–21. https://doi.org/10.1155/2023/2375223

Khaliq, T., Akhter, S., Sultan, P., & Hassan, Q. P., 2023. Critical Review on Rumexdentatus L. A Strong Pharmacophore and the Future Medicine: Pharmacology, Phytochemical Analysis and Traditional Uses. *Heliyon*, *9*(3), e14159. https://doi.org/10.1016/j.heliyon.2023.e14159

Latif, S., Shah, S. S. H., Qureshi, R., Mashwani, Z.-R., Jabeen, B., & Qaseem, M. F., 2023. Current Trends and Future Prospects of Medicinal Cannabis: An Underutilized Ancient Ethnomedicinal Plant for Human Wellbeing. *Journal of Xi'an Shiyou University, Natural Science Edition*, *19*(02), 1015–1025.

Legare, C. A., Raup-Konsavage, W. M., & Vrana, K. E., 2022. Therapeutic Potential of Cannabis, Cannabidiol, and Cannabinoid-Based Pharmaceuticals. *Pharmacology*, *107*(3–4), 131–149. https://doi.org/10.1159/000521683

Lema, F., Baptista, P., Oliveira, C., & Ramalhosa, E., 2023. Brown Rot Caused by Gnomoniopsissmithogilvyi (syn. Gnomoniopsiscastaneae) at the Level of the Chestnut Tree (Castanea sativa Mill.). *Applied Sciences*, *13*(6), 3969. https://doi.org/10.3390/app13063969

Macêdo, S. K. S., Lima, K. S. B., Silva, N. D. dos S., Campos, S. S. G., Araújo, B. R., Almeida, J. R. G. da S., Rolim, L. A., & Nunes, X. P., 2021. Genus Triplaris (Polygonaceae): A Review on Traditional Medicinal Use, Phytochemistry and Biological Activities. *Journal of Ethnopharmacology*, *277*, 114188. https://doi.org/10.1016/j.jep.2021.114188

Marx, Í. M. G., 2023. Co-Extraction Technique Improves Functional Capacity and Health-Related Benefits of Olive Oils: A Mini Review. *Foods*, *12*(8), 1667. https://doi.org/10.3390/foods12081667

Masyita, A., Mustika Sari, R., DwiAstuti, A., Yasir, B., RahmaRumata, N., Emran, T. Bin, Nainu, F., & Simal-Gandara, J., 2022. Terpenes and Terpenoids as Main Bioactive Compounds of Essential Oils, Their Roles in Human Health and Potential Application as Natural Food Preservatives. *Food Chemistry: X*, *13*, 100217. https://doi.org/10.1016/j.fochx.2022.100217

Mitra, A., Singh, M., Banga, A., Pandey, J., Tripathi, S. S., & Singh, D., 2023. Bioactive Compounds and Therapeutic Properties of Aloe Vera: A Review. *Plant Science Today*. https://doi.org/10.14719/pst.1715

Mohanty, S., Ray, A., Naik, P. K., Sahoo, A., Jena, S., Das, P. K., Patnaik, J., Panda, P. C., & Nayak, S., 2023. Variation in Yield, Chemical Composition and Biological Activities of Essential Oil of Three Curcuma Species: A Comparative Evaluation of Hydrodistillation and Solvent-Free Microwave Extraction Methods. *Molecules*, *28*(11), 4434. https://doi.org/10.3390/molecules28114434

Naikoo, S. H., Rashid, H., Gupta, R., Sharma, R. R., Kumar, S., Ahmad, R., Gudup, S., Singh, P. P., & Abdullah, S. T., 2023. A Novel Molecule 3-(1′-Methyltetrahydropyridinyl)-2,4-6-Trihydroxy Acetophenone Alleviates Ultraviolet-B-Induced Photoaging in Human Dermal Fibroblasts and BALB/c Mice. *Photochemistry and Photobiology*. https://doi.org/10.1111/php.13784

Nascimento, T., Gomes, D., Simões, R., & da Graça Miguel, M., 2023. Tea Tree Oil: Properties and the Therapeutic Approach to Acne – A Review. *Antioxidants*, *12*(6), 1264. https://doi.org/10.3390/antiox12061264

Rodrigues, R., Oliveira, M. B. P. P., & Alves, R. C., 2023. Chlorogenic Acids and Caffeine from Coffee By-Products: A Review on Skincare Applications. *Cosmetics*, *10*(1), 12. https://doi.org/10.3390/cosmetics10010012

Rui, Z., & Liu, J., 2020. Understanding of Free Radical Scavengers Used in Highly Durable Proton Exchange Membranes. *Progress in Natural Science: Materials International*, *30*(6), 732–742. https://doi.org/10.1016/j.pnsc.2020.08.013

Saewan, N., 2022. Effect of Coffee Berry Extract on Anti-Aging for Skin and Hair – In Vitro Approach. *Cosmetics*, *9*(3), 66. https://doi.org/10.3390/cosmetics9030066

Sharafan, M., Malinowska, M. A., Ekiert, H., Kwaśniak, B., Sikora, E., & Szopa, A., 2023. Vitisvinifera (Vine Grape) as a Valuable Cosmetic Raw Material. *Pharmaceutics*, *15*(5), 1372. https://doi.org/10.3390/pharmaceutics15051372

Sirgedaitė-Šėžienė, V., Čėsnienė, I., Leleikaitė, G., Baliuckas, V., & Vaitiekūnaitė, D., 2023. Phenolic and Antioxidant Compound Accumulation of Quercusrobur Bark Diverges Based on Tree Genotype, Phenology and Extraction Method. *Life*, *13*(3), 710. https://doi.org/10.3390/life13030710

Sun, W., & Shahrajabian, M. H., 2023. Therapeutic Potential of Phenolic Compounds in Medicinal Plants – Natural Health Products for Human Health. *Molecules*, *28*(4), 1845. https://doi.org/10.3390/molecules28041845

Upadhyay, P. R., Starner, R. J., Swope, V. B., Wakamatsu, K., Ito, S., & Abdel-Malek, Z. A., 2022. Differential Induction of Reactive Oxygen Species and Expression of Antioxidant Enzymes in Human Melanocytes Correlate with Melanin Content: Implications on the Response to Solar UV and Melanoma Susceptibility. *Antioxidants*, *11*(6), 1204. https://doi.org/10.3390/antiox11061204

Zhang, H. A., & Kitts, D. D., 2021. Turmeric and Its Bioactive Constituents Trigger Cell Signaling Mechanisms That Protect Against Diabetes and Cardiovascular Diseases. *Molecular and Cellular Biochemistry*, *476*(10), 3785–3814. https://doi.org/10.1007/s11010-021-04201-6

Zulfugarova, P., Zivari-Ghader, T., Maharramova, S., Ahmadian, E., Eftekhari, A., Khalilov, R., Turksoy, V. A., Rosić, G., & Selakovic, D., 2023. A Mechanistic Review of Pharmacological Activities of Homeopathic Medicine Licorice Against Neural Diseases. *Frontiers in Neuroscience*, *17*. https://doi.org/10.3389/fnins.2023.1148258

8 Skin Photoprotection Obtained with Natural Products

Ayşe Baldemir Kılıç, Sadi Yusufbeyoğlu, Mustafa Nisari, Ragip Ertas, Sule Ketenci Ertas, Muhammed Burak Yucel and Ismail Ocsoy

8.1 INTRODUCTION

In view of the stratospheric ozone hole and the effects of climate change, it is vital to protect human skin from the detrimental effects of increasing UV radiation. Life systems' exposure to visible light (400–700 nm) and terrestrial solar ultraviolet (UV) radiation (295–400 nm) has a significant effect. Absorption of photon energy by cellular chromophores can lead to either advantageous or detrimental outcomes. The wavelength dependency (effect spectrum) of a specific photobiological outcome is mostly controlled by the absorption spectrum of the chromophore. The skin is considered the primary organ subjected to sunlight exposure, as supported by various studies (Bickers and Athar, 2006; Wölfle et al., 2014; Lawrence et al., 2017).

The planet Earth is perpetually subjected to a continuous stream of photons that are emitted by the sun. The photons can be categorized into three primary types according to their wavelengths: infrared light photons, visible light photons, and ultraviolet (UV) light photons (McKenzie et al., 2007; Laikova et al., 2019). Approximately 56% of photons are classified within the infrared spectrum, which includes a range of wavelengths spanning from 780 to 5000 nm. The visible light spectrum, with wavelengths ranging from 400 to 780 nm, provides about 39% of the total composition in terms of photons. Ultraviolet (UV) light photons make up around 5% of the total population of photons, encompassing wavelengths ranging from 290 to 400 nm (Young et al., 1998; Bachelor and Bowden, 2004; Neale et al., 2023). UV light causes significant effects on several components present in the skin, such as melanin, DNA, RNA, proteins, lipids, water, tyrosine, tryptophan, and other aromatic amino acids, as well as trans-urocanic acid. The chromophores can absorb photons in the UV range, thereby initiating a range of photochemical reactions. These activities later cause interactions that include reactive oxygen species (ROS), which ultimately may lead to deleterious outcomes (Bachelor and Bowden, 2004; Widyarini, 2006; Correa et al., 2020). The molecular consequences of this exposure have been extensively studied and encompass various forms of harm to DNA, proteins, and lipids. For instance, the formation of DNA photoproducts has been observed as a direct result of this exposure (Madronich, 1992; McKenzie et al., 2007). Additionally, reactive oxygen species (ROS) generated during this exposure have been found to induce secondary damage (Afaq et al., 2002; Svobodová et al., 2003), affecting multiple molecular and cellular components. Initially, it was believed that the primary cause of damage was direct impact. Nevertheless, in recent years there has been a growing recognition of the significance of indirect and oxidative damage. The synthesis of oxidizing and light-sensitive compounds can be attributed to the oxidation of proteins, the DNA was subjected to damage, resulting in the inhibition of the DNA repair system (Katiyar, 2005; Norval et al., 2008; Mudit and Katiyar, 2010; Ahmady et al., 2020; Neale et al., 2023).

 DOI: 10.1201/9781003394846-8

There exist numerous strategies for mitigating the adverse impacts of UV rays on the skin. The application of sun protection creams containing active molecules capable of absorbing or reflecting UV radiation is a widely employed method to prevent the penetration of UV rays into the skin. The mitigating impact of sunscreen application on the skin's surface can reduce the deleterious consequences of UV radiation. Contemporary sunscreens are formulated using synthetic compounds, which may consist of one or many substances, which possess the ability to obstruct UV radiation and safeguard the integumentary system. The active chemicals in sunscreen consist of both inorganic and organic materials, which function to safeguard the skin. Inorganic sun protection possesses the ability to reflect and disperse UV and visible radiation, whereas organic sun protection can convert UV radiation into energy, manifesting as either heat or light. The quantity of these synthetic compounds is restricted due to regulatory measures (Korać and Khambholja, 2011; Mansuri et al., 2021). The compounds can be classified into two distinct categories, physical and chemical. Physical sunscreen can alter the trajectory and disperse ultraviolet B (UVB), ultraviolet A (UVA), and visible light wavelengths. In sunscreen formulations of physical filters, titanium dioxide (TiO_2) and zinc oxide (ZnO) are commonly used mineral filters. Contrary to popular belief, it is important to note that they primarily function using UVR absorption, with minimal impact resulting from reflection or scattering. According to a recent study, the typical level of ultraviolet radiation (UVR) reflected by certain nanoparticles (NPs) like TiO_2 and ZnO is within the range of 4% to 5%. This level of reflection corresponds to a sun protection factor (SPF) value of less than two. Ultraviolet radiation (UVR) is absorbed as a result of the acceleration of electrons transitioning from the valence band to the conduction band. These molecules do not exhibit light absorption within the visible spectrum. However, due to the structural arrangement of their particles, they impart a white appearance to the skin through a combination of scattering and reflection. This is not a desired situation for sunscreen, so these molecules are getting smaller and smaller to improve their facial qualities (Cayrol et al., 1999; Solish et al., 2020).

Chemical sunscreen functions by absorbing ultraviolet (UV) radiation and subsequently converting it into chemical energy in the form of heat or light (Chaudhuri et al., 2017; Lin et al., 2021). Organic chemical filters are made by binding an aromatic molecule to the carbonyl groups of the filter's chromophore. In general, the augmentation of conjugate double bonds and resonance structures confers stability to the excited state and induces a redshift in the absorption spectrum, resulting in a broader absorption band. It has been observed that UVB emitters, which produce light in the range of 280–315 nm, generally possess lower molecular weights compared to UVA emitters (315–400 nm) and wide-spectrum filters that encompass both UVB and UVA wavelengths (Goihman-Yahr, 1996; Krutmann, 2001).

Currently, there exists a variety of commercially produced chemical compounds that possess the ability to function as UV filters on the market. Nevertheless, the utilization of these substances is limited due to their potential adverse effects on human skin. In order to mitigate the aforementioned adverse consequences, there has been incredible scholarly attention directed toward the utilization of naturally occurring compounds as photoprotective agents. Natural products have had a significant impact on human medicine throughout many historical periods, spanning several centuries. These products have continuously demonstrated their efficacy in facilitating the discovery of novel therapeutic interventions owing to their extensive chemical variety and capacity to interact with diverse biological targets. In recent times, there has been an increasing scholarly focus on phytomedicine owing to its historical efficacy and extensive utilization of natural product combinations in traditional medicinal practices.

8.2 PHOTOPROTECTIVE

The initial response of the human body to safeguard against solar radiation involves the production of melanin, a pigment that effectively absorbs ultraviolet (UV) rays emitted by the sun, preventing potential damage to the skin cells. In certain instances, the production of melanin may be insufficient to adequately protect the skin. Various measures need to be undertaken to safeguard the facial region against the deleterious impacts of ultraviolet (UV) radiation (Svobodová et al., 2003; Sharma and Sharma, 2023). According to González et al. (2008), the most effective strategy for safeguarding oneself against the harmful effects of the sun is to minimize exposure during periods when ultraviolet (UV) rays are most intense (González et al., 2008). Additionally, it is recommended to utilize protective clothing, a wide-brimmed hat, and sunglasses. Another method of sun protection products contains substances that possess the ability to either absorb or reflect ultraviolet (UV) light when applied to the skin's surface. This serves to safeguard the skin against the detrimental consequences associated with exposure to the sun's UV rays.

Sun protection can be categorized into two primary classes based on their mechanisms of action: physical sun protection and chemical sun protection. Physical sunscreens contain inert particles, such as TiO_2 and ZnO, which effectively reflect UVA and UVB rays, hence preventing their penetration into the facial skin (Lowe, 1996; Shiffman, 1998; Solovchenko and Schmitz-Eiberger, 2003). Due to their ability to reflect visible photons, these particles frequently manifest on the surface of the skin, resulting in an unfavourable appearance for customers. One advantageous aspect of these substances is their chemical inertness, which renders them non-toxic and hypoallergenic (More, 2007; Cole et al., 2010). The majority of chemical sunscreens consist of organic aromatic compounds with a carbonyl functional group. The chromosphere facilitates the absorption of ultraviolet (UV) radiation by the molecule, leading to the emission of lower-energy radiation. This process serves to safeguard the skin against the detrimental impacts of UV radiation (Gilaberte and González, 2010; Rai et al., 2012; Garnacho Saucedo et al., 2020). Chemical sunscreens, such as oxybenzone, sulisobenzone, and octyl methoxycinnamate, serve as notable illustrations of this phenomenon. According to Rai et al. (2012), chemical sunscreen is visually preferable due to its ability to remain undetectable on the skin's surface. The absorption of ultraviolet (UV) radiation has the potential to activate organic sunscreen, leading to its interaction with cutaneous molecules and subsequent manifestation of adverse skin reactions (Rai et al., 2012). Certain chemical sunscreens, such as aminobenzoic acid and its esters, synamates, and oxybenzone, have been found to increase photosensitivity or provoke contact dermatitis in individuals (Dromgoole and Maibach, 1990). In recent years, there has been significant scholarly attention in the exploration of naturally occurring chemical compounds for their potential in sun protection.

8.3 NATURAL COMPOUNDS AND NMS-BASED PHOTOPROTECTIVES

The exploration of naturally produced chemicals from botanical sources that exhibit both safety and efficacy in the filtration of ultraviolet (UV) radiation holds great potential for the advancement of sun protection agents. When compared to synthetic analogues, natural materials demonstrate higher levels of ultraviolet (UV) absorption and antioxidant capacity (Cockell and Knowland, 1999).

Natural products, commonly known as secondary metabolites, are organic molecules that are biosynthesized by living organisms in their native habitats. The production of these metabolites often serves as a noticeable response to the environmental conditions encountered by organisms that possess defensive capabilities, thereby aiding in the survival and reproduction of their species. Several natural species, especially those exposed to significant levels of radiation, have developed various photo-adaptive mechanisms. These mechanisms include the production of antioxidants and the formation of secondary metabolites with UV-absorbing capabilities. The primary common feature across UV scanning molecules is the presence of the spp-electron system, which is seen in the conjugated bond configurations of linear chain molecules that display varying single and double bonds. Furthermore, this particular characteristic is also evident in a multitude of aromatic and cyclic compounds that exhibit electron delocalization (Nichols and Katiyar, 2010; Vostálová et al., 2019; Mansuri et al., 2021). The simultaneous use of these seconder metabolite combinations (phenolic acids, flavonoids, fluoride, lichen polyphenols, terpenoids, and microspore, etc.) can induce a decrease in inflammatory responses, alleviate oxidative stress, and limit the harmful effects on DNA integrity by obstructing the penetration of radiation through the layers of the skin. Moreover, it has been noted that specific naturally derived compounds exhibit the ability to absorb radiation and modulate various signaling pathways, thereby reducing oxidative stress, inflammation, and protecting the skin from the adverse consequences of prolonged exposure to UV radiation. Therefore, much scholarly focus has been dedicated to naturally occurring compounds due to their potential as effective agents in reducing photodamage in humans. These substances operate through various mechanisms and are involved in several phototoxic pathways Cockell and Knowland, 1999; Nichols and Katiyar, 2010; Kolbe, 2016).

It is worth mentioning that bio- or organic molecules incorporated products like sunscreens may not effectively reflect or scatter light compared to NMs-incorporated ones. In addition, some bio- or organic molecules are considered to be fewer stable agents against photodegradation, in which they can be easily oxidized or degraded to generate harmful materials like some radicals or toxic materials. To address this issue, the natural product-derived NMs can be a remedy to produce quite stable and efficient skin care products, especially towards UV exposure (Xia et al., 2010). Use of NMs like metal NMs, metal oxide NMs and lipid-based NMs have several advantages as photoprotection agents compared to organic molecule-based agents by increasing sun protection factor (SPF) and photostability with various protection mechanisms (Cole et al., 2016; Rajasekar et al., 2024; Some et al., 2020; Ekrikaya et al., 2021; Ocsoy et al., 2018; Demirbas et al., 2020; Karaagac et al., 2021; Temur et al., 2024; Ekrikaya et al., 2024). For instance, Souto et al. (2022) reported the use of lipid NPs in dermo-cosmetic products to provide effective photoprotection against solar radiation. Lipid NPs provide hydration and enhanced viscosity to deliver adjuvants to the applied surface, the skin is protected against the passage of UV radiation.

Inorganic NMs like TiO_2 NPs and ZnO NPs can be used in skin protection products for sun protection without accumulation in the inner layer of the skin. In addition, these NMs have acted as UVA and UVB filters with a broad spectrum of activity to prevent against even long-term solar radiation. Lignin-based NMs were used in both preparations of safer bio-based sunscreens and great UV absorbers in terms of sun protection factor (UVB-SPF) and UVA-UVB transmittance (Smijs and Pavel, 2011; Widsten et al., 2020). While nanodiamonds provide superior UV attenuation, Ivy NMs are considered non-toxic UV filters (Wu et al., 2015; Beasley and Meyer, 2010). Characterization of the UVA protection provided by avobenzone, zinc oxide, and titanium dioxide in broad-spectrum sunscreen products (Xia et al., 2010).

8.4 PHENOLIC COMPOUNDS

Phenolic compounds (PCs) are a highly prevalent and widely dispersed group of secondary metabolites in the kingdom of plants. PCs can be divided into three different groups: high-molecular-weight polyphenols, flavonoids, and phenolic acids. According to several sources (Lowe, 1996; Svobodová et al., 2003; Nichols and Katiyar, 2010; Mansuri et al., 2021), there are currently more than 8000 phenolic structures. PC use has a variety of effects on skin health, including photoprotection, anti-inflammatory qualities, anti-ageing effects, and photo-chemo-preventive capacities. The fundamental makeup of PCs can be responsible for these features' appearance. According to studies (Greenaway et al., 1990; Nichols and Katiyar, 2010), the presence of phenolic rings and hydroxyl groups exhibits strong free radical scavenging action. Multiple mechanisms of action, including (i) the inhibition of reactive oxygen species (ROS) biosynthesis, (ii) the capture of ROS, and (iii) the reduction of metal ions that serve as catalysts for ROS synthesis, can be used to explain the antioxidant activity displayed by PCs (Soobrattee et al., 2005; Alov et al., 2014; Ahumada-Fierro et al., 2021). According to data, it points to the possibility that PCs may exert their effects via multiple pathways. First, it has been demonstrated that phenolic compounds can reduce oxidative stress and inflammation by neutralizing free radicals such as reactive oxygen species (ROS) and reactive nitrogen species (RNs). Second, these substances can suppress lipid peroxidation and the synthesis of pro-inflammatory mediators including interleukin 6 (IL-6) and prostaglandin E2 (PGE2) and modify the activity of immune cells implicated in the inflammatory response. Thirdly, phenolic compounds have been discovered to regulate the activity of nuclear factors like nuclear factor B (NF-kB) and nuclear factor erythroid 2-related factor 2 (Nrf-2), which are crucial in both inflammatory and antioxidant responses, as well as the synthesis of eicosanoids, molecules involved in inflammation. The ability of PCs to regulate the expression of genes involved in processes connected to the skin can be attributed to their anti-ageing effects. PCs are in charge of controlling the genes involved in the processes of oxidative stress defence and skin cell renewal. PCs also help to maintain healthy skin tissue by promoting the production of elastin and collagen, inhibiting the activity of matrix metalloproteinases (MMPs) by activating activator protein-1 (AP-1) and nuclear factor-kappa B (NF-kB), and limiting the activity of collagenases and elastases (Dudonné et al., 2011; Abdul Karim et al., 2014; Działo et al., 2016). Additionally, it has been proposed that specific natural substances have the power to treat dermatological conditions by triggering DNA mutation repair mechanisms within damaged cellular structures (Kaur et al., 2019).

Multiple studies have found evidence that phenolic acids and flavonoids obtained from propolis, tea, grape, fern, and milk thistle exhibit photoprotective properties by decreasing oxidative stress, inflammation, and immunosuppression. Antioxidants, including vitamins such as Vitamin C and Vitamin E, as well as flavonoids and phenolic acids, are primarily responsible for combating free radical species, which are the primary contributors to many detrimental alterations in the skin (Saewan and Jimtaisong, 2015; Mansuri et al., 2021). Polyphenolic compounds are known to exert a substantial influence on the human diet. Flavonoids and phenolic acids are the primary categories of chemicals that are commonly detected in diverse food sources. Several compounds exhibit photoprotective properties as a result of their antioxidative, anti-inflammatory, and anticarcinogenic effects (González et al., 2008). Vanillic acid, *p*-coumaric acid, caffeic acid, phenolic acid, rosmarinic acid, chlorogenic acid, gallic acid, tannic acid, resveratrol, and curcumin are among the non-flavonoid phenolic compounds (PCs) that exhibit skin photo-protective activity (Hwang et al., 2014; Saewan and Jimtaisong, 2015; Daré et al., 2020). In the field of photoprotection, ferulic and caffeic acids have become the focus of intensive study. These two PCs' antioxidant properties have been proven by their ability to protect phospholipid membranes from UVR-induced peroxidation. Previous studies have shown that this protection is accomplished by inhibiting the lipid peroxides chain reaction. Both compounds have demonstrated photoprotective effects against oxidative stress, induction of matrix metalloproteinase-1 (MMP-1), and formation of reactive oxygen species (ROS) in human keratinocytes by modulating antioxidant defence mechanisms, such as the regulation of glutathione (GSH), catalase, and glutathione peroxidase (GPx) (Saewan and Jimtaisong, 2013; Hwang et al., 2014; Stevanato et al., 2014; Daré et al., 2020).

In a recent study, it was found that these two well-known photoprotective substances had the power to reduce UVA-induced melanogenesis. The nuclear factor erythroid similar to two pathways, a transcription factor in melanoma cells, is indirectly modulated to produce this effect (Chaiprasongsuk et al., 2016). Flavonoids are a category of secondary plant phenolics that possess notable antioxidant and chelating characteristics. The health benefits of these compounds were recognized prior to their isolation as natural products. A significant number of flavonoid variants, currently, there are more than 5000 known flavonoids, and they are found in a variety of plant species (Ross and Kasum, 2002). These variants play a crucial role in determining the visually appealing pigmentation observed in flowers, fruits, and foliage (Saewan and Jimtaisong, 2013) Specifically, there is a need to conduct novel studies that extend beyond *in vitro* assessments and progress towards *in vivo* investigations, in order to preserve and ascertain the sun protection factor values associated with flavonoids (José et al., 2016; Monsalve-Bustamante et al., 2020). When it comes to the preservation of photosensitive materials, flavonoids are a notable class of phenolic chemicals. The flavonoids constitute a class of naturally occurring compounds that function as catalysts in the light-dependent phase of photosynthesis. The aforementioned compounds have the capacity to absorb ultraviolet radiation (UVR) and so lessen the damaging effects of oxidative damage brought on by ROS. Due to the presence of double bonds, the flavonoid structure demonstrates a high propensity for UVR absorption. Additionally, flavonoids' ability to effectively scavenge ROS is influenced by the hydroxyl groups that are connected to their aromatic moieties (Stevanato et al., 2014; Saewan and Jimtaisong, 2015; Chaiprasongsuk et al., 2016). According to three main mechanisms – UVR absorption, direct and indirect antioxidant properties, including the induction of cytoprotective proteins, and modulation of various signal pathways – flavonoids have been found to exhibit protective effects against UVR damage (Ross and Kasum, 2002; Dunaway et al., 2018). Furthermore, it has been discovered that flavonoids use two different mechanisms – direct inhibition of MMPs and stimulation of the expression of the tissue inhibitor of MMPs – to successfully reduce the activity of matrix metalloproteinases (MMPs) in skin cells (Saewan and Jimtaisong, 2013). In addition, they serve the purpose of safeguarding plant cells against stress-induced damage by the elimination of reactive oxygen species (ROS). The protective effects of natural flavonoids against sun damage are attributed to their ability to absorb ultraviolet (UV) light, function as direct and secondary antioxidants, and possess anti-inflammatory and immune-modulating properties (Svobodová et al., 2003; Korać and Khambholja, 2011; Mansuri et al., 2021). The flavonoids can be classified into six distinct subclasses, namely flavones, flavonols, flavanonols, isoflavones, flavanols, and anthocyanidins. This classification is determined by the manner in which the B ring is connected to the C ring, as well as the oxidation state and functional groups present on the C ring (José et al., 2016).

Flavones provide beneficial characteristics in the prevention and treatment of atherosclerosis, osteoporosis, diabetes mellitus, and certain types of malignancies. These compounds are primarily found in cereals and plants conducted a study (Catarino et al., 2014; Li et al., 2022; Miao et al., 2022; Do Socorro Chagas et al., 2022) have identified apigenin and chrysin as the principal naturally occurring flavones. Apigenin, also known as 5,7,4'-trihydroxyflavone, is a plant flavone that is found in several sources such as cereal grains, aromatic herbs (such as parsley, rosemary, and thyme), fruits (including apples, cherries, and grapes), vegetables (such as beans, broccoli, celery, leeks, onions, barley, and tomatoes), and beverages (such as tea and wine) (Wu et al., 2011; Mani and Natesan, 2018). Chrysin, also known as 5,7-dihydroxyflavone, is a naturally occurring flavone compound that can be found in a variety of sources, including propolis and honey. It is considered an analogue of apigenin. Both compounds have anti-inflammatory properties, as well as the ability to scavenge free radicals. Additionally, they possess anticancer and antidiabetic properties and can be utilized for skin protection (Wu et al., 2011; Mani and Natesan, 2018).

Various plant-based foods, including onion, kale, broccoli, lettuce, tomato, apple, grape, berries, and tea, have been identified as notable sources of flavonols. According to some studies, the predominant flavonols identified in the natural environment include quercetin, kaempferol, myricetin, and isorhamnetin. Studies have proven that these substances have a photoprotective effect (Fahlman and Krol, 2009; Flamini et al., 2013; Han et al., 2015; Nagapan et al., 2018; Lin et al., 2023).

Anthocyanins represent a highly captivating category of polyphenolic compounds. They are found in a diverse array of plant species and fruits, playing a crucial role in determining their vibrant colouration, which spans the spectrum from red to blue. Six prevalent anthocyanidins are found in nature, namely pelargonidin, cyanidin, delphinidin, peonidin, petunidin, and malvidin. Anthocyanins have compounds that contain the capability to safeguard skin and ocular cells from light-induced harm and exhibit the capacity to absorb light within the ultraviolet (UV) and blue light spectrum, thereby counteracting the physiological processes linked to prolonged UV exposure. Consequently, these compounds possess the capability to safeguard skin and ocular cells from light-induced harm (Merzlyak and Chivkunova, 2000; Nagapan et al., 2018; Oliveira et al., 2020).

8.5 CAROTENOIDS

Carotenoids are tetraterpenoids with a central carbon chain consisting of alternative single and double bonds carrying different cyclical or acyclic end groups (Britton, 1995). Some types of carotenoids have dual-binding systems that provide photo-protective benefits, including the ability to absorb ultraviolet radiation and useful antioxidant properties. The second mentioned goal is accomplished by getting rid of molecules, known as peroxide and single-molecular oxygen, which are produced during photo-oxidation. These systems can also stop lipid peroxidation. Carotenoids can trigger phase 2 cytoprotective genes and prompt protection of cells. Expanded pair-binding systems make them better at protecting against sunlight, as they can absorb UV rays. They also contain antioxidants that can get rid of unstable molecules like peroxide and single-molecule oxygen generated during photo-oxidation. Additionally, they stop lipid peroxidation. Carotenoids can even stimulate cellular protective responses because they can turn on phase 2 cytoprotective genes. The protection of photos by dietary carotenoids has been assessed by Stahl and Sies earlier (Stahl et al., 2001; Stahl and Sies, 2012). Lycopene is among the most extensively examined carotenoids. It is deemed to be the most efficient against the pernicious ROS that forms in the skin after exposure to sunlight, namely the single oxygen radical. Topical use of lycopene protects against UVB damage by hindering UVB-induced ornithine decarboxylase (ODC) and myeloperoxidase activities. It also inhibits caspase-3 cleavage in the apoptotic pathway as well as inflammatory reactions (Stahl et al., 2001; Fazekas et al., 2003; Heinrich et al., 2003). A sunscreen made from tomato essence (which contains lycopene) has been successful. Lycopene also protects against the sun when consumed orally. Taking 16 mg of lycopene daily for ten weeks reduces skin redness caused by sun exposure by 40% (Stahl et al., 2001). Additionally, a diet enriched with a mix of carotenoids including β-carotene (24 mg/day) or β-carotene, lutein and lycopene (each 8 mg/day) for a duration of 12 weeks has illustrated the ability to protect against UV-related redness of the skin (Heinrich et al., 2003).

8.6 ALKALOIDS

A nitrogen atom in a heterocyclic ring refers to a group of substances known as alkaloids (Ziegler and Facchini, 2008). Caffeine, theophylline, and theobromine have been investigated for their photoprotective and antioxidant activities (Scapagnini et al., 2014; Cefali et al., 2016). In terms of photoprotection, caffeine is the alkaloid that has been examined the most. Caffeine treatment topically and orally has demonstrated anticancer benefits. In UVR-radiated rats, coffee applied topically decreased skin carcinogenesis (Zajdela and Latarjet, 1978). Caffeine treatment decreased tumour volume, tumour multiplication, and incidence. Additionally, it specifically promoted apoptosis in mouse skin cancers caused by UVB (Lu et al., 2001). According to epidemiological studies, increased caffeine consumption is linked to a lower incidence of skin cancer, such as BCC (basal cell carcinoma) (Lu et al., 2007; Song et al., 2012; Lou et al., 2013). In order to achieve this impact, faulty DNA cells undergo an increase in apoptosis, but surprisingly, normal epidermal cells do not experience this increase (Conney et al., 2008; Han et al., 2011). Some topical sunscreen formulations that contain caffeine have recently become the subject of discussion over their efficacy and safety. SPF values for caffeine formulations were greater according to *in vitro* functional characterization. Higher SPF values were also supported by *in vivo* experiments. As a photo-protector and photo-balancing agent, caffeine worked in concert with conventional

UV filters such as ethylhexyl methoxycinnamate, avobenzone, and titanium dioxide. So it has been established that caffeine is a useful and intriguing bioactive ingredient for use in the creation of sunscreen (Rosado et al., 2019). Certain metabolites can be produced by plants as a kind of UVR defence. As a result, these organisms constitute a special source of compounds with noteworthy photo-protective properties. Table 8.1 shows some studies on plant-derived products that have photoprotective activity.

TABLE 8.1
Studies on the Use of Natural Products as Photoprotective Agents

Plant Name/ Natural Product	Photoprotective Activity	Plant Extracts/ Active Compounds	Biological Model Used	Reference
Achyrocline satureioides (Lam.) D.C.	Antioxidant	Quercetin, luteolin and 3-O-methylquercetin	*in vitro*	(Morquio et al., 2005)
Aloe barbidensis Mill.	UV protection, Moisturizer, Immune enhance	Barbiloin, Aloe emodin, Oleic acid, Caprylic acid, Stearic acid	*in vitro*	(Prajakta and Shahu, 1970; Long, 2016; Bharadwaj et al., 2018)
Arctium lappa L.	Altering specific miRNA expression in NHDF cells	Arctiin	*in vitro*	(Lee et al., 2014)
Arnica montana L.	Anti-inflammatory	Volatile oil 5%, 56% fatty in extract	*in vitro* *in vivo*	(Kriplani et al., 2017; Da Silva Prade et al., 2020)
Azadirachta indica A. Juss.	Preventing photoaging Antioxidant	Azadirachtin and derivatives, nimbin and nimbidin	*in vivo* *iv vitro*	(Baby et al., 2022)
Caffeine	Sun protection Sunscreen adjuvant	Caffeine	*in vivo* *iv vitro*	(Rosado et al., 2019)
Calendula officinalis Hohen.	Prevent UV radiation-induced oxidative stress	Polyphenols flavonoids rutin, narcissin	*in vitro*	(Fonseca et al., 2010; Kurzawa et al., 2022)
Camellia sinensis L. (Kuntze)	Photostabilizing capacity, Anticarcinogenic, Anti-inflammatory, Antioxidant	Epicatechin Epicatechin-3-gallate Catechin	*in vivo* *in vitro*	(Zheng et al., 2008; Barg et al., 2014; Davis et al., 2022)
C. sinensis (L.) Kuntze *(Green tea)*	DNA damage prevention, Prevent UVA-Induced erythema	Polyphenols	*in vivo* *in vitro*	(Yusuf et al., 2007)
C. sinensis (L.) Kuntze *(Green tea)*	DNA damage prevention, Prevent UVA-Induced erythema	Polyphenols Epigallocatechin-3-Gallate (EGCG) and (-)-Epicatechin-3-gallate (ECG)	Clinical appearance, *in vitro*	(Elmets et al., 2001)
Capparis spinosa L.	Antioxidant, Photoprotection against UVB-Induced skin damage	Ethanol extract Caffeic acid, Ferulic acid, p-Cumaric acid, Cinnamic acid	*in vitro*	(Bonina et al., 2002)
C. spinosa L.	Antiallergic, Antioxidant, Antiaging,	Rutin and Gallic, Caffeic, Coumaric, Vanillic, Syringic, Ferulic, and Chlorogenic acids	*in vitro*	(Farhan et al., 2023; Harwansh and Deshmukh, 2023)

(Continued)

TABLE 8.1 *(Continued)*
Studies on the Use of Natural Products as Photoprotective Agents

Plant Name/ Natural Product	Photoprotective Activity	Plant Extracts/ Active Compounds	Biological Model Used	Reference
Carica papaya L.	Sun protection, Antioxidant, Anti-inflammatory, Photoprotective	Extract the flavonoid compounds in papaya seeds *Carica papaya* leaf extract (caffeic acid, rutin)	*in vitro*	(Seo et al., 2020; Budi Nursanto et al., 2023)
Citrus sp.	Absorb light in the UV-A	Furocoumarins on terpenes in lime, lemon, and bergamot essential oils	*in vitro*	(Bitterling et al., 2022)
Cocoa varieties	Cutaneous Photoprotective	Polyphenols, Flavanols, Caffeine, Theobromine, Theophylline	*in vivo* *in vitro*	(Scapagnini et al., 2014; Torres-Contreras et al., 2022)
Crocus sativus L.	Antioxidant, Antiaging, Anti-inflammatory	Safranal, Crocin	*in vivo*	(Ruszkiewicz et al., 2017; Sharma and Sharma, 2023)
Curcuma longa L.	Preventing photoaging, Anti-Oxidant	Polyphenol, Curcumin, Demethoxy Curcumin, Turmerone, Zingiberene	*in vivo*	(Kaur and Saraf, 2011; Feng et al., 2018; Son et al., 2023; Threskeia et al., 2023;)
Daucus carota L.	Free radical scavenger, Nourishes skin	Beta carotene Oxyacetylene	*in vivo* *in vitro*	(Müller et al., 2019; Aksu et al., 2020)
Epilobium parviflorum Schreb.	Antioxidant	Quercetin, Luteolin and 3-O-methylquercetin	*in vitro*	(Morquio et al., 2005)
Fragaria ananassa Duch.	Antioxidant, Anti-inflammatory	Anthocyanins, Hydrolysable tannins	*in vitro*	(Pombo et al., 2011; Giampieri et al., 2012)
Gentiana lutea L.	Antioxidant, UV-Protective, Antigenotoxicity	Gentiana lutea root extract	*in vitro*	(Cvetković et al., 2023)
Glycine max L.	Antioxidant, Anti-inflammatory	Hydroalcoholic extract isoflavones	*in vitro*	(Chiang et al., 2007)
Glycine max L.	Antioxidant, Anti-inflammatory	Hydroalcoholic extract isoflavones	*in vivo* (in mice)	(Huang et al., 2010)
Juglans regia L.	Keratin protection of skin, Antioxidant	Juglone Linoleic acid Linolenic acid	*in vivo* *in vitro*	(Muzaffer et al., 2018; Muzaffer et al., 2019)
Linum usitatissimum L.	Antioxidant, Antiapoptotic	β-carotene, vitamins A, C, and E	*in vivo*	(Tülüce et al., 2012)
Lycium ruthenicum Murr.	Photoprotective effects, Reducing apoptosis, Protective UVB irradiated	Anthocyanin	*in vitro*	(Yu et al., 2019; Wang et al., 2022)
Malus domestica Borkh.	Antioxidant, Protective UV irradiated	Quercetin Epicatechin Procyanidin ß-2	*in vivo* *in vitro*	(Solovchenko and Merzlyak, 2003; Solovchenko and Schmitz-Eiberger, 2003; Khayatan et al., 2022)

TABLE 8.1 *(Continued)*
Studies on the Use of Natural Products as Photoprotective Agents

Plant Name/ Natural Product	Photoprotective Activity	Plant Extracts/ Active Compounds	Biological Model Used	Reference
Moringa oleifera Lam.	Photoabsorption	Antioxidant	*in vitro*	(Baldisserotto et al., 2018; Gimenis et al., 2018)
Ocimum basilicum L.	Sun protection, Photoabsorption	Monoterpenes	*in vivo*	(Kale et al., 2010; Landi et al., 2014)
Panax ginseng M.	Prevent photodamaging effects, Anti-ageing	Gensenoside, Panaxoside, Glycoside	*in vitro* *in vitro*	(Lee et al., 2023; Dhanjal et al., 2020)
Piper longum L.	Antioxidant, UV protectant	Piperine, Methyl piperine, Piperinic acid	*in vitro*	(Choochana et al., 2015; Verma et al., 2017)
Polypodium leucotomos L.	Anti-ageing, DNA damage prevention, Antioxidant	Aqueous extract	*in vitro*	(Parrado et al., 2020)
Polypodium leucotomos L.	Antioxidant, Anti-inflammatory	Hydroalcoholic extract of Red Propolis	*in vivo* (rats)	(Batista et al., 2018)
Polypodium leucotomos L.	UV protectant, Photo protect	Oral and topical combination	Multi-Center prospective, randomized, parallel-group, assessor-blinded trial	(Pellacani et al., 2023)
Portulaca oleracea L.	Anti-Mutagenic Antioxidant, Anti-inflammatory	Carotenoids, Betaxanthins	*in vitro*	(Oh et al., 2019, 2023)
Propolis	Antioxidant	Hydroalcoholic propolis extract	*in vitro* (measurement SPF)	(Reis et al., 2009)
Propolis	Antioxidant, Against the oxidative stress in the skin	Green porolis extract, Coumaric acid, Drupanin, Baccharin, and Artepillin C	*in vitro* (measurement)	(Fonseca et al., 2011; Correa et al., 2020)
Propolis	Antioxidant, Anti-inflammatory	Ethanolic extract of Sydney propolis	*in vivo* (hairless mice)	(Cole et al., 2010)
Propolis	Antioxidants High UV absorption, Inhibited lipid peroxidation of linoleic acid (LA)	Ethanolic extract from Italian propolis were Caffeic acid, Caffeic acid phenethyl ester, and 1,1-dimethylallyl-caffeate, Kaempferol, Quercetin, and Galangin	*in vitro* (measurement)	(Gregoris et al., 2011)
Propolis	Antioxidant, Anti-inflammatory, Antiapoptotic, Antigenotoxic	Romanian propolis extract, Coumaric acid, Ferulic acid, and Quercetin	Topically (mouse model)	(Bolfa et al., 2013)
Propolis	Antioxidant, Anti-inflammatory	Hydroalcoholic extract of red propolis	*in vivo*	(Batista et al., 2018)

(Continued)

TABLE 8.1 *(Continued)*
Studies on the Use of Natural Products as Photoprotective Agents

Plant Name/ Natural Product	Photoprotective Activity	Plant Extracts/ Active Compounds	Biological Model Used	Reference
Propolis	Antioxidant, Anti-inflammatory	Ethanolic extracts of red propolis Liquiritigenin, Isoliquiritigenin, Pinocembrin, Formononetin, Naringenin, Vestitol, Biochanin A, and Daidzein	*in vitro* (Determination spf)	(Valverde et al., 2023)
Psoralea corylifolia L.	Anti-aging, Inhibiting oxidative stress	Flavonoid, Corylin, Psoralen	*in vitro* *in vivo*	(N. Li et al., 2022; Pandey et al., 2023)
Punica granatum L.	Protective effects against UVA- and UVB-Induced cell damage	Delphinidin, Cyanidin, Pelargonidin	*in vitro*	(Lisbeth et al., 2008)
Rubia codifo lia L.	Treat hyperpigmentation, Allergies, Sunburn	Pseudopurpurin, Manjistin, Puroxanthin,	*in vitro*	(Biswas et al., 2015; Do et al., 2023)
Sambucus nigra L.	UV protection, Sun protection, Antioxidant, Anti-inflammatory	Sambulin B, Quercetin, Kaempferol, Isorhamnetin-3-O-β-D-rutin glycoside, Quercetin-3-rutin glycoside, Keracyanin chloride, Kaempferol-3-rutinoside	*in vitro*	(Jarzycka et al., 2013; Stanković et al., 2017)
Silybum marianum L.	Inhibit UV-induced damage, Antioxidant	Silymarin, Silybin, Silydianin	*in vitro*	(Rojas et al., 2016; Rajnochová Svobodová et al., 2019)
Solanum lycopersicum L.	Antioxidant	Lycopene, Vitamins A, C, E, Anthocyanin	*in vitro*	(Kurzawa et al., 2022; Sopyan et al., 2017)
Terminalia chebula Retz	Antioxidant	Arjunolic acid, Ellegic acid	*in vivo*	(Mishra et al., 2011)
Terminalia chebula Retz	Antioxidant, Anti-inflammatory, Lipid peroxidation, Anti-aging	Chebulinic acid, Tannic acid, Ellagic acid	*in vitro* *in vivo* (clinical experiment)	(Gandhi and Krishnan Nair, 2005; Chakkalakal et al., 2023)
Trifolium pratense L.	Inhibiting DNA photodamage	Estrogenic isoflavones, Daidzein, Genistein, Formononetin, Biochanin	*in vivo* (hairless mice)	(Widyarini, 2006)
Vaccinium myrtillus L.	ROS formation stimulated by UV A and UV B reduced	Polyphenols	*in vivo*	(Calò and Marabini, 2014)
Vitis vinifera L.	Antioxidant Anti-inflammatory	Flavan-3-ol derivatives, Catechin, Oligomeric, Proanthocyanidins	*in vitro*	(Hübner et al., 2020)

8.7 CONCLUSION

Exposure to UV radiation from the sun is a recognized risk factor for various human diseases, including erythema, oedema, hyperpigmentation, the formation of sunburn cells, photo-immunosuppression, photo-induced aging, and skin cancer, through various mechanisms. To shield the skin from UV radiation, a topical application of sunscreen products is a recommended strategy. These products effectively block the UV radiation from reaching the epidermis. Numerous plant compounds have demonstrated beneficial effects on skin health and photoprotection in multiple studies. Secondary metabolites including alkaloids, carotenoids, and phenolic compounds have been found to shield the skin from damage caused by ultraviolet radiation. These compounds are capable of absorbing and blocking UVR while also impeding the mechanisms that lead to skin damage. Various of these compounds such as flavonoids, cinnamic acid derivatives, carotenoids, polyphenols, and anthocyanins have been incorporated into various herbal sunscreens and shown to be effective. These agents possess vital photoprotective and antioxidant properties that are necessary for optimal sun protection. There is a global movement towards natural cosmetics and the creation of high UV protection sunscreens with minimal chemical components. By combining natural and synthetic agents, a distinct approach to reducing the negative effects of UV radiation may be achieved. Due to their modulation of several signalling pathways in reducing the deleterious effects of UV radiation on the skin, natural compounds with broad biological activities are crucial in the production of sunscreen products. This chapter presents the advantages of natural compounds in offering efficient photoprotection through various pathways, thus supplying novel options for sunscreen formulations. The development of natural compounds as new commercial sunscreen agents in sun-care products must be safe and have a beneficial photoprotective effect. To ensure sun-care product efficacy, concern arises over the quantity of active constituents in natural extracts, as well as compatibility, concentration, and stability. It is also necessary to investigate the active compounds' stable dosage levels that produce the most beneficial effects.

This section comprises scientific studies investigating skin damage caused by sunlight to certain plant extracts and herbal active compounds within the natural products category. While protective preparations using natural products are currently available in the market, there exists a significant demand for materials produced with standardized natural products. Herbal sunscreens, utilizing diverse plant extracts, have the potential to effectively displace synthetic sunscreens justified by their minimal side effects and greater potency.

REFERENCES

Abdul Karim A, Azlan A, Ismail A, Hashim P, Abd Gani SS, Zainudin BH, Abdullah NA. 2014. Phenolic composition, antioxidant, anti-wrinkles and tyrosinase inhibitory activities of cocoa pod extract. *BMC Complement Altern Med.* 14(1), 1–13. https://doi.org/10.1186/1472-6882-14-381

Afaq F, Adhami VM, Ahmad N, Mukhtar H. 2002. Botanical antioxidants for chemoprevention of photocarcinogenesis. *Front Biosci.* 7, 784–792. https://doi.org/10.2741/afaq

Ahmady A, Amini MH, Zhakfar AM, Babak G, Sediqi MN. 2020. Sun protective potential and physical stability of herbal sunscreen developed from Afghan medicinal plants. *Turk J Pharm Sci.* 17(3), 285. https://doi.org/10.4274/tjps.galenos.2019.15428

Ahumada-Fierro NV, García-Mendoza E, Sandoval-Gil JM, Band-Schmidt CJ. 2021. Photosynthesis and photoprotection characteristics related to ROS production in three Chattonella (Raphidophyceae) species. *J Phycol.* 57(3). https://doi.org/10.1111/jpy.13138

Aksu M, Incegul Y, Kiralan SS, Kiralan M, Ozkan G. 2020. Cold pressed carrot (Daucus carota subsp. sativus) seed oil. In: *Cold Pressed Oils: Green Technology, Bioactive Compounds, Functionality, and Applications*. https://doi.org/10.1016/B978-0-12-818188-1.00030-X

Alov P, Tsakovska I, Pajeva I. 2014. Computational studies of free radical-scavenging properties of phenolic compounds. *Curr Top Med Chem.* 15(2), 85–104. https://doi.org/10.2174/1568026615666141209143702

Baby AR, Freire TB, Marques G de A, Rijo P, Lima FV, de Carvalho JCM, Rojas J, Magalhães WV, Velasco MVR, Morocho-Jácome AL. 2022. Azadirachta indica (Neem) as a potential natural active for dermocosmetic and topical products: A narrative review. *Cosmetics*. 9(3). https://doi.org/10.3390/cosmetics9030058

Bachelor MA, Bowden GT. 2004. UVA-mediated activation of signaling pathways involved in skin tumor promotion and progression. *Semin Cancer Biol.* 14(2). https://doi.org/10.1016/j.semcancer.2003.09.017

Baldisserotto A, Buso P, Radice M, Dissette V, Lampronti I, Gambari R, Manfredini S, Vertuani S. 2018. Moringa oleifera leaf extracts as multifunctional ingredients for "natural and organic" sunscreens and photoprotective preparations. *Molecules*. 23(3). https://doi.org/10.3390/molecules23030664

Barg M, Rezin GT, Leffa DD, Balbinot F, Gomes LM, Carvalho-Silva M, Vuolo F, Petronilho F, Dal-Pizzol F, Streck EL, Andrade VM. 2014. Evaluation of the protective effect of Ilex paraguariensis and Camellia sinensis extracts on the prevention of oxidative damage caused by ultraviolet radiation. *Environ Toxicol Pharmacol.* 37(1). https://doi.org/10.1016/j.etap.2013.11.028

Batista CM, Alves AVF, Queiroz LA, Lima BS, Filho RNP, Araújo AAS, de Albuquerque Júnior RLC, Cardoso JC. 2018. The photoprotective and anti-inflammatory activity of red propolis extract in rats. *J Photochem Photobiol B*. 180: 198–207. https://doi.org/10.1016/J.JPHOTOBIOL.2018.01.028

Beasley DG, Meyer TA. 2010. Characterization of the UVA protection provided by avobenzone, zinc oxide, and titanium dioxide in broad-spectrum sunscreen products. Am *J Clin Dermatol.* 11: 413–421. https://doi.org/10.2165/11536910-000000000-00000

Bharadwaj B, Vishnu Priya V, Balakrishna RN. 2018. Aloe vera – A review. *Drug Invention Today*. 10(Special Issue 4): 1797–1800.

Bickers DR, Athar M. 2006. Oxidative stress in the pathogenesis of skin disease. *J Invest Dermatol.* 126(12). https://doi.org/10.1038/sj.jid.5700340

Biswas R, Mukherjee PK, Dalai MK, Mandal PK, Nag M. 2015. Tyrosinase inhibitory potential of purpurin in Rubia cordifolia-A bioactivity guided approach. *Ind Crops Prod.* 74. https://doi.org/10.1016/j.indcrop.2015.04.066

Bitterling H, Mailänder L, Vetter W, Kammerer DR, Stintzing FC. 2022. Photo-protective effects of furocoumarins on terpenes in lime, lemon and bergamot essential oils upon UV light irradiation. *Eur Food Res Technol.* 248(4). https://doi.org/10.1007/s00217-021-03945-1

Bolfa P, Vidrighinescu R, Petruta A, Dezmirean D, Stan L, Vlase L, Damian G, Catoi C, Filip A, Clichici S. 2013. Photoprotective effects of Romanian propolis on skin of mice exposed to UVB irradiation. *Food Chem Toxicol.* 62. https://doi.org/10.1016/j.fct.2013.08.078

Bonina F, Puglia C, Ventura D, Aquino R, Tortora S, Sacchi A, Saija A, Tomaino A, Pellegrino ML, De Caprariis P. 2002. In vitro antioxidant and in vivo photoprotective effects of a lyophilized extract of Capparis spinosa L. buds. *J Cosmet Sci.* 53(6).

Britton G. 1995. Structure and properties of carotenoids in relation to function. *FASEB J.* 9(15). https://doi.org/10.1096/fasebj.9.15.8529834

Budi Nursanto E, Wijaya R, Permata Sari A. 2023. Flavonoid extraction from papaya (*Carica Papaya* L.) seed using ultrasound – assisted extraction method and determination of its SPF value. *J Emerg Supply Chain, Clean Energy Process Eng.* 2(1). https://doi.org/10.57102/jescee.v2i1.60

Calò R, Marabini L. 2014. Protective effect of *Vaccinium myrtillus* extract against UVA- and UVB-induced damage in a human keratinocyte cell line (HaCaT cells). *J Photochem Photobiol B.* 132. https://doi.org/10.1016/j.jphotobiol.2014.01.013

Catarino M, Alves-Silva J, Pereira O, Cardoso S. 2014. Antioxidant capacities of flavones and benefits in oxidative-stress related diseases. *Curr Top Med Chem.* 15(2). https://doi.org/10.2174/1568026615666141209144506

Cayrol C, Sarraute J, Tarroux R, Redoules D, Charveron M, Gall Y. 1999. A mineral sunscreen affords genomic protection against ultraviolet (UV) B and UVA radiation: In vitro and in situ assays. *Br J Dermatol.* 141(2). https://doi.org/10.1046/j.1365-2133.1999.02973.x

Cefali LC, Ataide JA, Moriel P, Foglio MA, Mazzola PG. 2016. Plant-based active photoprotectants for sunscreens. *Int J Cosmet Sci.* 38(4). https://doi.org/10.1111/ics.12316

Chaiprasongsuk A, Onkoksoong T, Pluemsamran T, Limsaengurai S, Panich U. 2016. Photoprotection by dietary phenolics against melanogenesis induced by UVA through Nrf2-dependent antioxidant responses. *Redox Biol.* 8. https://doi.org/10.1016/j.redox.2015.12.006

Chakkalakal M, Pan A, Nadora D, Gahoonia N, Chaudhuri RK, Burney W, Thacker S, Shakhbazova A, Subramanyam C, Chambers CJ, Sivamani RK. 2023. Randomized double-blind placebo-controlled supplementation with standardized terminalia chebula fruit extracts reduces facial sebum excretion, erythema, and wrinkle severity. *J Clin Med.* 12(4). https://doi.org/10.3390/jcm12041591

Chaudhuri RK, Ollengo MA, Singh P, Martincigh BS. 2017. 3-(3,4,5-Trimethoxybenzylidene)-2,4-pentanedione: Design of a novel photostabilizer with in vivo SPF boosting properties and its use in developing broad-spectrum sunscreen formulations. *Int J Cosmet Sci.* 39(1). https://doi.org/10.1111/ics.12344

Chiang HS, Wu W Bin, Fang JY, Chen BH, Kao TH, Chen YT, Huang CC, Hung CF. 2007. UVB-protective effects of isoflavone extracts from soybean cake in human keratinocytes. *Int J Mol Sci.* 8(7). https://doi.org/10.3390/i8070651

Choochana P, Moungjaroen J, Jongkon N, Gritsanapan W, Tangyuenyongwatana P. 2015. Development of piperic acid derivatives from Piper nigrum as UV protection agents. *Pharm Biol.* 53(4). https://doi.org/10.3109/13880209.2014.924020

Cockell Charles S., Knowland J. 1999. Ultraviolet radiation screening compounds. *Biol Rev.* 74(3). https://doi.org/10.1111/j.1469-185X.1999.tb00189.x

Cole C, Shyr T, Ou-Yang H. 2016. Metal oxide sunscreens protect skin by absorption, not by reflection or scattering. *Photodermatol Photoimmunol Photomed.* 32: 5–10. https://doi.org/10.1111/phpp.12214

Cole N, Sou PW, Ngo A, Tsang KH, Severino JAJ, Arun SJ, Duke CC, Reeve VE. 2010. Topical "sydney" propolis protects against UV-radiation-induced inflammation, lipid peroxidation and immune suppression in mouse skin. *Int Arch Allergy Immunol.* 152(2). https://doi.org/10.1159/000265530

Conney AH, Kramata P, Lou YR, Lu YP. 2008. Effect of caffeine on UVB-induced carcinogenesis, apoptosis, and the elimination of UVB-induced patches of p53 mutant epidermal cells in SKH-1 mice. *Photochem Photobiol.* 84(2). https://doi.org/10.1111/j.1751-1097.2007.00263.x

Correa L, De Carvalho Meirelles G, Balestrin L, De Souza PO, Moreira JCF, Schuh RS, Bidone J, Von Poser GL, Teixeira HF. 2020. In vitro protective effect of topical nanoemulgels containing Brazilian red propolis benzophenones against UV-induced skin damage. *Photochem Photobiol Sci.* 19(10). https://doi.org/10.1039/d0pp00243g

Cvetković S, Vuletić S, Vunduk J, Klaus A, Mitić-Ćulafić D, Nikolić B. 2023. The role of *Gentiana lutea* extracts in reducing UV-induced DNA damage. *Mutagenesis.* 38(1). https://doi.org/10.1093/mutage/geac006

da Silva Prade J, Bálsamo EC, Machado FR, Poetini MR, Bortolotto VC, Araújo SM, Londero L, Boeira SP, Sehn CP, de Gomes MG, et al. 2020. Anti-inflammatory effect of *Arnica montana* in a UVB radiation-induced skin-burn model in mice. *Cutan Ocul Toxicol.* 39(2). https://doi.org/10.1080/15569527.2020.1743998

Daré RG, Nakamura CV, Ximenes VF, Lautenschlager SOS. 2020. Tannic acid, a promising anti-photoaging agent: Evidences of its antioxidant and anti-wrinkle potentials, and its ability to prevent photodamage and MMP-1 expression in L929 fibroblasts exposed to UVB. *Free Radic Biol Med.* 160. https://doi.org/10.1016/j.freeradbiomed.2020.08.019

Davis SL, Marsh JM, Kelly CP, Li L, Tansky CS, Fang R, Simmonds MSJ. 2022. Protection of hair from damage induced by ultraviolet irradiation using tea (*Camellia sinensis*) extracts. *J Cosmet Dermatol.* 21(5). https://doi.org/10.1111/jocd.14387

Demirbas A, Karaytuğ T, Arabaci N, Yilmaz ES, Ocsoy I. 2020. Synthesis of metallic and metal oxide nanomaterials. In: *Green Synthesis of Nanomaterials for Bioenergy Applications* (pp. 99–123). John Wiley & Sons Ltd.

Dhanjal DS, Bhardwaj S, Sharma R, Bhardwaj K, Kumar D, Chopra C, Kuca K. 2020. Plant fortification of the diet for anti-ageing effects: A review. *Nutrients.* 12(10): 3008. https://doi.org/10.3390/nu12103008

Do KL, Su M, Mushtaq A, Ahsan T, Zhao F. 2023. Functionalization of silk with chitosan and *Rubia cordifolia* L. dye extract for enhanced antimicrobial and ultraviolet protective properties. *Text Res J.* 93(15–16): 3777–3789. https://doi.org/10.1177/00405175231167603

Do Socorro Chagas MS, Behrens MD, Moragas-Tellis CJ, Penedo GXM, Silva AR, Gonçalves-De-Albuquerque CF. 2022. Flavonols and flavones as potential anti-inflammatory, antioxidant, and antibacterial compounds. *Oxid Med Cell Longev.* 2022, Article ID 9966750, 21 pp. https://doi.org/10.1155/2022/9966750

Dromgoole SH, Maibach HI. 1990. Sunscreening agent intolerance: Contact and photocontact sensitization and contact urticaria. *J Am Acad Dermatol.* 22(6). https://doi.org/10.1016/0190-9622(90)70154-A

Dudonné S, Coutiãre P, Woillez M, Mérillon JM, Vitrac X. 2011. DNA macroarray study of skin aging-related genes expression modulation by antioxidant plant extracts on a replicative senescence model of human dermal fibroblasts. *Phytother Res.* 25(5). https://doi.org/10.1002/ptr.3308

Dunaway S, Odin R, Zhou L, Ji L, Zhang Y, Kadekaro AL. 2018. Natural antioxidants: Multiple mechanisms to protect skin from solar radiation. *Front Pharmacol.* 9(Apr). https://doi.org/10.3389/fphar.2018.00392

Działo M, Mierziak J, Korzun U, Preisner M, Szopa J, Kulma A. 2016. The potential of plant phenolics in prevention and therapy of skin disorders. *Int J Mol Sci.* 17(2). https://doi.org/10.3390/ijms17020160

Ekrikaya S, Danisman H, Baktir S, Arslan S, Ocsoy I. 2024. Bioinspired synthesis of nanocomposites. In: *Smart and Sustainable Applications of Nanocomposites* (pp. 36–66). IGI Global Scientific Publishing.

Ekrikaya S, Yilmaz E, Celik C, Demirbuga S, Ildiz N, Demirbas A, Ocsoy I. 2021. Investigation of ellagic acid rich-berry extracts directed silver nanoparticles synthesis and their antimicrobial properties with potential mechanisms towards *Enterococcus faecalis* and *Candida albicans*. *J Biotechnol.* 341: 155–162.

Elmets CA, Singh D, Tubesing K, Matsui M, Katiyar S, Mukhtar H. 2001. Cutaneous photoprotection from ultraviolet injury by green tea polyphenols. *J Am Acad Dermatol.* 44(3). https://doi.org/10.1067/mjd.2001.112919

Fahlman BM, Krol ES. 2009. UVA and UVB radiation-induced oxidation products of quercetin. *J Photochem Photobiol B.* 97(3). https://doi.org/10.1016/j.jphotobiol.2009.08.009

Farhan HS, Al-Azizz SA, Al-Jadaan SAN. 2023. Antibacterial and antioxidant activities of syzygium aromaticum capparis spinosa, and some novel acylselenourea and acylthourea derivatives. *Biomed Chem Sci.* 2(2). https://doi.org/10.48112/bcs.v2i2.449

Fazekas Z, Gao D, Saladi RN, Lu Y, Lebwohl M, Wei H. 2003. Protective effects of lycopene against ultraviolet B-induced photodamage. *Nutr Cancer.* 47(2). https://doi.org/10.1207/s15327914nc4702_11

Feng J, Du Z, Zhang L, Luo W, Zheng Y, Chen D, Xi L. 2018. Chemical composition and skin protective effects of essential oil obtained from ginger (*Zingiber officinale Roscoe*). *J Essent Oil Bear Plants.* 21(6): 1542–1549. https://doi.org/10.1080/0972060X.2018.1543448

Flamini R, Mattivi F, De Rosso M, Arapitsas P, Bavaresco L. 2013. Advanced knowledge of three important classes of grape phenolics: Anthocyanins, stilbenes and flavonols. *Int J Mol Sci.* 14(10). https://doi.org/10.3390/ijms141019651

Fonseca MJV, Fonseca YM, Marquele-Oliveira F, Vicentini FTMC, Furtado NAJC, Sousa JPB, Lucisano-Valim YM. 2011. Evaluation of the potential of Brazilian propolis against UV-induced oxidative stress. *Evid Based Complementary Altern Med.* 2011, Article ID 863917, 8 pp. https://doi.org/10.1155/2011/863917

Fonseca YM, Catini CD, Vicentini FTMC, Nomizo A, Gerlach RF, Fonseca MJV. 2010. Protective effect of Calendula officinalis extract against UVB-induced oxidative stress in skin: Evaluation of reduced glutathione levels and matrix metalloproteinase secretion. *J Ethnopharmacol.* 127(3). https://doi.org/10.1016/j.jep.2009.12.019

Gandhi NM, Krishnan Nair CK. 2005. Radiation protection by *Terminalia chebula*: Some mechanistic aspects. *Mol Cell Biochem.* 277(1–2). https://doi.org/10.1007/s11010-005-4819-9

Garnacho Saucedo GM, Salido Vallejo R, Moreno Giménez JC. 2020. Effects of solar radiation and an update on photoprotection. *An Pediatr (Engl Ed).* 92(6). https://doi.org/10.1016/j.anpedi.2020.04.014

Giampieri F, Alvarez-Suarez JM, Tulipani S, Gonzàles-Paramàs AM, Santos-Buelga C, Bompadre S, Quiles JL, Mezzetti B, Battino M. 2012. Photoprotective potential of strawberry (Fragaria × ananassa) extract against UV-A irradiation damage on human fibroblasts. *J Agric Food Chem.* 60(9). https://doi.org/10.1021/jf205065x

Gilaberte Y, González S. 2010. Update on photoprotection. *Actas Dermosifiliogr.* 101(8). https://doi.org/10.1016/S1578-2190(10)70696-X

Gimenis JM, Gomes AC, Dos Santos VHM, Ferreira PC, Oliveira CA, Baby AR, da Sılva RMG. 2018. Antioxidant and photoprotective potential of moringa oleifera lam (Moringaceae). *Biosci J.* 34(5). https://doi.org/10.14393/BJ-v34n5a2018-39845

Goihman-Yahr M. 1996. Skin aging and photoaging: An outlook. *Clin Dermatol.* 14(2). https://doi.org/10.1016/0738-081X(95)00150-E

González S, Fernández-Lorente M, Gilaberte-Calzada Y. 2008. The latest on skin photoprotection. *Clin Dermatol.* 26(6). https://doi.org/10.1016/j.clindermatol.2007.09.010

Greenaway W, Scaysbrook T, Whatley FR. 1990. The composition and plant origins of propolis: A report of work at oxford. *Bee World.* 71(3). https://doi.org/10.1080/0005772X.1990.11099047

Gregoris E, Fabris S, Bertelle M, Grassato L, Stevanato R. 2011. Propolis as potential cosmeceutical sunscreen agent for its combined photoprotective and antioxidant properties. *Int J Pharm.* 405(1–2). https://doi.org/10.1016/j.ijpharm.2010.11.052

Han W, Ming M, He YY. 2011. Caffeine promotes ultraviolet B-induced apoptosis in human keratinocytes without complete DNA repair. *J Biol Chem.* 286(26). https://doi.org/10.1074/jbc.M111.222349

Han X, Piao MJ, Kim KC, Madduma Hewage SRK, Yoo ES, Koh YS, Kang HK, Shin JH, Park Y, Yoo SJ, et al. 2015. Isorhamnetin protects human keratinocytes against ultraviolet B-induced cell damage. *Biomol Ther* (Seoul). 23(4). https://doi.org/10.4062/biomolther.2015.005

Harwansh RK, Deshmukh R. 2023. Recent Insight into UV-induced oxidative stress and role of herbal bioactives in the management of skin aging. *Curr Pharm Biotechnol.* 24. https://doi.org/10.2174/1389201024666230427110815

Heinrich U, Gärtner C, Wiebusch M, Eichler O, Sies H, Tronnier H, Stahl W. 2003. Supplementation with β-carotene or a similar amount of mixed carotenoids protects humans from UV-induced erythema. *J Nutr.* 133(1). https://doi.org/10.1093/jn/133.1.98

Huang CC, Hsu BY, Wu NL, Tsui WH, Lin TJ, Su CC, Hung CF. 2010. Anti-photoaging effects of soy isoflavone extract (aglycone and acetylglucoside form) from soybean cake. *Int J Mol Sci.* 11(12). https://doi.org/10.3390/ijms11124782

Hübner AA, Sarruf FD, Oliveira CA, Neto AV, Fischer DCH, Kato ETM, Lourenço FR, Baby AR, Bacchi EM. 2020. Safety and photoprotective efficacy of a sunscreen system based on grape pomace (*Vitis vinifera* l.) phenolics from winemaking. *Pharmaceutics.* 12(12). https://doi.org/10.3390/pharmaceutics12121148

Hwang E, Park SY, Lee HJ, Lee TY, Sun ZW, Yi TH. 2014. Gallic acid regulates skin photoaging in UVB-exposed fibroblast and hairless mice. *Phytother Res.* 28(12). https://doi.org/10.1002/ptr.5198

Jarzycka A, Lewińska A, Gancarz R, Wilk KA. 2013. Assessment of extracts of *Helichrysum arenarium, Crataegus monogyna, Sambucus nigra* in photoprotective UVA and UVB; Photostability in cosmetic emulsions. *J Photochem Photobiol B.* 128. https://doi.org/10.1016/j.jphotobiol.2013.07.029

José MT de AF, Pedrita AS, Emanuella CVP, Raimundo G de OJ, Fabrício SS, Jackson RG da SA, Larissa AR, Xirley PN, Edigênia C da CA. 2016. Flavonoids as photoprotective agents: A systematic review. *J Med Plants Res.* 10(47). https://doi.org/10.5897/jmpr2016.6273

Kale S, Sonawane A, Ansari A, Ghoge P, Waje A. 2010. Formulation and in- vitro determination of sun protection factor of *Ocimum basilicum*, Linn. leaf oils sunscreen cream. *Int J Pharm Pharm Sci.* 2(Suppl. 4).

Karaagac Z, Dadi S, Ekrikaya S, Ocsoy I. 2021. Plant extract guided nanomaterials and plant disease management. In: *Nanotechnology in Sustainable Agriculture* (pp. 1–20). CRC Press.

Katiyar SK. 2005. Silymarin and skin cancer prevention: Anti-inflammatory, antioxidant and immunomodulatory effects (Review). *Int J Oncol.* 26(1). https://doi.org/10.3892/ijo.26.1.169

Kaur CD, Saraf S. 2011. Topical vesicular formulations of Curcuma longa extract on recuperating the ultraviolet radiation-damaged skin. *J Cosmet Dermatol.* 10(4). https://doi.org/10.1111/j.1473-2165.2011.00586.x

Kaur P, Purewal SS, Sandhu KS, Kaur M. 2019. DNA damage protection: An excellent application of bioactive compounds. *Bioresour Bioprocess.* 6(1). https://doi.org/10.1186/s40643-019-0237-9

Khayatan D, Nilforoushzadeh MA, Ahmadi Ashtiani HR, Hashemian F. 2022. Effect of apple (Malus domestica) stem cells on UVB-induced damage skin with anti-inflammatory properties: An in vivo study. *Adv Mater Sci Eng.* 2022. https://doi.org/10.1155/2022/2417766

Kolbe L. 2016. Biological cell protection by natural compounds, a second line of defense against solar radiation. In: *Skin Stress Response Pathways: Environmental Factors and Molecular Opportunities.* https://doi.org/10.1007/978-3-319-43157-4_17

Korać RR, Khambholja KM. 2011. Potential of herbs in skin protection from ultraviolet radiation. *Pharmacogn Rev.* 5(10). https://doi.org/10.4103/0973-7847.91114

Kriplani P, Guarve K, Baghael US. 2017. Arnica montana L. – A plant of healing: Review. *J Pharm Pharmacol.* 69(8). https://doi.org/10.1111/jphp.12724

Krutmann J. 2001. The role of UVA rays in skin aging. *Eur J Dermatol.* 11.

Kurzawa M, Wilczyńska E, Brudzyńska P, Sionkowska A. 2022. Total phenolic content, antioxidant capacity and UV radiation protection properties of marigold (*Calendula officinalis*), carrot (*Daucus carota*), tomato (*Solanum lycopersicum*) and hop (*Humulus lupulus*) extracts. *Cosmetics.* 9(6). https://doi.org/10.3390/cosmetics9060134

Laikova KV, Oberemok VV, Krasnodubets AM, Gal'chinsky NV, Useinov RZ, Novikov IA, Temirova ZZ, Gorlov MV, Shved NA, Kumeiko VV, et al. 2019. Advances in the understanding of skin cancer: Ultraviolet radiation, mutations, and antisense oligonucleotides as anticancer drugs. *Molecules.* 24(8). https://doi.org/10.3390/molecules24081516

Landi M, Guidi L, Pardossi A, Tattini M, Gould KS. 2014. Photoprotection by foliar anthocyanins mitigates effects of boron toxicity in sweet basil (*Ocimum basilicum*). *Planta.* 240(5). https://doi.org/10.1007/s00425-014-2087-1

Lawrence KP, Long PF, Young AR. 2017. Mycosporine-like amino acids for skin photoprotection. *Curr Med Chem.* 25(40). https://doi.org/10.2174/0929867324666170529124237

Lee GT, Cha HJ, Lee KS, Lee KK, Hong JT, Ahn KJ, An IS, An S, Bae S. 2014. Arctiin induces an UVB protective effect in human dermal fibroblast cells through microRNA expression changes. *Int J Mol Med.* 33(3). https://doi.org/10.3892/ijmm.2014.1616

Lee HY, Kim EJ, Cho DY, Jung JG, Kim MJ, Lee JH, Kim W, Kang SS, Cho KM, Kang D. 2023. Photoprotective effect of fermented and aged mountain-cultivated ginseng sprout (*Panax ginseng*) on ultraviolet radiation-induced skin aging in a hairless mouse model. *Nutrients.* 15(7). https://doi.org/10.3390/nu15071715

Li N, Liu T, Zhu S, Yang Y, Wang Z, Zhao Z, Liu T, Wang X, Qin W, Yan Y, et al. 2022. Corylin from Psoralea fructus (*Psoralea corylifolia* L.) protects against UV-induced skin aging by activating Nrf2 defense mechanisms. *Phytother Res.* 36(8). https://doi.org/10.1002/ptr.7501

Li RL, Wang LY, Liu S, Duan HX, Zhang Q, Zhang T, Peng W, Huang Y, Wu C. 2022. Natural flavonoids derived from fruits are potential agents against atherosclerosis. *Front Nutr.* 9. https://doi.org/10.3389/fnut.2022.862277

Lin TC, Yang CY, Wu TH, Tseng CH, Yen FL. 2023. Myricetin nanofibers enhanced water solubility and skin penetration for increasing antioxidant and photoprotective activities. *Pharmaceutics.* 15(3). https://doi.org/10.3390/pharmaceutics15030906

Lin X, Zhang D, Ju L, Zhang Y, Jiang Y, Hou Q, Hu Z, Wang Y, Wang Z. 2021. Effect of high concentrate corn stalk particle size on physical state and production performance of lactating dairy cows. *Adv Biosci Biotechnol.* 12(2). https://doi.org/10.4236/abb.2021.122004

Lisbeth A, Noratto G, Hingorani L, Talcott ST, Mertens-Talcott SU. 2008. Protective effects of standardized pomegranate (*Punica granatum* L.) polyphenolic extract in ultraviolet-irradiated human skin fibroblasts. *J Agric Food Chem.* 56(18). https://doi.org/10.1021/jf8005307

Long V. 2016. Aloe vera in dermatology-the plant of immortality. *JAMA Dermatol.* 152(12). https://doi.org/10.1001/jamadermatol.2016.0077

Lou Y, Peng Q, Li T, Nolan B, Bernard JJ, Wagner GC, Lin Y, Shih WJ, Conney AH, Lu Y. 2013. Oral caffeine during voluntary exercise markedly inhibits skin carcinogenesis and decreases inflammatory cytokines in UVB-treated mice. *Nutr Cancer.* 65(7). https://doi.org/10.1080/01635581.2013.812224

Lowe NJ. 1996. *Sunscreens: Development: Evaluation, and Regulatory Aspects.* CRC Press.

Lu YP, Lou YR, Lin Y, Weichung Joe Shih, Huang MT, Yang CS, Conney AH. 2001. Inhibitory effects of orally administered green tea, black tea, and caffeine on skin carcinogenesis in mice previously treated with ultraviolet B light (high-risk mice): Relationship to decreased tissue fat. *Cancer Res.* 61(13).

Lu YP, Nolan B, Lou YR, Peng QY, Wagner GC, Conney AH. 2007. Voluntary exercise together with oral caffeine markedly stimulates UVB light-induced apoptosis and decreases tissue fat in SKH-1 mice. *Proc Natl Acad Sci U S A.* 104(31). https://doi.org/10.1073/pnas.0705839104

Madronich S. 1992. Implications of recent total atmospheric ozone measurements for biologically active ultraviolet radiation reaching the Earth's surface. *Geophys Res Lett.* 19(1). https://doi.org/10.1029/91GL02954

Mani R, Natesan V. 2018. Chrysin: Sources, beneficial pharmacological activities, and molecular mechanism of action. *Phytochemistry.* 145. https://doi.org/10.1016/j.phytochem.2017.09.016

Mansuri R, Diwan A, Kumar H, Dangwal K, Yadav D. 2021. Potential of natural compounds as sunscreen agents. *Pharmacogn Rev.* 15(29). https://doi.org/10.5530/phrev.2021.15.5

McKenzie RL, Aucamp PJ, Bais AF, Björn LO, Ilyas M. 2007. Changes in biologically-active ultraviolet radiation reaching the Earth's surface. *Photochem Photobiol Sci.* 6(3). https://doi.org/10.1039/b700017k

Merzlyak MN, Chivkunova OB. 2000. Light-stress-induced pigment changes and evidence for anthocyanin photoprotection in apples. *J Photochem Photobiol B.* 55(2–3). https://doi.org/10.1016/S1011-1344(00)00042-7

Miao L, Liu C, Cheong MS, Zhong R, Tan Y, Rengasamy KRR, Leung SWS, Cheang WS, Xiao J. 2022. Exploration of natural flavones' bioactivity and bioavailability in chronic inflammation induced-type-2 diabetes mellitus. *Crit Rev Food Sci Nutr.* https://doi.org/10.1080/10408398.2022.2095349

Mishra AK, Mishra A, Chattopadhyay P. 2011. Herbal cosmeceuticals for photoprotection from ultraviolet B radiation: A review. *Trop J Pharm Res.* 10(3). https://doi.org/10.4314/tjpr.v10i3.7

Monsalve-Bustamante YA, Puertas-Mejia MA, Mejia-Giraldo JC. 2020. Comparison of the photoprotective effect between hydrolyzed and aglycones flavonoids as sunscreen: A systematic review. *J Appl Pharm Sci.* 10(1). https://doi.org/10.7324/JAPS.2020.101016

More BD. 2007. Physical sunscreens: On the comeback trail. *Indian J Dermatol Venereol Leprol.* 73(2). https://doi.org/10.4103/0378-6323.31890

Morquio A, Rivera-Megret F, Dajas F. 2005. Photoprotection by topical application of Achyrocline satureioides ('Marcela'). *Phytother Res* [Internet]. 19(6): 486–490. [accessed 2023 Aug 11] https://doi.org/10.1002/PTR.1665

Mudit V, Katiyar SK. 2010. Molecular mechanisms of inhibition of photocarcinogenesis by silymarin, a phytochemical from milk thistle (*Silybum marianum* L. Gaertn.) (Review). *Int J Oncol.* 36(5). https://doi.org/10.3892/ijo-00000586

Müller R, Acosta-Motos JR, Großkinsky DK, Hernández JA, Lütken H, Barba-Espin G. 2019. UV-B exposure of black carrot (*Daucus carota* ssp. sativus var. atrorubens) plants promotes growth, accumulation of anthocyanin, and phenolic compounds. *Agronomy.* 9(6). https://doi.org/10.3390/agronomy9060323

Muzaffer U, Paul VI, Agilan B, Prasad NR. 2019. Protective effect of Juglans regia L., against ultraviolet-B induced photoaging in human epidermal keratinocytes. *Biomed Pharmacother.* 111. https://doi.org/10.1016/j.biopha.2018.12.129

Muzaffer U, Paul VI, Prasad NR, Karthikeyan R, Agilan B. 2018. Protective effect of *Juglans regia* L. against ultraviolet B radiation induced inflammatory responses in human epidermal keratinocytes. *Phytomedicine.* 42. https://doi.org/10.1016/j.phymed.2018.03.024

Nagapan TS, Ghazali AR, Basri DF, Lim WN. 2018. Photoprotective effect of stilbenes and its derivatives against ultraviolet radiation-induced skin disorders. *Biomed Pharmacol J.* 11(3). https://doi.org/10.13005/bpj/1481

Neale RE, Lucas RM, Byrne SN, Hollestein L, Rhodes LE, Yazar S, Young AR, Berwick M, Ireland RA, Olsen CM. 2023. The effects of exposure to solar radiation on human health. *Photochem Photobiol Sci.* 22(5). https://doi.org/10.1007/s43630-023-00375-8

Nichols JA, Katiyar SK. 2010. Skin photoprotection by natural polyphenols: Anti-inflammatory, antioxidant and DNA repair mechanisms. *Arch Dermatol Res.* 302(2). https://doi.org/10.1007/s00403-009-1001-3

Norval M, McLoone P, Lesiak A, Narbutt J. 2008. The effect of chronic ultraviolet radiation on the human immune system. *Photochem Photobiol.* 84. [place unknown]. https://doi.org/10.1111/j.1751-1097.2007.00239.x

Ocsoy I, Tasdemir D, Mazicioglu S, Tan W. 2018. Nanotechnology in plants. In: *Plant Genetics and Molecular Biology* (pp. 263–275). Springer.

Oh JH, Karadeniz F, Lee JI, Seo Y, Kong CS. 2023. *Oleracone C* from *Portulaca oleracea* attenuates UVB-induced changes in matrix metalloproteinase and type I procollagen production via MAPK and TGF-β/Smad pathways in human keratinocytes. *Int J Cosmet Sci.* 45(2). https://doi.org/10.1111/ics.12828

Oh JH, Seo Y, Kong CS. 2019. Anti-photoaging effects of solvent-partitioned fractions from Portulaca oleracea L. on UVB-stressed human keratinocytes. *J Food Biochem.* 43(4). https://doi.org/10.1111/jfbc.12814

Oliveira H, Correia P, Pereira AR, Araújo P, Mateus N, de Freitas V, Oliveira J, Fernandes I. 2020. Exploring the applications of the photoprotective properties of anthocyanins in biological systems. *Int J Mol Sci.* 21(20). https://doi.org/10.3390/ijms21207464

Pandey A, Agrawal M, Agrawal SB. 2023. Individual and combined effects of chromium and ultraviolet-B radiation on defense system, ultrastructural changes, and production of secondary metabolite psoralen in a medicinal plant *Psoralea corylifolia* L. *Environ Sci Pollut Res.* 30(2). https://doi.org/10.1007/s11356-022-22480-4

Parrado C, Nicolas J, Juarranz A, Gonzalez S. 2020. The role of the aqueous extract: Polypodium leucotomos in photoprotection. *Photochem Photobiol Sci.* 19(6). https://doi.org/10.1039/d0pp00124d

Pellacani G, Peris K, Ciardo S, Pezzini C, Tambone S, Farnetani F, Longo C, Chello C, González S. 2023. The combination of oral and topical photoprotection with a standardized Polypodium leucotomos extract is beneficial against actinic keratosis. *Photodermatol Photoimmunol Photomed.* https://doi.org/10.1111/phpp.12870

Pombo MA, Rosli HG, Martínez GA, Civello PM. 2011. UV-C treatment affects the expression and activity of defense genes in strawberry fruit (Fragaria×ananassa, Duch.). *Postharvest Biol Technol.* 59(1). https://doi.org/10.1016/j.postharvbio.2010.08.003

Prajakta S, Shahu K. 1970. Formulation and evaluation of vanishing herbal cream of crude drugs. *Asian J Pharm Res Dev.* 8(3). https://doi.org/10.22270/ajprd.v8i3.725

Rai R, Shanmuga S, Srinivas CR. 2012. Update on photoprotection. *Indian J Dermatol.* 57(5). https://doi.org/10.4103/0019-5154.100472

Rajasekar M, Mary J, Sivakumar M, Selvam M. 2024. Recent developments in sunscreens based on chromophore compounds and nanoparticles. *RSC Adv.* 14(4): 2529–2563. https://doi.org/10.1039/D3RA06107B

Rajnochová Svobodová A, Gabrielová E, Ulrichová J, Zálešák B, Biedermann D, Vostálová J. 2019. A pilot study of the UVA-photoprotective potential of dehydrosilybin, isosilybin, silychristin, and silydianin on human dermal fibroblasts. *Arch Dermatol Res.* 311(6). https://doi.org/10.1007/s00403-019-01928-7

Reis Gs Dos, Valadão Af, Lima Lrp De, Moreira Ml. 2009. Preparación de un protector solar y evaluación de la acción fotoprotectora del propóleo verde del Vale do Aço, Minas Gerais, Brasil. *Bol Latinoam Caribe Plantas Med Aromat.* 8(4).

Rojas J., Londoño C., Ciro Y. 2016. The health benefits of natural skin UVA photoprotective compounds found in botanical sources. *Int J Pharm Pharm Sci.* 8(3): 13–23. https://doi.org/10.22159/ijpps.2016v8i3.11238

Rosado C, Tokunaga VK, Sauce R, De Oliveira CA, Sarruf FD, Parise-Filho R, Maurício E, De Almeida TS, Velasco MVR, Baby AR. 2019. Another reason for using caffeine in dermocosmetics: Sunscreen adjuvant. *Front Physiol.* 10(May). https://doi.org/10.3389/fphys.2019.00519

Ross JA, Kasum CM. 2002. Dietary flavonoids: Bioavailability, metabolic effects, and safety. *Annu Rev Nutr.* 22. https://doi.org/10.1146/annurev.nutr.22.111401.144957

Ruszkiewicz JA, Pinkas A, Ferrer B, Peres TV, Tsatsakis A, Aschner M. 2017. Neurotoxic effect of active ingredients in sunscreen products, a contemporary review. *Toxicol Rep.* 4. https://doi.org/10.1016/j.toxrep.2017.05.006

Saewan N, Jimtaisong A. 2013. Photoprotection of natural flavonoids. *J Appl Pharm Sci.* 3(9). https://doi.org/10.7324/JAPS.2013.3923

Saewan N, Jimtaisong A. 2015. Natural products as photoprotection. *J Cosmet Dermatol.* 14(1). https://doi.org/10.1111/jocd.12123

Scapagnini G, Davinelli S, Di Renzo L, De Lorenzo A, Olarte HH, Micali G, Cicero AF, Gonzalez S. 2014. Cocoa bioactive compounds: Significance and potential for the maintenance of skin health. *Nutrients.* 6(8). https://doi.org/10.3390/nu6083202

Seo SA, Ngo HTT, Hwang E, Park B, Yi TH. 2020. Protective effects of Carica papaya leaf against skin photodamage by blocking production of matrix metalloproteinases and collagen degradation in UVB-irradiated normal human dermal fibroblasts. *S Afr J Bot.* 131. https://doi.org/10.1016/j.sajb.2020.03.019

Sharma M, Sharma A. 2023. A review on nature based sunscreen agents. *IOP Conf Ser Earth Environ Sci.* 1110. [place unknown]. https://doi.org/10.1088/1755-1315/1110/1/012047

Shiffman MA. 1998. Book review: Sunscreens: Development, evaluation, and regulatory aspects, second edition. *Am J Cosmet Surg.* 15(1). https://doi.org/10.1177/074880689801500121

Smijs TG, Pavel S. 2011. Titanium dioxide and zinc oxide nanoparticles in sunscreens: Focus on their safety and effectiveness. *Nanotechnol Sci Appl.* 4: 95–112. https://doi.org/10.2147/NSA.S19419

Solish N, Humphrey S, Waller B, Vanderveen S. 2020. Photoprotection with mineral-based sunscreens. *Dermatol Surg.* 46(12). https://doi.org/10.1097/DSS.0000000000002478

Solovchenko A, Merzlyak M. 2003. Optical properties and contribution of cuticle to UV protection in plants: Experiments with apple fruit. *Photochem Photobiol Sci.* 2(8). https://doi.org/10.1039/b302478d

Solovchenko A, Schmitz-Eiberger M. 2003. Significance of skin flavonoids for UV-B-protection in apple fruits. *J Exp Bot.* 54(389). https://doi.org/10.1093/jxb/erg199

Some S, Sarkar B, Biswas K, Jana TK, Bhattacharjya D, Dam P, Mondal R, Kumar A, Deb AK, Sadat A, Saha S, Kati A, Ocsoy I, Franco OL, Mandal A, Mandal S, Mandal AK, Ince AI. 2020. Bio-molecule functionalized rapid one-pot green synthesis of silver nanoparticles and their efficacy toward the multidrug resistant (MDR) gut bacteria of silkworms (Bombyx mori). *RSC Adv.* 10: 22742–22757.

Son D, Jun JS, Hong K. 2023. Photoprotection effect of Pu'er tea and *Curcuma longa* L. extracts against UV and blue lights. *J Appl Biol Chem.* 66. https://doi.org/10.3839/jabc.2023.016

Song F, Qureshi AA, Han J. 2012. Increased caffeine intake is associated with reduced risk of basal cell carcinoma of the skin. *Cancer Res.* 72(13). https://doi.org/10.1158/0008-5472.CAN-11-3511

Soobrattee MA, Neergheen VS, Luximon-Ramma A, Aruoma OI, Bahorun T. 2005. Phenolics as potential antioxidant therapeutic agents: Mechanism and actions. *Mutat Res Fund Mol Mech Mutagen.* 579(1–2). https://doi.org/10.1016/j.mrfmmm.2005.03.023

Sopyan I, Gozali D, Tiassetiana S. 2017. Formulation of tomato extracts (*Solanum lycopersicum* L.) as a sunscreen lotion. *Natl J Physiol Pharm Pharmacol.* 8(3): 453–458. https://doi.org/10.5455/njppp.2017.7.1039921112017

Souto EB, Jäger E, Jäger A, Štěpánek P, Cano A, Viseras C, de Melo Barbosa R, Chorilli M, Zielínska A, Severino P, et al. 2022. Lipid nanomaterials for targeted delivery of dermocosmetic ingredients: Advances in photoprotection and skin anti-aging. *Nanomaterials.* 12: 377. https://doi.org/10.3390/nano12030377

Stahl W, Heinrich U, Wiseman S, Eichler O, Sies H, Tronnier H. 2001. Dietary tomato paste protects against ultraviolet light-induced erythema in humans. *J Nutr.* 131(5). https://doi.org/10.1093/jn/131.5.1449

Stahl W, Sies H. 2012. β-Carotene and other carotenoids in protection from sunlight. *Am J Clin Nutr.* 96(5). https://doi.org/10.3945/ajcn.112.034819 https://doi.org/10.21175/RadProc.2017.48

Stanković M, Živković J, Tadić V, Arsić I. 2017. Skin protection against solar UV radiation by natural plant products: extracts from elder fruit (*Sambucus nigra* L.). In: The Fifth International Conference on Radiation and Applications in Various Fields of Research. *RAD Conf Proc.* 2: 231–236. https://doi.org/10.21175/RadProc.2017.47

Stevanato R, Bertelle M, Fabris S. 2014. Photoprotective characteristics of natural antioxidant polyphenols. *Regul Toxicol Pharmacol.* 69(1). https://doi.org/10.1016/j.yrtph.2014.02.014

Svobodová A, Psotová J, Walterová D. 2003. Natural phenolics in the prevention of UV-induced skin damage. A review. *Biomed Pap Med Fac Univ Palacky Olomouc Czech Repub.* 147(2). https://doi.org/10.5507/bp.2003.019

Temur N, Gundes NS, Korkmaz B, Ozkay ZA, Ocsoy I. 2024. Silver nanoparticles in dentistry. In: *Silver Nanoparticles for Drug Delivery* (pp. 265–288). Elsevier Inc.

Threskeia A, Sandhika W, Rahayu RP. 2023. Effect of turmeric (*Curcuma longa*) extract administration on tumor necrosis factor alpha and type 1 collagen expression in UVB-light radiated BALB/c mice. *J Appl Pharm Sci.* 13(5). https://doi.org/10.7324/JAPS.2023.19358

Torres-Contreras AM, Garcia-Baeza A, Vidal-Limon HR, Balderas-Renteria I, Ramírez-Cabrera MA, Ramirez-Estrada K. 2022. Plant secondary metabolites against skin photodamage: Mexican plants, a potential source of UV-radiation protectant molecules. *Plants.* 11(2), 220. https://doi.org/10.3390/plants11020220

Tülüce Y, Özkol H, Koyuncu I. 2012. Photoprotective effect of flax seed oil (*Linum usitatissimum* L.) against ultraviolet C-induced apoptosis and oxidative stress in rats. *Toxicol Ind Health.* 28(2). https://doi.org/10.1177/0748233711407239

Valverde TM, Soares BNG de S, Nascimento AM do, Andrade ÂL, Sousa LRD, Vieira PM de A, Santos VR, Seibert JB, Almeida TCS de, Rodrigues CF, et al. 2023. Anti-inflammatory, antimicrobial, antioxidant and photoprotective investigation of red propolis extract as sunscreen formulation in Polawax Cream. *Int J Mol Sci.* 24(6). https://doi.org/10.3390/ijms24065112

Verma A, Kushwaha HN, Srivastava AK, Srivastava S, Jamal N, Srivastava K, Ray RS. 2017. Piperine attenuates UV-R induced cell damage in human keratinocytes via NF-kB, Bax/Bcl-2 pathway: An application for photoprotection. *J Photochem Photobiol B.* 172. https://doi.org/10.1016/j.jphotobiol.2017.05.018

Vostálová J, Tinková E, Biedermann D, Kosina P, Ulrichová J, Svobodová AR. 2019. Skin protective activity of silymarin and its flavonolignans. *Molecules.* 24(6). https://doi.org/10.3390/molecules24061022

Wang L, Wan G, Wang G, Zhang M, Li N, Zhang Q, Yan H. 2022. Anthocyanin from *Lycium ruthenicum Murr.* in the Qaidam Basin alleviates ultraviolet-induced apoptosis of human skin fibroblasts by regulating the death receptor pathway. *Clin Cosmet Investig Dermatol.* 15: 2925–2932. https://doi.org/10.2147/CCID.S379486

Widsten P, Tamminen T, Liitiä T. 2020. Natural sunscreens based on nanoparticles of modified kraft lignin (CatLignin) petri. *ACS Omega.* 5: 13438–13446. https://doi.org/10.1021/acsomega.0c01742

Widyarini S. 2006. Protective effect of the isoflavone equol against DNA damage induced by ultraviolet radiation to hairless mouse skin. *J Vet Sci.* 7(3). https://doi.org/10.4142/jvs.2006.7.3.217

Wölfle U, Seelinger G, Bauer G, Meinke MC, Lademann J, Schempp CM. 2014. Reactive molecule species and antioxidative mechanisms in normal skin and skin aging. *Skin Pharmacol Physiol.* 27(6). https://doi.org/10.1159/000360092

Wu MS, Sun DS, Lin YC, Cheng CL, Hung SC, Chen PK, Chang HH. 2015. Nanodiamonds protect skin from ultraviolet B-induced damage in mice. *J Nanobiotechnol.* 13: 1–12. https://doi.org/10.1186/s12951-015-0127-4

Wu NL, Fang JY, Chen M, Wu CJ, Huang CC, Hung CF. 2011. Chrysin protects epidermal keratinocytes from UVA- and UVB-induced damage. *J Agric Food Chem.* 59(15). https://doi.org/10.1021/jf200931t

Xia L, Lenaghan SC, Zhang M, Zhang Z, Li Q. 2010. Naturally occurring nanoparticles from English ivy: An alternative to metal-based nanoparticles for UV protection. *J Nanobiotechnol.* 8: 12. https://doi.org/10.1186/1477-3155-8-12.

Young AR, Chadwick CA, Harrison GI, Nikaido O, Ramsden J, Potten CS. 1998. The similarity of action spectra for thymine dimers in human epidermis and erythema suggests that DNA is the chromophore for erythema. *J Invest Dermatol.* 111(6). https://doi.org/10.1046/j.1523-1747.1998.00436.x

Yu Q, Shen Y, Gan Y, Zheng L. 2019. Protective effect of total flavonoids from boxthorn leaf against UVB irradiation-induced skin injury. *Trop J Pharm Res.* 18(9): 1943–1947. https://doi.org/10.4314/tjpr.v18i9.5

Yusuf N, Irby C, Katiyar SK, Elmets CA. 2007. Photoprotective effects of green tea polyphenols. *Photodermatol Photoimmunol Photomed.* 23(1): 48–56. https://doi.org/10.1111/j.1600-0781.2007.00260.x

Zajdela F, Latarjet R. 1978. Inhibition of skin carcinogenesis in vivo by caffeine and other agents. *Natl Cancer Inst Monogr.* NO 50.

Zheng XQ, Jin J, Chen H, Du YY, Ye JH, Lu JL, Lin C, Dong JJ, Sun QL, Wu LY, Liang YR. 2008. Effect of ultraviolet B irradiation on accumulation of catechins in tea (*Camellia sinensis* (L) O. Kuntze). *Afr J Biotechnol.* 7(18).

Ziegler J, Facchini PJ. 2008. Alkaloid biosynthesis: Metabolism and trafficking. *Annu Rev Plant Biol.* 59. https://doi.org/10.1146/annurev.arplant.59.032607.092730.

9 Natural Dermal Preparation Used in the Therapy of Vitiligo

Sumitaksha Banerjee, Debankur Dey, Dwija Poddar, Harendra Kumar, Poulami Roy and Dattatreya Mukherjee

9.1 INTRODUCTION

Vitiligo (Leukoderma) is a chronic skin condition characterized by the development of white patches over the skin due to the loss of pigment-producing cells known as melanocytes (Ezzedine et al., 2015). Melanocytes produce melanin, the pigment that disperses colour to the skin, hair, and eyes. When these cells are destroyed or they stop functioning, depigmentation occurs, forming white or light-coloured patches over the skin. While vitiligo itself is not harmful or contagious, it can have significant social and psychological impacts on affected individuals due to its noticeable appearance. One piece of literature has shown that "vitiligo" comes from the word "vitilus," meaning calf (Abu Tahir et al., 2010). Another literature has shown that the word comes from the word "vitium," meaning defect or blemish (Gauthier and Benzekri, 2009). There have been several types of treatments of vitiligo, including steroid creams, psoralen and UV A light, narrow band UV B light, surgical treatments, etc. (Abu Tahir et al., 2010). Refractory vitiligo is always a cause of concern (Wang et al., 2024). As per Bellei et al. (2022), "The pathophysiology is complex and involves genetic predisposition, environmental factors, oxidative stress, intrinsic metabolic dysfunctions, and abnormal inflammatory/ immune responses." The psychological factors in a patient with vitiligo are shown in Figure 9.1 (Ezzedine et al., 2021).

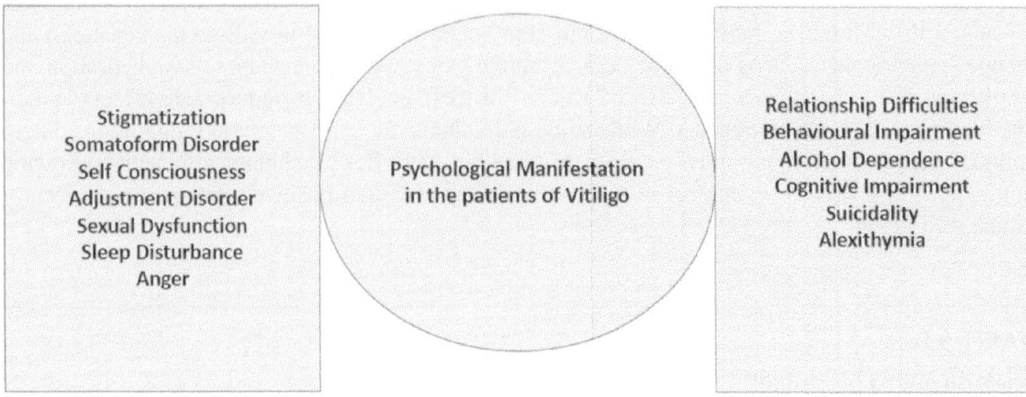

FIGURE 9.1 Psychological manifestations in the patients of vitiligo.

(Adapted with permission under Creative Commons CC-BY 4.0 license from Ref Ezzedine et al. (2021) Copyright 2021 the Authors).

DOI: 10.1201/9781003394846-9

Vitiligo is an autoimmune and genetic disease. In some cases, it is an inherent genetic defect which is getting expressed in melanocytes (Seneschal and Boniface, 2023; Bystryn, 1997). In the paper written by Jean Claude Bystryn (1997), it is mentioned that "Most authors regard the inheritance pattern as autosomal dominant with variable expression and incomplete penetrance. However, some favour a polygenic mode of inheritance. Several HLA abnormalities have been associated with vitiligo, including association with vitiligo, including association with Dr4, B13, BW35 and A30." There can be a relation between body stress factors and vitiligo. In stressful situations, the skin can release a chemical which is toxic to melanocytes. There are a couple of hypotheses present, but the exact aetiology of vitiligo is still unknown, that's why the treatment of vitiligo is very challenging. Classifications of Vitiligo is mentioned in Table 9.1 (Böhm et al., 2022).

Different types of vitiligo (Böhm et al., 2022):

1. Non-Segmental Vitiligo
2. Segmental Vitiligo
3. Mixed Vitiligo
4. Unclassifiable Vitiligo

Vitiligo is diagnosed based on skin discolouration. Skin biopsy is often not prescribed. Wood's light is used to diagnose a fair-skinned person. Different scales are used in the diagnosis procedure, VETF score, vitiligo area scoring index, and vitiligo extent score.

Natural preparations have gained significance in the therapeutic landscape of vitiligo, offering a complementary approach to conventional treatments. The appeal of natural remedies lies in their potential to address various aspects of the condition. Many natural ingredients, such as certain herbs and plant extracts, exhibit anti-inflammatory and immunomodulatory properties. Given that vitiligo often involves an autoimmune response, these natural substances may help regulate immune responses, potentially slowing the progression of the disorder. Additionally, some preparations are believed to stimulate melanin production or support the regeneration of melanocytes, contributing to repigmentation (Szczurko and Boon, 2008).

The antioxidant effects of certain natural preparations also play a crucial role. Oxidative stress has been implicated in the pathogenesis of vitiligo, and antioxidants found in vitamins C and E, polyphenols, and flavonoids may help counteract this stress, protecting melanocytes from damage (Wang, 2010). Moreover, maintaining overall skin health is essential in vitiligo management, and natural preparations such as oils, aloe vera, or herbal extracts can contribute to skin hydration and resilience. Natural remedies can also be valued for their potential to reduce side effects associated with conventional treatments. While some individuals may prefer natural preparations due to cultural beliefs or personal preferences, these remedies can offer psychological benefits, allowing individuals to feel more in control of their skincare regimen and positively impacting their emotional well-being.

TABLE 9.1
Classifications of Vitiligo

Severe Vitiligo	Moderate Vitiligo	Mild Vitiligo
BSA > 30%	BSA 10–30%	BSA < 10%

* BSA: Body Surface Area

Source: From Böhm et al., 2022.

9.2 UNDERSTANDING VITILIGO

9.2.1 Epidemiology

Vitiligo is a skin condition that transcends racial and gender boundaries, affecting individuals from diverse backgrounds. Globally the prevalence of vitiligo ranges from 0.5% to 2.0% but varies geographically (Picardo et al., 2015). While it commonly emerges in adolescence or early adulthood, its onset can occur at any age. The condition's origins are complex and not entirely elucidated, involving a combination of genetic predisposition, autoimmune mechanisms, and environmental factors. Genetics play a crucial role, as evidenced by the increased likelihood in individuals with a family history of vitiligo. The immune system's misguided assault on melanocytes, the pigment-producing cells, characterizes vitiligo as an autoimmune disorder. Environmental factors, such as chemical exposures, stress, and infections, may trigger or exacerbate the condition.

A South Korean study revealed that the prevalence of vitiligo has increased from 0.12% to 0.13% in three years among hospital-admitted patients (Lee et al., 2015). A Chinese study has shown that the prevalence is higher in males than females (Xiaoyan et al., 2013). Vitiligo is also associated with different comorbidities (Dahir and Thomsen, 2018).

9.2.2 Clinical Manifestations

The hallmark symptom of vitiligo is the emergence of depigmented or hypopigmented patches on the skin, creating a distinctive visual contrast. These patches, varying in size and shape, can manifest anywhere on the body, though they commonly appear on areas prone to friction or sun exposure. Notably affected regions include the face, hands, arms, elbows, knees, genitalia, and areas around body orifices. The unpredictable distribution of these patches underscores the unpredictable nature of vitiligo. Furthermore, in addition to its impact on the skin, vitiligo can extend its influence on the hair, resulting in premature greying or whitening. This not only accentuates the visible manifestations of the condition but also highlights the broader implications of vitiligo, which extends beyond a purely dermatological concern to affect aspects of an individual's overall appearance and self-image (Ezzedine et al., 2015).

9.2.3 Classification

Vitiligo exhibits a diverse spectrum of patterns and distributions, leading to the classification of different types of conditions. Focal vitiligo is characterized by scattered, isolated patches of depigmentation on specific areas of the body, while segmental vitiligo typically involves depigmented patches on one side of the body, often following a dermatomal or nerve distribution. These forms are more localized compared to the broader involvement seen in generalized vitiligo, where depigmented patches occur symmetrically on both sides of the body. Generalized vitiligo is the most common type and can affect various body parts, including the face, extremities, and trunk. In contrast, universal vitiligo represents the most extensive form, covering a significant portion of the body, and can result in nearly complete depigmentation. The classification of vitiligo types helps dermatologists tailor treatment approaches, as different patterns may respond differently to therapeutic interventions (Hu et al., 2023).

9.2.4 Pathogenesis

The pathogenesis of vitiligo involves a multifaceted interplay of factors, and while substantial progress has been made in understanding the condition, many aspects remain elusive. One key aspect of vitiligo's mechanism is its classification as an autoimmune disorder. In this context, the immune system, typically responsible for protecting the body against foreign invaders, mistakenly identifies melanocytes, the pigment-producing cells, as threats and launches an immune response against them. This immune assault leads to the destruction of melanocytes, causing depigmentation in affected areas of the skin. Genetic factors contribute significantly to the predisposition and development of vitiligo. Individuals with a family history of the condition are at a higher risk, suggesting a hereditary

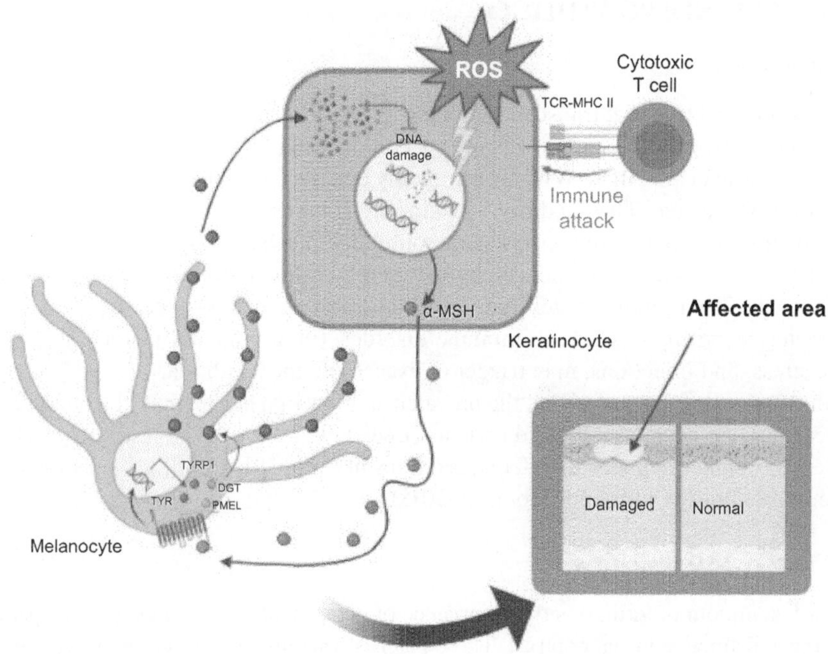

FIGURE 9.2 Pathogenic mechanisms behind vitiligo.

component. Specific genes associated with vitiligo susceptibility have been identified, further supporting the genetic influence on its onset. However, the inheritance pattern is complex and likely involves the interaction of multiple genes, making it challenging to pinpoint precise genetic mechanisms. The relationship between autoimmune responses and genetic predisposition in vitiligo is intricate and may be influenced by environmental factors. External triggers, such as exposure to certain chemicals, stress, or infections, are thought to play a role in initiating or exacerbating the autoimmune response. Additionally, the concept of oxidative stress, where an imbalance between reactive oxygen species and the body's ability to neutralize them occurs, is under investigation as a potential contributor to melanocyte dysfunction and destruction in vitiligo (Frisoli et al., 2020; Katz and Harris, 2021). The image of pathogenic mechanisms behind vitiligo has been highlighted in Figure 9.2.

9.2.5 Impact on Quality of Life

Vitiligo's impact on the quality of life extends beyond its visible manifestations, influencing various aspects of an individual's psychological, emotional, and social well-being. The noticeable depigmented patches on the skin often lead to self-consciousness and negative body image, affecting self-esteem and confidence. The societal emphasis on physical appearance can contribute to stigmatization and feelings of isolation, as individuals with vitiligo may experience societal misconceptions and prejudice. Psychologically, the condition can induce stress, anxiety, and depression, as individuals grapple with the emotional challenges of coping with a chronic, visible skin disorder. The unpredictability of the condition, with patches potentially spreading or changing in appearance over time, adds a layer of uncertainty and emotional strain. Social interactions may be impacted, as individuals with vitiligo may face stares, questions, or insensitive comments, leading to social withdrawal or avoidance of public settings. The emotional toll can be particularly pronounced during adolescence and early adulthood, critical periods for identity formation and social integration. Furthermore, the impact of vitiligo extends beyond the individual to their close relationships and support networks. Family members, friends, and romantic partners may also grapple with understanding and supporting their loved ones with vitiligo, contributing to a collective emotional burden (Vernwal et al., 2016; Morales-Sánchez et al., 2017).

9.2.6 Current Treatment Approaches and Limitations

Current treatment approaches for vitiligo aim to manage the condition and promote repigmentation, but it's important to note that there is no universal cure. Treatment plans are often individualized and based on factors such as the extent of depigmentation, the location of affected areas, and the patient's overall health. Common treatment modalities include topical corticosteroids, calcineurin inhibitors, phototherapy, NB-UVB monotherapy, immunosuppressive systemic therapies, targeted light therapies and surgical interventions (Patel et al., 2012).

Topical corticosteroids are frequently used to reduce inflammation and encourage repigmentation. However, their long-term use may be associated with side effects such as skin thinning and the development of striae. Calcineurin inhibitors, another class of topical medications, are considered for sensitive areas like the face and neck. They work by suppressing the immune response locally but may have limitations in terms of long-term safety. Phototherapy involves exposing the skin to ultraviolet (UV) light, either in the form of narrowband UVB or psoralen plus UVA (PUVA) therapy. These treatments stimulate melanocyte activity and can lead to repigmentation. However, the process requires regular sessions and may not be suitable for all skin types. Furthermore, the risk of UV exposure remains a concern. Surgical interventions, such as autologous melanocyte transplantation and skin grafting, are considered for stable vitiligo, where depigmentation has not progressed for a period. While these approaches can yield favourable results, they may not be suitable for all individuals, and the potential for scarring and colour mismatch exists.

Despite these treatment options, limitations persist in the field of vitiligo management. Response to treatment varies among individuals, and achieving complete repigmentation can be challenging. Additionally, treatments may need to be continued over the long term, and maintenance therapy is often required to sustain results. The unpredictable nature of vitiligo, with depigmentation potentially recurring or spreading, adds complexity to treatment planning. Moreover, the psychosocial impact of vitiligo remains a significant consideration. Even with successful repigmentation, individuals may continue to grapple with the emotional effects of the condition, emphasizing the need for holistic approaches that address both the physical and psychological aspects of vitiligo. Treatment of Vitiligo has been highlighted in Figure 9.3 and therapeutic approaches has been highlighted in Table 9.2. Treatment strategies and surgical treatment strategies are highlighted in Figures 9.4 and 9.5.

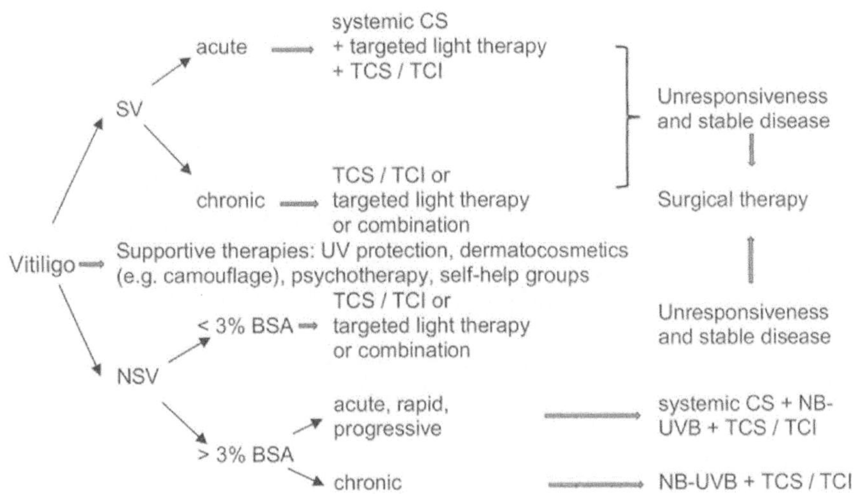

FIGURE 9.3 Treatment of vitiligo.

BSA, *body surface area*; CS, corticosteroids; NB-UVB, *narrowband* UVB; NSV, non-segmental vitiligo; SV, segmental vitiligo; TCI, topical calcineurin inhibitors; TCS, topical corticosteroids. (Reproduced with permission under Creative Common CC-BY 4.0 license from Ref (Böhm et al., 2022), Copyright 2022 the Authors).

TABLE 9.2
Novel Therapeutic Approaches of Vitiligo

Drug Name	Mechanism of Actions	Reference
Afamelanotide	synthetic analogue of alpha-melanocyte stimulating hormone	Böhm et al. (2022) Luger and Böhm (2015)
Ruxolitinib, Tofacitinib	JAK inhibitor	Böhm et al. (2022) Rothstein et al. (2017) Liu et al. (2017) Kim et al. (2018) Joshipura et al. (2018)

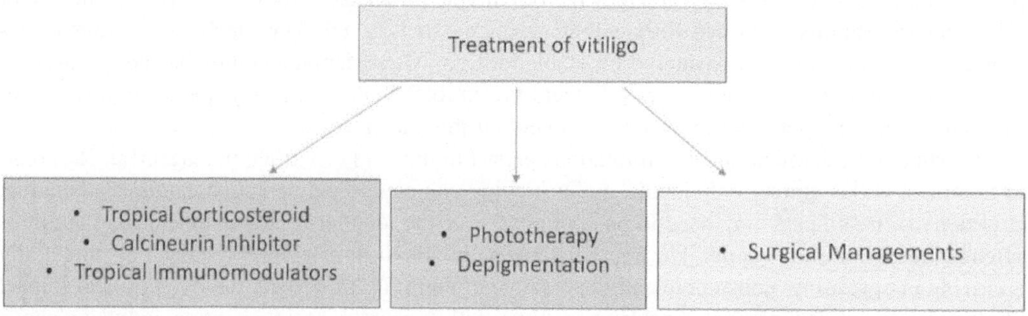

FIGURE 9.4 Treatment strategy of vitiligo.

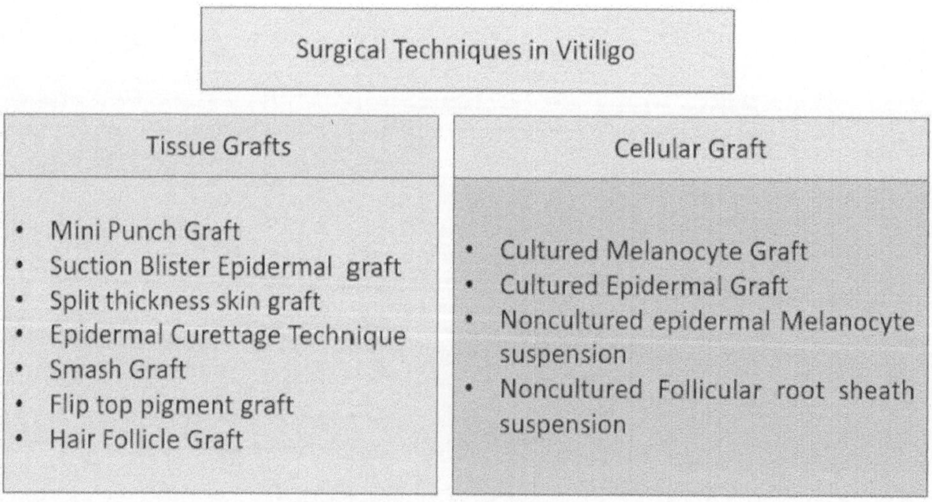

FIGURE 9.5 Surgical strategies in the treatment of vitiligo.

9.3 NATURAL PREPARATIONS FOR VITILIGO THERAPY

Natural preparations for vitiligo therapy have garnered attention for their potential benefits, often serving as complementary or alternative approaches to conventional treatments. While these remedies may not provide a cure, some individuals find them valuable in managing symptoms and promoting overall skin health. Several natural preparations have been explored in the context of vitiligo therapy (as summarized in Table 9.3):

1. **Ginkgo Biloba:** *Ginkgo biloba*, commonly known as the "maidenhair tree," has a rich history of medicinal use, with its extracts being employed for various health conditions. In recent years, *Ginkgo biloba* extracts have been explored for their potential in vitiligo treatment. Administered orally in tablet form, the exact mechanism of action in vitiligo remains unknown, but it is believed to be linked to the herb's anti-inflammatory, immunomodulatory, and antioxidant properties. Studies indicate that *Ginkgo biloba* can effectively control vitiligo activity and induce repigmentation when used alone or in conjunction with conventional therapies such as corticosteroids and phototherapies. However, variations in repigmentation outcomes may be influenced by factors such as genetic differences, *Ginkgo biloba* extract types, treatment duration, and daily dosage. The herb is generally safe and well-tolerated, with doses exceeding 240 mg/day potentially causing mild side effects. Caution is advised for individuals on anticoagulants due to the risk of over-thinning blood (Parsad et al., 2003).
2. **Cucumis Melo (Muskmelon):** *Cucumis melo,* also known as muskmelon, is a plant rich in antioxidants, particularly superoxide dismutase (SOD). Preliminary studies have investigated a topical preparation containing Cucumis melo SOD and catalase for vitiligo treatment (Naini et al., 2012). However, the results, when applied before natural or artificial UV irradiation, did not significantly differ from phototherapy alone. A more promising approach involves a topical formulation containing phenylalanine, Cucumis melo extract, and acetylcysteine, demonstrating safety and efficacy when combined with narrowband UVB target phototherapy. This combination appears to enhance repigmentation in cutaneous lesions (Yuksel et al., 2009).
3. **Khellin**: Khellin, derived from the plant *Ammi visnaga*, has been used historically in herbal medicine. Due to side effects, safer analogues of khellin have been developed for vitiligo treatment, showing good results in combination with UVA phototherapy. Khellin stimulates melanocytes' proliferation and melanogenesis and can be administered both systemically (oral) or topically. The oral intake of khellin in combination with UVA, known as KUVA therapy, is considered safe and clinically comparable to PUVA therapy. Topical khellin applications, combined with UVA or natural UV radiation, have also shown positive results, and recent studies suggest successful use in association with monochromatic excimer light (308 nm) for vitiligo treatment (Carlie et al., 2003; Morliere et al., 1988).
4. **Ayurvedic Medicine**: *Picrorhiza kurroa*, also known as Kutki, is an Ayurvedic herbal product with hepatoprotective properties. Recent studies have explored its antioxidant and immune-modulating activities. Administered orally with methoxsalen photochemotherapy, *Picrorhiza kurroa* has shown improved repigmentation results. Another Ayurvedic herbal product for vitiligo treatment involves the use of anarich (topical concentrated pharmaceutical preparations of plants in the Anacardiaceae family). However, more research is needed to fully understand its efficacy (Bedi et al., 1989).
5. **Polypodium Leucotomos:** *Polypodium leucotomos*, known as Calaguala, is a tropical fern with antioxidant and photoprotective properties. Used as adjuvant therapy for vitiligo patients undergoing phototherapy, studies indicate that combining PUVA therapy

TABLE 9.3
Summary of all the Natural Preparations Used in Vitiligo

Natural Preparations	Key Components	Administration	Mechanism of Action	Clinical Evidence and Considerations
Ginkgo Biloba	Extracts from *Ginkgo biloba*	Oral (tablet form)	Anti-inflammatory, immunomodulatory, antioxidant properties	Effective in controlling vitiligo, potential repigmentation, caution with anticoagulants (Parsad et al., 2003)
Cucumis Melo (Muskmelon)	Superoxide dismutase (SOD), catalase	Topical	Antioxidant-rich, combined with phototherapy for repigmentation	Topical formulation with phenylalanine, *Cucumis melo* extract, and acetylcysteine enhances repigmentation (Naini et al., 2012; Yuksel et al., 2009)
Khellin	Derived from *Ammi visnaga*	Oral, Topical	Stimulates melanocytes proliferation, safe KUVA therapy	Shows good results in combination with UVA phototherapy, systemic and topical applications (Carlie et al., 2003; Morliere et al., 1988)
Ayurvedic Medicine	*Picrorhiza kurroa*, anarich	Oral, Topical	Antioxidant, immune-modulating, improved repigmentation with methoxsalen photochemotherapy	Limited research, *Picrorhiza kurroa* shows promise (Bedi et al., 1989)
Polypodium Leucotomos	Extract from *Calaguala*	Oral	Antioxidant, photoprotective, enhances repigmentation	Combined with PUVA or narrowband UVB for higher repigmentation (Nestor et al., 2014)
Traditional Chinese Medicine	*Psoralen, Angelica Sinensis, Ligusticum wallichii, Tribulus Terrestris*	Oral, Topical	Standard PUVA therapy, "Barresi complex prescription" in Uyghur medicine	Good repigmentation results, long history of herbal product use (Hussain et al., 2016; Huang et al., 2016)
Green Tea Polyphenols	Polyphenols from green tea	Systemic, Topical	Anti-inflammatory, antioxidant, immunomodulatory properties	Counteracts oxidative stress in melanocyte unit (Jeong et al., 2005)
Capsaicin	Active component of chilli peppers	Topical	Anti-inflammatory, antioxidant properties, protects keratinocytes	Potential therapeutic tool, experimental studies support its use (Becatti et al., 2010)
Curcumin	Derived from turmeric	Topical	Antioxidant, anti-inflammatory effects, better repigmentation with narrowband UVB	Tetrahydrocurcumide cream shows slightly better results (Aggarwal and Harikumar, 2009)
Pyrostegia Venusta	Extract from cipó-de-são-joão (*Pyrostegia venusta*)	Topical	Antioxidant, anti-inflammatory, melanogenic properties	Under investigation for vitiligo therapy (Moreira et al., 2015)

with oral *Polypodium leucotomos* results in higher repigmentation compared to phototherapy alone. Similar results have been observed with the combination of narrowband UVB and oral *Polypodium leucotomos* (Nestor et al., 2014).

6. **Traditional Chinese Medicine (TCM) in Vitiligo Therapy:** TCM has a long history of using herbal products, and psoralen plus UVA (PUVA therapy) has been a standard vitiligo treatment. Psoralen, derived from *Psoralea Cordyfolia*, is combined with UVA exposure. Other TCM formulations for vitiligo include *Angelica Sinensis, Ligusticum wallichii, Tribulus Terrestris*, and others. The "Barresi complex prescription," widely used in Uyghur medicine, consists of *Psoralea corylifolia, Plumbago zeylanica, Brassica juncea, Nigella glandulifera*, and *Vernonia anthelmintic*, demonstrating good repigmentation results (Hussain et al., 2016; Huang et al., 2016).
7. **Green Tea Polyphenols:** Green tea polyphenols, known for their anti-inflammatory, antioxidant, and immunomodulatory properties, have been explored for vitiligo treatment. Administered systemically or topically, Green tea polyphenols are suggested to counteract oxidative stress in the melanocyte unit (Jeong et al., 2005).
8. **Capsaicin**: Capsaicin, an active component of chilli peppers, has anti-inflammatory and antioxidant properties, making it a potential therapeutic tool for vitiligo. Experimental studies have indicated its ability to protect keratinocytes from oxidative stress (Becatti et al., 2010).
9. **Curcumin:** Curcumin, derived from turmeric, has a broad range of properties, including antioxidant and anti-inflammatory effects. Tetrahydrocurcumide cream, containing curcumin, has been used in association with narrowband UVB for vitiligo treatment, showing slightly better repigmentation compared to phototherapy alone (Aggarwal and Harikumar, 2009).
10. **Pyrostegia Venusta:** *Pyrostegia venusta*, also known as cipó-de-são-joão, is an herb from the Bignoniaceae family. Although its mechanisms of action are still under investigation, it is considered effective for its antioxidant, anti-inflammatory, and melanogenic properties, making it a candidate for vitiligo therapy (Moreira et al., 2015).

9.4 SAFETY AND EFFICACY OF THE NATURAL PREPARATIONS

The efficacy and safety of natural preparations for vitiligo have been investigated across various clinical trials, revealing diverse outcomes and considerations. Among these, L-phenylalanine has demonstrated moderate evidence of efficacy when used as an adjunct to phototherapy. The studies reported beneficial effects on repigmentation; however, challenges in methodology, such as a lack of randomization and poor control, have been noted. Traditional Chinese Medicinal herbs (TCM) have shown weak evidence of usefulness, but limitations in study descriptions, treatment variations, and poor methodological quality underscore the need for caution in interpreting these results (Hussain et al., 2016; Moreira et al., 2015). Plants, including photosensitizing agents like *Picorrhiza kurroa* and *Polypodium leucotomos*, exhibited weak evidence when used in conjunction with phototherapy. Notably, *Ginkgo biloba* demonstrated moderate evidence of efficacy when administered alone. The trials, however, suffered from small sample sizes, inconsistent outcome measures, and poor study designs, warranting careful consideration of these findings. Vitamins, namely vitamin E and a combination of cobalamin and folic acid, displayed varied efficacy. Vitamin E, as an adjunct to phototherapy, showed weak evidence, while cobalamin and folic acid did not present a significant advantage over phototherapy alone. The limitations in statistical reporting, small sample sizes, and inconsistent outcome measures pose challenges in drawing definitive conclusions. On the safety front, adverse events reported across studies were generally minor, including erythema, pruritus, nausea, headaches, and mild gastrointestinal complaints. However, limited reporting and small sample sizes hindered a comprehensive understanding of the safety profiles. It is crucial to recognize that the heterogeneity in interventions, dosages, and outcome measures across studies makes it challenging to conduct a comprehensive meta-analysis. Recent clinical studies have been highlighted in Table 9.4.

TABLE 9.4
Clinical Studies

Study Name	Study Design	Sample Size (analyzed)	Main Significant Findings	Adverse Events
Szczurko et al. (2011)	Open labelled pilot trial, uncontrolled	G.biloba = 11	Total VASI score improved by +0.5 from baseline to w12	Nil
Parsad et al. (2003)	Double-blinded trial, placebo-controlled	G.biloba = 26(25); PLB = 26 (22)	Significant cessation of active progression of depigmentation in G.biloba arm	Well tolerated (2 = mild nausea, G.biloba arm)
Fenniche et al. (2018)	Open labelled trial, uncontrolled	Top.khellin + MEL = 20	9/20 pat.: > 75% repigmentation; 5/20 pat.: 50–75% repigmentation; 3/20 pat.: 25–50% repigmentation; 3/20 pat.: < 25% repigmentation	Erythema; transient burning with phototherapy
Saraceno et al. (2009)	Open labelled pilot trial, controlled	Top.khellin + MEL + VitE = 16; MEL + VitE = 16; VitE = 16	Significantly good (51–75%) response in (Top.khellin + MEL + VitE) and (MEL + VitE) compared to VitE; (Top.khellin + MEL + VitE) had significantly more excellent (> 75%) response than VitE	Increased sensitivity to light therapy
Hofer et al. (2001)	Retrospective study, uncontrolled	Oral khellin + UVA = 28	Re-pigmentation (response) of depigmented areas statistically significantly linked to the number of treatments and total cumulative UVA dose	8: Mild nausea
Ghorbanibirgani et al. (2014)	Randomized, double-blinded trial, controlled	Nigella sativa (black cumin) = 26; Fish oil = 26	4m: Significant improvements in VASI score in both arms; 6m: Significant improvement in VASI score in Nigella sativa arm	Insignificant in either arm
Colucci et al. (2015)	Open labelled, controlled	Phyllanthus emblica extract + VitE + carotenoids + topical a/o Phototherapy = 65; Topical a/o Phototherapy = 65	Significantly higher no. of patients with mild repigmentation on head/neck regions and on the trunk in active arm	Well tolerated
Bedi et al. (1989)	Blinded, self-controlled	P.kurroa (Picrolex) = 30	27 patients showed improved (accelerated) response to picrolex	Nil
Middelkamp-Hup et al. (2007)	Randomized, double-blinded trial, controlled	NB-UVB + P.leucotomos = 25; NB-UVB + PLB = 24	Significantly higher improvement in active arm (44% vs 27%) in head/meck	4(PL), 5(PLB): mild GI disturbances; mild itching and dryness
Reyes et al. (2006)	Randomized, double-blinded pilot trial, controlled	PUVA + PL = 10; PUVA + PLB (starch) = 9	Percentage of subjects with skin re-pigmentation > 50% was significantly higher in the arm of PUVA+PL than control	No significant phototoxic side effects or carcinogenesis were noted in the study from therapy

TABLE 9.4 *(Continued)*
Clinical Studies

Study Name	Study Design	Sample Size (analyzed)	Main Significant Findings	Adverse Events
Gupta et al. (1985)	Open labelled trial, uncontrolled	Canthaxanthin	Self-reported: 54% – not satisfied, 35% – satisfied, 10% – very satisfied; Physician reported: 42% – not satisfied, 37% 35% – satisfied, 21% – very satisfied	Orange palms, red stools

9.5 INTEGRATIVE APPROACHES

Phototherapy has historically served as a cornerstone in the therapeutic arsenal against vitiligo, encompassing a spectrum of modalities such as broadband UVB (BB-UVB), narrowband UVB (NB-UVB), excimer light, excimer laser, and psoralen plus ultraviolet A (PUVA) therapy. Within the stable phase of vitiligo, the pronounced repigmentation or "biostimulation" attributed to ultraviolet radiation potentially stems from the activation of functional melanocytes in perilesional integument or nascent melanocytes within hair follicles. Notably, NB-UVB also elicits the release of melanocyte-related growth factors from keratinocytes, thereby augmenting the amelioration of vitiligo. The targeted emission of monochromatic excimer light at 308 nm demonstrates efficacy in repigmenting vitiligo lesions. Comparative analyses underscore the superior efficacy of excimer lasers vis-à-vis NB-UVB, precipitating swifter and more substantial repigmentation alongside abbreviated treatment durations, thereby enhancing patient adherence (Seneschal et al., 2023). While excimer-based interventions bear promise for localized disease manifestations, their utility may be hindered by protracted treatment sessions, posing logistical challenges for both patients and healthcare providers. Despite its efficacy, excimer therapy remains underutilized owing to its exorbitant cost and limited adoption.

Conversely, photochemotherapy, or PUVA, entails the utilization of 8-methoxypsoralen as a photosensitizer concomitant with UV radiation exposure, primarily reserved for patients with generalized vitiligo. Topical PUVA emerges as a viable alternative for localized disease manifestations. Adjunctive to phototherapeutic interventions, systemic steroids function as immunomodulators in stabilizing vitiligo, with immunosuppressants like methotrexate and cyclosporine demonstrating efficacy. Notably, topical corticosteroids and calcineurin inhibitors, leveraging their immune-modulatory properties, are often concomitantly employed in the early disease phase (Esmat et al., 2017).

Expanding the therapeutic repertoire, transcutaneous drug delivery, and repigmentation facilitated by fractional CO_2 lasers have emerged as adjunctive strategies. Mechanistic insights into melanocyte dysfunction and destruction have engendered multifaceted approaches, encompassing genetic, autocytotoxic (oxidative stress), autoimmune, autoinflammatory, neuronal (neural), apoptotic, and adhesion disorder mechanisms (Grochocka et al., 2023). Surgical interventions, particularly tissue grafting methodologies like suction blister epidermal grafting (SBEG) and mini-punch grafting (MPG), as well as cellular grafting modalities such as epidermal cell suspension (ECS) and follicular cell suspension (FCS), have garnered attention. SBEG, leveraging negative pressure to induce blister formation, yields commendable colour matching and aesthetic outcomes, whereas MPG, characterized by puncture followed by graft transference, stands out as a rapid and efficacious technique for enhancing pigmentation (Seneschal et al., 2023).

Clinical assessment of disease activity often relies on longitudinal follow-up, incorporating methodologies like the Vitiligo Disease Activity Score (VDAS). Nonetheless, the efficacy of therapeutic interventions hinges on myriad factors, spanning from genetic predisposition to procedural proficiency. Synergistic combinations, such as phototherapy with oral antioxidants like polypodium leucotomos and gliadin-protected SuperOxide Dismutase (SOD), or systemic immunosuppressive therapies with phototherapy, exemplify therapeutic paradigms conducive to enhanced repigmentation. Post-Repigmentation, relapse rates approximate 44% within the first year post-treatment cessation, with biweekly application of 0.1% tacrolimus ointment emerging as a prophylactic measure against relapse, outperforming placebo (van Geel et al., 2023).

9.6 FUTURE DIRECTIONS

The landscape of treatment modalities evaluated in randomized controlled trials (RCTs) for vitiligo has expanded to encompass 84 interventions and continues to burgeon (Eleftheriadou et al., 2013). Amidst the profusion of interventions scrutinized, the paucity of robust evidence delineating their efficacy complicates the identification of a definitive or optimal treatment modality. A pressing imperative thus emerges for comprehensive evaluation of extant therapies, notably topical corticosteroids and calcineurin inhibitors, which stand as mainstays in the current armamentarium. Presently, topical corticosteroids reign supreme for managing localized vitiligo, while judicious employment of low-dose systemic corticosteroids serves to mitigate the onslaught of rapidly progressive disease. Tacrolimus and pimecrolimus, topical calcineurin inhibitors, and supersede corticosteroids in scenarios necessitating intervention on facial or high-risk areas vulnerable to cutaneous atrophy. Oral cyclosporine, exerting its immunomodulatory prowess through T cell activation constriction via IL-2 inhibition, assumes utility in the therapeutic schema.

Targeting oxidative stress constitutes a pivotal therapeutic endeavour, aspiring to modulate the intricate network orchestrating reactive oxygen species (ROS) scavenging within the cellular milieu. Pseudocatalase, manifesting the capacity to catalyze the conversion of hydrogen peroxide (H_2O_2) to oxygen (O_2) and water (H_2O) at an accelerated rate relative to catalase, aims to redress catalase deficiency in vitiligo-afflicted dermal terrain (Bellei et al., 2022). Moreover, ancillary medical interventions, psychological support interventions, and camouflage strategies merit scrutiny yet await meticulous evaluation within the realm of RCTs. Clinical trials probing interventions targeting interleukin-15 (IL-15) or its receptor CD122, aimed at impeding the generation and sustenance of cutaneous resident memory T cells in murine models of vitiligo, are underway. Inhibiting the disease's inception looms as a pivotal axis for future therapeutic endeavours in vitiligo.

Advancements in research have underscored a constellation of cytokines and pathways underpinning vitiligo pathogenesis. To engender optimal therapeutic outcomes, an integrative therapeutic paradigm targeting multifarious pathogenic cascades emerges as a harbinger of promise. Janus kinase (JAK)/signal transducer and activator of transcription (STAT) inhibitors, by abrogating interferon-gamma signalling pathways, potentiate Hedgehog and Wnt signalling cascades pivotal for epidermal pigmentation, thereby fostering melanocyte migration, proliferation, and differentiation (Wang et al., 2024). Ongoing research interrogating molecular and cellular alterations concomitant with vitiligo holds promise in addressing the unmet exigency for refractory vitiligo management. Furthermore, to potentiate repigmentation efficacy, synergistic amalgamation of topical and/or oral immunomodulators with strategies fostering melanocyte stem cell differentiation and proliferation in vitiligo lesions, particularly those domiciled in acral regions or areas beset by poliosis, assumes paramount significance [30]. Prior to the advent of novel monotherapeutic agents, the judicious formulation of rational combinatorial regimens predicated upon existing pharmacotherapeutic armamentarium stood as the vanguard of efficacious vitiligo management.

9.7 ACKNOWLEDGEMENT

We would like to extend our gratitude to Dr. Novonil Dev for his extensive support in the manuscript preparation.

REFERENCES

Abu Tahir, M., Pramod, K., Ansari, S.H., Ali, J. 2010. Current remedies for vitiligo. *Autoimmun Rev.* 9(7):516–520.

Aggarwal, B.B., Harikumar, K.B. 2009. Potential therapeutic effects of curcumin, the anti – inflammatory agent, against neurodegenerative, cardiovascular, pulmonary, metabolic, autoimmune and neoplastic diseases. *Int J Biochem Cell Biol.* 41(1):40–59.

Becatti, M., Prignano, F., Fiorillo, C., et al. 2010. The involvement of Smac/DIABLO, p53, NF – kB, and MAPK pathways in apoptosis of keratinocytes from perilesional vitiligo skin: Protective effects of curcumin and capsaicin. *Antioxid Redox Signal.* 13(9):1309–1321.

Bedi, K.L., Zutshi, U., Chopra, C.L., Amla, V. 1989. Picrorhiza kurroa, an ayurvedic herb, may potentiate photochemotherapy in vitiligo. *J Ethnopharmacol.* 27(3):347–352.

Bellei, B., Papaccio, F., Picardo, M. 2022. Regenerative medicine-based treatment for vitiligo: An overview. *Biomedicines.* 10(11):2744.

Böhm, M., Schunter, J.A., Fritz, K., Salavastru, C., Dargatz, S., Augustin, M., Tanew, A. 2022. S1 guideline: Diagnosis and therapy of vitiligo. *J Dtsch Dermatol Ges.* 20(3):365–378.

Bystryn, J.C. 1997. Immune mechanisms in vitiligo. *Clin Dermatol.* 15(6):853–861.

Carlie, G., Ntusi, N.B., Hulley, P.A., et al. 2003. KUVA (khellin plus ultraviolet A) stimulates proliferation and melanogenesis in normal human melanocytes and melanoma cells in vitro. *Br J Dermatol.* 149(4):707–717.

Colucci, R, Dragoni, F., Conti, R., Pisaneschi, L., Lazzeri, L., Moretti, S. 2015. Evaluation of an oral supplement containing Phyllanthus emblica fruit extracts, vitamin E, and carotenoids in vitiligo treatment. *Dermatol Ther.* 28(1):17–21.

Dahir, A.M., Thomsen, S.F. 2018. Comorbidities in vitiligo: Comprehensive review. *Int J Dermatol.* 57(10):1157–1164.

Eleftheriadou, V. 2013. Future horizons in vitiligo research: Focusing on the recommendations of the Cochrane systematic review 'Interventions for vitiligo' 2010. *Br J Dermatol.* 169(Suppl 3):67–70.

Esmat, S., Hegazy, R.A., Shalaby, S., Chu-Sung Hu, S., Lan, Cheng-Che E. 2017. Phototherapy and combination therapies for vitiligo. *Dermatol Clin.* 35(2):171–192.

Ezzedine, K., Eleftheriadou, V., Jones, H., Bibeau, K., Kuo, F.I., Sturm, D., Pandya, A.G. 2021. Psychosocial effects of vitiligo: A systematic literature review. *Am J Clin Dermatol.* 22(6):757–774.

Ezzedine, K., Eleftheriadou, V., Whitton, M., van Geel, N. 2015. Vitiligo. *Lancet* 386:74–84.

Fenniche, S., Zaouak, A., Tanfous, A.B., Jrad, M., Hammami, H. 2018. Successful treatment of refractory vitiligo with a combination of khellin and 308-nm excimer lamp: An open-label, 1-year prospective study. *Dermatol Ther* (Heidelb). 8(1):127–135.

Frączek, A., Kasprowicz-Furmańczyk, M., Placek, W., Owczarczyk-Saczonek, A. 2022. Surgical treatment of vitiligo. *Int J Environ Res Public Health.* 19(8):4812.

Frisoli, M.L., Essien, K., Harris, J.E. 2020. Vitiligo: Mechanisms of pathogenesis and treatment. *Annu Rev Immunol.* 38:621–648.

Gauthier, Y., Benzekri, L. 2009. Historical aspects. In Picardo, Mauro; Taïeb, Alain (eds.). *Vitiligo* (Online-Ausg ed.). Berlin: Springer. ISBN 978-3-540-69360-4.

Ghorbanibirgani, A., Khalili, A., Rokhafrooz, D. 2014. Comparing nigella sativa oil and fish oil in treatment of vitiligo. *Iran Red Crescent Med J.* 16(6):e4515.

Grochocka, M., Wełniak, A., Białczyk, A., Marek-Jozefowicz, L., Tadrowski, T., Czajkowski, R. 2023. Management of stable vitiligo – A review of the surgical approach. *J Clin Med.* 12:1984.

Gupta, A.K., Haberman, H.F., Pawlowski, D., Shulman, G., Menon, I.A. 1985. Canthaxanthin. *Int J Dermatol.* 24(1):528–532.

Hofer, A., Kerl, H., Wolf, P. 2001, Long-term results in the treatment of vitiligo with oral khellin plus UVA. *Eur J Dermatol.* 11(3):225–922.

Hu, Z., Wang, T. 2023. Beyond skin white spots: Vitiligo and associated comorbidities. *Front Med (Lausanne).* 10:1072837.

Huang, X., Ishikawa, M., Mansur, A., et al. 2016. The effects of bairesi complex prescription (a uyghur medicine prescription) and its five crude herbal extracts on melanogenesis in G-361 cells. *Evid Based Complement Alternat Med.* 2016:8415359.

Hussain I., Hussain, N., Manan, A., et al. 2016. Fabrication of anti-vitiligo ointment containing Psoralea corylifolia: In vitro and in vivo characterization. *Drug Des Devel Ther.* 10:3805–3816.

Jeong, Y.M., Choi, Y.G., Kim, D.S., et al. 2005. Cytoprotective effect of green tea extract and quercetin against hydrogen peroxide-induced oxidative stress. *Arch Pharm Res.* 28(11):1251–1256.

Joshipura, D., Alomran, A., Zancanaro, P., Rosmarin, D. 2018. Treatment of vitiligo with the topical Janus kinase inhibitor ruxolitinib: A 32-week open-label extension study with optional narrow-band ultraviolet B. *J Am Acad Dermatol* 78:1205–1207.

Katz, E.L., Harris, J.E. 2021. Translational research in Vitiligo. *Front Immunol.* 12:624517.

Kim, S.R., Heaton, H., Liu, L.Y., King, B.A. 2018. Rapid repigmentation of vitiligo using tofacitinib plus low-dose, narrowband UV-B phototherapy. *JAMA Dermatol.* 154:370–371.

Lee, H., Lee, M.H., Lee, D.Y., et al. 2015. Prevalence of vitiligo and associated comorbidities in Korea. *Yonsei Med J.* 56(3):719–725.

Liu, L.Y., Strassner, J.P., Refat, M.A., Harris, J.E., et al. 2017. Repigmentation in vitiligo using the Janus kinase inhibitor tofacitinib may require concomitant light exposure. *J Am Acad Dermatol.* 77:675–682.

Luger, T.A., Böhm, M. 2015. An α-MSH analog in erythropoietic protoporphyria. *J Invest Dermatol.* 135:929–931.

Middelkamp-Hup, M.A., Bos, J.D., Rius-Diaz, F., Gonzalez, S., Westerhof, W. 2007. Treatment of vitiligo vulgaris with narrow-band UVB and oral Polypodium leucotomos extract: A randomized double-blind placebo-controlled study. *J Eur Acad Dermatol Venereol.* 21(7):942–950.

Morales-Sánchez, M.A., Vargas-Salinas, M., Peralta-Pedrero, M.L., Olguín-García, M.G., Jurado-Santa Cruz, F. 2017. Impact of vitiligo on quality of life. *Actas Dermosifiliogr.* 108(7):637–642.

Moreira, C.G., Carrenho, L.Z.B., Pawloski, P.L., et al. 2015. Pre-clinical evidences of Pyrostegia venusta in the treatment of vitiligo. *J Ethnopharmacol.* 168:315–325.

Morliere, P., Hönigsmann, H., Averbeck, D., et al. 1988. Phototherapeutic, photobiologic, and photosensitizing properties of khellin. *J Invest Dermatol.* 90(5):720–724.

Naini, F.F., Shooshtari, A.V., Ebrahimi, B., et al. 2012. The effect of pseudocatalase/superoxide dismutase in the treatment of vitiligo: A pilot study. *J Res Pharm Pract.* 1(2):77–80.

Nestor, M., Bucay, V., Callender, V., et al. 2014. Polypodium leucotomos as an adjunct treatment of pigmentary disorders. *J Clin Aesthet Dermatol.* 7(3):13–17.

Parsad, D., Pandhi, R., Juneja, A. 2003. Effectiveness of oral Ginkgo biloba in treating limited, slowly spreading vitiligo. *Clin Exp Dermatol.* 28(3):285–287.

Patel, N.S., Paghdal, K.V., Cohen, G.F. 2012. Advanced treatment modalities for vitiligo. *Dermatol Surg.* 38(3):381–391.

Picardo, M., Dell'Anna, M.L., Ezzedine, K., Hamzavi, I., Harris, J.E., Parsad, D., et al. 2015. Vitiligo. *Nat Rev Dis Primers.* 1:15011.

Reyes, E., Jaén, P., de las Heras, E., Carrión, F., Alvarez-Mon, M., de Eusebio, E., Alvare, M., Cuevas, J., González, S., Villarrubia, V.G. 2006. Systemic immunomodulatory effects of Polypodium leucotomos as an adjuvant to PUVA therapy in generalized vitiligo: A pilot study. *J Dermatol Sci.* 41(3):213–216.

Rothstein, B., Joshipura, D., Saraiya, A., Abdat, R., et al. 2017. Treatment of vitiligo with the topical Janus kinase inhibitor ruxolitinib. *J Am Acad Dermatol.* 76:1054–1060.

Saraceno, R., Nisticò, S.P., Capriotti, E., Chimenti, S. 2009. Monochromatic excimer light 308 nm in monotherapy and combined with topical khellin 4% in the treatment of vitiligo: A controlled study. *Dermatol Ther.* 22(4):391–394.

Seneschal, J., Boniface, K. 2023. Vitiligo: Current therapies and future treatments. *Dermatol Pract Concept.* 13(4)S2:e2023313S.

Seneschal, J., Speeckaert, R., Taïeb, A., Wolkerstorfer, A., Passeron, T., Pandya, A.G., Lim, H.W., Ezzedine, K., Zhou, Y., Xiang, F., Thng, S., Tanemura, et al. 2023. Worldwide expert recommendations for the diagnosis and management of vitiligo: Position statement from the international Vitiligo Task Force-Part 2: Specific treatment recommendations. *J Eur Acad Dermatol Venereol.* 37(11):2185–2195.

Szczurko, O., Boon, H.S. 2008. A systematic review of natural health product treatment for vitiligo. *BMC Dermatol.* 22(8):2.

Szczurko, O., Shear, N., Taddio, A., Boon, H. 2011. Ginkgo biloba for the treatment of vitilgo vulgaris: An open label pilot clinical trial. *BMC Complement Altern Med.* 11:21.

van Geel, N., Speeckaert, R., Taïeb, A., Ezzedine, K., Lim, H.W., Pandya, A.G., Passeron, T., Wolkerstorfer, A., Abdallah, M., Alomar, A., Bae, J.M., Bekkenk, M., Benzekri, L., Böhm, M., Eleftheriadou, V., Esmat, S., et al. 2023. Worldwide expert recommendations for the diagnosis and management of vitiligo: Position statement from the International Vitiligo Task Force Part 1: Towards a new management algorithm. *J Eur Acad Dermatol Venereol.* 37(11):2173–2184.

Vernwal, D., Sharma, D., Agrawal, S., Sharma, D., Sushil, C., Vijayvergia, D. 2016. A study to know correlation among DE-pigmentation of body areas and sex of vitiligo patients with their self esteem and impact of vitiligo on quality of life. *Int J Contemp Med Res.* 3(8):2194–2197.

Wang, T. 2010. Mechanism of oxidative stress and antioxidant therapy in vitiligo. *Int J Dermatol Venereol.* 36:101–103.

Wang, X., Wu, W., Chen, J., Li, C. Li, S. 2024. Management of the refractory vitiligo patient: Current therapeutic strategies and future options. *Front Immunol.* 14:1294919.

Xiaoyan, W., Juan, D., Tinglin, W., Cheng, Z., Yiwei, S., Xiaolan, D., . . . Zhang, J. 2013. Prevalence and clinical profile of vitiligo in China: A community-based study in six cities. *Acta dermato-venereologica* 93(1).

Yuksel, E.P., Aydin, F., Senturk, N., et al. 2009. Comparison of the efficacy of narrow band ultraviolet B and narrow band ultraviolet B plus topical catalase-superoxide dismutase treatment in vitiligo patients. *Eur J Dermatol.* 19(4):341–344.

10 The Action Mechanisms of Polyphenols in the Prevention and Treatment of Skin Diseases

Jülide Secerli, Şevval Çelikten, Onur Erdem and Merve Güdül Bacanlı

10.1 INTRODUCTION

Skin is the first line of protection against illness. Therefore, agents that disrupt the integrity of the skin can also damage general health (Ndiaye et al., 2011). Skin diseases and cancer occur as a result of frequent skin exposure to physical, biological and chemical genotoxic agents (Oyetakinwhite et al., 2012). For example, ultraviolet (UV) radiation, a physical agent, is considered to be the main reason for early skin ageing and carcinogenesis, causing a range of skin disorders, including non-melanoma and melanoma skin cancers. UV radiation also directly causes DNA damage, immunosuppression, and oxidative stress (Ndiaye et al., 2011; Nichols and Katiyar, 2010).

Ultraviolet (UV) radiation occurs in three different forms: UVA, UVB, and UVC. With its longest wavelength, UVA radiation is associated with many skin conditions, including skin ageing and skin carcinogenesis. Skin cancer is thought to occur mostly as a result of UVB radiation. UVC, which has the shortest wavelength, can cause DNA damage, leading to mutagenesis. For these reasons, UV radiation can cause or worsen skin ageing such as wrinkles, dryness, and discolouration (Ndiaye et al., 2011). There are various methods for the treatment and prevention of UV-induced skin disorders. These approaches are generally seen as (I) physical and chemical protection (preventing the interaction of UV with skin components), (II) biological protection (reducing UV-induced damage), or (III) repair and regeneration (reversing existing symptoms) (Kostyuk et al., 2018). Besides oxidative stress, inflammation, DNA damage, deregulation of cellular signalling pathways, immunosuppression, and exposure to UV radiation from the sun are the main causes of skin cancer (Afaq and Katiyar, 2011).

The process of accumulating harmful alterations in cells and tissues with age is known as ageing. There are many studies and theories on mechanisms of aging, and most of these are concord on oxidative stress theory. Oxidative stress-induced free radicals accumulate the more damage there is, thus the ageing process and the repair mechanisms cannot function correctly (Pandey and Rizvi, 2009).

Skin ageing is characterized by wrinkling, loss of elasticity and increased roughness due to loss of hydration, hyaluronic acid, collagen, and elastin. Throughout the ageing process, cells retain the capacity to form tropoelastin, the precursor monomer of elastin, while the cross-linking process of extracellular tropoelastin decreases with age (Chowdhury et al., 2021). Photooxidative damage, which is brought on by very reactive free radicals that can harm DNA and proteins, is another factor that causes ageing. Additionally, free radicals activate matrix metalloproteinases (MMPs), which can break down the skin's collagen and elastin (Chowdhury et al., 2021).

UV exposure affects the transcription factor pathways activator protein-1 (AP-1) and nuclear factor-kappa B (NF-κB) and binds inhibitory κB (IκB) to NF-κB, forming an inactive complex. In addition, UV exposure induces inflammation by inducing IκB proliferation and the nuclear translocation of NF-κB and the upregulation of many cytokines, including cyclooxygenase (COX)-2,

 DOI: 10.1201/9781003394846-10

interleukins (ILs), and inducible nitric oxide synthase (i-NOS). Furthermore, MMP-1 is formed when NF-κB activation is stimulated, and this causes the dermal collagen structure to be disrupted (Lin et al., 2019). The action mechanism of UV radiation in skin is given in Figure 10.1.

The most prevalent cells in the epidermis, keratinocytes, as they develop, collect melanin pigments and effectively prevent UV radiation from penetrating the skin. However, melanin synthesis does not occur in keratinocytes, but in melanocytes occurring in both the dermis and epidermis (D'Orazio et al., 2013). Melanogenesis is a process of melanin synthesis that protects against UV exposure and prevents skin damage and skin cancer development. Tyrosinase is an enzyme involved in the enzymatic processes that produce melanin, which is responsible for catalyzing the rate-limiting phases of melanogenesis in melanocytes and melanoma cells. Up-regulation of tyrosinase is thought to increase melanin formation. Cyclic AMP (cAMP), p38 mitogen-activated protein kinase (MAPK) and phosphatidylinositol 3-kinase (PI3K) are also involved in the signal transduction pathways of melanogenesis (Jiang et al., 2011).

In 2020, it is projected that 325,000 new instances of melanoma and about 57,000 deaths will result from malignant melanomas. By 2040, these numbers are predicted to rise to 510,000 new cases and 96,000 deaths (Ashique et al., 2023). In reported cases, melanoma is the least frequent type of skin cancer in humans, however, it ranks 18th globally and seventh in Europe in terms of frequency of occurrence (Pontes et al., 2023).

According to epidemiologic research, sun exposure is the primary cause of skin cancers such as melanoma, squamous cell carcinoma (SCC), and basal cell carcinoma (BCC). Those with fair skin and those who are sun-sensitive have a higher incidence of cancer. While the incidence of SCC is higher with occupational exposure, the incidence of melanoma and BCC is higher with other exposures (Souto et al., 2019).

Since the pathophysiology of melanoma involves apoptosis, angiogenesis, and autophagy, it is important to investigate complementary therapeutic approaches that can reduce the resistance and side effects of treatment (Pourhanifeh et al., 2019). For this reason, there is an increasing interest in polyphenolic compounds, which have antioxidant, chemopreventive, and immunomodulatory effects, especially in the protection and treatment of the skin. Most of these secondary metabolites have various interactions with UV rays. For example, phenylpropanoids and bioflavonoids are synthesized via the phenylpropanoid pathway immediately after UV exposure in plant cells. The OH groups in their structure effectively absorb UVA+UVB without causing additional photochemical

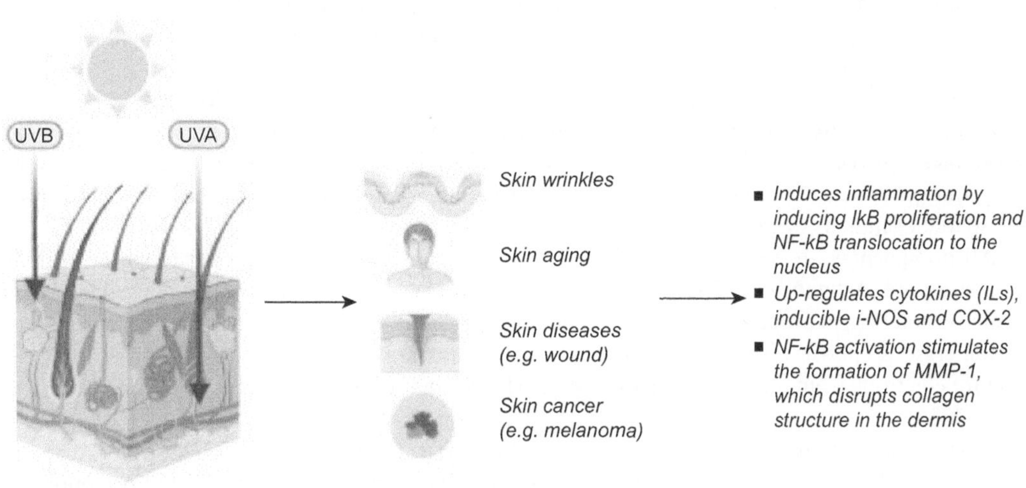

FIGURE 10.1 The action mechanism of UV radiation in skin (created with BioRender.com).

reactions. Therefore, it is considered safer than classical synthetic sunscreens, which are aromatic polyphenols. In addition, other secondary metabolites (stilbenes, catechins, etc.) protect against undesirable reactions (type II photoreactions) caused by reactive oxygen species (ROS) resulting from UV reaction with organic material in the presence of molecular oxygen (Kostyuk et al., 2018).

Bioactive substances such as polyphenols have been associated with numerous health advantages. These phytochemicals are a varied group that function as secondary metabolites of plants and are primarily engaged in defence against harsh weather, UV rays, herbivores, infections, and other environmental stresses. They can also be responsible for the bitter, astringent, colour, taste, odour, and oxidative stability of food (Pandey and Rizvi, 2009; Souto et al., 2019).

Since the positive effects of phytochemical compounds on the skin are known, studies on this subject and interest in these compounds are increasing day by day. Therefore, this chapter is aimed to focus on the role of phenolic compounds in the protection and treatment of the skin and the studies on this subject.

10.2 POLYPHENOLIC COMPOUNDS

Polyphenols are secondary metabolites found in many plant foods such as fruits, vegetables, cereals, and red wine, and contain at least two phenyl cycles or phenol groups bearing hydroxyl groups (W. Li et al., 2023). Mostly, polyphenols are found in foods in complex mixtures. Phenolic chemicals are more prevalent in the outer layers of plants than they are in the inside. Many conditions affect the polyphenolic content of plants, such as harvest time, processing and storage conditions, also environmental factors, for example, stress exposure to the plant, exposure to sun, weather, air, water, and soil conditions, and content (Pandey and Rizvi, 2009).

Polyphenols constitute a large group of compounds containing multiple phenolic hydroxyl groups attached to one or more benzene cycles (Zillich et al., 2015). To present, more than 8000 polyphenols have been identified (Ashwin et al., 2021). All plant-based phenolic compounds derive from the same source, phenylalanine or shikimic acid. They are mainly formed by conjugation with one or more sugar radicals (glucose, rhamnose, galactose, arabinose, xylose, glucuronic acid, etc.) bound to hydroxyl groups. However, they can also be formed by direct linkage between aromatic carbons and sugars without binding to hydroxyl groups. Phenols are also known to form linkages with other compounds such as carboxylic acids, amines, lipids, etc. (Pandey and Rizvi, 2009; Souto et al., 2019). It is divided into groups according to the number of phenols in its structure, the substitution group or the type of connection between phenol units (Zekrumah et al., 2023). The two major classes of polyphenolic compounds are the flavonoids and the non-flavonoids (Figure 10.2). Non-flavonoids include phenolic acids, stilbenes, and lignans (Souto et al., 2019; W. Li et al., 2023).

Plant foods include phenolic acids (ferulic acid, caffeic acid, etc.) more frequently than other substances. They are typically present in cereals (rice bran and wheat bran) as well as fruits and vegetables. Quercetin is a flavonoid that is often found in plants, particularly onions. Furthermore, the primary sources of resveratrol are grapes and peanuts, whereas cereals (such as rye and oats) are the source of lignans (W. Li et al., 2023).

All polyphenols are scavengers of free radicals because they have reducing properties due to their structure (Ashwin et al., 2021). Phenolic compounds have been the focus of studies, due to their numerous biological properties, which include photoprotective, anti-inflammatory, antioxidant, and anti-allergic properties. The effects of these compounds are due to the phenolic group that can capture electrons from reactive nitrogen species (RNS) and reactive oxygen species (ROS) (Pontes et al., 2023).

In vivo and *in vitro* studies support the beneficial effects of polyphenols on chronic diseases including cardiovascular diseases, cancer, neurodegenerative diseases, diabetes, and osteoporosis. However, in order to increase the accuracy of the studies, the daily intake of these compounds should be known, and studies should be supported accordingly. While 100 g of fruit contains up

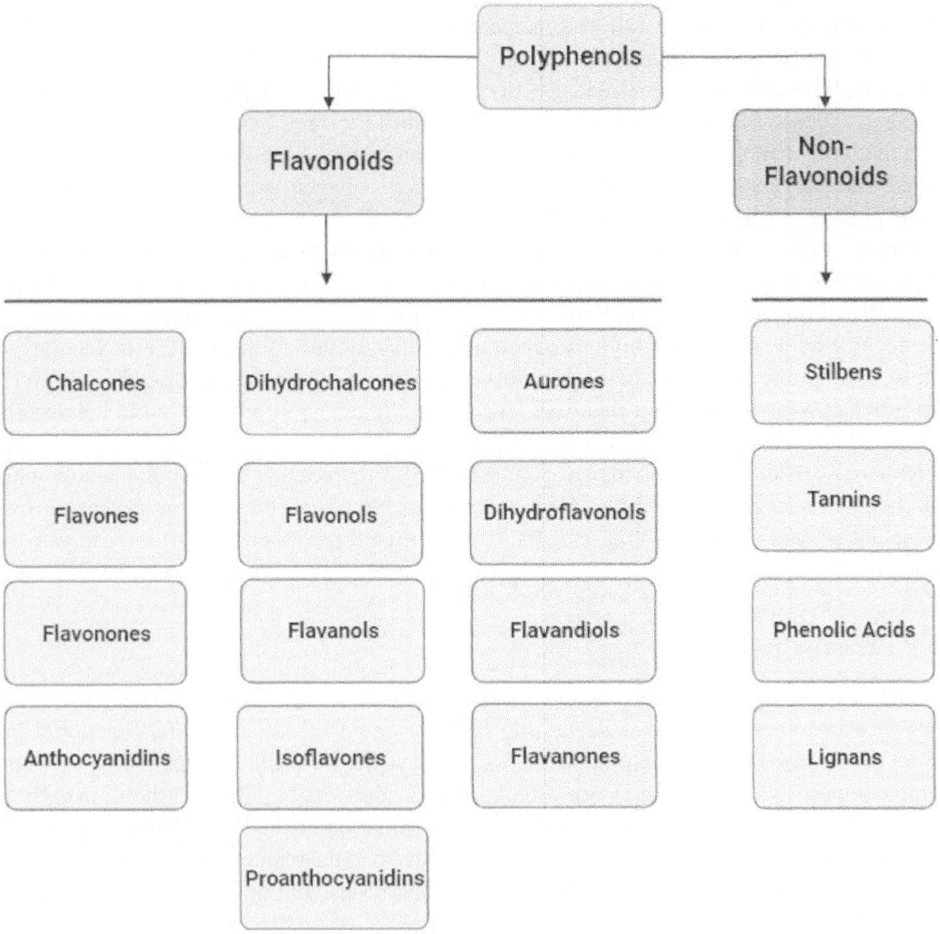

FIGURE 10.2 Classification of polyphenolic compounds (Created with BioRender.com).

to 200–300 mg, a cup of coffee/tea and a glass of red wine contains 100 mg of polyphenols. Daily dietary intake is approximately 1 g, which is higher than other antioxidants (Scalbert et al., 2005).

Since phenolic compounds have anti-inflammatory activity, they are successful in recruiting immune cells to the skin and stopping the growth of secondary infections that arise when the skin barrier is compromised. They also reduce cellular damage caused by free radicals during an allergic reaction due to their antioxidant activity (Singh et al., 2011). Since phenolic compounds have biological activities and low toxicity, they show promise in reducing the causes and effects of ageing, illnesses, damage, and malignancies of the skin (Działo et al., 2016; Nichols and Katiyar, 2010). In addition, these phenolic compounds extracted from plants inhibit proteinase enzyme activity, which is responsible for the rapid degradation of skin proteins (Fares et al., 2023).

Polyphenols are known to show chemoprevention and skin protective activity. Chemoprevention refers to the inhibition, inversion, or delay of the damaging events that occur in skin exposed to UV light. Chemoprevention offers a workable approach to reducing the risk of cancer in this sense, because the use of polyphenols for chemopreventive purposes means that individuals can take precautions against the risk of skin cancer by making changes in their diet and lifestyle (Ichihashi et al., 2003; J. Kim et al., 2001; Nichols and Katiyar, 2010). Also, many studies on polyphenols suggest that their use improves and speeds up wound healing (Amber et al., 2014).

Polyphenol compounds have elastin binding properties and therefore can help the coacervation of tropoelastin secreted by the cell and increase the amount of elastin. In a study, the effects of flavonols on skin were investigated in human dermal fibroblasts (HDF) under UV exposure. For this purpose, HDF cell lines were treated with epigallocatechin 3-gallate (EGCG) and pentagalloyl glucose and exposed to UVA (three times a week, 12 min) for three, seven, and 14 days. The results showed that polyphenols increased elastin accumulation compared to the control group. It was also observed that polyphenol compounds caused increased collagen accumulation in UVA-exposed cells and prevented cell oxidation during UVA exposure (Chowdhury et al., 2021).

In studies, it has been reported that the permeability of polyphenol compounds through the skin varies according to the molecular weight and polarity of the compounds. In a study, the penetration capacity of polyphenols through the skin was investigated using pig skin. It was reported that protocatechuic acid showed the highest penetration capacity due to its smallest molecular weight and most hydrophilic properties. Catechin, resveratrol, rutin, quercetin, and EGCG followed that, in that order. It was also reported that high levels of polyphenols were observed in the dermis and epidermis layers of the skin (Zillich et al., 2013, 2015).

Polyphenols differ in their biological activities due to there being many polyphenols with different chemical structures. When the plant extracts are prepared, they are in a complex form of different polyphenols, thus studying is difficult with mixed polyphenols, as they can affect each other's activities.

10.2.1 Flavonoids

Flavonoids are phytotherapeutic compounds that are abundant in foods, plants, and fruits and have various pharmacological activities depending on their chemical structure and functional groups (Islam et al., 2022). The diphenylpropane (C6-C3-C6) structure of flavonoids, the biggest class of phenolic compounds, is made up of two aromatic rings connected by three carbons, usually forming an oxygenated heterocycle. Based on the level of oxidation of the heterocyclic ring, flavonoids are classified into 13 subclasses: anthocyanidins, isoflavones, flavonones, aurones, chalcones, dihydrochalcones, flavones, flavanols, dihydroflavonols, flavanones, and flavandiols (Souto et al., 2019; Tresserra-Rimbau et al., 2018).

Most of the time, polyphenols are present in food as glycosides, esters, etc. Thus, they cannot be absorbed without modification (D'Archivio et al., 2007). Before being absorbed, these substances must be hydrolyzed by intestinal enzymes or colonic microbiota. Many factors can affect the absorption process, and the site of absorption is one of them. Some polyphenols are absorbed better in the gastrointestinal tract, intestines, or other parts of the digestive tract. Gastric level of absorption is also possible (Gee et al., 1998). After the course of absorption, polyphenol metabolites are found in the blood bound to carrier proteins, especially to albumin (Day and Williamson, 2001; Pandey and Rizvi, 2009). During the absorption process, polyphenols undergo methylation, sulfation, and glucuronidation by phase two enzymes. Through this process, metabolites are formed. After distribution, polyphenols accumulate in tissues is of importance as the concentration of this accumulation may be the cause of the biological effects of polyphenols (Pandey and Rizvi, 2009). Excretion of polyphenols occurs through bile and urine (D'Archivio et al., 2007).

The flavonoid's antioxidant capacity varies mainly depending on the substitutions in the B and C rings (Figure 10.3). For example, the quantity and locations of hydroxyl groups on the B ring, as well as the double bonds and substitutions on the C9 ring, affect antioxidant activity (Zillich et al., 2015).

Isoflavones are known to show estrogen-like activity, therefore they may have antiandrogenic effects and affect the reproductive system regardless of sex (Mennen et al., 2005). In a study, it was found that the antiandrogenic activity of soybean isoflavones is important in treating acne vulgaris and regulating dihydrotestosterone (DHT) enzyme levels (Riyanto et al., 2015). According to many studies, isoflavones can be used for the relief of menopausal symptoms due to their estrogen-like and antiandrogenic effects (Chen et al., 2019). In an *in vitro* study, isoflavones were found to work

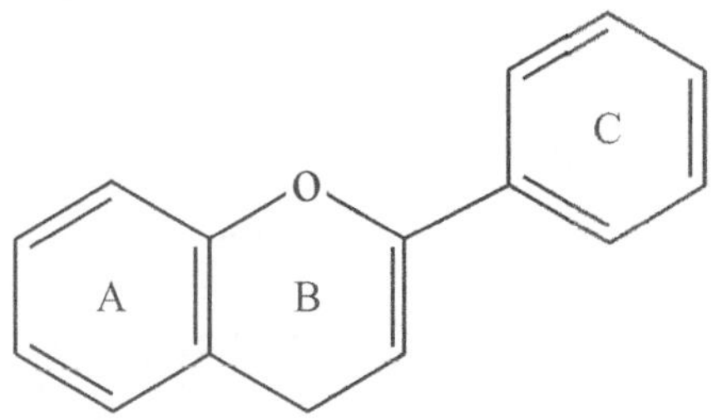

FIGURE 10.3 Basic flavonoid structure.

as estrogen antagonists and estrogen agonists according to endogenous estrogen levels in the body before and after menopause (Russell et al., 2002). In a study conducted in mice with estrogen deficiency and lack of aromatase activity, isoflavones were found to regulate morphology by acting like estrogen (Britt et al., 2005).

The hydroxyl groups in their structure scavenge free radicals and show antioxidant activity by mediating their antioxidant effects by chelating metal ions. Flavonoids are also known to have antibacterial, hepatoprotective, anti-inflammatory, anticancer, and antiviral effects (S. Kumar and Pandey, 2013). The first flavonoid isolated from orange is rutin and more than 4000 flavonoids have been identified to date (Islam et al., 2022). Some examples of flavonoids are given in Figure 10.4. These compounds show structural similarity due to the derivation of silibinin by the addition of a lignan to the quercetin structure and EGCG by the addition of a gallate (Bhatia et al., 2001).

Teas such as oolong, white, black, and green teas, which are one of the most important sources of antioxidants, are the leading plants containing polyphenol compounds (Zillich et al., 2015). Up to 30% of the dry weight of green tea is made up of polyphenols such as phenolic acids, flavondiols, and flavanols while catechin, which accounts for about 10%, is the most important flavonoid (Komes et al., 2010). The main polyphenols in teas are EGCG, epigallocatechin, epicatechin, epicatechin-3-gallate (ECG) in flavan-3-ol structure (Zillich et al., 2015). Studies have reported that catechins extracted from green teas have strong antioxidant effects and improve skin health by inhibiting collagenase and tyrosinase activity (Hong et al., 2014). It was shown in a study that applying extracts of white or green tea on human skin can protect it from UV exposure from the sun. It was found that these extracts reduced the harmful impact of UV radiation on skin response (Afaq and Katiyar, 2011; Camouse et al., 2009).

The antioxidant activity of green tea polyphenols (GTPs) is due to scavenging free radicals by preventing the activation of MAPKs, a serine/threonine protein involved in cellular functions in the skin (Oyetakinwhite et al., 2012). In a study, it was reported that topical application of these polyphenols inhibited UVB-induced phosphorylation of the expression of MAPKs such as c-Jun NH2-terminal kinases (JNK), extracellular signal-regulated kinases (ERK), and p38 in mice (Afaq et al., 2003; Oyetakinwhite et al., 2012).

Many studies have demonstrated that the administration of GTPs or catechins in mice's drinking water provides protection against the development of skin tumours. This was demonstrated by a decrease in the number, size, and incidence of tumours per group when compared to animals who were not given GTP (Nichols and Katiyar, 2010). Another study of GTPs showed that drinking water with GTPs provided to mice protected them against UVB radiation tumourigenesis and partially regressed existing skin papillomas. GTPs also have the ability to inhibit lipid peroxidation (Katiyar et al., 1994, 1997; Katiyar and Elmets, 2001; Nichols and Katiyar, 2010).

In a study, it was stated that dietary using EGCG increases skin tolerance (Jeon et al., 2009; Nichols and Katiyar, 2010). In another study, oral administration of green tea extract (contains EGCG etc.) before and In mice, psoralen with UVA therapy reduced skin conditions such as hyperplasia, hyperkeratosis, erythema, and oedema in mice (Nichols and Katiyar, 2010; Zhao et al., 1999).

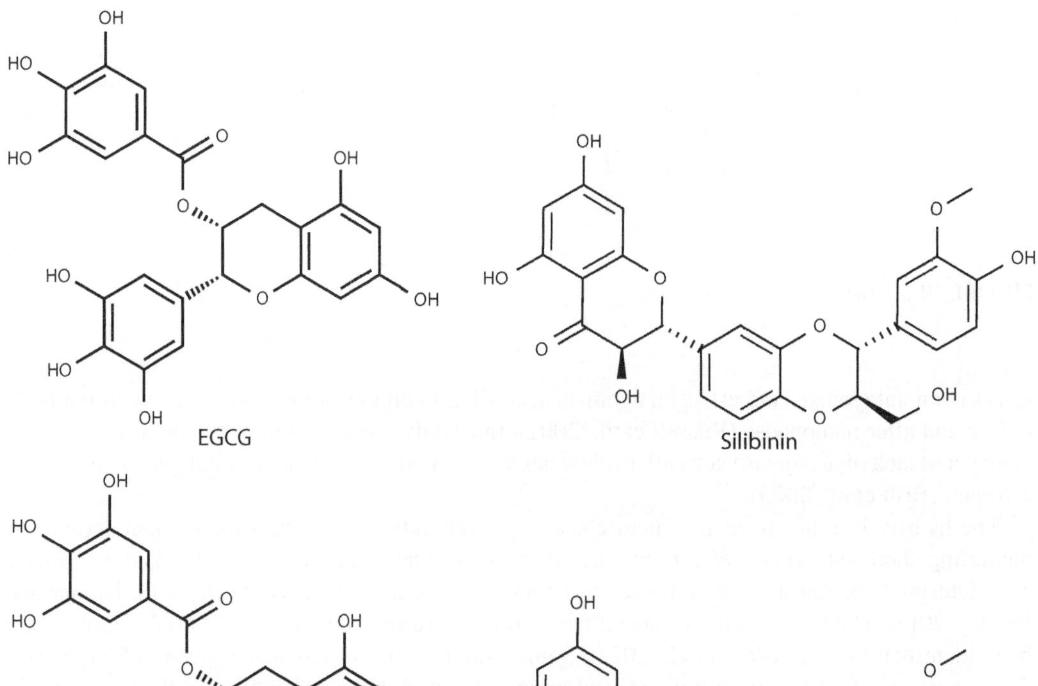

FIGURE 10.4 Structures of some flavonoid compounds.

It was discovered that topical application of green tea extract reduced the number of apoptotic keratinocytes and UV-induced p53 expression in human skin, indicating its potential in daily use as photochemopreventive (Mnich et al., 2009; Nichols and Katiyar, 2010).

EGCG has been found to block UVB-induced leukocyte infiltration in mice and human skin, potentially inhibiting UVB-induced ROS production. EGCG usage in treatment also resulted in the reduction of nitric oxide and hydrogen peroxide in UVB irritated dermis and epidermis, inducible nitric oxide synthase (iNOS) expressing cells (S. Katiyar et al., 2007; Nichols and Katiyar, 2010). Studies suggest that using GTPs, EGCG, or GSPs topically can prevent acute or chronic UV-induced protein oxidation in mice skin (Nichols and Katiyar, 2010).

Flavonoids are known for their antioxidant properties, which can be enhanced by interacting with other antioxidants. However, the extent to which a flavonoid acts as an oxidant or antioxidant depends on its molecular structure and the cellular redox state. For instance, EGCG can autoxidize in culture media, generating hydrogen peroxide (H_2O_2) in cell cultures and promoting cytotoxicity. This makes EGCG a promising antitumor agent, as it can produce H_2O_2 near tumour cells and have anticancer effects, particularly on the skin, when used topically or in the diet (Halliwell, 2008; H. S. Kim et al., 2014; Souto et al., 2019).

When the skin is damaged, it cannot prevent bacteria from invading the tissues and this can cause infected wounds and severe tissue damage. In a study, an antibacterial composite hydrogel system with controlled release of Cu-Cu-EGCG nanocapsules was developed. This hydrogel showed inhibitory activity on *Escherichia coli* and *Staphylococcus aureus* and promoted the proliferation and migration of L929 fibroblasts. In *in vivo* wound healing experiments, the angiogenesis and inflammation-regulating effect of these hydrogels was confirmed and collagen deposition and wound healing rate were reported to be accelerated (N. Liu et al., 2023).

Fungi, yeasts, and bacteria produce the enzyme known as tannase (tannin acyl hydrolase), which is distinguished by its hydrolyzing activity on depside bonds of substrates like tannic acid, ECGG, and ECG (Hong et al., 2014). According to a study, green tea's catechins and the tannase enzyme worked together to enhance the compounds' capacity to scavenge radicals including superoxide anions and hydrogen peroxide, as well as their antioxidant properties and the chelation of metal ions (Hong et al., 2014; Lu and Chen, 2007). In another study, the effects of green tea extract formulations in the presence of tannase on enzyme markers (collagenase, elastase and tyrosinase activities) that have important effects on the skin were investigated. When the formulation prepared as a result of the study was compared with normal green tea extracts, no significant difference was observed in collagenase activity inhibition, while elastase and tyrosinase activity inhibition was found to be higher. The study suggests that the tannase-treated formulation can be used as an additional ingredient in skin care products such as anti-ageing creams (Hong et al., 2014).

Quercetin is one of the most abundant flavonoids in plants, found mainly in green and black tea, apples, and onions. However, its use is limited especially in cancer treatment due to its low water solubility, low cellular bioavailability, and high instability (Sun et al., 2014). Therefore, a new drug delivery system is needed for the effective administration of quercetin. Quercetin has antiallergic, anti-inflammatory, anti-melanogenesis and anticarcinogenic effects. These activities of quercetin are due to the fact that it is an antioxidant substance that mediates anti-melanogenesis by inactivating the enzyme tyrosinase (Choi and Shin, 2016).

Normal human epidermal melanocytes (NHEM) and human melanoma (HMVII) cells were used in a study to examine the impact of quercetin on melanogenesis. It was shown that exposure to quercetin stimulated melanogenesis in NHEM cells and increased the amount of melanin in HMVII cells in a time- and dose-dependent manner. It was found that by raising tyrosinase activity and reducing elements such as melanogenic inhibitors, quercetin increased melanogenesis in both lines (Nagata et al., 2004).

In another study, it was reported that quercetin only slightly affected the stimulation of tyrosinase activity expression in melanoma (B16F10) cells at low concentrations (<20 μm), while tyrosinase activity, synthesis, and the expression of TRP-2 and TRP-related protein-1 were all reduced after exposure to 50 μm quercetin. It has also been reported to decrease phosphorylated protein kinase-A and intracellular cAMP levels in cells stimulated with α-melanocyte stimulating hormone. It was also reported that low concentrations of quercetin decreased tyrosinase expression and activity and melanin content in normal human epidermal melanocytes (Yang et al., 2011).

In vitro and *in vivo* skin penetration were examined in a study assessing the potential of quercetin-loaded nanostructured lipid carriers as a topical administration system. Anti-inflammatory and antioxidant effects were also studied, and light microscopy was used to evaluate the application's effect on the skin. The study's findings suggested that quercetin's antioxidant and anti-inflammatory properties, as well as its penetration and retention in the epidermis and dermis, might all be enhanced by the created nanoparticles (NPs) (Guo et al., 2012).

Silymarin is a mixture of lignans extracted from the milk thistle plant. There are studies suggesting that silymarin slows the development of radiodermatitis and restraints cutaneous toxicity in patients receiving radiotherapy. Silymarin's ability to eliminate free radicals and anti-inflammatory activity has been stated in *in vitro* studies. When used on the skin or in the diet, the main component of silymarin, silibinin, has an inhibitory effect on photocarcinogenicity (Amber et al., 2014; Nichols and Katiyar, 2010). Di Franco et. al. observed progress with the usage of aloe vera in radiodermatitis, however later it was demonstrated that it was the result of repetitive moisturization of the skin (Amber et al., 2014).

In a study based on the fact that silibinin, quercetin, and EGCG show skin cancer preventive properties through several different mechanisms, it was aimed to evaluate the potential effects of these compounds on mitogenic signalling molecules and cell cycle regulators.

Exposure of these compounds to human epidermoid carcinoma (A431) cells was shown to inhibit the activation of epidermal growth factor receptor and downstream adaptor protein Shc. Silibinin has been demonstrated to exert an inhibitory effect on the activation of mitogen-activated protein kinase (MAPK) – extracellular signal-regulated kinase (ERK) 1 and ERK 2. It was also reported that silibinin caused a significant decrease in cyclin-dependent kinase (CDK)-4, CDK-2 and cyclin D1 and induction of Cip1/p21 and Kip1/p27 in the cell cycle. Additionally, it was noted that silibinin remarkably increased Cip1/p21 and Kip1/p27 in the cell cycle and remarkably decreased cyclin-dependent kinase (CDK)-4, CDK-2, and cyclin D1. It was demonstrated that whereas EGCG produced induction of Cip1/p21 and a decrease in CDK-4 at low dosages, quercetin generated a moderate increase in Cip1/p21 and a decrease in CDK-4 and cyclin D1 without affecting Kip1/p27. The study's findings included the observation that silibinin at high doses, together with quercetin and EGCG at low and high concentrations, produced apoptosis in cells. The findings imply that quercetin, EGCG, and silibinin all affect cancer in various ways (Bhatia et al., 2001).

In some *in vitro* and *in vivo* studies, it has been reported that quercetin and resveratrol have a combined effect on different types of cancer and cooperate in inducing apoptosis, inhibiting cell proliferation and preventing ageing in glioma cell lines. A study was conducted to investigate the bioactivities of liposomes containing quercetin and resveratrol on oxidative stress and inflammation in pre-cancerous and cancerous skin. According to the study's findings, cellular absorption was higher, quercetin and resveratrol in liposomes effectively scavenged ROS in fibroblasts, and the presence of these compounds did not alter the antioxidant activity in comparison to the control. In addition, because of the evaluation of the *in vivo* effectiveness of liposomes in a mouse skin lesion model (topical), a remarkable decrease in oedema and leukocyte infiltration and a significant improvement in tissue damage were reported (Caddeo et al., 2016).

In a study, a nanostructured lipid carrier (NLC) gel was developed to increase the distribution of quercetin and resveratrol in dermal and epidermal layers and its cytotoxicity was investigated. As a result of the 3-[4,5-dimethylthiazol-2-yl]-2,5 diphenyl tetrazolium bromide (MTT) assay performed on human epidermoid carcinoma (A431) cell lines, the IC_{50} values of the NLC gel and the control group gel were found to be 86.50 μM and 123.64 μM, respectively. According to the study, gels created using nanotechnology may be able to penetrate the skin's deeper layers and offer a promising approach to treating skin cancer (Imran et al., 2020).

In a study, it was shown that giving grape seed proanthocyanidins to SKH-1 hairless mice in a UVB-induced photocarcinogenesis procedure reduced the size and number of tumours. Additionally, the authors discovered that there was a delay in the development of papillomas into carcinomas (Mittal et al., 2003; Souto et al., 2019).

Pomegranate fruit extract (PFE) is a natural source of antioxidants such as anthocyanins, ellagitannins and hydrolysable tannins. These compounds have potent antioxidant properties. According to research by Afaq et al., PFE products can help repair DNA damage caused by UVB radiation. The research was carried out on reconstituted human skin. The results showed that PFE products can improve DNA repair (Afaq et al., 2009, 2010; Afaq and Katiyar, 2011).

Pycnogenol (PYC®) is a brand-name product of Horphag Research Ltd. that is sold in the United Kingdom and Geneva, Switzerland. It is a trademarked blend of procyanidins that is extracted from pine bark (Pinus pinaster, formerly known as Pinus maritima). Studies suggest that it can be used to treat chronic venous insufficiency for skin changes, and skin ulcers, as well as melasma (chloasma) for the hyperpigmentation of the skin in women exposed to sunlight and is known to help reduce erythema. It also has a smoothening and elasticizing effect on the skin (D'Andrea, 2010; Ruddock and Facey, 2023).

Chrysin is a flavonoid that is commonly utilized in formulations for dark circles, puffiness under the eyes, wrinkles, and ageing skin. Recently, research has concentrated on this compound. Due to its antioxidant, anti-inflammatory, and antiallergic qualities, the chemical chrysin – which is frequently present in passion fruit, vegetables, honey, propolis, and mushrooms – is beneficial for treating skin conditions (Islam et al., 2022). It is reported to have chemopreventive properties, especially against various cancers due to its anti-proliferative and pro-apoptotic effects (H. Liu et al., 2013). It is reported to have chemopreventive properties, especially against various cancers due to its anti-proliferative and pro-apoptosis effects (H. Liu et al., 2013). In addition, although chrysin is reported to be non-toxic and biocompatible, its usage is limited by its water solubility and poor bioavailability (Nagaraja et al., 2021). *In vivo* studies have reported that the anticancer activity of chrysin is limited (H. Liu et al., 2013). Therefore, there is a need to improve its physicochemical properties by using methods such as nanotechnology in order to increase solubility and bioavailability (Nagaraja et al., 2021). The results of a study with chrysin nanoemulsions demonstrated that the therapeutic effect in human melanoma (A375 and SK-MEL-2) cell lines was significantly improved and resulted in a dose reduction. To evaluate the safety of the product, cell viability testing in mouse fibroblast (L929) cell lines showed that the nanoemulsion did not affect cell growth and was reported to be biocompatible and safe for topical application (Nagaraja et al., 2021).

10.2.2 Stilbenes

Stilbenes, which are 1,2-diphenylethylene-containing secondary compounds, are typically identified as trans-resveratrol derivatives (3,5,4'-trihydroxy-trans-stilbene) (Chong et al., 2009). These compounds are found in plants such as grapes, pine, peanuts, and sorghum and are produced using the phenylpropanoid pathway (Chong et al., 2009; Dubrovina and Kiselev, 2017). Resveratrol and its derivatives are being investigated for having cardioprotective, potent antioxidant, anti-inflammatory and anticancer effects (De Filippis et al., 2017). The structure of some stilbene compounds is given in Figure 10.5.

Resveratrol is a stilbene containing three hydroxyl groups found especially in grapevine skin, Japanese knotweed (*Polygonum cuspidatum*), and Mojave yucca plant (*Yucca schidigera*) and is characterized by its anticancer properties (Chinembiri et al., 2014; Gatouillat et al., 2010). As demonstrated by assays including the three primary phases of carcinogenesis, resveratrol possesses chemopreventive activity in addition to antioxidant, antimutagenic, anti-inflammatory, antiproliferative, and photoprotective properties (Jang et al., 1997; Ndiaye et al., 2011). Resveratrol shows its anticancer effect by inducing apoptosis in amelanotic and melanotic cells. To prove this activity, apoptosis induced in doxorubicin-resistant mouse cells was investigated (Chinembiri et al., 2014; Gatouillat et al., 2010). By blocking the NF-κB signalling pathway, resveratrol is hypothesized to have the antimetastatic ability to prevent the induction of lipopolysaccharide epithelial to mesenchymal transition (M.-C. Chen et al., 2012; Chinembiri et al., 2014).

In an *in vitro* study, it was reported that the combination of resveratrol and temozolomide showed cytotoxic activity against melanoma cells. However, the same effect was not observed in *in vivo* studies (Chinembiri et al., 2014; Niles et al., 2006; Osmond et al., 2012).

Resveratrol is also used as a natural pigment remover due to its anti-hyperpigmentation effect in skin disorders (Sheweita et al., 2022). However, due to the low oral bioavailability, it has been

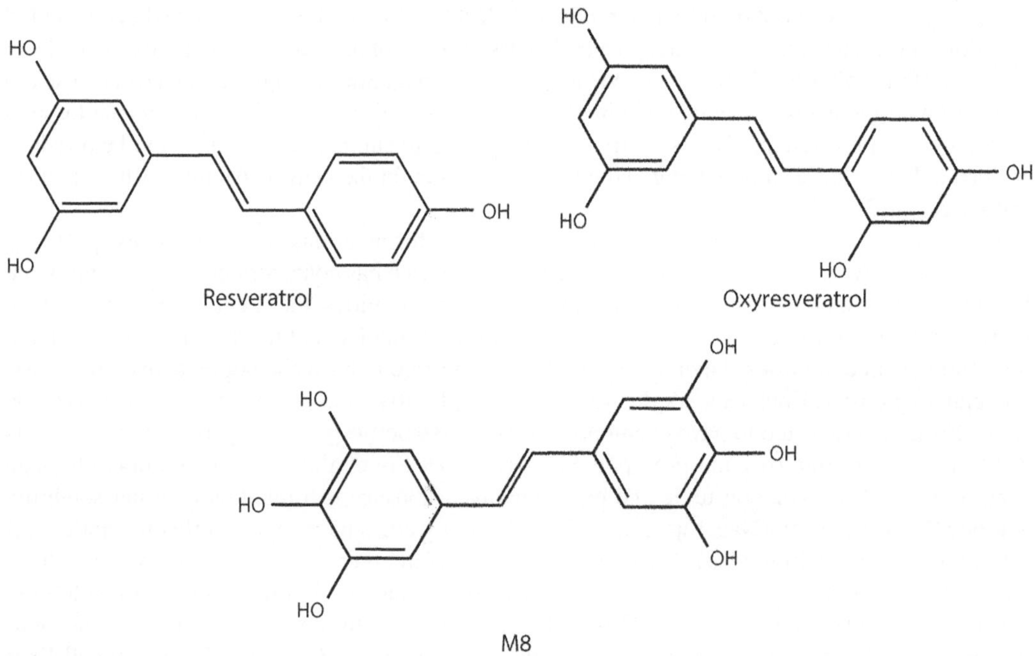

FIGURE 10.5 Structure of some stilbene compounds.

reported to give positive/negative results in the *in vivo* studies, and some studies have reported that it has no effect or little effect (Su et al., 2013).

In a study, the permeability of resveratrol 24 hours after topical application was investigated by *in vitro* and *in vivo* methods. By gradually removing the *stratum corneum* layers by tape stripping the permeability of human skin was examined. Using α,α-diphenyl-β-picrylhydrazyl (DPPH) test and Raman spectroscopy, the *in vitro* profile of resveratrol was determined, along with its ability to scavenge free radicals. As a result of the study, Raman spectra showed that resveratrol penetrated deep into the skin. Furthermore, large levels of resveratrol were found in several *stratum corneum* layers close to the surface, whereas a consistent low level was found in the upper layers of the live epidermis. Following *in vivo* administration, the concentration of resveratrol in the outermost *stratum corneum* layers was measured by tape stripping, which demonstrated that resveratrol was mostly found in the human *stratum corneum* layers. Resveratrol was found to have more antioxidant activity in the inner layers of the *stratum corneum* and to retain its antiradical action following topical administration (Alonso et al., 2017).

In a study conducted due to the low bioavailability of resveratrol, piceid, the precursor of resveratrol, was used since it was thought to have the same bioactivity. The MTT assay was used to assess the anti-proliferation effects of resveratrol and piceid in human liver tumour (HepG2) and human breast cancer (MDA-MB-231, MCF-7) cells. Flow cytometry was used to assess the effects of these compounds on cell cycle and apoptosis. The study showed that while resveratrol showed a strong protective effect against H_2O_2-induced cell damage, piceid demonstrated better scavenging capacity against hydroxyl radicals. Additionally, it was noted that only at high concentrations did resveratrol and piceid exhibit notable cytotoxicity in tumour cells (≥ 50 μmol/L). However, low resveratrol concentration (< 30 μmol/L) was reported to increase cell viability. It has been reported that the effect of these two compounds on tumour cells is due to cell cycle arrest and the effect on apoptosis is relatively low (Su et al., 2013).

Low solubility and isomerization when exposed to UV properties of resveratrol limits its use in skin products. In a study for this purpose, a nanoformulation form was prepared as a skin-whitening agent on the back skin of guinea pigs by combining hydrophilic surfactants and oils to increase the solubility of resveratrol. When Fontana Mountain dye was applied to the back skin of guinea pigs, it was found that the melanin levels in the NPs treated group were much lower than those in the control group. The safety of topical application of the nanoformulation was supported by the fact that it did not cause changes in free radical levels and oxidative stress markers in the livers and kidneys of guinea pigs. The study demonstrated that the resveratrol NPs were safe and non-toxic (Sheweita et al., 2022).

In a study using resveratrol with lipoic acid, which can enhance the effects of antioxidant compounds, lipid-core nanocapsules suitable for cutaneous administration were formed to improve chemical stability and photostability, controlled release and antioxidant effects. The cytotoxicity of NPs was determined by MTT method using BALB/c-3T3 fibroblasts and antioxidant activity was determined by lipid peroxidation method. As a result of the study, it was found that the nanoformulation controlled the release of the substance and increased the stability and photostability of the substance in UVA exposure. Co-nanocapsulation of resveratrol and lipoic acid was reported to potentiate antioxidant activity without increasing cytotoxicity in fibroblasts (Davies et al., 2020).

Oxiresveratrol (OXY) is a derivative of resveratrol, a substance that has an extra hydroxyl group in its aromatic ring, which gives it greater water solubility. It has been reported that OXY inhibits tumour necrosis factor-alpha (TNF-α), which lowers the incidence of psoriasis. However, OXY does not have anti-inflammatory and anti-ageing properties. There is also little research on its antioxidant mechanisms. In a study with this purpose, the effects of OXY on Kelch-like ECH-associated protein 1 – the nuclear factor erythroid 2-related factor 2 – antioxidant response elements (Keapl-Nrf2-ARE) signalling pathway were evaluated. According to the study's findings, OXY can bind with Nrf2 to cause nuclear translocation using Western blotting, and luciferase reporter gene analysis can demonstrate its agonistic actions on ARE. It was also reported that OXY not only increased the mRNA expression levels of ARE-mediated genes NQO1, HO-1 and GCLM but also increased their protein expression levels. This indicates that the substance reduces oxidative damage caused by H_2O_2 and demonstrates its potential as an antioxidant via stimulating the Keap-Nrf2-ARE signalling pathway (Ma et al., 2022).

M8 (3,3',3',4,4',4',5,5'-hexahydroxystilbene) is a stilbene derivative. According to a study, it inhibits cancer growth and metastasis development *in vivo* and *in vitro*. It has also been found to have a stronger anticancer effect than resveratrol. M8 shows its anticancer effect by inhibiting signaling pathways involved in apoptosis, affecting the cell cycle, and inhibiting metastasis, and DNA repair mechanisms (Paulitschke et al., 2010).

Research showed that M8 was superior to resveratrol in primary human melanoma cells obtained from cutaneous and lymph node metastases, the metastatic 1205Lu cell line, and M24met melanoma cells separated *ex vivo*. Besides its strong anticancer effect, M8 does not show any cytotoxicity against healthy tissue. Because M8 neither shows an antiproliferative effect in melanocytes nor stimulates mismatch repair (MMR) proteins from DNA repair systems in healthy cells. This means M8 selectively shows an anticancer effect (Paulitschke et al., 2010).

10.2.3 Lignans

Lignans are polyphenolic compounds commonly found in fibre-rich plants, legumes and vegetables, especially flax and sesame seeds, pumpkin seeds, wheat, barley, rye, and oats (Imran et al., 2015). Secoisolariciresinol, matairesinol, lariciresinol, and pinoresinol are the most often consumed lignans in conjunction with diets. As phytoestrogens, lignans are known for their estrogen agonist and antagonistic characteristics (Pathak et al., 2018).

A study aimed to develop a formulation of flaxseed extracts as a wound-healing agent. Using the Folin-Ciocalteu method, the total phenolic content was assessed, and the DPPH test was used to assess the antioxidant properties. The phenolic cream's ability to diffuse and its ability to promote wound healing was assessed in Wistar rats. The formulation may help these compounds' antioxidant ability, as seen by the high correlation found between antioxidant capacity and total phenolic content. The number of phenolic compounds released by transdermal diffusion of the cream in the abdominal skin of rats for 24 hours was reported to be very high. As a result of macroscopic evaluation of epidermal lesions, it was reported to increase the effectiveness of the treatment and wound contraction, and decrease epithelialization and scar area (Draganescu et al., 2015).

Sesame (*Sesamum indicum L.*) is one of the sources of lignan compounds such as sesamolin, sesamin, sesamol, sesaminol, sesaminol, sesamolinol (Figure 10.6) (Rosalina and Weerapreeyakul, 2021). In a study, the antimelanogenesis properties of sesamol, sesaminol, and sesamolin, sun protection function and enzyme inhibition in melanoma cell lines were investigated. As a result of the study, it was reported that all three compounds exhibited sun protection function by absorbing UVB. It was also reported that the compounds inhibit melanin synthesis by blocking melanin induction and inhibiting melanogenic enzyme formation (Srisayam et al., 2017). In another study, the intracellular uptake pathways, and cytotoxicity of these three compounds were investigated in melanoma and healthy cells. According to the results, esamin, and sesamoline were discovered to have higher

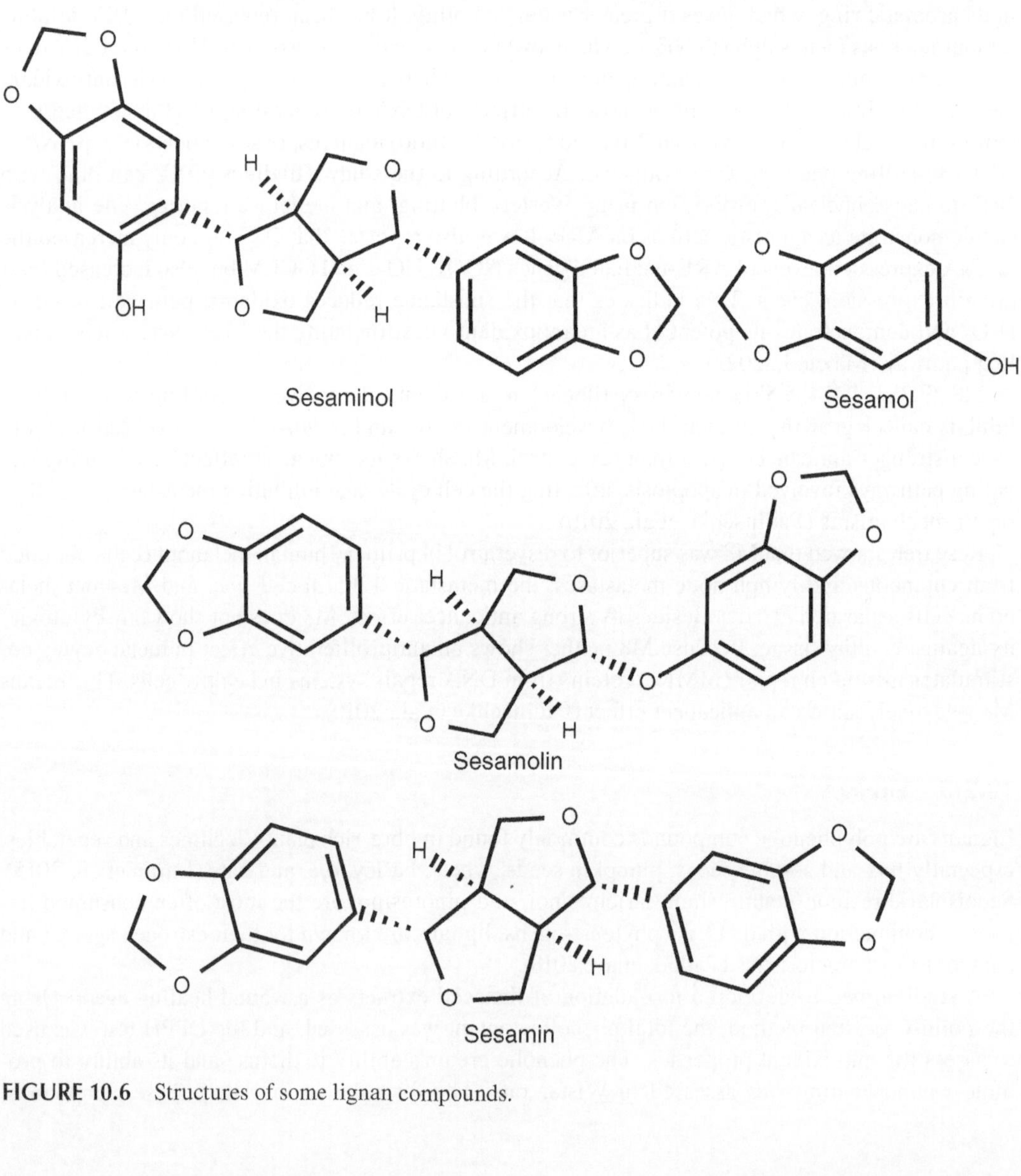

FIGURE 10.6 Structures of some lignan compounds.

intracellular concentrations than sesamol. Sesamol was the only medication that reduced melanoma cell viability. Sesamol's intracellular concentration should be increased by enhancing its physicochemical characteristics because it has the potential to prevent the proliferation of melanoma cells (Srisongkram and Weerapreeyakul, 2022).

Sesamin, the most widely researched of these compounds, is a lipid-soluble lignan that shows inflammation and free radical scavenging activity. According to studies, sesamin controls several signal transduction pathways involved in inflammation, angiogenesis, apoptosis, oxidative stress, proliferation, and autophagy (Majdalawieh et al., 2017).

In a study, sesamin has been shown to raise tyrosinase mRNA and protein levels, as well as melanin content and activity in a dose-dependent manner in B16 melanoma cells. Furthermore, through a cAMP-dependent pathway, it was discovered to be able to activate protein kinase-A and the cAMP response element binding protein; it did not affect the activation of p38 MAPK or Akt. According to the study, sesamin can promote melanogenesis in B16 cells by activating cAMP and upregulating tyrosinase (Jiang et al., 2011).

In a study, the effect of sesamin against UVB-induced photodamage and photoinflammation in human skin fibroblasts (Hs68) and mouse skin was investigated. Following UVB irradiation, Hs68 cells showed a decrease in intracellular ROS generation, an increase in total collagen content, and a decrease in MAPK phosphorylation and matrix metalloproteinase (MMPs) overexpression. Additionally, it was discovered to prevent NF-κB translocation and reduce the overexpression of COX-2 and inducible nitric oxide synthase (i-NOS). Additionally, it was discovered to suppress the overexpression of COX-2 and inducible nitric oxide synthase (i-NOS), as well as to block NF-κB translocation. In addition, it was found to lower mouse skin levels of MMP-1, IL-1, i-NOS, and NF-κB (Lin et al., 2019).

In a study investigating the preventive effects of sesamin on UVA-exposed keratinocytes, cell viability and oxidative stress indicators, apoptosis and intracellular ROS levels were evaluated using HaCaT keratinocytes. The study's findings included the following: sesamin did not exhibit cytotoxicity up to 160 μmol/L; pretreatment with sesamin at concentrations of 20 and 40 μM reversed suppressed cell viability; increased levels of lactate dehydrogenase (LDH) and malondialdehyde (MDA); and decreased levels of glutathione (GSH) and superoxide dismutases (SOD). Furthermore, it has been demonstrated to phosphorylate apoptotic signal-regulating kinase-1 (ASK-1), c-Jun-NH(2)-terminal kinase (JNK), and p38, while stimulating the production of Bcl-2. It also inhibits the expression of Bax, MMP-1, and MMP-9. The study shows that sesamin attenuates UVA-induced keratinocyte damage (H. Li et al., 2024).

Studies on the pharmacological activity of sesamolin are more limited and these studies have reported that sesamolin showed less antioxidant activity than sesamol (Rosalina and Weerapreeyakul, 2021; Suja et al., 2004).

10.2.4 Phenolic Acids

Phenolic acids are widely present and are known for their various health benefits, including antimicrobial, anticancer, anti-inflammatory, neuroprotective, antimutagenic, and high antioxidant effects. They are phenolic compounds with the addition of a carboxylic acid group. They too are secondary metabolites of plants and vital human nutrients.

Phenolic acids can be derived from seeds, skins of fruits, leaves, and vegetables. They can be found bound to different chemical functional groups. Hydroxybenzoic acid and hydroxycinnamic acid are two subtypes of phenolic acids. Benzoic acid is the source of hydroxybenzoic acids, which are present in black radish, red fruits, etc. Hydroxycinnamic acids are present in food and originate from cinnamic acid. The hydroxycinnamic acids ferulic acid, caffeic acid, and p-coumaric acid are the most prevalent ones (Kumar and Goel, 2019).

Numerous studies suggest that polyphenolic acids are direct antioxidants, thus consumption of polyphenolic acid-rich foods can lead to improved health and reduced oxidative stress. They especially have beneficial effects on diseases like heart disease and cancer (Kumar and Goel, 2019; Ness and Powles, 1997).

Cinnamic acid derivatives induce apoptosis by causing irreversible DNA damage to cancer cells leading to their death, thus showing anticancer activity. According to some studies, cinnamic acid shows antiproliferative activity against melanoma cells. A study stated that cinnamic acid induces melanoma cell differentiation and in certain concentrations reduces melanoma cell proliferation (L. Liu et al., 1995).

Some natural compounds exert their cytotoxic activity by interfering with cell division in the mitotic phase. As a result, the aneugenic pathway is stimulated. According to a study, the aneugenic potential may lead to carcinogenicity (Efthimiou et al., 2007; Niero and MacHado-Santelli, 2013).

It was found that an increase in the quantity of micronucleated cells was observed even at low concentrations of cinnamic acid. Therefore, cinnamic acid is considered to have genotoxic properties. Despite its genotoxic activity, it is thought that cinnamic acid may be used in the treatment of melanoma due to its antiproliferative effect at genotoxic concentrations (Niero and MacHado-Santelli, 2013).

Vanillic acid is a polyphenolic acid used for pharmaceutical, cosmetic, flavouring, etc. purposes. It shows many activities such as antioxidant activity because of its ability to scavenge free radicals and anti-inflammatory activity. It is an active and potent antioxidant and has a protective effect on the skin from active radicals that increase due to UV exposure and prevent premature ageing. There is data that vanillic acid inhibits the activity of tyrosinase in a dose-dependent manner. Although it shows cytotoxicity at some concentrations, it is widely used especially in cosmetics due to its anti-tyrosinase activity (Girawale et al., 2022).

Caffeic acid is a well-known polyphenolic acid found in coffee. There are studies stating that caffeic acid shows anticancer activity in various types of cancer cells. Many previous studies have suggested that caffeic acid is effective against skin cancer and preventive against UVB-induced skin damage (G. Yang et al., 2014). Yang et al. reported that caffeic acid acts as a chemopreventive against skin cancer caused by sun UV radiation by preventing human skin cancer cells from colonizing. At the same time, as a result of long-term topical application of caffeic acid to mouse dorsal skin, they revealed a decrease and suppression of solar UV-induced skin cancer formation (Yang et al., 2014). According to a study on SK-MEL-28 cells, a decrease in the cell viability, colony formation, reproduction amount, and proliferation of cells was observed when caffeic acid was used in treatment concentrations for a certain period. In the study, data on the immunomodulatory and antioxidant activities of caffeic acid were also obtained. It has been shown that caffeic acid causes a rise in the quantity of early apoptotic cells, pushes the cells to apoptosis, and as a result, causes the death of the cells resulting in anticancer effects (Pelinson et al., 2019).

10.2.5 Tannins

As with other polyphenols tannins are also secondary metabolites of plants, which are used for tanning or converting animal skin to leather. Tannins are known to have antityrosinase activity and are actively used in cosmetics for this purpose Tannins are water soluble and can precipitate alkaloids, gelatins, etc. They may be derived from quinic acid or naringenin chalcone. Tannins can either be categorized according to their resistance to hydrolysis in hot water/or to be in contact with tannases as, hydrolysable, and nonhydrolysable. Or their structural characteristics as, gallotannins, ellagitannins, complex tannins, and condensed tannins (Khanbabaee and van Ree, 2001; Ruddock and Facey, 2023; Desmedt et al., 2016; D'Mello et al., 2016).

According to studies on *Acacia nilotica (L.)* and *Acacia catechu (L.)*, derived tannins from these plants have healing effects on skin ulceration and diseases as they show antioxidant activity (Karim and Azlan, 2012; Khanbabaee and van Ree, 2001; Shen et al., 2006).

Studies state that *Monochaetum multiflorum (Bonpl.) Naudin* contains higher ellagitannins which show less antitumor activity and potency than dimeric ellagitannins. However, some studies state that ellagitannins are used to heal skin injuries (Isaza et al., 2004; Okuda and Ito, 2011; Ruddock and Facey, 2023).

A study states that *Phyllanthus muellerianus (Kuntze) Exell* contains the ellagitannins: geraniin, corilagin, and furosin. The plant's main component is gallic acid and the authors found out that gallic acid stimulates tissue regeneration, barrier formation, cellular matrix regeneration, reduction of oxidative stress-originated cellular damages, and inhibition of inflammation thus helping the healing process (Agyare et al., 2011; Ruddock and Facey, 2023).

A study demonstrated that pentagalloylglucose extracted from the tea leaves of *Anacardium occidentale L.* shows more cytotoxicity to cancerous cells than healthy cells, rapidly decreases tumour cells, and generates oxidative stress in cancer cells in high concentrations (Taiwo et al., 2020).

Green tea tannins are remarkable for their ability to capture and neutralize free radicals and radical scavenging properties without causing undesirable effects on the skin (Fares et al., 2023). In a study, anti-ageing creams were formulated with green tea tannins (including avocado oil) to investigate the anti-ageing effects, nourishing effects, and protective effects against solar UVB rays. The sunscreen effect was found by calculating the *in vitro* sun protection factor (SPF). In the study, it was reported that the cream showed a medium SPF value (6–16), which can absorb UVB up to high content, as a result of using different tannin concentrations. It was also found that the cream showed anti-ageing skin care properties with antibacterial and antifungal effects. To sum up, these components seem to play a significant part in nourishing and shielding the skin from UVB rays from the sun (Fares et al., 2023).

In a study of brown algae polyphenols, a special type of polyphenols called phlorotannins, can reduce tumour multiplicity and volume in SKH-1 mice after UVB radiation-induced skin carcinogenesis. In addition to suppressing COX-2 activity and cell proliferation and 6% and 21% lower tumour incidence, the polyphenols also prevented UVB-induced oxidative stress, inflammation, and skin cell proliferation (Hwang et al., 2006; Souto et al., 2019).

In a study, it was suspected that persimmon reduces skin pigmentation and conducted a study using arbutin, which is used to treat skin pigmentation, as a control. As a result of this study, it was found that persimmon tannins (PT) reduced melanin density more than arbutin. PT physiologically shows pigmentation inhibition through its effects on tyrosinase activity, oxidative stress, inflammatory response and signalling pathways. PT is a promising compound in the treatment of skin pigmentation (Sun et al., 2021).

A study conducted on sunscreen using tannin microcapsules, considering the protection of tannins against the risks of solar UV such as skin burns, dermatitis, and cancer formation. According to the study, the beneficial effects of sunscreens containing tannin microcapsules were especially effective against the activities of UV radiation, especially against smaller wavelength and more harmful rays such as UVB and were effective in increasing the effectiveness and duration of protection and reducing the amount of sunscreen use with synergistic effect (Alfonsi et al., 2023).

10.3 CONCLUSION

The skin serves as a barrier against a variety of illnesses, including frequent exposure to xenobiotics. All of the exposures to the skin affect skin structure and integrity, resulting in skin ageing, skin diseases and cancer. UV radiation specifically damages DNA, suppresses the immune system, disrupts cellular signalling pathways, and increases oxidative stress. These effects result in early skin ageing and a variety of skin conditions, including melanoma and non-melanoma skin malignancies. Apoptosis, angiogenesis, and autophagy play a role in the pathophysiology of melanoma, a global health threat. For this reason, complementary therapeutic approaches are being focused on reducing treatment resistance and side effects of melanoma. Polyphenolic compounds, which are secondary metabolites, have antioxidant, chemopreventive, and immunomodulatory effects.

Polyphenol compounds are found in many plant foods and contain multiple phenolic hydroxyl groups bound to one or more benzene rings. The two major classes of polyphenolic compounds are flavonoids and non-flavonoids (phenolic acids, stilbenes, tannins, and lignans). These compounds have reducing properties due to their structure and thus show scavenging properties of free radicals. They also show many biological activities such as anti-inflammatory, antioxidant, antiallergic, and photoprotective. According to reports, these compounds' anti-inflammatory activity effectively draws immune cells to the skin and stops secondary infections from developing after the skin barrier is disrupted; their antioxidant activity lessens the damage that free radicals cause to cells during allergic reactions. Studies have reported that polyphenols have strong antioxidant effects, scavenge free radicals by preventing the activation of MAPKs, a serine/threonine protein, reduce the number of apoptotic keratinocytes and UV-induced p53 expression in human skin, and improve skin health by inhibiting collagenase and tyrosinase activity. However, the undesirable properties of these compounds such as low water solubility, inadequate cellular bioavailability and high instability limit their use. Therefore, in order to overcome these limitations, recent studies have focused on new delivery systems. The genotoxicity and carcinogenicity of nanoparticles, which are the most important of these, need to be well evaluated.

The main topics of this chapter are the protective and therapeutic effects of phenolic compounds, as well as the toxicity and activity investigations in this field, Investigating the effects of xenobiotics and UV radiation on skin is important for many skin diseases, especially melanoma. Therefore, more *in vitro* and *in vivo* studies are needed to determine the mechanisms, cytotoxic, and genotoxic effects of secondary compounds such as polyphenols on skin.

REFERENCES

Afaq, F., Ahmad, N., & Mukhtar, H. 2003. Suppression of UVB-induced phosphorylation of mitogen-activated protein kinases and nuclear factor kappa B by green tea polyphenol in SKH-1 hairless mice. *Oncogene*, *22*(58), 9254–9264. https://doi.org/10.1038/sj.onc.1207035

Afaq, F., & Katiyar, S. K. 2011. Polyphenols: Skin photoprotection and inhibition of photocarcinogenesis. *Mini Reviews in Medicinal Chemistry*, *11*(14), 1200–1215. https://doi.org/10.2174/138955711804586739

Afaq, F., Khan, N., Syed, D. N., & Mukhtar, H. 2010. Oral feeding of pomegranate fruit extract inhibits early biomarkers of UVB radiation-induced carcinogenesis in SKH-1 hairless mouse epidermis. *Photochemistry and Photobiology*, *86*(6), 1318–1326. https://doi.org/10.1111/j.1751-1097.2010.00815.x

Afaq, F., Zaid, M. A., Khan, N., Dreher, M., & Mukhtar, H. 2009. Protective effect of pomegranate-derived products on UVB-mediated damage in human reconstituted skin. *Experimental Dermatology*, *18*(6), 553–561. https://doi.org/10.1111/j.1600-0625.2008.00829.x

Agyare, C., Lechtenberg, M., Deters, A., Petereit, F., & Hensel, A. 2011. Ellagitannins from Phyllanthus muellerianus (Kuntze) Exell.: Geraniin and furosin stimulate cellular activity, differentiation and collagen synthesis of human skin keratinocytes and dermal fibroblasts. *Phytomedicine*, *18*(7), 617–624. https://doi.org/10.1016/j.phymed.2010.08.020

Alfonsi, E., Lange, H., Zongo, L., Poce, G., Sgarzi, M., & Crestini, C. 2023. Tannin microcapsules for synergy-enhanced sunscreen formulations. *Industrial Crops and Products*, *192*. https://doi.org/10.1016/j.indcrop.2022.116105

Alonso, C., Martí, M., Barba, C., Carrer, V., Rubio, L., & Coderch, L. 2017. Skin permeation and antioxidant efficacy of topically applied resveratrol. *Archives of Dermatological Research*, *309*(6), 423–431. https://doi.org/10.1007/S00403-017-1740-5/FIGURES/9

Amber, K. T., Shiman, M. I., & Badiavas, E. V. 2014. The use of antioxidants in radiotherapy-induced skin toxicity. *Integrative Cancer Therapies*, *13*(1), 38–45. https://doi.org/10.1177/1534735413490235

Ashique, S., Afzal, O., Hussain, A., Zeyaullah, M., Altamimi, M. A., Mishra, N., Ahmad, M. F., Dua, K., Altamimi, A. S. A., & Anand, K. 2023. It's all about plant derived natural phytoconstituents and phytonanomedicine to control skin cancer. *Journal of Drug Delivery Science and Technology*, *84*, 104495. https://doi.org/10.1016/J.JDDST.2023.104495

Ashwin, K., Pattanaik, A. K., & Howarth, G. S. 2021. Polyphenolic bioactives as an emerging group of nutraceuticals for promotion of gut health: A review. *Food Bioscience*, *44*, 101376. https://doi.org/10.1016/J.FBIO.2021.101376

Bhatia, N., Agarwal, C., & Agarwal, R. 2001. Differential responses of skin cancer-chemopreventive agents silibinin, quercetin, and epigallocatechin 3-gallate on mitogenic signaling and cell cycle regulators in human epidermo . . . differential responses of skin cancer-chemopreventive agents silibinin, quercetin, and epigallocatechin 3-gallate on mitogenic signaling and cell cycle regulators in human epidermoid carcinoma A431 cells. *Nutrition and Cancer*, 39(2). https://doi.org/10.1207/S15327914nc392_20

Britt, K. L., Simpson, E. R., & Findlay, J. K. 2005. Effects of phytoestrogens on the ovarian and pituitary phenotypes of estrogen-deficient female aromatase knockout mice. *Menopause*, *12*(2). https://journals.lww.com/menopausejournal/fulltext/2005/12020/effects_of_phytoestrogens_on_the_ovarian_and.12.aspx

Caddeo, C., Nacher, A., Vassallo, A., Armentano, M. F., Pons, R., Fernàndez-Busquets, X., Carbone, C., Valenti, D., Fadda, A. M., & Manconi, M. 2016. Effect of quercetin and resveratrol co-incorporated in liposomes against inflammatory/oxidative response associated with skin cancer. *International Journal of Pharmaceutics*, *513*(1–2), 153–163. https://doi.org/10.1016/J.IJPHARM.2016.09.014

Camouse, M. M., Domingo, D. S., Swain, F. R., Conrad, E. P., Matsui, M. S., Maes, D., Declercq, L., Cooper, K. D., Stevens, S. R., & Baron, E. D. 2009. Topical application of green and white tea extracts provides protection from solar-simulated ultraviolet light in human skin. *Experimental Dermatology*, *18*(6), 522–526. https://doi.org/10.1111/j.1600-0625.2008.00818.x

Chen, L. R., Ko, N. Y., & Chen, K. H. 2019. Isoflavone supplements for menopausal women: A systematic review. *Nutrients*, *11*(11). MDPI AG. https://doi.org/10.3390/nu11112649

Chen, M.-C., Chang, W.-W., Kuan, Y.-D., Lin, S.-T., Hsu, H.-C., & Lee, C.-H. 2012. Resveratrol inhibits LPS-induced epithelial-mesenchymal transition in mouse melanoma model. *Innate Immunity*, *18*(5), 685–693. https://doi.org/10.1177/1753425912436589

Chinembiri, T. N., Du Plessis, L. H., Gerber, M., Hamman, J. H., & Du Plessis, J. 2014. Review of natural compounds for potential skin cancer treatment. *Molecules*, *19*(8), 11679–11721. MDPI AG. https://doi.org/10.3390/molecules190811679

Choi, M. H., & Shin, H. J. 2016. Anti-melanogenesis effect of quercetin. *Cosmetics*, *3*(2), 18. https://doi.org/10.3390/COSMETICS3020018

Chong, J., Poutaraud, A., & Hugueney, P. 2009. Metabolism and roles of stilbenes in plants. *Plant Science*, *177*(3), 143–155. https://doi.org/10.1016/J.PLANTSCI.2009.05.012

Chowdhury, A., Nosoudi, N., Karamched, S., Parasaram, V., & Vyavahare, N. 2021. Polyphenol treatments increase elastin and collagen deposition by human dermal fibroblasts; Implications to improve skin health. *Journal of Dermatological Science*, *102*(2), 94–100. https://doi.org/10.1016/J.JDERMSCI.2021.03.002

D'Andrea, G. 2010. Pycnogenol: A blend of procyanidins with multifaceted therapeutic applications? *Fitoterapia*, *81*(7), 724–736. https://doi.org/10.1016/j.fitote.2010.06.011

D'Archivio, M., Filesi, C., Benedetto, R., Gargiulo, R., Giovannini, C., & Masella, R. 2007. Polyphenols, dietary sources and bioavailability. *Annali Dell'Istituto Superiore Di Sanità*, *43*, 348–361.

Davies, S., Contri, R. V., Guterres, S. S., Pohlmann, A. R., & Guerreiro, I. C. K. 2020. Simultaneous nanoencapsulation of lipoic acid and resveratrol with improved antioxidant properties for the skin. *Colloids and Surfaces B: Biointerfaces*, *192*, 111023. https://doi.org/10.1016/J.COLSURFB.2020.111023

Day, A. J., & Williamson, G. 2001. Biomarkers for exposure to dietary flavonoids: A review of the current evidence for identification of quercetin glycosides in plasma. *British Journal of Nutrition*, *86*(S1), S105–S110. https://doi.org/DOI: 10.1079/BJN2001342

De Filippis, B., Ammazzalorso, A., Fantacuzzi, M., Giampietro, L., Maccallini, C., & Amoroso, R. 2017. Anticancer activity of stilbene-based derivatives. *ChemMedChem*, *12*(8), 558–570. https://doi.org/10.1002/CMDC.201700045

Desmedt, B., Courselle, P., De Beer, J. O., Rogiers, V., Grosber, M., Deconinck, E., & De Paepe, K. 2016. Overview of skin whitening agents with an insight into the illegal cosmetic market in Europe. *Journal of the European Academy of Dermatology and Venereology*, *30*(6), 943–950. https://doi.org/10.1111/jdv.13595

D'Mello, S. A. N., Finlay, G. J., Baguley, B. C., & Askarian-Amiri, M. E. 2016. Signaling pathways in melanogenesis. *International Journal of Molecular Sciences*, *17*(7). https://doi.org/10.3390/ijms17071144

D'Orazio, J., Jarrett, S., Amaro-Ortiz, A., & Scott, T. 2013. UV radiation and the skin. *International Journal of Molecular Sciences*, *14*(6), 12222–12248. https://doi.org/10.3390/IJMS140612222

Draganescu, D., Ibanescu, C., Tamba, B. I., Andritoiu, C. V., Dodi, G., & Popa, M. I. 2015. Flaxseed lignan wound healing formulation: Characterization and in vivo therapeutic evaluation. *International Journal of Biological Macromolecules*, *72*, 614–623. https://doi.org/10.1016/J.IJBIOMAC.2014.09.012

Dubrovina, A. S., & Kiselev, K. V. 2017. Regulation of stilbene biosynthesis in plants. *Planta*, *246*(4), 597–623. https://doi.org/10.1007/S00425-017-2730-8

Działo, M., Mierziak, J., Korzun, U., Preisner, M., Szopa, J., & Kulma, A. 2016. The potential of plant phenolics in prevention and therapy of skin disorders. *International Journal of Molecular Sciences*, *17*(2), 160. https://doi.org/10.3390/IJMS17020160

Efthimiou, M., Andrianopoulos, C., Stephanou, G., Demopoulos, N. A., & Nikolaropoulos, S. S. 2007. Aneugenic potential of the nitrogen mustard analogues melphalan, chlorambucil and p-N,N-bis(2-chloroethyl)aminophenylacetic acid in cell cultures in vitro. *Mutation Research – Fundamental and Molecular Mechanisms of Mutagenesis*, *617*(1–2), 125–137. https://doi.org/10.1016/j.mrfmmm.2007.01.009

Fares, M. M., Radaydeh, S. K., & AlAmeen, H. M. 2023. Green tannins /Avocado oil composites; suncare and skincare materials. *Arabian Journal of Chemistry*, *16*(6), 104764. https://doi.org/10.1016/J.ARABJC.2023.104764

Gatouillat, G., Balasse, E., Joseph-Pietras, D., Morjani, H., & Madoulet, C. 2010. Resveratrol induces cell-cycle disruption and apoptosis in chemoresistant B16 melanoma. *Journal of Cellular Biochemistry*, *110*(4), 893–902. https://doi.org/10.1002/jcb.22601

Gee, J. M., Dupont, M. S., Rhodes, M. J. C., & Johnson, I. T. 1998. Quercetin glucosides interact with the intestinal glucose transport pathway. *Free Radical Biology and Medicine*, *25*(1), 19–25. https://doi.org/10.1016/S0891-5849(98)00020-3

Girawale, S. D., Meena, S. N., Nandre, V. S., Waghmode, S. B., & Kodam, K. M. 2022. Biosynthesis of vanillic acid by Ochrobactrum anthropi and its applications. *Bioorganic and Medicinal Chemistry*, *72*. https://doi.org/10.1016/j.bmc.2022.117000

Guo, C. Y., Yang, C. F., Li, Q. L., Tan, Q., Xi, Y. W., Liu, W. N., & Zhai, G. X. 2012. Development of a Quercetin-loaded nanostructured lipid carrier formulation for topical delivery. *International Journal of Pharmaceutics*, *430*(1–2), 292–298. https://doi.org/10.1016/J.IJPHARM.2012.03.042

Halliwell, B. 2008. Are polyphenols antioxidants or pro-oxidants? What do we learn from cell culture and in vivo studies? *Archives of Biochemistry and Biophysics*, *476*(2), 107–112. https://doi.org/10.1016/J.ABB.2008.01.028

Hong, Y.-H., Jung, E. Y., Noh, D. O., & Suh, H. J. 2014. Physiological effects of formulation containing tannase-converted green tea extract on skin care: Physical stability, collagenase, elastase, and tyrosinase activities. *Integrative Medicine Research*, *3*(1), 25–33. https://doi.org/10.1016/J.IMR.2013.12.003

Hwang, H., Chen, T., Nines, R. G., Shin, H. C., & Stoner, G. D. 2006. Photochemoprevention of UVB-induced skin carcinogenesis in SKH-1 mice by brown algae polyphenols. *International Journal of Cancer*, *119*(12), 2742–2749. https://doi.org/10.1002/ijc.22147

Ichihashi, M., Ueda, M., Budiyanto, A., Bito, T., Oka, M., Fukunaga, M., Tsuru, K., & Horikawa, T. 2003. UV-induced skin damage. *Toxicology*, 189(1–2), 21–39. Elsevier Ireland Ltd. https://doi.org/10.1016/S0300-483X(03)00150-1

Imran, M., Ahmad, N., Anjum, F. M., Khan, M. K., Mushtaq, Z., Nadeem, M., & Hussain, S. 2015. Potential protective properties of flax lignan secoisolariciresinol diglucoside. *Nutrition Journal*, *14*(1), 1–7. https://doi.org/10.1186/S12937-015-0059-3/PEER-REVIEW

Imran, M., Iqubal, M. K., Imtiyaz, K., Saleem, S., Mittal, S., Rizvi, M. M. A., Ali, J., & Baboota, S. 2020. Topical nanostructured lipid carrier gel of quercetin and resveratrol: Formulation, optimization, in vitro and ex vivo study for the treatment of skin cancer. *International Journal of Pharmaceutics*, *587*, 119705. https://doi.org/10.1016/J.IJPHARM.2020.119705

Isaza, J. H., Ito, H., & Yoshida, T. 2004. Oligomeric hydrolyzable tannins from Monochaetum multiflorum. *Phytochemistry*, *65*(3), 359–367. https://doi.org/10.1016/j.phytochem.2003.11.017

Islam, M. M., Nagaraja, S., Hafsa, N. E., Meravanige, G., Asdaq, S. M. B., & Anwer, M. K. 2022. Polyphenol chrysin for management of skin disorders: Current status and future opportunities. *Journal of King Saud University – Science*, *34*(5), 102026. https://doi.org/10.1016/J.JKSUS.2022.102026

Jang, M., Cai, L., Udeani, G. O., Slowing, K. V., Thomas, C. F., Beecher, C. W. W., Fong, H. H. S., Farnsworth, N. R., Kinghorn, A. D., Mehta, R. G., Moon, R. C., & Pezzuto, J. M. 1997. Cancer chemopreventive activity of resveratrol, a natural product derived from grapes. *Science*, *275*(5297), 218–220. https://doi.org/10.1126/SCIENCE.275.5297.218

Jeon, H. Y., Kim, J. K., Kim, W. G., & Lee, S. J. 2009. Effects of oral epigallocatechin gallate supplementation on the minimal erythema dose and uv-induced skin damage. *Skin Pharmacology and Physiology*, *22*(3), 137–141. https://doi.org/10.1159/000201562

Jiang, Z., Li, S., Liu, Y., Deng, P., Huang, J., & He, G. 2011. Sesamin induces melanogenesis by microphthalmia-associated transcription factor and tyrosinase up-regulation via cAMP signaling pathway. *Acta Biochimica et Biophysica Sinica*, *43*(10), 763–770. https://doi.org/10.1093/ABBS/GMR078

Karim, A. A., & Azlan, A. 2012. Fruit pod extracts as a source of nutraceuticals and pharmaceuticals. *Molecules*, *17*(10), 11931–11946. https://doi.org/10.3390/MOLECULES171011931

Katiyar, S. K., Agarwal, R., & Mukhtar, H. 1994. Inhibition of spontaneous and photo-enhanced lipid peroxidation in mouse epidermal microsomes by epicatechin derivatives from green tea. *Cancer Letters*, *79*(1), 61–66

Katiyar, S. K., & Elmets, C. A. 2001. Green tea polyphenolic antioxidants and skin photoprotection (Review). *International Journal of Oncology*, *18*.

Katiyar, S. K., Elmets, C. A., & Katiyar, S. K. 2007. Green tea and skin cancer: Photoimmunology, angiogenesis and DNA repair. *Journal of Nutritional Biochemistry*, *18*(5), 287–296. https://doi.org/10.1016/j.jnutbio.2006.08.004

Katiyar, S. K., Mohan, R. R., & Agarwal, R. 1997. Protection against induction of mouse skin papillomas with low and high risk of conversion to malignancy by green tea polyphenols. *Carcinogenesis*, *18*(3).

Khanbabaee, K., & van Ree, T. 2001. Tannins: Classification and definition. *Natural Product Reports*, *18*(6), 641–649. https://doi.org/10.1039/b101061l

Kim, H. S., Quon, M. J., & Kim, J. A. 2014. New insights into the mechanisms of polyphenols beyond antioxidant properties; lessons from the green tea polyphenol, epigallocatechin 3-gallate. *Redox Biology*, *2*(1), 187–195. https://doi.org/10.1016/J.REDOX.2013.12.022

Kim, J., Hwang, J.-S., Cho, Y.-K., Han, Y., Jeon, Y.-J., Yang, K.-H., & Yang, K. H. 2001. Protective effects of (-)-epigallocatechin-3-gallate on UVA-and UVB-induced skin damage. *Original Research Article Skin Pharmacol Appl Skin Physiol*, *14*. www.karger.com/journals/sph

Komes, D., Horžić, D., Belščak, A., Ganić, K. K., & Vulić, I. 2010. Green tea preparation and its influence on the content of bioactive compounds. *Food Research International*, *43*(1), 167–176. https://doi.org/10.1016/J.FOODRES.2009.09.022

Kostyuk, V., Potapovich, A., Albuhaydar, A. R., Mayer, W., De Luca, C., & Korkina, L. 2018. Natural substances for prevention of skin photoaging: Screening systems in the development of sunscreen and rejuvenation cosmetics. *Rejuvenation Research*, *21*(2), 91. https://doi.org/10.1089/REJ.2017.1931

Kumar, N., & Goel, N. 2019. Phenolic acids: Natural versatile molecules with promising therapeutic applications. *Biotechnology Reports*, *24*. Elsevier B.V. https://doi.org/10.1016/j.btre.2019.e00370

Kumar, S., & Pandey, A. K. 2013. Chemistry and biological activities of flavonoids: An overview. *The Scientific World Journal*, *2013*. https://doi.org/10.1155/2013/162750

Li, H., Zhu, L., Weng, Z., Fu, H., Liu, J., Mao, Q., Li, W., Ding, B., & Cao, Y. 2024. Sesamin attenuates UVA-induced keratinocyte injury via inhibiting ASK-1-JNK/p38 MAPK pathways. *Journal of Cosmetic Dermatology*, *23*(1). https://doi.org/10.1111/JOCD.15951

Li, W., Chen, H., Xu, B., Wang, Y., Zhang, C., Cao, Y., & Xing, X. 2023. Research progress on classification, sources and functions of dietary polyphenols for prevention and treatment of chronic diseases. *Journal of Future Foods*, *3*(4), 289–305. https://doi.org/10.1016/J.JFUTFO.2023.03.001

Lin, T. Y., Wu, P. Y., Hou, C. W., Chien, T. Y., Chang, Q. X., Wen, K. C., Lin, C. Y., & Chiang, H. M. 2019. Protective effects of sesamin against UVB-induced skin inflammation and photodamage in vitro and in vivo. *Biomolecules*, *9*(9), 479. https://doi.org/10.3390/BIOM9090479

Liu, H., Liu, K., Huang, Z., Park, C. M., Thimmegowda, N. R., Jang, J. H., Ryoo, I. J., He, L., Kim, S. O., Oi, N., Lee, K. W., Soung, N. K., Bode, A. M., Yang, Y., Zhou, X., Erikson, R. L., Ahn, J. S., Hwang, J., Kim, K. E., . . . Kim, B. Y. 2013. A chrysin derivative suppresses skin cancer growth by inhibiting cyclin-dependent kinases. *Journal of Biological Chemistry*, *288*(36), 25924–25937. https://doi.org/10.1074/JBC.M113.464669

Liu, L., Hudgins, W. R., Shack, S., Yin, M. Q., & Samid, D. 1995. Cinnamic acid: A natural product with potential use in cancer intervention. *International Journal of Cancer*, *62*(3), 345–350. https://doi.org/10.1002/IJC.2910620319

Liu, N., Zhu, S., Deng, Y., Xie, M., Zhao, M., Sun, T., Yu, C., Zhong, Y., Guo, R., Cheng, K., Chang, D., & Zhu, P. 2023. Construction of multifunctional hydrogel with metal-polyphenol capsules for infected full-thickness skin wound healing. *Bioactive Materials*, *24*, 69–80. https://doi.org/10.1016/J.BIOACTMAT.2022.12.009

Lu, M. J., & Chen, C. 2007. Enzymatic tannase treatment of green tea increases in vitro inhibitory activity against N-nitrosation of dimethylamine. *Process Biochemistry*, *42*(9), 1285–1290. https://doi.org/10.1016/J.PROCBIO.2007.06.003

Ma, X., Li, Y., Lv, C., Liu, B., Yuan, C., Huang, W., Luo, Q., Xiao, Y., Sun, C., Li, T., & Zhang, J. 2022. Modulation of Keap1-Nrf2-ARE signaling pathway by oxyresveratrol, a derivative of resveratrol from grape skin. *Food Bioscience*, *50*, 102162. https://doi.org/10.1016/J.FBIO.2022.102162

Majdalawieh, A. F., Massri, M., & Nasrallah, G. K. 2017. A comprehensive review on the anti-cancer properties and mechanisms of action of sesamin, a lignan in sesame seeds (Sesamum indicum). *European Journal of Pharmacology*, *815*, 512–521. https://doi.org/10.1016/J.EJPHAR.2017.10.020

Mennen, L. I., Walker, R., Bennetau-Pelissero, C., & Scalbert, A. 2005. Risks and safety of polyphenol consumption. *The American Journal of Clinical Nutrition*, *81*(1 Suppl). https://doi.org/10.1093/ajcn/81.1.326s

Mittal, A., Elmets, C. A., & Katiyar, S. K. 2003. Dietary feeding of proanthocyanidins from grape seeds prevents photocarcinogenesis in SKH-1 hairless mice: Relationship to decreased fat and lipid peroxidation. *Carcinogenesis*, *24*(8), 1379–1388. https://doi.org/10.1093/carcin/bgg095

Mnich, C. D., Hoek, K. S., Virkki, L. V., Farkas, A., Dudli, C., Laine, E., Urosevic, M., & Dummer, R. 2009. Green tea extract reduces induction of p53 and apoptosis in UVB-irradiated human skin independent of transcriptional controls. *Experimental Dermatology*, *18*(1), 69–77. https://doi.org/10.1111/j.1600-0625.2008.00765.x

Nagaraja, S., Basavarajappa, G. M., Attimarad, M., & Pund, S. 2021. Topical nanoemulgel for the treatment of skin cancer: Proof-of-technology. *Pharmaceutics*, *13*(6), 902. https://doi.org/10.3390/PHARMACEUTICS13060902

Nagata, H., Takekoshi, S., Takeyama, R., Homma, T., & Osamura, R. Y. 2004. Quercetin enhances melanogenesis by increasing the activity and synthesis of tyrosinase in human melanoma cells and in normal human melanocytes. *Pigment Cell Research*, *17*(1), 66–73. https://doi.org/10.1046/J.1600-0749.2003.00113.X

Ndiaye, M., Philippe, C., Mukhtar, H., & Ahmad, N. 2011. The grape antioxidant resveratrol for skin disorders: Promise, prospects, and challenges. *Archives of Biochemistry and Biophysics*, *508*(2), 164–170. https://doi.org/10.1016/J.ABB.2010.12.030

Ness, A. R., & Powles, J. W. 1997. Fruit and vegetables, and cardiovascular disease: A review. *International Journal of Epidemiology*, *26*(1), 1–13. https://doi.org/10.1093/ije/26.1.1

Nichols, J. A., & Katiyar, S. K. 2010. Skin photoprotection by natural polyphenols: Anti-inflammatory, antioxidant and DNA repair mechanisms. *Archives of Dermatological Research*, *302*(2), 71–83. https://doi.org/10.1007/S00403-009-1001-3/FIGURES/1

Niero, E. L. D. O., & MacHado-Santelli, G. M. 2013. Cinnamic acid induces apoptotic cell death and cytoskeleton disruption in human melanoma cells. *Journal of Experimental and Clinical Cancer Research*, *32*(1). https://doi.org/10.1186/1756-9966-32-31

Niles, R. M., Cook, C. P., Meadows, G. G., Fu, Y. M., McLaughlin, J. L., & Rankin, G. O. 2006. Resveratrol is rapidly metabolized in athymic (Nu/Nu) mice and does not inhibit human melanoma xenograft tumor growth. *The Journal of Nutrition*, *136*(10), 2542–2546. https://doi.org/10.1093/JN/136.10.2542

Okuda, T., & Ito, H. 2011. Tannins of constant structure in medicinal and food plants-hydrolyzable tannins and polyphenols related to tannins. *Molecules*, *16*(3), 2191–2217. https://doi.org/10.3390/molecules16032191

Osmond, G. W., Augustine, C. K., Zipfel, P. A., Padussis, J., & Tyler, D. S. 2012. Enhancing melanoma treatment with resveratrol. *Journal of Surgical Research*, *172*(1), 109–115. https://doi.org/10.1016/j.jss.2010.07.033

Oyetakinwhite, P., Tribout, H., & Baron, E. 2012. Protective mechanisms of green tea polyphenols in skin. *Oxidative Medicine and Cellular Longevity*. https://doi.org/10.1155/2012/560682

Pandey, K. B., & Rizvi, S. I. 2009. Plant polyphenols as dietary antioxidants in human health and disease. *Oxidative Medicine and Cellular Longevity*, *2*(5), 270–278. https://doi.org/10.4161/oxim.2.5.9498

Pathak, S., Kesavan, P., Banerjee, A., Banerjee, A., Celep, G. S., Bissi, L., & Marotta, F. 2018. Metabolism of dietary polyphenols by human gut microbiota and their health benefits. https://doi.org/10.1016/B978-0-12-813006-3.00025-8

Paulitschke, V., Schicher, N., Szekeres, T., Jäger, W., Elbling, L., Riemer, A. B., Scheiner, O., Trimurtulu, G., Venkateswarlu, S., Mikula, M., Swoboda, A., Fiebiger, E., Gerner, C., Pehamberger, H., & Kunstfeld, R. 2010. 3,3′,4,4′,5,5′-hexahydroxystilbene impairs melanoma progression in a metastatic mouse model. *Journal of Investigative Dermatology*, *130*(6), 1668–1679. https://doi.org/10.1038/jid.2009.376

Pelinson, L. P., Assmann, C. E., Palma, T. V., da Cruz, I. B. M., Pillat, M. M., Mânica, A., Stefanello, N., Weis, G. C. C., de Oliveira Alves, A., de Andrade, C. M., Ulrich, H., Morsch, V. M. M., Schetinger, M. R. C., & Bagatini, M. D. 2019. Antiproliferative and apoptotic effects of caffeic acid on SK-Mel-28 human melanoma cancer cells. *Molecular Biology Reports*, *46*(2), 2085–2092. https://doi.org/10.1007/s11033-019-04658-1

Pontes, J. F., Fonseca, M., Macedo, A. S., Grenha, A., & Fonte, P. 2023. New and revisited approaches on the transdermal delivery of polyphenols-loaded nanoparticles for melanoma prevention and treatment. In *Phytopharmaceuticals and Herbal Drugs: Prospects and Safety Issues in the Delivery of Natural Products*, 341–373. https://doi.org/10.1016/B978-0-323-99125-4.00001-9

Pourhanifeh, M. H., Abbaszadeh-Goudarzi, K., Goodarzi, M., Piccirillo, S. G. M., Shafiee, A., Hajighadimi, S., Moradizarmehri, S., Asemi, Z., & Mirzaei, H. 2019. Resveratrol: A new potential therapeutic agent for melanoma? *Current Medicinal Chemistry*, *28*(4), 687–711. https://doi.org/10.2174/0929867326666191212101225

Riyanto, P., Subchan, P., & Lelyana, R. 2015. Advantage of soybean isoflavone as antiandrogen on acne vulgaris. *Dermato-Endocrinology*, *7*(1). https://doi.org/10.1080/19381980.2015.1063751

Rosalina, R., & Weerapreeyakul, N. 2021. An insight into sesamolin: Physicochemical properties, pharmacological activities, and future research prospects. *Molecules (Basel, Switzerland)*, *26*(19). https://doi.org/10.3390/MOLECULES26195849

Ruddock, P. L., & Facey, P. 2023. Tannins. In *Pharmacognosy: Fundamentals, Applications, and Strategies, Second Edition*, 211–251. Elsevier. https://doi.org/10.1016/B978-0-443-18657-8.00015-3

Russell, L., Hicks, G. S., Low, A. K., Shepherd, J. M., & Brown, C. A. 2002. Phytoestrogens: A viable option? *The American Journal of the Medical Sciences*, *324*(4), 185–188. https://doi.org/10.1097/00000441-200210000-00004

Scalbert, A., Manach, C., Morand, C., Rémésy, C., & Jiménez, L. 2005. Dietary polyphenols and the prevention of diseases. *Critical Reviews in Food Science and Nutrition*, *45*(4), 287–306. https://doi.org/10.1080/1040869059096

Shen, D., Wu, Q., Wang, M., Yang, Y., Lavoie, E. J., & Simon, J. E. 2006. Determination of the predominant catechins in Acacia catechu by liquid chromatography/electrospray ionization-mass spectrometry. *Journal of Agricultural and Food Chemistry*, *54*(9), 3219–3224. https://doi.org/10.1021/jf0531499

Sheweita, S. A., El-Masry, Y. M., Zaghloul, T. I., Mostafa, S. K., & Elgindy, N. A. 2022. Preclinical studies on melanogenesis proteins using a resveratrol-nanoformula as a skin whitener. *International Journal of Biological Macromolecules*, *223*, 870–881. https://doi.org/10.1016/J.IJBIOMAC.2022.11.036

Singh, A., Holvoet, S., & Mercenier, A. 2011. Dietary polyphenols in the prevention and treatment of allergic diseases. *Clinical & Experimental Allergy*, *41*(10), 1346–1359. https://doi.org/10.1111/J.1365-2222.2011.03773.X

Souto, E. B., Sampaio, A. C., Campos, J. R., Martins-Gomes, C., Aires, A., & Silva, A. M. 2019. Polyphenols for skin cancer: Chemical properties, structure-related mechanisms of action and new delivery systems. *Studies in Natural Products Chemistry*, *63*, 21–42. https://doi.org/10.1016/B978-0-12-817901-7.00002-2

Srisayam, M., Weerapreeyakul, N., & Kanokmedhakul, K. 2017. Inhibition of two stages of melanin synthesis by sesamol, sesamin and sesamolin. *Asian Pacific Journal of Tropical Biomedicine*, *7*(10), 886–895. https://doi.org/10.1016/J.APJTB.2017.09.013

Srisongkram, T., & Weerapreeyakul, N. 2022. Route of intracellular uptake and cytotoxicity of sesamol, sesamin, and sesamolin in human melanoma SK-MEL-2 cells. *Biomedicine & Pharmacotherapy*, *146*, 112528. https://doi.org/10.1016/J.BIOPHA.2021.112528

Su, D., Cheng, Y., Liu, M., Liu, D., Cui, H., Zhang, B., Zhou, S., Yang, T., & Mei, Q. 2013. Comparision of piceid and resveratrol in antioxidation and antiproliferation activities in vitro. *PLoS ONE*, *8*(1), e54505. https://doi.org/10.1371/JOURNAL.PONE.0054505

Suja, K. P., Jayalekshmy, A., & Arumughan, C. 2004. Free radical scavenging behavior of antioxidant compounds of sesame (Sesamum indicum L.) in DPPH• system. *Journal of Agricultural and Food Chemistry*, *52*(4), 912–915. https://doi.org/10.1021/JF0303621

Sun, M., Nie, S., Pan, X., Zhang, R., Fan, Z., & Wang, S. 2014. Quercetin-nanostructured lipid carriers: Characteristics and anti-breast cancer activities in vitro. *Colloids and Surfaces B: Biointerfaces*, *113*, 15–24. https://doi.org/10.1016/J.COLSURFB.2013.08.032

Sun, W., Chen, X., Nan, X., Zhang, J., Dong, L., Ji, W., Sheng, G., & Zhou, Q. 2021. Inhibition of persimmon tannin extract on guinea pig skin pigmentation. *Journal of Cosmetic Dermatology*, *20*(8), 2648–2656. https://doi.org/10.1111/jocd.13915

Taiwo, B. J., Popoola, T. D., Van Heerden, F. R., & Fatokun, A. A. 2020. Pentagalloylglucose, isolated from the leaf extract of Anacardium occidentale L., could elicit rapid and selective cytotoxicity in cancer cells. *BMC Complementary Medicine and Therapies*, *20*(1). https://doi.org/10.1186/s12906-020-03075-3

Tresserra-Rimbau, A., Lamuela-Raventos, R. M., & Moreno, J. J. 2018. Polyphenols, food and pharma. Current knowledge and directions for future research. In *Biochemical Pharmacology*, Vol. 156, 186–195. Elsevier Inc. https://doi.org/10.1016/j.bcp.2018.07.050

Yang, G., Fu, Y., Malakhova, M., Kurinov, I., Zhu, F., Yao, K., Li, H., Chen, H., Li, W., Lim, D. Y., Sheng, Y., Bode, A. M., Dong, Z., & Dong, Z. 2014. Caffeic acid directly targets ERK1/2 to attenuate solar UV-induced skin carcinogenesis. *Cancer Prevention Research*, *7*(10), 1056–1066. https://doi.org/10.1158/1940-6207.CAPR-14-0141

Yang, Y. M., Son, Y. O., Lee, S. A., Jeon, Y. M., & Lee, J. C. 2011. Quercetin inhibits α-MSH-stimulated melanogenesis in B16F10 melanoma cells. *Phytotherapy Research: PTR*, *25*(8), 1166–1173. https://doi.org/10.1002/PTR.3417

Zekrumah, M., Begua, P., Razak, A., Wahab, J., Moffo, N., Ivane, A., Oman, M., Elrashied, H., Zou, X., & Zhang, D. 2023. Role of dietary polyphenols in non-communicable chronic disease prevention, and interactions in food systems: An overview. *Nutrition*, *112*, 112034. https://doi.org/10.1016/J.NUT.2023.112034

Zhao, J. F., Zhang, Y. J., Jin, X. H., Athar, M., Santella, R. M., Bickers, D. R., & Wang, Z. Y. 1999. Green tea protects against psoralen plus ultraviolet a-induced photochemical damage to skin1. *Journal of Investigative Dermatology*, *113*(6), 1070–1075. https://doi.org/10.1046/J.1523-1747.1999.00806.X

Zillich, O. V., Schweiggert-Weisz, U., Eisner, P., & Kerscher, M. 2015. Polyphenols as active ingredients for cosmetic products. *International Journal of Cosmetic Science*, *37*(5), 455–464. https://doi.org/10.1111/ICS.12218

Zillich, O. V., Schweiggert-Weisz, U., Hasenkopf, K., Eisner, P., & Kerscher, M. 2013. Release and in vitro skin permeation of polyphenols from cosmetic emulsions. *International Journal of Cosmetic Science*, *35*(5), 491–501. https://doi.org/10.1111/ICS.12072

11 Phytochemical Compounds with a Role in Wound Healing and the Proliferation of Dermal Fibroblasts

Muhammed Burak Yucel, Ragip Ertas, Mustafa Nisari, Sule Ketenci Ertas, Ayşe Baldemir Kılıç, Sadi Yusufbeyoğlu and Ismail Ocsoy

11.1 INTRODUCTION

11.1.1 Definition and Overview of Phytochemicals

Phytochemicals, a term coined from the Greek word 'phyto' meaning plant, are naturally occurring compounds in plants, known for their role in plant defence and human health. These compounds, produced as a part of plants' protective mechanisms against various environmental stresses, pathogens, and predators, have garnered significant attention for their potential health benefits when consumed by humans (Craig, 1999; Liu, 2004). Phytochemicals encompass a wide array of compounds, each with distinct chemical structures and functions. They are broadly categorized into several major groups, polyphenols, flavonoids, terpenoids, alkaloids, and sulfides; for instance, polyphenols, known for their antioxidant properties, play a crucial role in combating oxidative stress, a contributor to chronic diseases. Flavonoids, a subclass of polyphenols, are abundantly found in fruits, vegetables, and beverages like tea and wine, and are known for their anti-inflammatory and cardioprotective effects (Pandey et al., 2009). The impact of phytochemicals on human health has been a focus of research, linking their consumption to reduced risks of several chronic diseases. Studies have shown that diets rich in phytochemical-containing foods are associated with lower risks of cancer, cardiovascular diseases, diabetes, and neurodegenerative disorders. This is attributed to their ability to modulate various biological processes including antioxidant defence, enzyme activity, cell signalling, and gene expression (Liu, 2004).

Recent research has expanded to explore the potential of phytochemicals in dermatology, particularly their role in wound healing and the proliferation of dermal fibroblasts, essential for skin repair and regeneration. This new area of study suggests promising therapeutic applications for phytochemicals in skin health, capitalizing on their natural healing properties (Chainani-Wu, 2003; Thangapazham et al., 2007). In essence, phytochemicals are considered a natural member of the intersection between plant biology and human health. Their various biological activities which are rooted in their natural origin, have made them a subject of intense study in nutrition and medical science. The exploration of their role in processes like wound healing and skin regeneration underscores the profound and beneficial relationship between plants world and human health and with this, some phytochemicals already have been used in traditional medicine for centuries.

DOI: 10.1201/9781003394846-11

In summary, phytochemicals represent a fascinating and vital aspect of human nutrition and health. Their wide-ranging biological and cellular activities, coupled with their natural origin make them a thrilling subject of study in the quest for maintaining and enhancing human health. The exploration of their role in specific processes like wound healing and skin regeneration opens new avenues in both nutritional science and medical research, highlighting the intricate and beneficial relationship between the plant kingdom and human health.

11.1.2 Importance of Wound Healing and Skin Regeneration

Wound healing and skin regeneration are complex processes that require a well-orchestrated series of events involving cell migration, proliferation, and differentiation. Phytochemicals have emerged as pivotal agents in facilitating these processes, primarily due to their anti-inflammatory, antioxidant, and antimicrobial properties. Inflammation is a critical phase in wound healing, but excessive inflammation can impede the process and lead to scarring and poor wound healing. Phytochemicals like flavonoids and curcumin have been shown to exert significant anti-inflammatory effects, modulating the inflammatory response and promoting better wound-healing outcomes. Oxidative stress can impair wound healing. Phytochemicals, particularly polyphenols, are potent antioxidants that neutralize free radicals and protect tissues from oxidative damage. This action accelerates wound closure and prevents cell death around the wound site (Pandey and Rizvi, 2014). The risk of infection is a major concern in wound healing. Several phytochemicals have antimicrobial properties that are beneficial in preventing wound infections. For example, allicin from garlic and catechins from green tea have been reported to exhibit broad-spectrum antimicrobial activity against wound pathogens (Reygaert, 2014).

Dermal fibroblasts play a key role in wound healing since they are responsible for collagen production and tissue formation. Phytochemicals have been found to promote dermal fibroblast proliferation and collagen synthesis, thereby enhancing the structural integrity of the regenerated tissue (Varani et al., 2006). Clinical studies have started to validate the efficacy of phytochemicals in wound healing and skin regeneration. For instance, topical applications containing phytochemicals have shown improved wound healing in both animal models and human trials (Thangapazham et al., 2016). Various properties of phytochemicals, such as anti-inflammatory, antioxidant, antimicrobial and fibroblast stimulating, position them as key components in the field of wound healing and skin regeneration. Their natural origin, coupled with these beneficial properties, underscores the potential of phytochemicals in developing new therapeutic strategies for skin repair and regeneration.

11.1.3 Brief Overview of Dermal Fibroblasts in Wound Healing

In the dynamic world of cellular biology, dermal fibroblasts play a key role in wound healing and skin regeneration. Understanding their function is key to developing advanced therapeutic strategies for skin injuries and disorders. Dermal fibroblasts are specialized cells located within the dermis, the skin's second layer. They are primarily responsible for maintaining the structural integrity and biomechanical properties of the skin by producing and organizing the extracellular matrix (ECM), which includes collagen, elastin, and glycosaminoglycans (Sorrell and Caplan, 2004). Wound healing is a complex process that involves multiple phases which are hemostasis, inflammation, proliferation, and remodeling (Figure 11.1). Dermal fibroblasts play a crucial role, particularly in the latter two phases:

1. **Proliferation Phase:** After the inflammatory phase begins, fibroblasts migrate to the wound site, proliferate, and begin depositing new ECM components. This phase is characterized by the formation of granulation tissue, which provides a temporary matrix for cell migration and further tissue development (Gurtner et al., 2008).

2. **Remodeling Phase**: In this final phase, fibroblasts continue to synthesize and remodel the ECM, contributing to the strength and flexibility of the newly formed tissue. They differentiate from myofibroblasts, which have contractile properties that help in wound contraction, a vital aspect of wound closure. The longest step of wound healing is the remodelling step, which can last for years (Hinz, 2007).

Beyond wound healing, dermal fibroblasts are central to skin regeneration and anti-ageing therapies. Their ability to produce collagen is particularly important in maintaining skin elasticity and reducing the appearance of wrinkles. Age-related decline in fibroblast function is a significant factor in skin ageing, leading to reduced collagen production and skin laxity. Understanding the biology of dermal fibroblasts has led to advancements in skin repair therapies. This includes the development of fibroblast-based treatments for chronic wounds, burn injuries, and cosmetic procedures that are aimed at skin rejuvenation (Varani et al., 2006). To sum up; dermal fibroblasts are more than just structural cells; they are active participants in wound healing and skin regeneration. Their multifaceted role in ECM production, wound contraction, and tissue remodelling underscores their importance in both natural healing processes and clinical applications for skin health.

11.2 PHYTOCHEMICALS AND THEIR PROPERTIES

11.2.1 Classification of Phytochemicals

Phytochemicals, an array of bioactive compounds found in plants, are classified into various categories based on their chemical structure and the function they perform.

11.2.1.1 Polyphenols

Polyphenols are a large and diverse group of phytochemicals characterized by the presence of multiple phenol units. They are known for their antioxidant properties and are divided into subcategories such as flavonoids, phenolic acids, stilbenes, and lignans. Among them, flavonoids which are the most studied group, include flavonols, flavones, isoflavones, and anthocyanins, found in various fruits, vegetables, and teas. Polyphenols are linked with reduced risk of chronic diseases due to their ability to mitigate oxidative stress and inflammation (Kroon et al., 2004).

11.2.1.2 Flavonoids

Flavonoids are a large group of polyphenols with over 6,000 identified compounds. They are known for their anti-inflammatory, antioxidant, and anti-carcinogenic properties. Common flavonoids include quercetin, kaempferol, catechins, and anthocyanins, which are abundant in a variety of plant foods like berries, apples, and onions (Middleton et al., 2000).

11.2.1.3 Carotenoids

Carotenoids are pigments that give plants their red, orange, and yellow colours. They are divided into two categories xanthophylls and carotenes. Beta-carotene, lycopene, and lutein are well-known carotenoids, recognized for their role as antioxidants and their contribution to human health, particularly in eye health and disease prevention. Carotenoids are converted into vitamin A in the body, which is essential for vision, growth, and immune function (Krinsky and Johnson, 2005).

11.2.1.4 Saponins

Saponins are a group of phytochemicals that form foam in aqueous solutions and are secondary metabolites of plants. It contains terpenoids and steroid alkaloids (Rex et al., 2018). These compounds have anti-inflammatory and antimicrobial effects, which are essential in the initial stages of wound healing to reduce inflammation and prevent infection (Vitale et al., 2022).

11.2.1.5 Terpenoids

Terpenoids are the largest class of phytochemicals, consisting of more than 40,000 compounds. They are responsible for the aromatic features of plants, and they have significant therapeutic properties, including anti-inflammatory, antibacterial, and antiviral effects. Some terpenoids are also known to inhibit tumour growth. For instance, while terpenes such as gentiopicroside, sweroside, and swertiamarin support the mitotic activities of fibroblasts, borneol and thymol terpenes contribute to wound healing with an increase in angiogenic and collagen synthesis (Budovsky et al., 2015).

11.2.1.6 Alkaloids

Alkaloids are nitrogen-containing compounds that are often used for their pharmacological effects. They have a wide range of therapeutic properties that include analgesics and stimulation of the central nervous system. Common alkaloids include caffeine, nicotine, and morphine. For example, nicotine has wound-healing and angiogenesis-stimulating effects because it can affect many intracellular processes even in small doses. In addition, Turcz alkaloids such as mesaconitine and hypaconitine have positive effects on wound healing by stimulating fibroblast proliferation and TFG-β1 secretion from fibroblasts which leading to greater amounts of growth factor and secretion by skin stromal cells (Budovsky et al., 2015).

11.2.2 General Properties and Functions of Phytochemicals in the Body

Phytochemicals are natural compounds found in plants and play significant roles in human health due to their multiple properties and biological activities. Some of these features have been tried to be explained in this section.

11.2.2.1 Antioxidant Activity

One of the most notable properties of many phytochemicals is their antioxidant activity. Compounds like flavonoids and carotenoids act as antioxidants, neutralizing free radicals and preventing oxidative stress, a contributor to chronic diseases and ageing. This antioxidant property is crucial in protecting cells and tissues from oxidative damage, thereby reducing the risk of various diseases, including cancer and cardiovascular diseases (Liu, 2004; Pandey et al., 2009).

11.2.2.2 Anti-Inflammatory Effects

Phytochemicals, particularly polyphenols like curcumin and resveratrol, exhibit strong anti-inflammatory properties. They modulate the inflammatory response by inhibiting enzymes and cytokines such as TNF-α and IL-1 that are involved in inflammation, thus playing a role in preventing and managing inflammatory diseases (Chainani-Wu, 2003).

11.2.2.3 Modulation of Detoxification Enzymes

Biocompounds found in cruciferous vegetables such as broccoli, which are highly rich in glucosinolates, which are converted into isothiocyanates by the intestinal flora, may affect the activity of enzymes involved in detoxification processes, activate enzymes that help detoxify carcinogens, and thus offer protective effects against certain types of cancer (Talalay and Fahey, 2001).

11.2.2.4 Hormonal Regulation

Isoflavones, a class of phytochemicals found in soy, are known to mimic or modulate hormones, particularly estrogen. They can bind to estrogen receptors and they may show estrogenic or anti-estrogenic effects, which is important in the context of hormone-related cancers and menopausal symptoms (Cassidy et al., 2006).

11.2.2.5 Antimicrobial Properties

Phytochemicals like allicin from garlic and catechins from green tea have demonstrated antimicrobial properties. They can inhibit the growth of bacteria, fungi, and viruses, thus contributing to the prevention and treatment of infections (Reygaert, 2014).

11.2.2.6 Enhancing Immune Function

Certain phytochemicals can strengthen the immune system. For example, beta-glucans found in mushrooms are known to enhance the immune response by stimulating the activity of macrophages and natural killer cells (Kozarski et al., 2023).

Phytochemicals, with their heterogeneous biological activities, play a significant role in the maintenance of health and the prevention of diseases. Their function in the body encompasses antioxidant activities, anti-inflammatory effects, modulation of detoxification enzymes, hormonal regulation, antimicrobial properties, and immune system enhancement that exhibit the potential of these natural compounds in contributing to overall health and well-being.

11.2.3 Mechanisms of Action at the Cellular Level

Phytochemicals, the bioactive compounds found in plants show a variety of effects at the cellular level. These effects are crucial for their health-promoting properties.

11.2.3.1 Antioxidant Mechanisms

Phytochemicals like flavonoids and carotenoids act as antioxidants primarily by scavenging reactive oxygen species (ROS) and free radicals. They neutralize these harmful molecules, thereby preventing oxidative damage to cellular components such as DNA, proteins, and lipids. This action is crucial in protecting against cellular damage and chronic diseases associated with oxidative stress (Liu, 2004; Pandey et al., 2009).

11.2.3.2 Anti-Inflammatory Pathways

Many phytochemicals, such as curcumin and resveratrol, show anti-inflammatory effects by modulating signalling pathways involved in inflammation. They can inhibit key inflammatory mediators like nuclear factor-kappa B (NF-κB) and cytokines such as TNF-α and IL-1, thereby reducing the inflammatory response at the cellular level. It can even be said that the main effect of phytochemicals on wound healing is by reducing nuclear factor-kappa B (NF-κB) levels and reactive oxygen radicals. This property is significant in the prevention and treatment of various inflammatory conditions (Chainani-Wu, 2003; Thangapazham et al., 2016).

11.2.3.3 Modulation of Enzyme Activity

Certain phytochemicals influence the activity of various enzymes. For example, compounds in cruciferous vegetables, such as glucosinolates, can increase the body's ability to detoxify and eliminate carcinogens by inducing phase II detoxification enzymes. Similarly, phytochemicals can inhibit enzymes involved in the synthesis of cholesterol, impacting lipid metabolism (Rodríguez-Cantú et al., 2011; Talalay and Fahey, 2001).

Phytochemicals can regulate cell proliferation and induce apoptosis (programmed cell death) in cancer cells. Compounds like genistein (an isoflavone) and sulforaphane (found in cruciferous vegetables) have been shown to interrupt the cell cycle in cancer cells, leading to their death and preventing tumour growth (Khan et al., 2008; Zhang and Tang, 2007).

11.2.3.4 Hormonal Actions

Isoflavones, such as those found in soy, can bind to estrogen receptors and show estrogenic or anti-estrogenic effects. The interaction of phytoestrogens such as α-naphthoflavone (ANF) with hormone receptors can affect intracellular enzyme systems such as hormone-related aromatase and is particularly relevant for breast and prostate cancer preventive effects (Campbell and Kurzer, 1993; Cassidy et al., 1998).

11.2.3.5 Enhancement of Immune Function

A variety of different amounts of phytochemicals found in echinacea and some mushrooms may enhance immune function by stimulating the activity of immune cells such as macrophages and lymphocytes. These effects may strengthen the body's defence mechanism against pathogens and diseases (Block and Mead, 2003).

Eventually, the mechanisms of action of phytochemicals at the cellular level are multiple and impactful. They include antioxidant activities, modulation of inflammatory responses, enzyme regulation, control of cell proliferation and apoptosis, hormonal interactions, and immune system enhancement. These cellular actions form the basis of the health benefits associated with phytochemicals and highlight their potential in disease prevention and therapy.

11.3 WOUND HEALING PROCESS

11.3.1 Phases of Wound Healing

Wound healing is a complex and dynamic process, essential for the restoration of tissue integrity following injury. It encompasses four distinct and at the same time overlapping stages hemostasis, inflammation, proliferation, and remodelling.

11.3.1.1 Hemostasis

The hemostasis phase is the body's immediate response to injury, initiating within minutes. It involves vasoconstriction to reduce blood loss and the formation of a fibrin clot. Platelets play a critical role in this stage by aggregating at the wound site, releasing clotting factors, and forming a temporary plug to stop bleeding. This sets the stage for the inflammatory phase (Velnar et al., 2009).

11.3.1.2 Inflammation

Following hemostasis, the inflammatory phase begins, typically lasting a few days. It is characterized by the recruiting of neutrophils and macrophages to the wound site. These immune cells clear the wound of debris, bacteria, and damaged tissue. They also release cytokines and growth factors that are crucial for the healing process. While inflammation is necessary for healing, prolonged inflammation can hinder the process and lead to chronic wounds (Eming et al., 2007)

11.3.1.3 Proliferation

The proliferation phase overlaps with inflammation and can last several weeks. This phase is characterized by angiogenesis (formation of new blood vessels), collagen deposition, granulation tissue formation, and re-epithelialization. Fibroblasts play a major role in this stage, proliferating and synthesizing collagen and extracellular matrix proteins to provide strength and structure to the healing tissue (Gurtner et al., 2008).

11.3.1.4 Remodeling

The final phase of wound healing is remodelling, which can last up to a year or more. During this phase, the newly formed tissue gradually gains strength and flexibility. Collagen is remodelled from type III to type I, and the extracellular matrix is reorganized. This phase is crucial for restoring the functional integrity of the skin and reducing scar formation (Wolfram et al., 2009).

Understanding these stages of wound healing is essential for developing effective treatment strategies for various types of wounds and for addressing complications that can arise during the healing process.

11.3.2 Role of Dermal Fibroblasts in Wound Healing

Dermal fibroblasts are key players in the process of wound healing, significantly contributing to tissue repair and regeneration. Their role encompasses various aspects of the healing process, from the formation of new tissue to the remodelling of the wound area.

11.3.2.1 Activation and Migration

Upon injury, dermal fibroblasts are activated and migrate to the wound site. This migration is crucial for initiating the repair process. Activated fibroblasts begin to proliferate and produce extracellular matrix (ECM) components, preparing the ground for new tissue formation (Tomasek et al., 2002).

11.3.2.2 Collagen Synthesis

One of the primary functions of fibroblasts during wound healing is the synthesis of collagen, the main structural protein in the skin. Collagen provides strength and structural integrity to the newly formed tissue. Fibroblasts initially produce type III collagen, which is later replaced by the stronger type I collagen during the remodelling phase (Gelse et al., 2003).

11.3.2.3 Extracellular Matrix Modulation

Besides collagen, fibroblasts produce and remodel other ECM components such as elastin and glycosaminoglycans. This ECM forms the scaffolding necessary for cell migration, proliferation, and differentiation, essential for effective wound healing (Sorrell and Caplan, 2004).

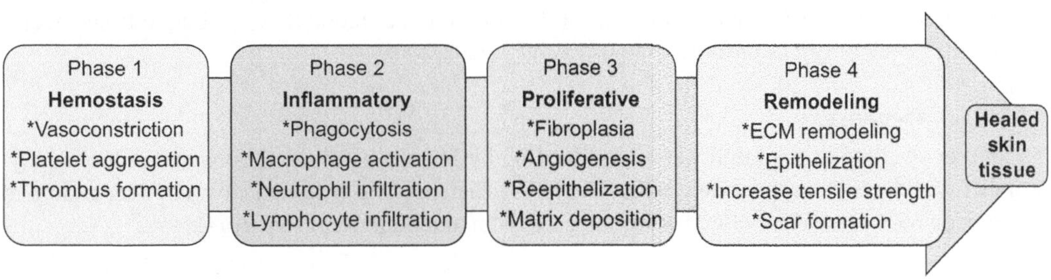

FIGURE 11.1 Phases of wound healing (El-Sherbeni and Negm, 2023).

11.3.2.4 Wound Contraction

Fibroblasts differentiate into myofibroblasts, acquiring contractile properties. Myofibroblasts play a significant role in wound contraction, reducing the wound size and facilitating closure. This is particularly important in the healing of large wounds (Hinz, 2007).

11.3.2.5 Regulation of Inflammatory Response

Fibroblasts also modulate the inflammatory response in the wound area. They interact with immune cells and release cytokines and growth factors that regulate inflammation and aid in tissue repair (Eming et al., 2007).

Dermal fibroblasts are essential for wound healing, with their roles extending from the initial response to injury through to the final stages of wound closure and tissue remodelling. Understanding these roles is critical for developing treatments for wound healing and skin regeneration.

11.3.3 Common Challenges and Complications in Wound Healing

Wound healing is a complex process, and many various factors can lead to challenges and complications, impeding proper healing. Understanding these challenges is crucial for effective wound management and prevention of chronic wound conditions.

11.3.3.1 Infection

One of the most common complications in wound healing is infection. Bacterial contamination can delay healing, exacerbate inflammation, and lead to further tissue damage. Chronic wounds are particularly susceptible to infections, which can lead to systemic complications if not properly managed (Bowling et al., 2009).

11.3.3.2 Chronic Inflammation

While inflammation is a necessary and important phase of healing, prolonged or chronic inflammation can hinder the process. It can lead to excessive tissue damage and prevent progression to later stages of healing. Chronic inflammation is often seen in wounds of diabetic patients and can lead to diabetic ulcers (Guo and DiPietro, 2010).

11.3.3.3 Poor Blood Circulation

Adequate blood flow is essential for delivering nutrients and oxygen to the wound site, as well as for removing waste products. Conditions that disrupt the circulation, such as diabetes and peripheral arterial disease, can significantly slow wound healing and increase the risk of complications (Sen, 2009).

11.3.3.4 Pressure Ulcers

Pressure ulcers, also known as bedsores, are a significant challenge, particularly in immobile patients. Constant pressure on certain areas of the body reduces blood flow, leading to tissue necrosis and the development of chronic wounds (Reddy et al., 2008).

11.3.3.5 Oedema

Edema, or swelling, can complicate wound healing by increasing tissue tension and reducing oxygen diffusion. It is often associated with venous insufficiency and lymphatic dysfunction, and managing oedema is crucial for effective wound care (Mortimer and Rockson, 2014).

11.3.3.6 Nutritional Deficiencies

Nutrition plays a vital role in wound healing. Deficiencies in proteins, vitamins, and minerals can impair immune function, collagen synthesis, and tissue regeneration, leading to delayed wound healing (Molnar et al., 2014).

Briefly, the challenges and complications in wound healing can vary from infections and chronic inflammation to issues related to blood circulation, pressure ulcers, oedema, and nutritional deficiencies. Addressing these factors carefully is essential for promoting effective wound healing and preventing the development of chronic wounds.

11.4 PHYTOCHEMICALS IN DERMAL FIBROBLAST FUNCTION

11.4.1 Detailed Analysis of how Phytochemicals Affect Dermal Fibroblast Proliferation

Phytochemicals, naturally occurring compounds in plants, have been increasingly recognized for their significant effects on dermal fibroblast proliferation, a crucial aspect of skin repair and wound healing. The well-known effects of phytochemicals on dermal fibroblasts have been tried to be explained here.

11.4.1.1 Stimulation of Fibroblast Proliferation

Several phytochemicals have been shown to stimulate the proliferation of dermal fibroblasts. For instance, flavonoids like quercetin and genistein can promote fibroblast growth, enhancing their capacity to repair and regenerate skin tissue. This effect is particularly beneficial in wound healing, where rapid fibroblast proliferation is needed for effective tissue repair (Huang et al., 2018)

In recent years, the extract prepared from the sap of the Lepiniopsis ternatensis plant has been found to significantly reduce M1 macrophage TNF-α secretion and also increase dermal fibroblast proliferation without affecting keratinocytes (Moses et al., 2020).

Moreover, considering the need for new therapeutic agents that can increase the proliferative ability of mesenchymal stem cells (MSCs) and at the same time reduce ageing processes, compounds produced from phytochemicals such as Panax ginseng, liquorice root and Salvia miltiorrhiza have been studied.

These phytocompounds have been observed to increase MSC numbers and delay cellular ageing *in vitro*, especially by upregulating the β-catenin pathway and reducing ROS levels (Yuan et al., 2022).

Again, a study conducted in 2020 revealed that progress has been made in wound healing with the new generation green methods used to obtain phytochemicals and that phytochemicals, especially those containing high amounts of phenol, have high antioxidant and fibroblast proliferation-enhancing effects (Addis et al., 2020).

11.4.1.2 Enhancement of Collagen Synthesis

Phytochemicals also play a role in enhancing collagen synthesis by dermal fibroblasts. Compounds like epigallocatechin gallate (EGCG) found in green tea have been observed to increase collagen production. Collagen is a critical component of the extracellular matrix and is essential for maintaining skin strength and elasticity (Nakamuta et al., 2005).

11.4.1.3 Antioxidant Effects

Oxidative stress can impair fibroblast function and delay wound healing. Phytochemicals with antioxidant properties, such as resveratrol and curcumin, protect fibroblasts from oxidative damage. By neutralizing free radicals, these compounds maintain fibroblast viability and promote healthy skin regeneration (Bisht et al., 2010).

In addition, the effects of phenolic acid and flavonoid phytochemical compounds on chronic wound healing have been demonstrated with their chondrocyte and fibroblast protective antioxidant and anti-bacterial activities by suppressing inflammatory cytokines in chronic osteoarthritis (Salem et al., 2023).

In addition, it is available in the literature that Aesculus hippocastanum green fruit mesocarps (AHFME), whose methanol extracts contain abundant phenol, flavonoid and alkaloid phytochemicals, show significant antioxidant activity, and that the seed extracts of the plant are also antioxidant and cause an increase in fibroblast proliferation (Çelik and Aslantürk, 2023).

In another study conducted in recent years, Avenanthramide C (Avn C), a polyphenolic phytochemical found only in oats (Avena sativa L.), prevented H_2O_2-induced intracellular oxidative stress in dermal human fibroblasts by scavenging reactive oxygen radicals and increasing Heme Oxygenase-1 (HO-1) levels. It also showed an anti-inflammatory effect by reducing the levels of pro-inflammatory cytokines, especially TNF-α, by reducing phosphorylated nuclear factor-κB (NF-κB), and a cytoprotective effect was observed on human dermal fibroblasts (Wang and Eskiw, 2019).

11.4.1.4 Modulation of Growth Factors and Cytokines

Certain phytochemicals can modulate the release of growth factors and cytokines are vital for fibroblast function. For example, saponins have been reported to increase the expression of growth factors like TGF-β, which is important for fibroblast proliferation and differentiation (Choi et al., 2019).

11.4.1.5 Anti-Inflammatory Properties

Inflammation can negatively affect fibroblast activity. Phytochemicals with anti-inflammatory properties can create a more conducive environment for fibroblast proliferation and function. By reducing inflammation, these compounds facilitate the healing process (Chainani-Wu, 2003).

In conclusion, phytochemicals significantly influence dermal fibroblast proliferation and activity, impacting processes such as collagen synthesis, protection against oxidative stress, modulation of growth factors, and inflammation. These effects underscore the potential of phytochemicals in enhancing skin health and wound healing.

11.4.2 Specific Phytochemicals Known to Enhance Wound Healing

A lot of phytochemicals have been identified for their potent effects in enhancing wound healing. These compounds, derived from plants, exhibit a range of activities that accelerate the healing process.

11.4.2.1 Curcumin

Curcumin, the active component of turmeric, is notorious for its anti-inflammatory and antioxidant properties. It promotes wound healing by reducing inflammation, oxidative stress, and enhancing collagen synthesis. Curcumin has been shown to accelerate wound closure and improve tissue regeneration in several studies (Bisht et al., 2010).

11.4.2.2 Epigallocatechin Gallate (EGCG)

Found in green tea, EGCG is a powerful antioxidant. It aids wound healing by reducing oxidative damage and inflammation. EGCG also enhances the proliferation and migration of skin cells, which are crucial for wound repair (Nakamuta et al., 2005).

11.4.2.3 Quercetin

Quercetin, a flavonoid present in many fruits and vegetables, exhibits antioxidant, anti-inflammatory, and wound-healing properties. It supports wound healing by promoting fibroblast proliferation and collagen synthesis, which are essential for tissue repair and regeneration (Huang et al., 2018).

11.4.2.4 Resveratrol

Resveratrol, found in grapes and berries, is another potent phytochemical that enhances wound healing. It improves healing by modulating inflammation, promoting angiogenesis (formation of new blood vessels), and stimulating the migration and proliferation of skin cells (Thangapazham et al., 2005).

11.4.2.5 Aloe Vera

Aloe vera contains multiple phytochemicals, including glycoproteins and polysaccharides, which have been used traditionally for wound healing. Aloe vera gel accelerates wound healing by its anti-inflammatory activity, hydrating properties, and stimulating collagen production (Richardson et al., 2005).

These phytochemicals, each with unique properties, contribute significantly to wound healing by modulating key processes such as inflammation, cell proliferation, collagen synthesis, and angiogenesis. Their natural origin and broad range of activities make them appealing for therapeutic applications in skin repair and regeneration.

11.4.2.6 Centella Asiatica

It has been used as a wound healer for centuries in Asia and traditional medicine. It has been shown to be effective in healing both normal, delayed and bad wounds, thanks to the saponins and sugar esters such as asiaticoside and madecassic it contains In particular, asiaticoside is known for its healing effects on diabetic and chronic wounds and has accelerating effects on wound contraction, collagen synthesis and angiogenesis processes (Gorain et al., 2022; Nadar and Selvakumar, 2018).

11.4.2.7 Malva Neglecta Wallr

Malva neglecta Wallr, commonly known as common mallow and belonging to the Malvaceae family, has traditionally been used in herbal medicine for wound healing. Recent studies have sought to scientifically validate these traditional uses and explore the phytochemical properties of the plant in relation to wound healing and dermal fibroblast proliferation (Saleem et al., 2020).

11.4.2.8 Alocasia Longiloba

The plant, generally found in Malaysia and known as 'Keladi Candik', has been used in traditional medicine methods in wound healing for many years. The effects of the plant, especially its ethanol extract, which contains many phytochemicals, including alkaloids, flavonoids, tannins, phenols and saponins, on wound healing have been previously described in rat models in the literature (Hamzah et al., 2022).

11.4.3 Case Studies or Research Findings Demonstrating These Effects

Numerous studies and research findings have highlighted the efficacy of various phytochemicals in enhancing wound healing. Some of these are summarized below.

11.4.3.1 Curcumin in Wound Healing

A study by Bisht et al. (2010) demonstrated that curcumin enhances wound healing in both *in vitro* and *in vivo* settings. The study revealed that curcumin treatment led to increased fibroblast proliferation and collagen synthesis, crucial factors for effective wound healing. The research suggested that curcumin's antioxidant properties played a significant role in its wound-healing effects (Bisht et al., 2010).

11.4.3.2 Green Tea Polyphenols and Skin Regeneration

Nakamuta et al. (2005) explored the role of green tea polyphenols, particularly epigallocatechin gallate (EGCG), in skin regeneration. The study found that EGCG promoted the proliferation and differentiation of keratinocytes, essential for skin repair. Moreover, EGCG's anti-inflammatory and antioxidant properties were beneficial in reducing wound inflammation and oxidative stress (Nakamuta et al., 2005).

11.4.3.3 Aloe Vera for Wound Healing

In a clinical trial, Richardson et al. (2005) examined the efficacy of aloe vera in preventing skin reactions in patients undergoing radiation therapy. The study found that topical application of aloe vera gel reduced the severity of radiation-induced skin reactions, suggesting its potential to promote wound healing and skin regeneration (Richardson et al., 2005).

11.4.3.4 Resveratrol in Wound Healing

A research study by Thangapazham et al. (2005) investigated the effects of resveratrol on cutaneous wound healing. The findings indicated that topical application of resveratrol accelerated wound closure, enhanced epithelialization, and increased angiogenesis in the wound area. These effects were attributed to resveratrol's anti-inflammatory and pro-angiogenic properties (Thangapazham et al., 2005).

11.4.3.5 Quercetin's Role in Wound Repair

Huang et al. (2018) conducted a study focusing on the effects of quercetin on wound healing. The results showed that quercetin facilitated wound closure, improved collagen deposition, and enhanced fibroblast proliferation. The study highlighted quercetin's antioxidant and anti-inflammatory activities as key factors in its wound-healing properties (Huang et al., 2018).

11.4.3.6 Centella Asiatica in Wound Healing

The plant is traditionally used for its antimicrobial, antioxidant, and anti-inflammatory capabilities. These properties are critical in wound healing as they help prevent infection, reduce oxidative stress, and minimize inflammation at the wound site. Also, various extracts of Centella asiatica, including asiatic acid, β-sitosterol, asiaticoside, and madecassoside, have been identified to facilitate the wound healing process in both incisions and burn wounds. Asiatic acid, in particular, is noted for its effectiveness in healing wounds (Arribas-López et al., 2022).

11.4.3.7 Malva Neglecta Wallr in Wound Repair

Malva neglecta's crude methanolic extract (Mn.Cme) was chemically characterized using HPLC-DAD and GCMS analysis. The analysis revealed 25 phenolic compounds, with significant amounts of hydrotyrosol, coumaroylhexoside, kaempferol-3-(p-coumaroyldiglucoside)-7-glucoside, quercetin-3-O-rutinoside, and epicatechin-3-O-(4-O-methyl)-gallate. Oleic acid, taurine, ethylene dimercaptan, isoeugenol, patchoulane, methyl 12-methyltetradecanoate, and isopropyl myristate were also found in high abundance.

The wound healing properties of Mn.Cme were evaluated using an excision wound model in rats. The extract, formulated as an ointment, was applied topically to wound areas once daily for 14 days. Wound contraction was measured regularly, and results showed a considerable wound healing tendency at all doses, with absolute healing observed after applying 2 g of Mn.Cme ointment. The healing process was dose-related, with a reduced epithelialization period.

The extract exhibited significant anti-radical potential in various assays, indicating its antioxidant properties which are beneficial in the wound healing process (Saleem et al., 2020).

These case studies and research findings collectively demonstrate the significant potential of phytochemicals in wound healing. Through various mechanisms, including anti-inflammatory, antioxidant, and cell-proliferative activities, phytochemicals like curcumin, EGCG, aloe vera, resveratrol, and quercetin offer promising avenues for enhancing wound repair and skin regeneration.

11.4.3.8 Alocasia Longiloba for Wound Repair

The extract of A. longiloba prepared with 6% chloroform has wound healing properties. They play a positive role in wound healing, especially with the various phytochemicals they contain, and their anti-microbial, antioxidant and anti-bacterial properties in the wound area (Hamzah et al., 2022).

11.5 THERAPEUTIC APPLICATIONS

11.5.1 Use of Phytochemicals in Topical Treatments for Wound Healing

The application of phytochemicals in topical wound healing treatments has gained significant attention due to their natural origin and potent biological activities. These can be examined as follows:

11.5.1.1 Aloe Vera in Wound Healing

Aloe vera, rich in glycoproteins and polysaccharides, is widely used in topical formulations for wound healing. Studies like those by Maenthaisong et al. (2007) have shown that aloe vera accelerates wound contraction, increases collagen synthesis and angiogenesis and reduces inflammation, making it effective for treating burns and surgical wounds (Maenthaisong et al., 2007).

11.5.1.2 Curcumin-Based Topical Formulations

Curcumin, derived from turmeric, is recognized for its anti-inflammatory and antioxidant properties. Research by Sidhu et al. demonstrated that topical application of curcumin enhances wound healing by promoting fibroblast proliferation and collagen deposition and proliferation (Sidhu et al., 1998).

11.5.1.3 Tea Tree Oil for Antimicrobial Action

Tea tree oil, which contains compounds such as Terpinen-4-ol, a monoterpene, is used in topical treatments for its antimicrobial properties. It is effective against a range of pathogens that commonly infect wounds (Carson et al., 2006).

11.5.1.4 Green Tea Extracts in Skin Regeneration

Green tea polyphenols, especially EGCG, have been incorporated into topical treatments due to their ability to promote skin regeneration and protect against UV damage, as shown in studies by Yamamoto et al. (2003). These extracts aid in wound healing by enhancing cell proliferation and differentiation (Yamamoto et al., 2003).

11.5.1.5 Chamomile Extracts for Anti-Inflammatory Effect

Chamomile extracts, containing bisabolol and chamazulene, are used in topical applications for their anti-inflammatory and skin-soothing properties. They are beneficial in treating skin irritations and promoting wound healing (Srivastava et al., 2010).

The use of phytochemicals in topical treatments offers a natural and effective approach to wound healing. Aloe vera, curcumin, tea tree oil, green tea extracts, and chamomile are some of the key phytochemicals that have been studied and applied with success in various wound-healing formulations. These phytochemicals provide diverse mechanisms of action, including antimicrobial, anti-inflammatory, antioxidant, and cell-proliferative properties, contributing to their effectiveness in wound care.

11.5.1.6 Centella Asiatica in Wound Healing and Repair

Centella asiatica extracts improve wound healing by enhancing collagen synthesis and microcirculatory function, this is crucial for the repair and regeneration of damaged skin tissue. Compounds like madecassoside, madecassic acid, asiaticoside, and asiatic acid present in the plant contribute to this healing process. Asiatic acid, in particular, has been identified as a highly active component for wound healing. The extracts of Centella asiatica are effective in both incision and burn wounds, and the plant has also been used traditionally for managing keloids, which are a type of scar. Moreover, the plant might enhance wound healing through improved angiogenesis and the formation of new blood vessels, which is a critical aspect of tissue repair and regeneration. This effect is possibly due to its stimulating impact on collagen I, fibroblast growth factor (FGF), and vascular endothelial growth factor (VEGF) production. In summary, Centella asiatica's effectiveness in topical wound healing is attributed to its ability to enhance collagen synthesis, improve microcirculatory function, and provide antimicrobial, antioxidant, and anti-inflammatory effects. Additionally, its role in promoting angiogenesis further supports its use in wound care and tissue regeneration (Arribas-López et al., 2022).

11.5.1.7 Portulaca Gandiflora in Wound Repairing

Topical application of 10% and 20% extracts of the purslane (Portulaca grandiflora) plant, which contains large amounts of flavonoids, twice a day in the rabbit excisional wound model showed high wound healing activity on the seventh and 11th days (Budiawan et al., 2023).

In addition, the wound healing effect of at least 20% ethanol extract prepared from the leaves of the Indian almond plant known as Terminalia catappa L., when applied topically once a day in *in vivo* mouse experiments, was found to be significantly higher than the simple ointment and control group.

11.5.2 Synergistic Effects of Phytochemicals with Other Treatments

Phytochemicals often exhibit enhanced therapeutic effects when combined with other treatments, offering synergistic benefits in various medical applications. These effects are examined in this section (Nugroho et al., 2019).

11.5.2.1 Phytochemicals and Antibiotics

Phytochemicals like those found in tea tree oil, garlic, and curcumin have been shown to enhance the efficacy of antibiotics. A study by Betts et al. demonstrated that certain phytochemicals can reduce antibiotic resistance in bacteria, making the antibiotics more effective. This synergy is particularly significant in the era of increasing antibiotic resistance (Betts and Wareham, 2014).

11.5.2.2 Cancer Treatment Enhancement

In oncology, phytochemicals such as curcumin and resveratrol have been found to enhance the effectiveness of chemotherapy and radiation therapy. Studies like those by Patel et al. have shown that these phytochemicals can sensitize cancer cells to conventional treatments, potentially reducing required dosages and side effects (Patel et al., 2010).

11.5.2.3 Phytochemicals and Skin Care Products

In dermatology, phytochemicals like green tea polyphenols and flavonoids are often combined with skin care products to enhance their antioxidant and anti-ageing effects. The study by Yamamoto et al. (2003) suggests that the addition of EGCG to skin care formulations can improve their protective effects against UV damage and skin ageing.

The synergy between phytochemicals and many medical treatment options, ranging from antibiotics and chemotherapy to NSAIDs and skin care products, highlights the potential of these natural compounds in enhancing treatment efficacy, reducing side effects, and overcoming resistance issues. This approach represents an evolving area of research with significant implications for the development of more effective and holistic treatment strategies.

11.5.3 Limitations and Potential Side Effects

While phytochemicals offer significant benefits in wound healing, limitations and potential side effects must be taken into account.

11.5.3.1 Allergic Reactions

Some individuals may experience allergic reactions to certain phytochemicals when applied topically. For example, tea tree oil, commonly used for its antimicrobial properties, can cause allergic contact dermatitis in some individuals (Carson et al., 2006).

11.5.3.2 Interaction with Medications

Certain phytochemicals can interact with prescription medications, potentially altering their efficacy. For instance, curcumin has been shown to interact with anticoagulant and antiplatelet drugs, potentially increasing the risk of bleeding (Gurley, 2012).

11.5.3.3 Cytotoxicity

High concentrations of some phytochemicals can be cytotoxic. A study by Lü et al. (2010) indicated that high doses of resveratrol could have cytotoxic effects on cells, underscoring the need for careful dosing in therapeutic applications.

11.5.3.4 Delayed Wound Healing

Certain phytochemicals, if not properly used, can potentially delay wound healing. For example aloe vera, while beneficial in many cases, can inhibit the healing of deep surgical wounds (Schmidt and Greenspoon, 1991).

11.5.3.5 Phototoxicity

Some phytochemicals, when applied topically, can make the skin more sensitive to sunlight, leading to phototoxicity. Compounds such as furocoumarins are found in some essential oils that have this effect (Korać and Khambholja, 2011).

While phytochemicals are promising agents in wound healing, their use is associated with potential side effects and interactions. Allergic reactions, interactions with medications, cytotoxicity at high concentrations, potential for delayed healing, and phototoxicity are among the key considerations. These factors stress the importance of cautious and informed use of phytochemicals in therapeutic contexts.

11.6 FUTURE DIRECTIONS AND RESEARCH

11.6.1 Current Trends in Research on Phytochemicals and Wound Healing

Recent research trends in phytochemicals for wound healing focus on understanding their mechanisms, enhancing their efficacy, and integrating them into modern medical practices. It is obvious that the importance of this issue is increasing, especially if the burdens of chronic wound healing on the healthcare system and patient care are taken into account.

11.6.1.1 Nanoformulations of Phytochemicals

A significant trend is the development of nanoformulations to improve the delivery and efficacy of phytochemicals in wound healing. These nanomaterials can mimic the extracellular matrix (ECM) environment of wounds, and which type of nanomaterial should be used in which type of wound depends on the type and depth of the wound. Studies have explored the use of nanocarriers for the controlled release of phytochemicals, enhancing their therapeutic effects and minimizing side effects (Contri et al., 2016).

Various nanocarriers can be produced with these nanoformulations and nanocarriers; they can be examined as nanoparticles, metal nanoparticles, vesicles, nanofibers, nanogels and nanoemulsions, and as drug delivery systems, they can play a role in vital aspects such as controlled release of therapeutics, biological and synthetic drugs in the target tissue, and maintenance of steady-state concentrations and reduction of toxicity through controlled and continuous dosing (Gorain et al., 2022).

Nanoparticles offer a versatile platform for the delivery of phytochemicals especially aiming for enhancing their stability, solubility, and bioavailability. A study highlighted the use of polymeric nanoparticles for controlled release, improving the therapeutic outcomes of phytochemicals in wound healing (Contri et al., 2016).

Metal nanoparticles, particularly silver and gold, have been utilized for their antimicrobial properties in wound healing. The integration of phytochemicals with these metal nanoparticles can synergistically enhance their efficacy. A study discussed has shown the potential of silver nanoparticles (Ag NPs) as a vehicle for phytochemicals, providing sustained release while maximizing antimicrobial activity (Rai et al., 2009; Some et al., 2020; Ekrikaya et al., 2021, 2024; Ocsoy et al., 2018; Demirbas et al., 2020; Karaagac et al., 2021; Temur et al., 2024).

Liposomal vesicles, composed of phospholipid bilayers, are effective in encapsulating both hydrophilic and lipophilic phytochemicals. Their biocompatibility makes them ideal for wound care and wound healing applications. This issue has been particularly beneficial in delivering curcumin to the wound site with liposomal nanosystems (Ashtikar et al., 2016).

Nanofibers, often used in the form of mats or dressings, provide a conducive environment for wound healing. According to our literature knowledge, aloe vera-loaded electrospun nanofibers have been shown to effectively deliver phytochemicals to the wound, promote healing and provide a protective barrier (Unnithan et al., 2014).

Nanogels are hydrogel nanoparticles that can absorb significant amounts of water or biological fluids, making them suitable for moist wound environments. Their ability to encapsulate and release phytochemicals in response to environmental stimuli (like pH or temperature) can be advantageous in wound healing (Gupta et al., 2016).

Nanoemulsions provide a stable means of delivering lipophilic phytochemicals. Their small droplet size allows for better penetration and retention at the wound site. Studies on this nanocarrier system highlighted the use of nanoemulsions for enhancing the therapeutic efficacy of phytochemicals in wound management (Sutradhar and Amin, 2013).

The development of advanced drug delivery systems like nanoparticles, metal nanoparticles, liposomal vesicles, nanofibers, nanogels, and nanoemulsions has revolutionized the controlled release of phytochemicals in wound healing. These systems not only improve the bioavailability and stability of phytochemicals but also provide sustained and targeted delivery, enhancing the overall efficacy of wound treatment.

In recent studies, many promising results have begun to emerge regarding the use of phytochemicals through nanocarriers in wound healing and dermal fibroblast proliferation.

One of them is AuNP (Gold Nanoparticles), prepared at a concentration of 20 mg, using leaf extracts of the impatiens balsamina plant; A positive and significant contribution to wound healing, superior to traditional drugs, was observed in both excisional and thermal wounds as early as the eighth day (Nie et al., 2020).

In addition, total ethanol extract (TEE) and polysaccharide (PS), prepared from the seeds of the Lepidium Salivatum plant, which contains large amounts of phytochemical phenolic acids, flavonoids, and tannins in its structure, are loaded on polyvinyl alcohol (PVA) nanofibers produced by the electrospinning method and are delivered to the wound area, providing wound healing properties. It has been described in the literature in recent years that it can be used as a dressing and that its effectiveness is higher in PVA nanofibers containing TEE than in PS nanofibers. This again reveals the importance of the future role of nanocarriers (Amer et al., 2022).

Another study on the importance of nanoparticles in wound healing shows that Ag NPs prepared through green synthesis with methanolic extracts of the seeds of the Pongamia Pinnata plant, which contains many phytochemicals in its structure (such as flavonoids, alkaloids), have antioxidant, anti-bacterial, anti-inflammatory properties. It showed that they have inflammatory, angiogenic and wound-closing mitogenic and collagen synthesis potential (Paul and Londhe, 2019).

Regarding the use of nanomaterials, Ag NP-loaded hydrogels offer promising developments with their effectiveness in chronic wounds, their wound closure accelerating and tissue regeneration-improving activities, their biocompatibility, their lower toxicity compared to synthetic substances, and their features such as controlled and continuous drug delivery in the long term (Dam et al., 2023).

Again, the anti-bacterial activity of Ag NPs, produced by the green synthesis method and obtained using Salvia Nubicola extract, against the phytopathogen ralstonia solanacearum, which is a plant pathogen and causes vascular wilt of tomato, has been demonstrated in *in vitro* experiments, making it an alternative to classical pesticides. It emphasizes the importance of diversity in the usage areas of nanoparticles (Dilbar et al., 2023).

Drug delivery nanosystems; nanofibers can be examined as polymeric nanoparticles (PNP), metallic nanoparticles, nanohydrogels, liposomal vesicular nanosystems and lipid nanoparticles.

Such nanotechnological systems, which can target specific areas in the body with biological therapeutics and distribute drug release and delivery in a controlled release and increase the permeation of therapeutics to the underlying skin layers, together with the phytochemicals loaded into these nanosystems, are beneficial in wound healing both *in vivo* and *in vitro*, and are found to be anti-inflammatory, anti-inflammatory, and anti-inflammatory. They have been shown to have oxidative and anti-bacterial effects and stimulate angiogenesis and wound contraction by increasing vascular endothelial growth factor (VEGF) and fibroblast growth factor (FGF) in the wound area. Moreover, the new generation nanosystem treatments mentioned in detail previously, along with 3D bioprinting methods and stem cell treatments, are expected to take their place among the new generation wound treatments in the near future (Arifin et al., n.d.; Hajialyani et al., 2018; Qadir et al., 2021).

11.6.1.2 Antimicrobial Properties

With the rise of antibiotic-resistant infections, the antimicrobial properties of phytochemicals are a growing area of interest. Research showed that compounds like tea tree oil and curcumin highlight their potential as alternatives or adjuncts to antibiotics in wound care (Pazyar et al., 2014).

Phytochemicals are known for their broad-spectrum antimicrobial properties, which are crucial in preventing and treating wound infections. Compounds such as allicin from garlic and catechins from green tea exhibit strong antimicrobial effects against a range of bacteria, fungi, and viruses. A study in the literature highlighted the antimicrobial potency of allicin against various microbial pathogens (Ankri and Mirelman, 1999).

Furthermore, the rise of antibiotic-resistant bacteria has prompted research into alternative treatments like phytochemicals. Many plant-derived compounds, including essential oils and flavonoids, have been shown in the literature to be effective against antibiotic-resistant strains, making them valuable in modern wound care (Cowan, 1999).

Biofilms in wounds can hinder healing and increase infection risk. Phytochemicals, such as those found in cranberry extracts, have been shown to prevent biofilm formation, as noted in research in the previous literature knowledge, thus aiding in effective wound management (Pazyar et al., 2014).

Phytochemicals play a significant role in wound healing due to their antimicrobial effects. They offer a promising alternative or adjunct to traditional antibiotics, especially in the era of increasing antibiotic resistance. Their broad-spectrum activity, ability to combat resistant strains, synergistic effects with antibiotics, and prevention of biofilm formation underline their potential in effective wound care.

11.6.1.3 Phytochemicals in Chronic Wound Healing

There is increasing focus on the effectiveness of phytochemicals in treating chronic wounds, especially in diabetic patients. The latest research has shown that certain phytochemicals can address the underlying pathologies of chronic wounds like diabetic ulcers (Yeung et al., 2021).

11.6.1.4 Natural Product-Based Bioactive Dressings

The development of bioactive wound dressings infused with phytochemicals is a notable trend. These dressings not only provide physical protection but also deliver therapeutic agents directly to the wound site (Boateng and Catanzano, 2015).

Current research in phytochemicals and wound healing is multifaceted, focusing on innovative delivery systems, combination therapies, antimicrobial effects, chronic wound management, and bioactive dressings, as detailed previously. These trends demonstrate the growing integration of traditional knowledge with modern technology and medicine, expanding the potential of phytochemicals in effective wound care solutions.

11.6.2 Potential for New Discoveries and Therapeutic Applications

The field of phytochemicals is ripe with possibilities for new discoveries and therapeutic applications, particularly in treating diseases and enhancing health. These application areas are examined.

11.6.2.1 Neurodegenerative Diseases

Recent research has demonstrated the potential of phytochemicals in the treatment of neurodegenerative diseases. Compounds such as genistein resveratrol and curcumin are being studied for their neuroprotective properties.

Phytochemicals are known for their antioxidant and anti-inflammatory effects. Through these effects, phytochemicals may play a role in regaining glial activity, correcting mitochondrial dysfunction and preventing ROS accumulation by controlling oxidative damage and neuroinflammation present in the pathogenesis of neurodegenerative disease (Tripathi et al., 2021).

11.6.2.2 Cancer Therapy

Phytochemicals have attracted attention in oncology due to their potential in cancer treatment. In particular, epigallocatechin gallate (EGCG) has been shown to inhibit tumour growth by inhibiting the methyltransferase enzyme and to exert anti-proliferative effects through Janus Kinase inhibition. Additionally, thymoquinone, a phytochemical obtained from the Nigella sativa plant, has been studied to clear existing superoxides and have anticancer properties. Further studies should focus on the anticancer properties of phytochemicals, including enhancing the effectiveness of chemotherapy drugs (Amin et al., 2009).

11.6.2.3 Cardiovascular Health

The cardiovascular benefits of phytochemicals such as flavonoids and polyphenols are well documented. These phytochemicals play a role in reducing the risk of cardiovascular disease, especially through the antioxidant and anti-inflammatory effects of the stilbenes and lignans they contain (Tresserra-Rimbau et al., 2014).

11.6.2.4 Antimicrobial Resistance

With concerns about antimicrobial resistance increasing, phytochemicals offer a promising alternative. Research in the literature has identified several phytochemicals from the group of phenols, terpenoids, alkaloids, lectins, gallic acid, and polyacetylenes with potent antimicrobial properties that could be developed as new treatments for resistant bacterial strains (Cowan, 1999).

11.6.2.5 Diabetes Management

Phytochemicals are also being investigated for diabetes management. In particular, flavonoid-containing plant extracts of the Smilax moranensis plant have the potential to regulate blood sugar levels and increase insulin sensitivity (Andrade-Cetto, 2011).

The potential of phytochemicals in new therapeutic applications is wide, covering areas such as neurodegenerative diseases, cancer therapy, cardiovascular health, antimicrobial resistance, and diabetes management. Ongoing research continues to uncover the various roles of these natural compounds, offering promising thoughts for innovative treatments and health solutions.

11.6.3 The Need for Further Research and Clinical Trials

Despite promising findings, the field of phytochemicals in wound healing requires more comprehensive research and clinical trials to fully understand their potential and ensure safe, effective use.

11.6.3.1 Efficacy and Mechanisms of Action

While numerous *in vitro* and animal studies have demonstrated the wound-healing benefits of phytochemicals, there is a need for more clinical trials to confirm these effects in humans. A study by Moosavi et al. emphasizes the importance of understanding the precise mechanisms of action of phytochemicals in the wound healing process (Moosavi et al., 2018).

11.6.3.2 Optimal Dosage and Formulation

Determining the optimal dosage and formulation of phytochemicals for wound healing is crucial. Research has shown us that nanoformulations have the potential to improve the bioavailability and efficacy of phytochemicals, but more studies are needed to optimize these parameters for clinical use and to achieve optimization, nanoformulations and nanocarriers may be a promising option, as reviewed previously (Contri et al., 2016).

11.6.3.3 Long-Term Safety and Side Effects

Long-term safety and potential side effects of phytochemicals in wound healing remain areas needing further exploration. Clinical trials are essential to assess any possible adverse effects, particularly when used over extended periods, as discussed in the context of antimicrobial phytochemicals (Ulbricht and Chao, 2010).

11.6.3.4 Effectiveness in Chronic and Non-Healing Wounds

The effectiveness of phytochemicals in treating chronic and non-healing wounds, especially in conditions like diabetes and vascular insufficiencies, requires more investigation. Clinical trials focusing on these specific wound types are necessary to develop targeted phytochemical-based treatments (Boateng and Catanzano, 2015).

11.6.3.5 Comparison with Conventional Treatments

The potential of phytochemicals in wound healing is evident but further research and clinical trials are essential to establish their efficacy, optimal dosing, safety, effectiveness in chronic wounds, and comparison with existing treatments. Such studies will be crucial in transitioning phytochemicals from traditional and experimental use to evidence-based clinical applications.

11.7 CONCLUSION

11.7.1 Summary of Key Findings and Their Implications

Research on phytochemicals has provided insightful findings on their role in wound healing and dermal fibroblast proliferation.

11.7.1.1 Enhancement of Wound Healing

Studies have consistently shown that certain phytochemicals, such as curcumin, resveratrol, and epigallocatechin gallate (EGCG), enhance wound healing. They accelerate wound closure, promote collagen synthesis, and reduce inflammation. For instance, curcumin has been observed to enhance fibroblast proliferation and increase collagen deposition (Phan et al., 2001).

11.7.1.2 Dermal Fibroblast Proliferation

Phytochemicals have been found to stimulate the proliferation of dermal fibroblasts, key cells in wound healing. Compounds like quercetin and flavonoids in green tea are known to promote fibroblast growth and function, aiding in skin repair and regeneration (Yamamoto et al., 2003).

11.7.1.3 Antioxidant and Anti-Inflammatory Properties

The antioxidant and anti-inflammatory properties of phytochemicals play a significant role in their wound-healing efficacy. They help mitigate oxidative stress and reduce inflammation at the wound site, thereby facilitating the healing process (Kumar and Pandey, 2013).

11.7.1.4 Promising Potential in Chronic Wounds

Phytochemicals show promise in treating chronic wounds, particularly in diabetic patients, due to their ability to address underlying issues such as prolonged inflammation and impaired angiogenesis (Moosavi et al., 2018).

11.7.1.5 Need for Further Research

Despite promising findings, further research and clinical trials are essential to fully understand the potential of phytochemicals in wound healing, determine optimal dosages, and ensure safety and efficacy in diverse patient populations (Contri et al., 2016)

11.7.2 The Importance of Phytochemicals in Wound Healing and Skin Health

Phytochemicals are emerging as potent agents in enhancing wound healing and promoting dermal fibroblast proliferation, primarily due to their anti-inflammatory, antioxidant, and cell-stimulative properties. While current findings are promising, more comprehensive research is needed to translate these benefits effectively into clinical practice.

The following features, which we have tried to examine throughout this chapter, support our opinion.

11.7.2.1 Accelerating Wound Healing

Phytochemicals such as curcumin and resveratrol have been shown to accelerate the wound-healing process by acting protectively against hydrogen peroxide and increasing VEGF levels. As shown in studies, they improve the various stages of healing, including inflammation, tissue formation, and remodelling (Phan et al., 2001; Shah and Amini-Nik, 2017).

11.7.2.2 Promoting Dermal Fibroblast Proliferation

Dermal fibroblasts are very important for skin regeneration. Phytochemicals such as quercetin and green tea polyphenols have been found to stimulate dermal fibroblast proliferation and reduce ROS levels, thereby contributing to wound healing and skin regeneration (Yamamoto et al., 2003).

11.7.2.3 Anti-Inflammatory and Antioxidant Properties

The anti-inflammatory and antioxidant properties of phytochemicals play a critical role in wound healing and skin health. Phytochemicals, especially curcumin, help alleviate damage to skin cells and improve the healing process by reducing inflammation and oxidative stress by inhibiting enzymes and cytokines such as phospholipase, cyclooxygenase 2, nitric oxide and prostaglandins, and TNF (Chainani-Wu, 2003).

11.7.2.4 Antimicrobial Activity

Some phytochemicals, such as those found in tea tree oil and garlic, exhibit antimicrobial properties, making them effective in preventing and treating skin infections, a crucial aspect of wound care (Cowan, 1999).

11.7.2.5 Skin Ageing and Photoprotection

Phytochemicals also offer benefits in terms of skin ageing and photoprotection. Compounds like flavonoids and carotenoids protect the skin from UV radiation and combat signs of ageing (Svobodová et al., 2003).

The importance of phytochemicals in wound healing and skin health is underpinned by their ability to accelerate healing, promote cell proliferation, and provide anti-inflammatory, antioxidant, and antimicrobial effects. These properties make phytochemicals valuable in various dermatological applications, from wound care to skin protection and anti-ageing treatments.

11.7.3 Final Thoughts and Future Outlook

11.7.3.1 Present Insights

Current research has established a substantial role for phytochemicals in enhancing wound healing and stimulating dermal fibroblast proliferation. Key phytochemicals like curcumin, resveratrol, flavonoids and epigallocatechin gallate (EGCG) have been shown to promote wound closure, modulate inflammation, and enhance collagen synthesis (Phan et al., 2001; Yamamoto et al., 2003).

11.7.3.2 Future Perspectives

The future vision of phytochemical research in wound healing focuses on understanding its molecular mechanisms, optimizing delivery methods, and enhancing efficacy. The development of nanoformulations and controlled-release systems appears promising. A particularly promising area is the application of phytochemicals in the management of chronic wounds, especially in diabetic patients. Additionally, personalized medicine approaches could tailor phytochemical treatments to individual genetic profiles and specific wound-healing responses (Kumar and Pandey, 2013).

While the therapeutic potential is significant, the safety profiles, optimal dosages, and long-term effects of phytochemicals need thorough evaluation through clinical trials. Establishing these parameters is crucial for the safe and effective transition of phytochemicals from bench to bedside (Moosavi et al., 2018). An integrative approach, combining phytochemicals with conventional wound healing treatments, could offer more comprehensive care strategies. Understanding how phytochemicals interact with other drugs and their synergistic potential will be key in this integration (Contri et al., 2016).

In addition, the promising effect of phytochemicals on wound healing and dermal fibroblast proliferation compared to normal drugs will increase even more in the future with the use of nanocarriers as new-generation drug delivery systems. In this way, satisfactory improvements can be achieved, especially in chronic wound care and fibroblasts, by ensuring more effective delivery of therapeutics to the target tissue, maintaining effective concentration and reducing toxicity rates and the most effective treatment methods will be offered to patients while reducing patient care times and costs. That is, the regular and sustained release properties of active ingredients into the wound site and new-generation treatments will form the basis for future studies.

The exploration of phytochemicals in wound healing and skin health is an exciting and evolving field. With their natural origin and potent bioactivities, phytochemicals hold promise for innovative and effective wound care solutions. Future research, focused on clinical applications, safety, and integration with conventional therapies, will be instrumental in harnessing the full potential of phytochemicals in wound healing.

REFERENCES

Addis, R., Cruciani, S., Santaniello, S., Bellu, E., Sarais, G., Ventura, C., Maioli, M., & Pintore, G. (2020). Fibroblast proliferation and migration in wound healing by phytochemicals: Evidence for a novel synergic outcome. *International Journal of Medical Sciences*, *17*(8), 1030.

Amer, A. A., Mohammed, R. S., Hussein, Y., Ali, A. S. M., & Khalil, A. A. (2022). Development of lepidium sativum extracts/PVA electrospun nanofibers as wound healing dressing. *ACS Omega*, *7*(24), 20683–20695.

Amin, A., Gali-Muhtasib, H., Ocker, M., & Schneider-Stock, R. (2009). Overview of major classes of plant-derived anticancer drugs. *International Journal of Biomedical Science: IJBS*, *5*(1), 1.

Andrade-Cetto, A. (2011). Hypoglycemic effect of Smilax moranensis root on N5-STZ diabetic rats. *Pharmacologyonline*, *1*, 111–115.

Ankri, S., & Mirelman, D. (1999). Antimicrobial properties of allicin from garlic. *Microbes and Infection*, *1*(2), 125–129.

Arifin, A. F., Rijal, S., & Handayani Idrus, H. (n.d.). Phytochemicals in medicinal plants in wound healing. *International Journal of Medical Science and Dental Research*, *6*(5), 107–113.

Arribas-López, E., Zand, N., Ojo, O., Snowden, M. J., & Kochhar, T. (2022). A systematic review of the effect of Centella asiatica on wound healing. *International Journal of Environmental Research and Public Health*, *19*(6), 3266.

Ashtikar, M., Nagarsekar, K., & Fahr, A. (2016). Transdermal delivery from liposomal formulations–Evolution of the technology over the last three decades. *Journal of Controlled Release*, *242*, 126–140.

Betts, J. W., & Wareham, D. W. (2014). In vitro activity of curcumin in combination with epigallocatechin gallate (EGCG) versus multidrug-resistant Acinetobacter baumannii. *BMC Microbiology*, *14*, 1–5.

Bisht, K., Wagner, K.-H., & Bulmer, A. C. (2010). Curcumin, resveratrol and flavonoids as anti-inflammatory, cyto-and DNA-protective dietary compounds. *Toxicology*, *278*(1), 88–100.

Block, K. I., & Mead, M. N. (2003). Immune system effects of echinacea, ginseng, and astragalus: A review. *Integrative Cancer Therapies*, *2*(3), 247–267.

Boateng, J., & Catanzano, O. (2015). Advanced therapeutic dressings for effective wound healing – a review. *Journal of Pharmaceutical Sciences*, *104*(11), 3653–3680.

Bowling, F. L., Jude, E. B., & Boulton, A. J. M. (2009). MRSA and diabetic foot wounds: Contaminating or infecting organisms? *Current Diabetes Reports*, *9*(6), 440–444.

Budiawan, A., Purwanto, A., Puradewa, L., Cahyani, E. D., & Purwaningsih, C. E. (2023). Wound healing activity and flavonoid contents of purslane (Portulaca grandiflora) of various varieties. *RSC Advances*, *13*(15), 9871–9877.

Budovsky, A., Yarmolinsky, L., & Ben-Shabat, S. (2015). Effect of medicinal plants on wound healing. *Wound Repair and Regeneration*, *23*(2), 171–183.

Campbell, D. R., & Kurzer, M. S. (1993). Flavonoid inhibition of aromatase enzyme activity in human preadipocytes. *The Journal of Steroid Biochemistry and Molecular Biology*, *46*(3), 381–388.

Carson, C. F., Hammer, K. A., & Riley, T. V. (2006). Melaleuca alternifolia (tea tree) oil: A review of antimicrobial and other medicinal properties. *Clinical Microbiology Reviews*, *19*(1), 50–62.

Cassidy, A., Albertazzi, P., Nielsen, I. L., Hall, W., Williamson, G., Tetens, I., Atkins, S., Cross, H., Manios, Y., & Wolk, A. (2006). Critical review of health effects of soyabean phyto-oestrogens in post-menopausal women. *Proceedings of the Nutrition Society*, *65*(1), 76–92.

Cassidy, A., Faughnan, M., Hughes, R., Fraser, C., Cathcart, A., Taylor, N., Setchell, K., & Bingham, S. (1998). Hormonal effects of phytoestrogens in postmenopausal women and middle-aged men. *American Journal of Clinical Nutrition*, *68*(6), 1531S–1531S.

Çelik, T. A., & Aslantürk, Ö. S. (2023). Phytochemical screening and in vitro assessments of antioxidant and cytotoxic potentials of extracts from Aesculus hippocastanum L. green fruit mesocarps. *International Journal of Secondary Metabolite*, *10*(2), 245–256.

Chainani-Wu, N. (2003). Safety and anti-inflammatory activity of curcumin: A component of tumeric (Curcuma longa). *The Journal of Alternative & Complementary Medicine*, *9*(1), 161–168.

Choi, J. H., Kim, S. M., Lee, G. H., Jin, S. W., Lee, H. S., Chung, Y. C., & Jeong, H. G. (2019). Platyconic acid A, platycodi radix-derived saponin, suppresses TGF-β1-induced activation of hepatic stellate cells via blocking SMAD and activating the PPARγ signaling pathway. *Cells*, *8*(12), 1544.

Contri, R. V, Fiel, L. A., Alnasif, N., Pohlmann, A. R., Guterres, S. S., & Schäfer-Korting, M. (2016). Skin penetration and dermal tolerability of acrylic nanocapsules: Influence of the surface charge and a chitosan gel used as vehicle. *International Journal of Pharmaceutics*, *507*(1–2), 12–20.

Cowan, M. M. (1999). Plant products as antimicrobial agents. *Clinical Microbiology Reviews*, *12*(4), 564–582.

Craig, W. J. (1999). Health-promoting properties of common herbs. *The American Journal of Clinical Nutrition*, *70*(3), 491s–499s.

Dam, P., Celik, M., Ustun, M., Saha, S., Saha, C., Kacar, E. A., Kugu, S., Karagulle, E. N., Tasoglu, S., & Buyukserin, F. (2023). Wound healing strategies based on nanoparticles incorporated in hydrogel wound patches. *RSC Advances*, *13*(31), 21345–21364.

Demirbas, A., Karaytuğ, T., Arabaci, N., Yilmaz, E. S., & Ocsoy, I. (2020). Synthesis of metallic and metal oxide nanomaterials. In *Green Synthesis of Nanomaterials for Bioenergy Applications*, 99–123. John Wiley & Sons Ltd.

Dilbar, S., Sher, H., Ali, H., Ullah, R., Ali, A., & Ullah, Z. (2023). Antibacterial efficacy of green synthesized silver nanoparticles using salvia nubicola extract against ralstonia solanacearum, the causal agent of vascular wilt of tomato. *ACS Omega*, *8*(34), 31155–31167.

Ekrikaya, S., Danisman, H., Baktir, S., Arslan, S., & Ocsoy, I. (2024). Bioinspired synthesis of nanocomposites. In *Smart and Sustainable Applications of Nanocomposites*, 36–66. IGI Global Scientific Publishing.

Ekrikaya, S., Yilmaz, E., Celik, C., Demirbuga, S., Ildiz, N., Demirbas, A., & Ocsoy, I. (2021). Investigation of ellagic acid rich-berry extracts directed silver nanoparticles synthesis and their antimicrobial properties with potential mechanisms towards *Enterococcus faecalis* and *Candida albicans*. *Journal of Biotechnology*, *341*, 155–162.

El-Sherbeni, S. A., & Negm, W. A. (2023). The wound healing effect of botanicals and pure natural substances used in in vivo models. *Inflammopharmacology*, *31*(2), 755–772.

Eming, S. A., Krieg, T., & Davidson, J. M. (2007). Inflammation in wound repair: Molecular and cellular mechanisms. *Journal of Investigative Dermatology*, *127*(3), 514–525.

Gelse, K., Pöschl, E., & Aigner, T. (2003). Collagens – structure, function, and biosynthesis. *Advanced Drug Delivery Reviews*, *55*(12), 1531–1546.

Gorain, B., Pandey, M., Leng, N. H., Yan, C. W., Nie, K. W., Kaur, S. J., Marshall, V., Sisinthy, S. P., Panneerselvam, J., & Molugulu, N. (2022). Advanced drug delivery systems containing herbal components for wound healing. *International Journal of Pharmaceutics*, *617*, 121617.

Guo, S. al, & DiPietro, L. A. (2010). Factors affecting wound healing. *Journal of Dental Research*, *89*(3), 219–229.

Gupta, A., Low, W. L., Radecka, I., Britland, S. T., Mohd Amin, M. C. I., & Martin, C. (2016). Characterisation and in vitro antimicrobial activity of biosynthetic silver-loaded bacterial cellulose hydrogels. *Journal of Microencapsulation*, *33*(8), 725–734.

Gurley, B. J. (2012). Pharmacokinetic herb-drug interactions (part 1): Origins, mechanisms, and the impact of botanical dietary supplements. *Planta Medica*, *78*(13), 1478–1489.

Gurtner, G. C., Werner, S., Barrandon, Y., & Longaker, M. T. (2008). Wound repair and regeneration. *Nature*, *453*(7193), 314–321.

Hajialyani, M., Tewari, D., Sobarzo-Sánchez, E., Nabavi, S. M., Farzaei, M. H., & Abdollahi, M. (2018). Natural product-based nanomedicines for wound healing purposes: Therapeutic targets and drug delivery systems. *International Journal of Nanomedicine*, 5023–5043.

Hamzah, N. H. C., Mohammed, A., Sirajudin, K. N. M. S., Hamzah, Z., & Wei, L. S. (2022). Phytochemical screening and wound healing effect of Alocasia longiloba Miq. petiole ethanolic extract and its fractions in excision wound rat model. *Borneo International Journal of Biotechnology (BIJB)*, *2*, 108–124.

Hinz, B. (2007). Formation and function of the myofibroblast during tissue repair. *Journal of Investigative Dermatology*, *127*(3), 526–537.

Huang, Y.-H., Wu, P.-Y., Wen, K.-C., Lin, C.-Y., & Chiang, H.-M. (2018). Protective effects and mechanisms of Terminalia catappa L. methenolic extract on hydrogen-peroxide-induced oxidative stress in human skin fibroblasts. *BMC Complementary and Alternative Medicine*, *18*(1), 1–9.

Karaagac, Z., Dadi, S., Ekrikaya, S., & Ocsoy, I. (2021). Plant extract guided nanomaterials and plant disease management. In *Nanotechnology in Sustainable Agriculture*, 1–20. CRC Press.

Khan, N., Adhami, V. M., & Mukhtar, H. (2008). Apoptosis by dietary agents for prevention and treatment of cancer. *Biochemical Pharmacology*, *76*(11), 1333–1339.

Korać, R. R., & Khambholja, K. M. (2011). Potential of herbs in skin protection from ultraviolet radiation. *Pharmacognosy Reviews*, *5*(10), 164.

Kozarski, M., Klaus, A., van Griensven, L., Jakovljevic, D., Todorovic, N., Wan, W. A. A. Q. I., & Vunduk, J. (2023). Mushroom β-glucan and polyphenol formulations as natural immunity boosters and balancers: Nature of the application. *Food Science and Human Wellness*, *12*(2), 378–396.

Krinsky, N. I., & Johnson, E. J. (2005). Carotenoid actions and their relation to health and disease. *Molecular Aspects of Medicine*, *26*(6), 459–516.

Kroon, P. A., Clifford, M. N., Crozier, A., Day, A. J., Donovan, J. L., Manach, C., & Williamson, G. (2004). How should we assess the effects of exposure to dietary polyphenols in vitro? *The American Journal of Clinical Nutrition*, *80*(1), 15–21.

Kumar, S., & Pandey, A. K. (2013). Chemistry and biological activities of flavonoids: An overview. *The Scientific World Journal*. Article ID 162750, 16 pp.

Liu, R. H. (2004). Potential synergy of phytochemicals in cancer prevention: Mechanism of action. *The Journal of Nutrition*, *134*(12), 3479S–3485S.

Lü, J., Lin, P. H., Yao, Q., & Chen, C. (2010). Chemical and molecular mechanisms of antioxidants: Experimental approaches and model systems. *Journal of Cellular and Molecular Medicine*, *14*(4), 840–860.

Maenthaisong, R., Chaiyakunapruk, N., Niruntraporn, S., & Kongkaew, C. (2007). The efficacy of aloe vera used for burn wound healing: A systematic review. *Burns*, *33*(6), 713–718.

Middleton, E., Kandaswami, C., & Theoharides, T. C. (2000). The effects of plant flavonoids on mammalian cells: Implications for inflammation, heart disease, and cancer. *Pharmacological Reviews*, *52*(4), 673–751.

Molnar, J. A., Underdown, M. J., & Clark, W. A. (2014). Nutrition and chronic wounds. *Advances in Wound Care*, *3*(11), 663–681.

Moosavi, M. A., Haghi, A., Rahmati, M., Taniguchi, H., Mocan, A., Echeverría, J., Gupta, V. K., Tzvetkov, N. T., & Atanasov, A. G. (2018). Phytochemicals as potent modulators of autophagy for cancer therapy. *Cancer Letters*, *424*, 46–69.

Mortimer, P. S., & Rockson, S. G. (2014). New developments in clinical aspects of lymphatic disease. *The Journal of Clinical Investigation*, *124*(3), 915–921.

Moses, R. L., Dally, J., Lundy, F. T., Langat, M., Kiapranis, R., Tsolaki, A. G., Moseley, R., & Prescott, T. A. K. (2020). Lepiniopsis ternatensis sap stimulates fibroblast proliferation and down regulates macrophage TNF-α secretion. *Fitoterapia*, *141*, 104478.

Nadar, M. S. A. M., & Selvakumar, P. M. (2018). Plant-derived compounds for wound healing: A review. *Organic & Medicinal Chemistry International Journal*, *5*(1), 13–17.

Nakamuta, M., Higashi, N., Kohjima, M., Fukushima, M., Ohta, S., Kotoh, K., Kobayashi, N., & Enjoji, M. (2005). Epigallocatechin-3-gallate, a polyphenol component of green tea, suppresses both collagen production and collagenase activity in hepatic stellate cells. *International Journal of Molecular Medicine*, *16*(4), 677–681.

Nie, S., Wei, R., Zhou, H., Zhang, L., Chen, Z., & Hou, L. (2020). Eco-friendly synthesis of AuNPs for cutaneous wound-healing applications in nursing care after surgery. *Green Processing and Synthesis*, *9*(1), 366–374.

Nugroho, R. A., Utami, D., Aryani, R., Nur, F. M., Sari, Y. P., & Manurung, H. (2019). In vivo wound healing activity of ethanolic extract of Terminalia catappa L. leaves in mice (Mus musculus). *Journal of Physics: Conference Series*, *1277*(1), 012031.

Ocsoy, I., Tasdemir, D., Mazicioglu, S., & Tan, W. (2018). Nanotechnology in plants. In *Plant Genetics and Molecular Biology*, 263–275. Springer, Cham.

Pandey, K. B., & Rizvi, S. I. (2014). Role of red grape polyphenols as antidiabetic agents. *Integrative Medicine Research*, *3*(3), 119–125.

Patel, V. B., Misra, S., Patel, B. B., & Majumdar, A. P. N. (2010). Colorectal cancer: Chemopreventive role of curcumin and resveratrol. *Nutrition and Cancer*, *62*(7), 958–967.

Paul, M., & Londhe, V. Y. (2019). Pongamia pinnata seed extract-mediated green synthesis of silver nanoparticles: Preparation, formulation and evaluation of bactericidal and wound healing potential. *Applied Organometallic Chemistry*, *33*(3), e4624.

Pazyar, N., Yaghoobi, R., Rafiee, E., Mehrabian, A., & Feily, A. (2014). Skin wound healing and phytomedicine: A review. *Skin Pharmacology and Physiology*, *27*(6), 303–310.

Phan, T.-T., See, P., Lee, S.-T., & Chan, S.-Y. (2001). Protective effects of curcumin against oxidative damage on skin cells in vitro: Its implication for wound healing. *Journal of Trauma and Acute Care Surgery*, *51*(5), 927–931.

Qadir, A., Jahan, S., Aqil, M., Warsi, M. H., Alhakamy, N. A., Alfaleh, M. A., Khan, N., & Ali, A. (2021). Phytochemical-based nano-pharmacotherapeutics for management of burn wound healing. *Gels*, *7*(4), 209.

Rai, M., Yadav, A., & Gade, A. (2009). Silver nanoparticles as a new generation of antimicrobials. *Biotechnology Advances*, *27*(1), 76–83.

Reddy, M., Gill, S. S., Kalkar, S. R., Wu, W., Anderson, P. J., & Rochon, P. A. (2008). Treatment of pressure ulcers: A systematic review. *Jama*, *300*(22), 2647–2662.

Rex, J. R. S., Muthukumar, N., & Selvakumar, P. M. (2018). Phytochemicals as a potential source for anti-microbial, anti-oxidant and wound healing-a review. *MOJ Bioorganic & Organic Chemistry*, *2*(2), 61–70.

Reygaert, W. C. (2014). The antimicrobial possibilities of green tea. *Frontiers in Microbiology*, *5*, 434.

Richardson, J., Smith, J. E., McIntyre, M., Thomas, R., & Pilkington, K. (2005). Aloe vera for preventing radiation-induced skin reactions: A systematic literature review. *Clinical Oncology*, *17*(6), 478–484.

Rodríguez-Cantú, L. N., Gutiérrez-Uribe, J. A., Arriola-Vucovich, J., Díaz-De La Garza, R. I., Fahey, J. W., & Serna-Saldivar, S. O. (2011). Broccoli (Brassica oleracea var. italica) sprouts and extracts rich in glucosinolates and isothiocyanates affect cholesterol metabolism and genes involved in lipid homeostasis in hamsters. *Journal of Agricultural and Food Chemistry*, *59*(4), 1095–1103.

Saleem, U., Khalid, S., Zaib, S., Anwar, F., Ahmad, B., Ullah, I., Zeb, A., & Ayaz, M. (2020). Phytochemical analysis and wound healing studies on ethnomedicinally important plant Malva neglecta Wallr. *Journal of Ethnopharmacology*, *249*, 112401.

Salem, M. A., Mohamed, O. G., Mosalam, E. M., Elberri, A. I., Abdel-Bar, H. M., Hassan, M., Al-Karmalawy, A. A., Tripathi, A., Ezzat, S. M., & Abo Mansour, H. E. (2023). Investigation of the phytochemical composition, antioxidant, antibacterial, anti-osteoarthritis, and wound healing activities of selected vegetable waste. *Scientific Reports*, *13*(1), 13034.

Schmidt, J. M., & Greenspoon, J. S. (1991). Aloe vera dermal wound gel is associated with a delay in wound healing. *Obstetrics and Gynecology*, *78*(1), 115–117.

Sen, C. K. (2009). Wound healing essentials: Let there be oxygen. *Wound Repair and Regeneration*, *17*(1), 1–18.

Shah, A., & Amini-Nik, S. (2017). The role of phytochemicals in the inflammatory phase of wound healing. *International Journal of Molecular Sciences*, *18*(5), 1068.

Sidhu, G. S., Singh, A. K., Thaloor, D., Banaudha, K. K., Patnaik, G. K., Srimal, R. C., & Maheshwari, R. K. (1998). Enhancement of wound healing by curcumin in animals. *Wound Repair and Regeneration*, *6*(2), 167–177.

Some, S., Sarkar, B., Biswas, K., Jana, T. K., Bhattacharjya, D., Dam, P., Mondal, R., Kumar, A., Deb, A. K., Sadat, A., Saha, S., Kati, A., Ocsoy, I., Franco, O. L., Mandal, A., Mandal, S., Mandal, A. K., & Ince, A. I. (2020). Bio-molecule functionalized rapid one-pot green synthesis of silver nanoparticles and their efficacy toward the multidrug resistant (MDR) gut bacteria of silkworms (Bombyx mori). *RSC Advances*, *10*, 22742–22757.

Sorrell, J. M., & Caplan, A. I. (2004). Fibroblast heterogeneity: More than skin deep. *Journal of Cell Science*, *117*(5), 667–675.

Srivastava, J. K., Shankar, E., & Gupta, S. (2010). Chamomile: A herbal medicine of the past with a bright future. *Molecular Medicine Reports*, *3*(6), 895–901.

Sutradhar, K. B., & Amin, M. L. (2013). Nanoemulsions: Increasing possibilities in drug delivery. *European Journal of Nanomedicine*, *5*(2), 97–110.

Svobodová, A., Psotová, J., & Walterová, D. (2003). Natural phenolics in the prevention of UV-induced skin damage. A review. *Biomedical papers of the Medical Faculty of the University Palacky, Olomouc, Czech Republic*, *147*(2), 137–145.

Talalay, P., & Fahey, J. W. (2001). Phytochemicals from cruciferous plants protect against cancer by modulating carcinogen metabolism. *The Journal of Nutrition*, *131*(11), 3027S–3033S.

Temur, N., Gundes, N. S., Korkmaz, B., Ozkay, Z. A., & Ocsoy, I. (2024). Silver nanoparticles in dentistry. In *Silver Nanoparticles for Drug Delivery*, 265–288. Elsevier Inc.

Thangapazham, R. L., Sharad, S., & Maheshwari, R. K. (2016). Phytochemicals in wound healing. *Advances in Wound Care*, *5*(5), 230–241.

Thangapazham, R. L., Sharma, A., & Maheshwari, R. K. (2005). Biomarkers of angiogenesis in wound healing and cancer: Role of botanicals and phytochemicals. *IV International Conference on Quality and Safety Issues Related to Botanicals*, *720*, 129–137.

Thangapazham, R. L., Sharma, A., & Maheshwari, R. K. (2007). Beneficial role of curcumin in skin diseases. In *The Molecular Targets and Therapeutic Uses of Curcumin in Health and Disease*, 343–357. Springer.

Tomasek, J. J., Gabbiani, G., Hinz, B., Chaponnier, C., & Brown, R. A. (2002). Myofibroblasts and mechano-regulation of connective tissue remodelling. *Nature Reviews Molecular Cell Biology*, *3*(5), 349–363.

Tresserra-Rimbau, A., Rimm, E. B., Medina-Remón, A., Martínez-González, M. A., López-Sabater, M. C., Covas, M. I., Corella, D., Salas-Salvadó, J., Gómez-Gracia, E., & Lapetra, J. (2014). Polyphenol intake and mortality risk: A re-analysis of the PREDIMED trial. *BMC Medicine*, *12*(1), 1–11.

Tripathi, S. S., Kumar, R., Pandey, P. K., Singh, A. K., & Gupta, N. (2021). Role of phytochemicals in neurodegenerative disorders. *Recent Advances in the Treatment of Neurodegenerative Disorders*, *50*, 50.

Ulbricht, C. E., & Chao, W. (2010). Phytochemicals in the oncology setting. *Current Treatment Options in Oncology*, *11*(3–4), 95–106.

Unnithan, A. R., Sasikala, A. R. K., Sathishkumar, Y., Lee, Y. S., Park, C. H., & Kim, C. S. (2014). Nanoceria doped electrospun antibacterial composite mats for potential biomedical applications. *Ceramics International*, *40*(8), 12003–12012.

Varani, J., Dame, M. K., Rittie, L., Fligiel, S. E. G., Kang, S., Fisher, G. J., & Voorhees, J. J. (2006). Decreased collagen production in chronologically aged skin: Roles of age-dependent alteration in fibroblast function and defective mechanical stimulation. *The American Journal of Pathology*, *168*(6), 1861–1868.

Velnar, T., Bailey, T., & Smrkolj, V. (2009). The wound healing process: An overview of the cellular and molecular mechanisms. *Journal of International Medical Research*, *37*(5), 1528–1542.

Vitale, S., Colanero, S., Placidi, M., Di Emidio, G., Tatone, C., Amicarelli, F., & D'Alessandro, A. M. (2022). Phytochemistry and biological activity of medicinal plants in wound healing: An overview of current research. *Molecules*, *27*(11), 3566.

Wang, C., & Eskiw, C. H. (2019). Cytoprotective effects of Avenathramide C against oxidative and inflammatory stress in normal human dermal fibroblasts. *Scientific Reports*, *9*(1), 2932.

Wolfram, D., Tzankov, A., Pülzl, P., & Piza-Katzer, H. (2009). Hypertrophic scars and keloids – a review of their pathophysiology, risk factors, and therapeutic management. *Dermatologic Surgery*, *35*(2), 171–181.

Yamamoto, T., Hsu, S., Lewis, J., Wataha, J., Dickinson, D., Singh, B., Bollag, W. B., Lockwood, P., Ueta, E., & Osaki, T. (2003). Green tea polyphenol causes differential oxidative environments in tumor versus normal epithelial cells. *Journal of Pharmacology and Experimental Therapeutics*, *307*(1), 230–236.

Yeung, A. W. K., Tzvetkov, N. T., Durazzo, A., Lucarini, M., Souto, E. B., Santini, A., Gan, R.-Y., Jozwik, A., Grzybek, W., & Horbańczuk, J. O. (2021). Natural products in diabetes research: Quantitative literature analysis. *Natural Product Research*, *35*(24), 5813–5827.

Yuan, T.-J., Xu, X.-H., Zhou, N., Yan, G., Gu, T.-W., & Peng, L.-H. (2022). Phytochemicals as new therapeutic candidates simultaneously stimulate proliferation and counteract senescence of stem cells. *Biomedicine & Pharmacotherapy*, *151*, 113170.

Zhang, Y., & Tang, L. (2007). Discovery and development of sulforaphane as a cancer chemopreventive phytochemical 1. *Acta Pharmacologica Sinica*, *28*(9), 1343–1354.

12 Role of Dietary Phytochemicals in Skin Cancer

Debjyoti Paul, Kankana Das, Jatindra Nath Bhakta and Krishnendu Manna

12.1 INTRODUCTION

As per the latest report published by the American Cancer Society for the year 2023, skin cancer ranks as the 17th most prevalent cancer globally (2013). In broad terms, skin cancer manifests itself through several types, including melanoma (originating from melanocytes), non-melanoma (often basal cell carcinoma, BCC, and squamous cell carcinoma, SCC), and the less common yet extremely aggressive neuroendocrine skin cancer known as Merkel cell carcinoma (MCC). Melanoma is a significant global health concern, with approximately 150,000 new cases diagnosed annually, highlighting the pressing need for comprehensive and innovative medical interventions worldwide. It is also pertinent to note that since non-melanoma cancer cells are not tracked by clinical registries, the incidence reports of non-melanoma cancer cases are incomplete. Often non-melanoma skin cancers are effectively managed through surgical procedures or ablation, resulting in significantly improved long-term outlooks (Simoes et al., 2015). Hence, there is a probability that the documented global occurrences of skin cancer are lower than the actual figures. However, in every form of cancer treatment and management, the impact on patients' quality of life is profound, presenting a significant contributor to the escalating burden of diseases on a global scale (Figure 12.1).

Skin cancer is particularly widespread among the Caucasian populations of the world and among the most frequently diagnosed malignancies (Erb et al., 2005). Nevertheless, individuals with a higher level of melanin in their skin are frequently identified with skin cancer at more advanced stages, which poses challenges for effective treatment (Mangione et al., 2023). High levels of UV exposure from the sun, resulting in frequent sunburns, weakened immune systems, individuals who have undergone organ transplants, and viral infections like human immune virus (HIV) and human papillomavirus (HPV), are among the well-established causes of skin cancer manifestations (Saladi and Persaud, 2005; Trager et al., 2022). The involvement of UV exposure in causing cancer is also now established. Skin malignancies occur because of a multifaceted series of stochastic occurrences triggered and facilitated by ultraviolet (UV) radiation, encompassing both UVB and UVA wavelengths, and potentially extending to visible light and infrared radiation. UVB radiation mostly induces DNA lesions in the form of neighbouring pyrimidine dimers, specifically involving the nucleotide combinations of TT, CT, TC, and CC. There are two primary forms of lesions that are formed: cyclobutane pyrimidine dimers (CPD) and 6-4 photoproducts [6-4PPs] (Mullenders et al., 1993). The lesions are typically healed through the utilization of the nucleotide excision repair (NER) mechanism (Hoeijmakers, 2001; Stary and Sarasin, 2002). Nucleotide Excision Repair (NER) is a process that restores the integrity of a damaged DNA strand through a sequential series of steps. These steps mainly consist of recognizing the lesion, unwinding the DNA helix, marking the lesion, making dual incisions in the damaged DNA strand, accurately resynthesizing the

DOI: 10.1201/9781003394846-12

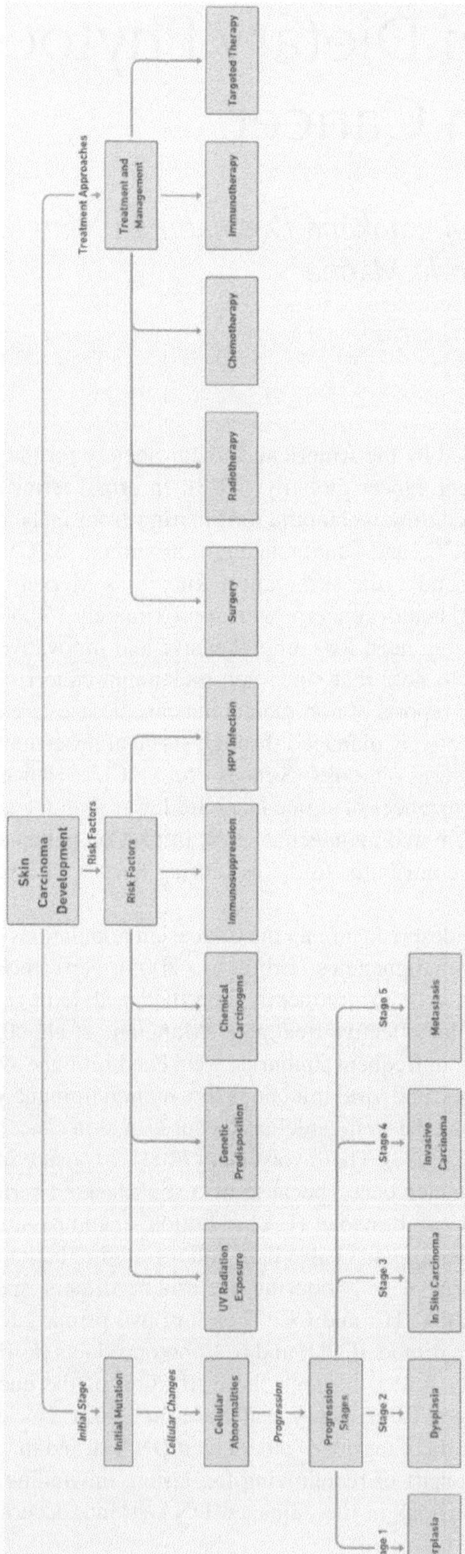

FIGURE 12.1 Representative illustration of the multistage progression of skin carcinoma, beginning with initial mutations and spanning through various cellular changes to metastasis. It also highlights the risk factors associated with skin carcinoma development and outlines the primary treatment approaches.

resulting gap, and ligating the newly synthesized DNA strand. Despite the diverse and sometimes stochastic nature of skin cancer causes, focusing on established factors such as sunburn prevention and administering HPV vaccines has demonstrated significant effectiveness in reducing skin cancer incidences (Trager et al., 2022).

The National Comprehensive Cancer Network (NCCN) and the American Joint Committee on Cancer (AJCC) have collaboratively developed a set of evidence-based guidelines to facilitate the identification and optimal care of skin cancer. According to the established recommendations, the initial step in the clinical inquiry process involves gathering the patient's medical history and doing a thorough physical examination. It is highly recommended to perform thorough full-body inspections, as patients commonly present several cases of skin cancer. The risk factors associated with skin cancer encompass various dimensions. These include advanced age, a greater quantity of benign moles, particularly melanoma, and individuals with red hair or fair skin who have a history of exposure to ultraviolet (UV) radiation. Strong intermittent exposure to UV radiation has been linked to melanoma, while chronic cumulative exposure is associated with squamous cell carcinoma. Additionally, individuals with compromised immune systems are also at risk. The presence of a personal medical background involving skin precancers or a familial medical history pertaining to such illnesses is regarded as a substantial risk factor (Nataren et al., 2022). From this standpoint, it is crucial to recognize that although the general outlook for the treatment of skin cancer is positive, it is essential to emphasize that melanoma cancer, which represents just 2% of all instances of skin cancer, is responsible for an alarming 80% of deaths (Banach et al., 2021). The prevailing consensus suggests that the development of skin cancer is influenced by a variety of genetic mechanisms, encompassing modifications in tumour suppressor genes, proto-oncogenes, and signalling pathways. The incidence rates of the principal molecular processes associated with distinct types of skin cancer are provided below.

12.1.1 The Pathogenesis of Melanoma

In the context of melanoma, it is frequently noticed that mutations in the p53 gene tend to become apparent towards the later stages of the illness, thereby serving as an indicator of its advanced progression. Nevertheless, prior research has observed the existence of mutations associated with non-melanocytic skin malignancies (NMSC) in premalignant lesions such as actinic keratosis (AK), indicating their potential involvement in the early stages of NMSC development (Arafeh et al., 2015). The incidence rates of p53 mutations in melanomas exhibit considerable variety, as evidenced by multiple publications. Numerous studies have documented diverse incidence rates pertaining to the issue under examination. Various studies have reported divergent findings on the occurrence of the phenomenon under investigation. While certain studies indicate a prevalence rate spanning from 92% to 100%, alternative estimates propose a narrower range of 7% to 27%. Approximately two-thirds of patients diagnosed with AK (actinic keratosis) display p53 mutations (Xu et al., 2013). Moreover, the P16 gene, which is responsible for the synthesis of the p16 protein (Zhang and Rosdahl, 2004), is commonly inactivated in both melanocytic and non-melanocytic cancers. The origin of tumours can be partially attributable to the disruption of cell cycle regulation caused by the functional loss of p16.

Recent research findings have unveiled novel insights into the tumour suppressor activities associated with certain genes. The protein phosphatase 2A regulatory subunit PR70 has been identified as an example of an illustration, demonstrating its role as a suppressor of gonosomal melanoma tumours (Banach et al., 2021). Similarly, it has been observed that the RASA2 gene exhibits a significant prevalence of inactivating mutations in the context of melanoma, where it is responsible for encoding the Ras GAP protein. The presence of these genetic anomalies results in heightened activation of the RAS pathway, augmented cellular proliferation and migration, and diminished survival rates in those affected (Arafeh et al., 2015).

Proto-oncogenes and NRAS have garnered considerable attention within the domain of cancer. Approximately 15% of melanomas exhibit genetic modifications in the proto-oncogene RAS, with a specific emphasis on NRAS. The role of NRAS mutations in the disruption of signalling pathways and subsequent advancement of melanoma has been extensively documented.

BRAF mutations have been detected in around 50% of melanoma cases that have been diagnosed. The mutations manifest inside the BRAF gene, which encodes a kinase enzyme that plays a pivotal role in the activation of the MAP kinase/ERK signalling pathway (Figure 12.2). The BRAF V600E mutation is well acknowledged for its significant role in the hyperactivation of the mitogen-activated protein kinase (MAPK) pathway, resulting in enhanced cellular proliferation, angiogenesis, and cytokine production (Kulkarni et al., 2017).

The occurrence of mutations in proto-oncogenes, namely GNA11 and GNAQ, which encode G-protein alpha subunits, is commonly observed in cases of uveal melanoma and cutaneous melanoma (Metz et al., 2013). It is widely hypothesized that mutations contribute to the regulation of melanoblast populations and the facilitation of neoplastic progression.

Mutations in the proto-oncogene C-KIT have been observed in melanomas situated in acral regions and mucosae (Pham et al., 2020). Around 11% of melanomas that are found in acral areas and mucosae display identifiable alterations. The aforementioned mutations have been associated with the emergence of resistance towards targeted therapy, such as anti-BRAF or anti-MEK treatments (Ponti et al., 2017).

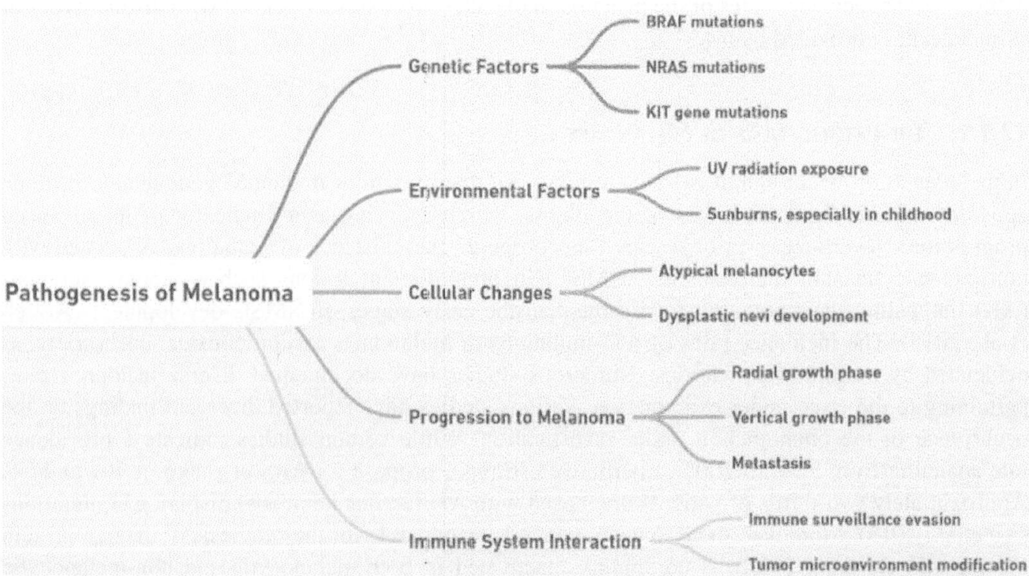

FIGURE 12.2 Representative image provides a structured overview of melanoma pathogenesis, broken down into several interconnected categories. Mutations in the BRAF, NRAS, and KIT genes are important contributing factors that lead to melanoma. Environmental factors such as UV radiation exposure and sunburn also play a critical role in the progression of melanoma. These factors lead to cellular changes characterized by the presence of atypical melanocytes and the development of dysplastic nevi. The progression of melanoma is characterized by a series of stages, beginning with the radial growth phase, advancing to the vertical growth phase, and culminating in metastasis. It also addresses the interaction with the immune system, outlining how melanoma cells can evade immune surveillance and modify the tumour microenvironment to facilitate their own survival and proliferation.

12.1.2 Non-Melanocytic Skin Cancer Pathogenesis

Non-melanocytic skin cancer is a type of skin cancer that arises from cells other than melanocytes (Scherer and Kumar, 2010). The gene BCORL1, recognized for its tumour-suppressive characteristics, has been reported to exhibit an association with resistance to pharmaceutical therapies. The process of reducing the expression of E-cadherin is facilitated by its interaction with CtBP. Mutations in the BCORL1 gene have been identified in approximately 10% of instances of cutaneous melanoma (Mologni et al., 2018).

Approximately 23% of patients diagnosed with malignant melanoma have been shown to exhibit mutations in the CTNNB1 gene. The CTNNB1 gene has been identified as having a significant association with the process of tumour suppression. The gene assumes a crucial function as a fundamental constituent within the Wnt signalling pathway, which governs the mechanisms of cellular proliferation and differentiation (Cerami et al., 2012).

The aetiology of cutaneous malignancy is distinguished by a diverse array of genetic alterations, encompassing mutations in many tumour suppressor genes such as p53, P16, PR70, RASA2, BCORL1, and CTNNB1. Moreover, this process also involves proto-oncogenes such as RAS, BRAF, GNA11, GNAQ, and C-KIT. It is imperative to acknowledge that the progression of skin cancer is significantly impacted by pivotal signalling pathways, particularly MAPK and Wnt. Through the identification and specific targeting of distinct molecular processes accountable for the onset and advancement of skin cancer, healthcare practitioners can shift away from a uniform treatment strategy. In contrast, it is possible to customize drugs to accommodate the unique genetic and molecular characteristics of each patient. Hence, it is crucial to acquire a thorough understanding of these molecular pathways to facilitate the implementation of focused therapeutic approaches and improve the overall treatment of different forms of skin cancer.

Furthermore, it is worth noting that molecular markers that are linked to skin cancer possess the capability to function as early diagnostic indicators. The identification of these indicators enables timely detection and action, potentially resulting in enhanced prognostic outcomes and increased patient survival rates.

In addition, it is important to acknowledge that conventional methods of treating and managing skin cancer should be adapted to incorporate the expanding knowledge regarding the potential advantages of phytochemicals. These compounds specifically target molecular pathways and regulate the expression of the aforementioned genes implicated in the development of skin cancer. Phytochemicals have the potential to enhance the overall effectiveness of conventional medicines through their complementary actions. One potential approach to enhance tumour response rates and improve patient outcomes is the integration of phytochemical-rich meals or supplements with radiation or chemotherapy treatments. Furthermore, phytochemicals exhibit a high degree of accessibility and cost-effectiveness in comparison to certain expensive medications, hence enhancing their availability to a wider spectrum of patients and perhaps alleviating the economic strain associated with the treatment of skin cancer. Significantly, phytochemicals have the ability to ameliorate chronic inflammation and oxidative stress, both of which play crucial roles in the development of skin cancer. Consequently, they have the capacity to decelerate or even halt the advancement of precancerous lesions towards malignancy.

The focus of this chapter is to examine the molecular basis of skin cancer symptoms and conduct a comprehensive investigation of naturally occurring phytochemical substances. The primary objective of this study is to offer a thorough comprehension of the molecular mechanisms that are crucial for effectively addressing the fundamental factors responsible for the development of skin cancer. This study presents a meta-analysis of the molecular aetiology of skin cancer, alongside a comprehensive review of the current literature on the use of phytochemicals for the management of various manifestations. Furthermore, this discussion will highlight the benefits of employing phytochemicals as opposed to standard therapeutic methods.

12.2 PHYTOCHEMICALS IN THE MANAGEMENT OF SKIN CANCER MANIFESTATIONS

Phytochemicals, derived from the Greek word "*phyto*" meaning "plant", are bioactive non-nutrient compounds that exist in a wide range of plant-based food sources including fruits, vegetables, cereals, and other similar things. In nature, phytochemicals play a crucial role in the plant's defence mechanisms against pathogens and ultraviolet radiation, frequently serving as pigments. When consumers partake in vegetables, herbs, seeds, nuts, and fruits, they are ingesting a diverse range of phytochemicals, both nutritive and nonnutritive, some of which have the potential to demonstrate pharmacological effects. In several studies, the potential of these chemicals to mitigate the risk of severe chronic disorders has also been established (Zhang et al., 2015). Phytochemicals, owing to their varied bioactive components, demonstrate resistance against bacterial, fungal, and pesticidal ailments, and possess various modes of operation, including antioxidant attributes as well as the ability to modulate gene expression and signal transduction pathways (Radi et al., 2021; Chen and Liu, 2018; Thakur et al., 2014). Consequently, phytochemicals exert significant influence on the cancer process through a multitude of gene-modulating mechanisms, hence displaying promise as viable agents for cancer prevention owing to their minimal toxicity profiles (Thakur et al., 2014).

Phytochemicals encompass polyphenols, terpenoids, alkaloids, phytosterols, and organosulfur compounds that exhibit diverse levels of antioxidant and prooxidant effects contingent upon their respective concentrations (Somani et al., 2015). As an illustration, polyphenols, which belong to an extensively researched group of phytochemicals, consist of approximately 8000 compounds (Rathod et al., 2023). These compounds have been found to exert various effects on cellular processes, such as inducing apoptosis, inhibiting cell proliferation, and suppressing migration and invasion in cancer cells, particularly when administered at elevated concentrations (Rizeq et al., 2020). On the other hand, when administered in lesser quantities, polyphenols demonstrate scavenging capabilities by effectively counteracting the presence of free radicals within cellular structures (Maisuthisakul et al., 2007). The protective benefits of phytochemicals with antioxidative properties are crucial in mitigating lipid peroxidation. These compounds effectively counteract the detrimental impact of free radicals on cellular membranes by removing electrons from lipids (Figure 12.3). Significantly, a considerable body of research conducted in controlled laboratory settings (*in vitro*) as well as in living creatures (*in vivo*) has demonstrated that specific substances produced from plants (phytochemicals) can impede the binding of mutagens present in the environment to DNA. The process efficiently impedes the start of cancer by inhibiting the enzymatic activity of phase I enzymes, which are responsible for the activation of carcinogens. The enzymes in question encompass cytochrome P450 (CYP) enzymes and epoxide hydrolases (Wang et al., 2012; George et al., 2021).

Furthermore, phytochemicals play a role in the chemopreventive process through their ability to decrease the activity of phase I enzymes and enhance the activation or expression of phase II enzymes. Phase II enzymes play a crucial role in the detoxification process by attaching polar groups to activated carcinogens. These enzymes, such as glutathione S-transferases (GST), uridine 5'-diphospho-glucuronosyltransferase (UGT), NAD(P)H dehydrogenase (quinone)-1 (NQO1), sulfotransferases (SULT), N-acetyltransferases (NAT), and methyltransferases (e.g., S-methyl transferase and catechol O-methyl transferase, COMT), facilitate the removal of carcinogens from the body and their subsequent excretion. This enzymatic activity ultimately reduces the risk of cancer initiation (Wang et al., 2012; Warwick et al., 2012).

Several phytochemicals have been shown to inhibit tumour promotion by decreasing the viability and proliferation of existing cancer cells. The observed consequences encompass the suppression of cellular proliferation and the initiation of programmed cell death. These outcomes are achieved through the induction of cell cycle arrest, which is facilitated by the overexpression of

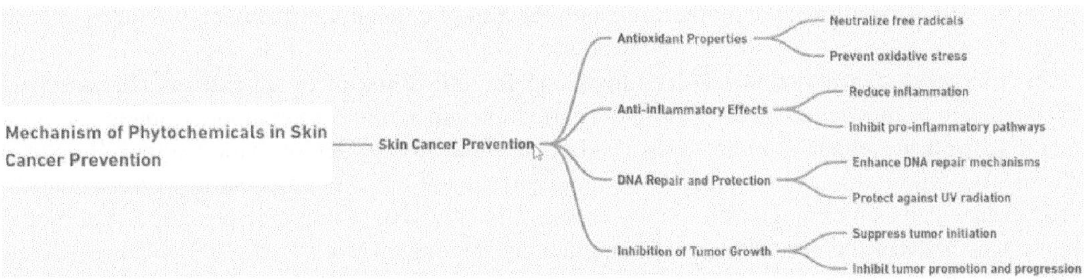

FIGURE 12.3 The figure illustrates the various mechanisms by which phytochemicals contribute to the prevention of skin cancer. The pivotal strategies for skin cancer prevention branch out into four key areas of action. Firstly, the antioxidant properties of phytochemicals help neutralize free radicals and prevent oxidative stress. Secondly, their anti-inflammatory effects serve to reduce inflammation by inhibiting pro-inflammatory pathways. Thirdly, they aid in DNA repair and protection by enhancing DNA repair mechanisms and shielding against UV radiation. Lastly, these compounds play a role in the inhibition of tumour growth by suppressing tumour initiation and inhibiting both tumour promotion and progression.

the tumour suppressor P53 or the inhibition of signalling pathways involving oncogenes like Akt and Ras. Furthermore, it has been observed that phytochemicals could hinder the advancement and dissemination of tumours by suppressing enzymes associated with inflammatory mechanisms, such as cyclooxygenase-2 (COX-2), as well as those involved in the formation of new blood vessels, such as vascular endothelial growth factor (VEGF). It is crucial to acknowledge that phytochemicals and other exogenous substances can regulate epigenetic changes, such as modifications in DNA methylation, histone acetylation, and regulation of microRNA synthesis. These variations ultimately influence the expression of genes, even restoring normalcy in cancerous cells (Shankar et al., 2016).

12.3 EXPLORING THE IMPORTANCE OF NATURAL COMPOUNDS FOR SKIN CANCER MANAGEMENT

The examination and acknowledgement of the significance of investigating natural chemicals for the management of skin cancer are imperative, given the multifaceted and frequently stochastic aetiology of skin cancer. The aforementioned elements involve genetic predisposition, epigenetic changes, and environmental impacts. Moreover, the management of rare skin cancers, particularly melanoma cancer and MCC types, is further complicated by the occurrence of significant adverse effects and the development of resistance to conventional therapeutic approaches (Kalal et al., 2017). Patients undergoing chemotherapy frequently find secondary treatments to be both highly toxic and undesirable (Luke and Schwartz, 2013).

Natural chemicals obtained from many plant components, including leaves, bark, stems, roots, and bulbs, exhibit considerable potential as pharmaceutical agents for combating cancer. The investigation of these phytochemicals as primary components in the production of novel pharmaceuticals is currently underway (Cragg and Newman, 2013). Despite the use of tactics such as the employment of nanostructured micellar drug delivery vehicles to address drug resistance in traditional chemotherapy, the overall prognosis for patients diagnosed with uncommon skin malignancies continues to be unfavourable. Consequently, there has been an increasing scholarly focus on the utilization of complementary and alternative medicines, which comprise a range of natural therapeutic approaches. These modalities are being explored as supplementary treatments to conventional chemotherapy, with the aim of enhancing treatment outcomes (Keene et al., 2019).

In contrast to conventional practices involving the utilization of crude extracts characterized by undefined compositions and suboptimal extraction procedures, contemporary methodologies facilitate the achievement of targeted distribution, accurate dosage determination, and well-defined chemical composition of phytochemicals. The advancement in formulation strategies has contributed to the increasing recognition of phytochemicals as viable options for prescription formulations, especially in the later stages of chemotherapy treatment (More et al., 2021). However, it is imperative to conduct additional research on a clinical scale to verify the safety and effectiveness of administered phytochemical formulations and establish them as recognized and standardized clinical procedures in the field of cancer treatment.

It is noteworthy to observe that a considerable piece of pharmaceuticals currently on the market are derived from natural sources, with a majority of anticancer medications (almost 70%) originating from these sources (Cragg and Newman, 2005). Natural sources comprise a diverse range of entities, including flora, fauna, microorganisms, and aquatic organisms. Plants, due to their widespread availability and abundance, serve as a significant natural resource in the field of drug development, offering a multitude of primary chemicals that can be utilized in the creation of novel pharmaceuticals. However, it is worth noting that the current market offers a limited selection of organically derived medications specifically designed to combat skin-related cancers, and none of these have received approval for topical administration. This limitation arises from the recognized adverse effects associated with the topical application of these drugs to the skin (Chinembiri et al., 2014). A thorough comprehension of the molecular and cellular consequences of diverse phytochemicals obtained from natural origins, coupled with investigations on the cytotoxicity and preferential absorption of these substances by malignant and non-malignant cells, has the potential to facilitate the development of efficacious topical interventions for the management of skin malignancies. In order to facilitate the utilization of herbal products in cancer therapies by licensed healthcare practitioners, it is imperative to establish rigorous pharmacovigilance, conduct large, randomized studies, and provide extensive online material that has been confirmed by regulatory organizations across the globe.

12.4 POTENT PHYTOCHEMICALS AND THEIR FORMULATIONS FOR SKIN CANCER PREVENTION AND THERAPY

The significance of phytochemicals in the prevention of skin cancer is of utmost importance, as they demonstrate diverse molecular mechanisms of therapeutic efficacy. Phytochemicals can be classified into three distinct categories according to their efficacy in inhibiting the progression of skin cancer.

I. ANTIPROLIFERATIVE PHYTOCHEMICALS

Phytochemicals with antiproliferative properties, obtained from botanical sources, are of paramount importance in the suppression and restraint of cancer cell growth and proliferation. These chemicals play a role in the regulation of important cellular signalling pathways, the initiation of programmed cell death (apoptosis), and the inhibition of angiogenesis, which is a critical mechanism for the provision of nutrients to tumours. Prominent instances encompass curcumin, derived from turmeric, resveratrol, obtained from grapes, peanuts, and berries, ursolic acid, sourced from citrus fruits, apples, and basil; apigenin, extracted from various fruits and vegetables; luteolin, found in carrots, peppers, and celery; and silymarin, derived from milk thistle (Zhao et al., 2016; Fang et al., 2013; Liu et al., 2019; Zhao et al., 2017; Yao et al., 2019; Vaid et al., 2015). Research indicates that these phytochemicals exhibit inhibitory properties against cancer cells, underscoring their potential as medicines for chemoprevention. The combination of these phytochemicals has the potential to enhance their therapeutic effectiveness, presenting a promising strategy for the management of cancer.

II. PHYTOCHEMICALS WITH ANTI-METASTATIC PROPERTIES

Phytochemicals with antimetastatic properties, found in several plant species, exhibit the capacity to hinder the dissemination of malignant cells within the human body. This characteristic holds considerable potential in the realm of skin cancer prevention and treatment. The previously mentioned phytochemicals selectively interact with pathways that are implicated in the metastatic cascade, including cellular adhesion, migration, and invasion. Antimetastatic phytochemicals are characterized by their distinct features, which set them apart from antiproliferative and apoptosis-inducing phytochemicals. These properties include anti-inflammatory, antioxidant, and anti-angiogenic effects. These compounds demonstrate the ability to affect many cellular signalling pathways, hence exerting inhibitory effects on both cell proliferation and angiogenesis. Several noteworthy examples, such as luteolin, resveratrol, curcumin, β-carotene, and sulforaphane, have demonstrated the potential to inhibit the progression of melanoma (Jones and Katiyar, 2013).

III. AGENTS THAT INDUCE APOPTOSIS

Several phytochemicals have been identified as agents that increase apoptosis, a process that induces programmed cell death, specifically in cancer cells. This is particularly noteworthy in the context of controlling advanced stages of skin cancer. Several phytochemicals have been identified as important regulators of apoptosis, a process of programmed cell death. These include indole-3-carbinol (I3C) (Kim et al., 2006), genistein (Cui et al., 2017), quercetin (Soll et al., 2020), resveratrol (Zhao et al., 2021), and curcumin (Zhao et al., 2016). These compounds elicit apoptosis by stimulating the activation of caspase-3 and caspase-9, while concurrently suppressing the production of Bcl-2, a protein that exerts an inhibitory effect on apoptosis. In addition, phytochemicals have been observed to inhibit the proliferation of skin cancer by stimulating apoptosis, suppressing anti-apoptotic proteins such as Bcl-XL and X-IAP, modulating the cell cycle and microRNA, and promoting apoptosis-autophagic cell death in cancerous cells (Choudhari et al., 2023). Several phytochemicals, such as Chrysin, silymarin, Resveratrol, and Kaempferol, have been observed to impede cellular functioning by inducing pauses in the G1, G2, and M phases. Several phytochemicals have been found to have a notable impact on cell division, leading to programmed cell death by either apoptotic or antiapoptotic activities. This is supported by the presence of substances such as ursolic acid, quercetin, kaempferol, cordycepin, and fisetin (Lagoa et al., 2020).

In addition to the multifarious anticarcinogenic attributes, phytochemicals have also been found to counter multidrug resistance (MDR). Tumourous cells frequently demonstrate resistance to chemotherapy, leading to the development of MDR. The resistance observed might manifest as either inherent or acquired, resulting from molecular modifications that occur either during or after the treatment. Plant-derived natural extracts have exhibited the capacity to enhance the efficacy of chemotherapeutic drugs, while concurrently mitigating toxicity and the need for frequent administration. Various strategies have been developed to address the issue of drug resistance, which involves the inhibition of proteins, signalling pathways, ligands, or overexpressed amino acids and enzymes that are associated with resistance in particular cell types. An alternative methodology is the development of chemotherapeutic drugs that effectively bypass resistance mechanisms and utilize novel tactics for the purpose of selectively targeting cancer cells. Phytochemicals derived from plants are known to have a significant impact on the inhibition of malignant tumours and the reduction of resistance, hence improving the effectiveness of therapeutic interventions. The combination of these phytochemicals in therapeutic applications has demonstrated notable advantages, including the reduction of dosages and dosing intervals, as well as the efficient inhibition of cancer cells (Sun et al., 2019).

Furthermore, phytochemicals exert their effects through the inhibition of key signalling pathways, including CDC2, CDK2, CDK4 kinases, topoisomerase, mTOR, TNK, PI3K, Akt, Bcl-2, Bcl-1, cytokines, and interleukins. In addition, it is possible for them to decrease the level of gene expression during the processes of mitosis or meiosis, specifically targeting genes that have had mutations such as p21 and p53. In addition, they may also participate in DNA repair mechanisms. Flavones, which are prominent components of various natural products, have been found to induce the synthesis of defensive enzymes. The regulation of the activation processes is mediated by proteins like Bax, Bid, and Bak. Moreover, it has been observed that antioxidant enzymes such as glutathione (GSH), glutathione S-transferase (GST), and glutathione peroxidase (GPxn) exhibit significant anticancer properties (Lagoa et al., 2020; Iqbal et al., 2017).

Formulations using concentrated phytochemicals, either separately or in combination as part of pharmacological therapy, have been extensively investigated due to the multifaceted processes by which these compounds combat the progression of skin cancer.

12.4.1 Curcumin (CUR)

CUR is a potent phytochemical that has been extracted from *Curcumin longa* rhizomes. It has demonstrated promising anti-inflammatory and antioxidant characteristics in a range of inflammatory conditions, including psoriasis, as indicated by previous studies (Abbasi et al., 2018). Additionally, CUR possesses the capacity to inhibit cancer development through the modulation of various molecular targets, including cyclooxygenase-2 (COX-2), 5-lipoxygenase (5-LOX), signal transducer and activator of transcription 3 (STAT3, nuclear factor kappa-light-chain-enhancer of activated B cells (NF-κB), phosphorylase kinase, and multiple apoptotic cytokines (Abbasi et al., 2018). Kuttan et al. presented the preliminary findings on the anticancer potential of CUR (1987). It was discovered that CUR has exhibited a reduction in the size of cancer lesions in a cohort of more than 62 patients. Furthermore, it has been demonstrated that CUR exhibits potential efficacy in various types of cancer, either as a standalone treatment or in combination with other therapeutic drugs, resulting in a significant synergistic impact. Furthermore, extensive research has been conducted to examine the potential of CUR in offering protective effects against several types of cancer, including head and neck squamous cell carcinoma (SCC), prostate cancer, multiple myeloma, pancreatic cancer, lung cancer, and colorectal cancer (Abbasi et al., 2018). Additionally, the current study demonstrates that CUR effectively inhibits melanoma growth in a mouse model (C57BL/6 mice) implanted with melanoma cells (B78H1 cells) by upregulating miRNA-2015-5p expression. This upregulation of miRNA-2015-5p plays a crucial role in controlling apoptosis and proliferation (Dahmke et al., 2013). In a separate investigation, the potential anti-inflammatory properties of CUR were examined in relation to the SRB12-p9 skin cancer cell line inside a mouse skin model. The injection of CUR to immunodeficient mice has shown potential in suppressing the formation of cutaneous squamous cell carcinoma (SCC) via downregulating the pS6 biomarker. Curcumin (CUR) exhibits the capacity to effectively suppress cell proliferation in RB12-p9 cells at a concentration of 20μM or higher. This observation suggests that CUR possesses promising therapeutic potential for the treatment of skin cancer. Cucurbitacin (CUR) can be delivered via many ways, and both oral and topical administration have demonstrated comparable potential in the mouse skin model. The safety and efficacy of CUR have been thoroughly investigated through many clinical trials, establishing it as a potent phytocompound for the advancement of pharmaceutical interventions targeting skin cancer (Gupta et al., 2012) [60].

12.4.2 Caffeic Acid (CA)

CA is a polyphenolic chemical that exhibits promise and is found in vast quantities in coffee, as well as in several fruits and vegetables. CA is a bioactive phytocompound that exhibits promising antioxidant, anti-inflammatory, and anticancer properties (Fernandez et al., 1998; Moon et al., 2009;

Tsai et al., 2011). Moreover, Yang et al. (2013) discovered in their investigation that the application of CA has the potential to impede the formation of colonies and suppress EGF-induced tumour growth in keratinocyte cells. Furthermore, it has been observed that CA can diminish the migratory capacity of cancer stem cells through the upregulation of p38 phosphorylation and the inhibition of the NF-κB/snail signalling pathway. The administration of p38 has been seen to decrease the DNA-binding capacity of NF-κB to the promoter region of the snail gene, leading to the subsequent inhibition of snail gene transcription. Additionally, a disruption in the process of epithelial-mesenchymal transition (EMT) has been seen in human keratinocyte tumours treated with CA. Epithelial-mesenchymal transition (EMT) is a biological process by which epithelial cells undergo a loss of polarity and acquire migratory and invasive characteristics. In malignant human keratinocyte cells treated with CA, there is an observed rise in the level of E-cadherin, while the levels of N-cadherin and vimentin drop. The findings of this study demonstrate that CA confers a protective effect on cancer cells, namely in terms of inhibiting invasion and migration (Yang et al., 2013; Esmaeili et al., 2016). The scientific community maintains a positive outlook on the potential development of novel formulations derived from these chemicals, aiming to offer skin cancer patients the most effective therapy alternative.

The application of CA topically on the dorsal surface of mice with UV-induced skin cancer was found to effectively limit the occurrence of cancer, as reported in a study (Esmaeili et al., 2016). The Ras-Raf-MEK-extracellular signal-regulated kinase (ERK) 1/2 cascade of the MAPK signalling system is the most regulated molecular pathway in human malignancies. This molecular mechanism governs numerous other cellular activities, including cell development and cell proliferation (McCubrey et al., 2007). In addition, it has been observed that CA also inhibits the functions of ERK1/2 in *in vitro* experiments and exhibits chemotherapeutic effects against UV-induced skin cancer (Esmaeili et al., 2016). Furthermore, the researchers observed that COX-2 and Fyn kinase expressions were upregulated during UV-induced skin carcinogenesis, in addition to MAPK. Caffeic acid (CA) interferes with the activity of NF-kB and AP-1 leading to the down-modulation of COX-2 levels that are otherwise found to be elevated due to UVB radiation. This interference, which lead to the inhibition of prostaglandin E2 and the blocking of Fyn kinase activity was reported by Kang et al. (2009) in their landmark study that included a mouse model of skin cancer.

Additionally, Chao et al. (2009) discovered that CA exhibits anti-inflammatory properties by reducing the expression of IL-1β mRNA, TNF-α, IL-6 protein in the cardiac tissue of mice with diabetes. Furthermore, in a separate study conducted by Song et al., it was revealed that CA exhibits anti-inflammatory properties through the reduction of phospholipase A2 and myeloperoxidase (MPO) activities in a mouse model with induced skin inflammation (Song et al., 2008).

CA has the capacity to reduce the expression of mRNA and proteins associated with IL-1β, TNF-α, and IL-6 at the specific site of action and even in human keratinocyte cell cultures (Khan et al., 2012). The results of this study suggest that CA exhibits potent anti-inflammatory properties by inhibiting the production of cytokines and reducing MPO activity. In general, these findings can serve as a valuable resource for informing the design and evaluation of computer-aided interventions, given their significant therapeutic promise in clinical settings for the treatment of skin cancer.

12.4.3 Caffeic Acid Phenethyl Ester (CAPE)

CAPE is a potentially significant bioactive compound that has been extracted from several honey products. CAPE exhibits robust inhibitory capabilities *in vitro* and *in vivo* against various types of malignancies, including pancreatic, gastric, lung, colon, liver, melanoma, and breast cancer (Xiang et al., 2006; Kudugunti et al., 2011; Onori et al., 2009). *In vitro* investigations have demonstrated that CAPE exhibits robust anticarcinogenic, anti-mitogenic, and anti-inflammatory effects (Natarajan et al., 1996). Moreover, the proliferation, invasion, and migration of the cutaneous papilloma induced by exposure to 12-O-tetradecanoylphorbol-13-acetate (TPA) have been significantly impeded by CAPE. It has been observed that CAPE exhibits the ability to down-regulate the expression of vascular endothelial growth factors (VEGF) and multidrug resistance 1 (MDR-1) membrane protein. Cape Aloe-Emodin (CAPE) has been found to possess the capability to regulate the cell cycle and trigger apoptosis through the NF-κB pathway (Wu et al., 2011).

Moreover, it has been observed that in leukaemia cells, CAPE has the ability to trigger apoptosis by upregulation of Bax and downregulation of Bcl-2 expression (Chen et al., 2001). At lower concentrations, CAPE has demonstrated antioxidant activity on mouse skin cells (Frenkel et al., 1993). There has been a scarcity of studies conducted on the human clinical trials of CAPE in relation to skin cancer, indicating a need for additional investigation in this area. Nevertheless, the compound's antioxidant properties, ability to block certain processes, and its capacity to inhibit VEGF show potential as a viable therapy option for skin cancer.

12.4.4 Ursolic Acid (UA)

UA is a phytochemical with significant potential that is found in several plant sources such as thyme, basil, and rosemary, among others (Iqbal et al., 2018a). The compound UA has demonstrated encouraging properties in terms of inhibiting cell proliferation, preventing the onset of cancer, acting as an antioxidant, and reducing inflammation. UA has been shown to trigger cell death in multiple myeloma cell lines through the activation of caspase-3 via the mitochondria intrinsic signalling cascade. Additionally, UA has been found to up-regulate the expression of p53 and caspase-3, while down-regulating Bcl-2 (Harmand et al., 2005; Manu and Kuttan, 2008). Ursolic acid has also demonstrated potential anti-proliferative effects through its ability to modulate the G1 phase of the cell cycle and regulate the expression of p21 WAF1, which is involved in cell cycle regulation (Es-Saady et al., 1996; Harmand et al., 2003). Moreover, in a separate investigation, it has been demonstrated that UA can hinder the phosphorylation of p65 and IκBα, resulting in the suppression of the NF-κB signalling pathway (Shishodia et al., 2003). The inhibition of NF-κB by UA mediates a reduction in the production of cyclin D1, MMP-9, and COX-2 enzymes. Ursolic acid (UA) has been found to have many effects on human lymphocytes exposed to UVB irradiation. In addition to inducing apoptosis and cell cycle arrest, pre-treatment with UA has been observed to confer a UV antioxidant effect. This is evidenced by a reduction in lipid hydroperoxide levels and an enhancement in antioxidant levels (Ramachandran et al., 2008). A study conducted on three healthy volunteers utilized a liposome-coating formulation to examine its effects on the ceramide contents of human skin. The results indicated an increase in ceramide levels (Both et al., 2002). Nevertheless, the sample size utilized in this experiment was limited and consisted solely of non-cancerous skin specimens.

12.4.5 Genistein (GEN)

Soybeans are rich sources of the isoflavone genistein (GEN) (Singh et al., 2014). Interestingly, in the management of cancer, cardiovascular disorders, and osteoporosis soybean-rich diets have proven to be very useful (Iqbal et al., 2018b). The therapeutic benefits found in such diets have been attributed to the wide distribution of phytoestrogens such as GEN. Soya-phytoestrogens have potent anti-inflammatory, antioxidant, anti-inhibitory, and anti-proliferative activities, as supported by many studies (Rusin et al., 2010; Wei et al., 1998; Sarkar and Li, 2002). In such studies, GEN has demonstrated profound chemoprotective action in various forms of malignancies, including breast cancer, neuroblastoma, MM (multiple myeloma) and NMSC (non-melanoma skin cancer) (Afaq and Katiyar, 2011; Li et al., 2012). GEN has been shown to have potent anti-angiogenic properties that contribute significantly to diminishing tumour proliferation and metastasis. Additionally, it has been found to induce cell cycle arrest and facilitate cell death through caspase-mediated mechanisms (Rusin et al., 2011; Chandra et al., 2016). According to a study, GEN has demonstrated a reduction in UV-induced sunburn in humans thereby providing protection against both UV-induced photoageing and skin cancer (Wei et al., 2002). Furthermore, GEN exhibits photoprotective effects on human skin through the inhibition of pyrimidine dimer production generated by UVB radiation.

Additionally, in several experimental models involving xenografts, GEN has been found to play a direct role in the disruption of the cell cycle in proliferating cancer cells and thereby stemming their metastatic hallmarks (Danciu et al., 2013; Ji et al., 2012). The reduction of malondialdehyde (MDA) (marker of lipid peroxidation) and H_2O_2 levels of the epidermis played a mechanistic role in the photoprotective attributes of GEN in the experimental animals treated with GEN prior to UV exposure (Wei et al., 2002). In a separate investigation, the progression of the MM cell cycle was impeded by GEN through its interaction with many molecular targets, including p53, p21, checkpoint kinase, and Chk2 (Casagrande and Darbon, 2000; Rauth et al., 1997). The gene expression network (GEN) not only plays a role in regulating the cell cycle but also functions to limit the process of angiogenesis (Kiguchi et al., 1990; Sjoberg et al., 1995; Yan and Han, 1997). There exists a substantial body of scientific information that provides support for the utilization of genetic engineering (GEN) in the prevention and treatment of human skin cancers such as multiple myeloma (MM) and non-melanoma skin cancer (NMSC). Both *in vitro* and *in vivo* research are still required to assess the route of administration, toxicity, dosing schedule, and efficacy of gene editing nucleases (GEN) for the treatment and management of cancer.

12.4.6 Luteolin

In a wide range of plant sources such as carrots, celery, olives, and peppers, phytochemical luteolin is found in significant proportions. Luteolin demonstrates significant potential in exhibiting anti-inflammatory, antioxidant, and anticancer characteristics. Several studies involving luteolin have corroborated its abilities to block angiogenesis, facilitate caspase-mediated cell death, and enhance the sensitivity of cells to anticancer therapies across several types of malignancies (López-Lázaro, 2009). Numerous research studies have demonstrated that luteolin possesses the ability to stimulate melanogenesis while concurrently mitigating the aggressive behaviour of skin cancer cells. This impact is achieved through the modulation of focal adhesion kinase (FAK) signalling cascade and β3 integrin as evidenced by several investigations (Horibe et al., 2013; Ruan et al., 2012). In addition, previous studies have demonstrated that luteolin possesses the ability to initiate programmed cell death (apoptosis) and inhibit the proliferation of skin cancer cells through the upregulation of Bax, downregulation of Bcl-2, and attenuation of ERK1/2 signalling pathways (Iwashita et al., 2000; Nakashima et al., 2010). Prior investigations have demonstrated promising outcomes in the field of cancer therapies for this phytochemical. Nevertheless, additional *in vitro, in vivo*, and human clinical trials are required to establish the effectiveness and pharmacokinetics of this prospective phytocompound.

12.4.7 Indole-3-Carbinol (I3C)

Phytochemical profile of various cruciferous vegetables, such as broccoli, Brussels sprouts, and cauliflower comprises I3C that demonstrates promising application against skin cancer manifestations (Abbasi et al., 2018). In addition, the anticancer effects of indole-3-carbinol (I3C) have also been observed and reported in several published reports against other types of cancer manifestations such as breast, cervical, gastrointestinal, and lung cancer (Kuttan et al., 1987). The studies exploring the potential of I3C against UVB-sensitized MM cells have provided evidence that indicates that I3C augments cell cycle arrest and induces apoptosis by inhibiting Bcl-2 and down-regulating microphthalmia-associated transcription factor (MITF) (Dahmke et al., 2013; Gupta et al., 2012). Additionally, studies reveal that I3C can modulate the degradation of phosphatase and tensin homolog (PTEN) and therefore are able to suppress the growth of human MM cells (Safe et al., 2008). The administration of dietary supplements containing indole-3-carbinol (I3C) has been observed to enhance the responsiveness to chemotherapy in a mouse model (Weng et al., 2007). The study of I3C hitherto has been limited to only cellular and animal models. These preliminary findings necessitate rigorous scientific inquiry to establish its safety and effectiveness.

12.4.8 Resveratrol (RV)

RV is a naturally occurring stilbene polyphenol that exhibits a wide distribution in various plant sources such as mulberries, peanuts, and grapes. The topical administration of this substance has significant inhibitory effects on all three distinct stages of carcinogenesis in a mouse model (Kim et al., 2006; Kim et al., 2011). The compound has demonstrated significant anticancer properties due to its anti-proliferative, anti-inflammatory, and antioxidant capabilities (Aronchik et al., 2014). Additionally, it has been observed to effectively scavenge reactive oxygen species (ROS) (Christensen and LeBlanc, 1996) and potentially decrease ROS levels in human skin fibroblast cells *in vivo* (Jang et al., 1997). In addition to its antioxidant capacity, resveratrol (RV) has been found to counteract the anti-inflammatory impact by impeding the activity of COX-1 in living organisms (Aziz et al., 2005) and COX-2 in the skin of mice (Ndiaye, et al., 2011). This is mostly achieved through the inhibition of NF-κB expression and the suppression of both p38 MAPK and ERK (Guthrie et al., 2017; Jagdeo et al., 2010; Subbaramaiah et al., 1998).

Resveratrol (RV) has demonstrated promising efficacy in suppressing carcinogenesis and lowering murine epidermal hyperplasia. Additionally, RV has been found to decrease the expression of proteins such as COX-2, Bcl-2, p21 and enzymes, suggesting its potential in modulating various phytocompounds (Kundu et al., 2006). Rhein (RV) has been seen to potentially serve as an adjuvant alongside other chemotherapeutic treatments in the management of multiple myeloma (MM) with distant metastatic disease. In this context, RV has demonstrated the potential to reduce the viability of skin cancer cells and enhance the cytotoxic effects of temozolomide on cancer cells (Niles et al., 2006). Moreover, it has been observed that RV possesses the capacity to inhibit the activity of redox factor-1 (Ref-1), hence increasing the susceptibility of skin cancer cells to the alkylating agent dacarbazine (Kowalczyk et al., 2010). Moreover, the upregulation of Akt/PKB proteins in MM cells is an additional mechanism through which RV exhibits chemotherapeutic actions on MM skin cancer (Osmond et al., 2012).

Additionally, it has been observed that RV administered orally exhibits limited *in vivo* bioavailability due to its rapid elimination by the liver and intestine, resulting in decreased levels inside the human body (Yang et al., 2005). The administration of RV orally to mice with skin cancer has been found to have little contact with skin cells and tumours, hence demonstrating its incapacity to effectively inhibit cancer proliferation (Bhattacharya et al., 2011). Hence, the application of RV by topical administration holds great potential in the fields of chemoprevention and chemotherapeutics. Currently, the administration of topical creams containing RV has demonstrated promising enhancements in skin elasticity, hydration, and brightness, without any adverse effects observed in individuals with normal skin (Moyano-Mendez et al., 2014). Furthermore, another study conducted on a sample size of 55 patients showed that the combination of vitamin E conjugated RV with baicalin resulted in significant improvement in photodamaged skin within a period of around 12 weeks (Farris et al., 2014). The clinical trials conducted in this study utilized a limited sample size and focused on the examination of non-malignant skin conditions. Although extensive clinical trials are required to determine their effectiveness and safety in the prevention and treatment of both MM and NMSC, preliminary findings have indicated promising outcomes. The scientific community remains actively involved in the investigation of the further pharmacological potential of these substances, with the anticipation of discovering new opportunities.

12.4.9 Gingerol

Phenolic phytochemical gingerol with a strong taste is derived from the root of *Zingiber officinale*. The initial investigation on [6]-gingerol was conducted by Park et al. (1998), wherein the topical application of this compound inhibited skin papilloma formation. Furthermore, the compound known as [6]-gingerol has been found to exhibit anti-inflammatory properties through many mechanisms. These include reducing the activity of epidermal ornithine decarboxylase, inhibiting the enzyme COX-2, and suppressing the activity of NF-κB through modifying the activity of p38 mitogen-activated protein kinase (MAPK) (Park et al., 1998; Kim et al., 2004). In addition, it has been observed that [6]-gingerol provides an antioxidant effect by counter-modulating intracellular reactive oxygen species (ROS) levels that were elevated by UV radiation stimulation. This compound has also been found to activate caspases 3, 8, and 9 (Kim et al., 2007).

Numerous additional ways encompass the induction of AP-1 DNA binding activity (Bode et al., 2001) and the modulation of proteins such as Bax, Bcl-2, p53, and survivin (Nigam et al., 2010). Hitherto, no scholarly publications have been released regarding human trials in this area. However, ongoing research is focused on the integration of [6]-gingerol as stable nanoparticle formulations for topical administration, with the aim of improving their safety and efficacy (Ratcharin et al., 2012). This form of delivery technique for [6]-gingerol has the potential to offer a more convenient and stable alternative for further human clinical investigations.

12.4.10 Capsaicin

Red pepper and red chilli pepper are potent sources of the bioactive compound capsaicin with immense and proven therapeutic benefits. There exists a scientific debate about the potential carcinogenic properties of capsaicin, as well as its potential applications in chemotherapy (Bode and Dong, 2011). In their study, Hwang et al. (2010) provide evidence that the application of capsaicin topically on a mouse model treated with TPA resulted in the stimulation of skin cancer. This finding suggests that capsaicin acts as a pro-carcinogen by activating the tyrosine kinase EGFR and COX-2. Contrarily, previous studies have yielded contrasting findings, indicating that the application of capsaicin topically did not result in a significant escalation in the development of skin cancer in mice when compared to control groups. In fact, it was shown that capsaicin administration exhibited a noteworthy inhibition of papilloma formation in mice, hence impeding the progression of skin cancer (Park and Surh, 1997). Capsaicin has demonstrated robust chemopreventive attributes through its ability to halt the cell cycle, induce apoptosis, and impede the growth of cancer cells by counteracting the expression of NF-κB, AP-1, STAT3, and COX-2 (Oyagbemi et al., 2010). Additionally, capsaicin has demonstrated its therapeutic potential by triggering cell death in a human cutaneous squamous cell carcinoma cell line through the inhibition of mitochondrial function, mediated by caspase activation (Hail and Lotan, 2002). Other research studies have demonstrated that capsaicin exhibits an anti-mitogenic effect on metastatic MM cells through the downregulation of phosphatidylinositol 3-kinase (PI3-K) expression (Shin et al., 2008). Capsaicin exhibits synergistic effects through the induction of caspase-mediated cell death in MM cell lines when co-administered with HA14-1 (Marques et al., 2013). Researchers are currently engaged in additional investigations and epidemiological studies to establish the potential therapeutic applications of capsaicin more effectively in the context of cancer treatment.

Despite the absence of prior research elucidating the precise mechanisms underlying the utilization of capsaicin through topical administration for the treatment of skin cancer, insights can be gleaned from the application of capsaicin in other contexts. A comprehensive analysis revealed that a significant proportion of patients (one out of three) who received capsaicin experienced a greater incidence of side effects, such as stinging, erythema, and burning, as compared to those who received a placebo (Mason et al., 2004). The potential adverse effects mentioned above pose as downsides for the topical application of capsaicin in the prevention and treatment of skin cancer using chemopreventive and chemotherapeutic approaches. Hence, the development of a unique drug delivery system that involves the design and formulation of capsaicin in combination with other medicines, which exhibit minimal or no adverse effects, is a promising approach for the treatment of skin cancer.

12.4.11 Eugenol

Eugenol is a phytochemical compound classified as a phenol, mostly found in various plant sources such as cloves, nutmeg, bay leaves, basil, and cinnamon. The administration of this substance can be achieved via many pathways within the body. The application of eugenol topically and the administration of aqueous infusions of cloves orally in mice with skin cancer resulted in a decrease in the occurrence of papilloma formation. In addition, eugenol exhibits anti-proliferative and antioxidant properties via distinct pathways. Eugenol exhibits its antioxidant activity through the quick scavenging of reactive oxygen species (ROS), as well as by inhibiting the production of ROS and lipid peroxidation (Sukumaran et al., 1994). Moreover, the application of eugenol in a localized manner has been shown to potentially mitigate inflammation through the inhibition of the COX-2 enzyme, stimulation of nitric oxide synthase (iNOS) expression, reduction in the levels of TNF-α, IL-6, and PGE2, as well as modulation of the expression of NK-κB (Kaur et al., 2010). In addition, eugenol has been observed to reduce the expression of oncogenes H-ras and c-Myc, regulate the expression of p53, and promote apoptosis through the downregulation of the E2F1 transcriptional factor production (Ghosh et al., 2005; Pal et al., 2010).

In a recent study, eugenol was developed into a nanoemulsion for the purpose of topical anti-inflammatory application on murine skin. The results indicated that eugenol formulations containing 2% concentration exhibited superior anti-inflammatory effects in comparison to topical piroxicam, specifically after a duration of 1.5 hours (Esmaeili et al., 2016). Further *in vitro* and *in vivo* investigations are required to provide a more comprehensive understanding of the anti-inflammatory, antioxidant, and anticancer properties of eugenol, as well as to facilitate the development of novel formulations.

12.4.12 Epigallocatechin-3-Gallate

Epigallocatechin-3-Gallate (EGCG) is a phytocompound that has been extracted from Camellia sinensis and extensively investigated for its chemopreventive and chemotherapeutic properties. It is a constituent of green tea phenols (GTP) and has demonstrated anti-inflammatory, anti-proliferative, and antioxidant capabilities. Katiyar et al. (2000) conducted a significant study on GTP, focusing on its anti-inflammatory properties. Their findings suggest that GTP exerts its anti-inflammatory effects by suppressing the activity of COX and lipoxygenase enzymes. Additionally, GTP was found to reduce the incidence of skin cancer by decreasing epidermal hyperplasia and oedema. The researchers subsequently investigated the antioxidant properties of EGCG on human skin cancer through topical application, which resulted in a considerable reduction of reactive oxygen species (ROS) products caused by UV radiation (Katiyar et al., 1992). The radiations have been found to be linked with the suppression of MAPK signalling cascades (Katiyar et al., 2001). Additional anti-proliferative activities encompass the control of NF-κB signalling cascades (Afaq et al., 2003; Ahmad et al., 2000), suppression of tumour-induced activator protein (AP-1) (Dong et al., 1997), inhibition of angiogenesis, and recruitment of T cells (Mantena et al., 2005).

Additionally, Nihal et al. (2009) have provided evidence indicating that EGCG enhances the susceptibility of MM cells to growth inhibition triggered by interferon, diminishes the proliferation of malignancy cells, and triggers cell death mediated by caspase. The researchers discovered novel findings indicating that the combined administration of EGCG and interferon resulted in a synergistic impact, which was far more effective than the individual treatment of either compound alone. EGCG has been found to exhibit the ability to downregulate inflammation, resulting in a decrease in the release of interleukin (IL)-1β and the downregulation of NF-κB activity. This, in turn, leads to a reduction in the proliferation of cancer cells (Ellis et al., 2011). Recent research has shown that scientists have proven the inhibitory effects of EGCG on MM cell invasion through the downregulation of tumour necrosis factor (TNF) receptor-associated factor 6 (TRAF6) function (Zhang et al., 2016).

Various human trials have demonstrated the robust therapeutic potential of EGCG in the chemoprevention and chemotherapeutics of skin cancer. The problem of topical versus oral application continues to be of significant importance. Researchers have noticed that mice who were provided GTP orally or through injection demonstrated inhibition or reversal of UV-induced cutaneous papilloma (Wang et al., 1992). However, another study has shown that tumour inhibition in mice was shown only when EGCG was applied topically, rather than administered orally (Gensler et al., 1996). A research investigation involving human subjects who were given topical GTP has substantiated its efficacy in providing protection against erythema generated by UV radiation (Mukhtar and Ahmad, 1999). However, a separate single-blind randomized clinical experiment including a sample size of 50 persons found that the administration of GTP orally, in conjunction with vitamin C, did not result in a significant reduction in skin leukocyte infiltration and erythema when compared to a placebo group (Farrar et al., 2015). It was hypothesized by the scientific community that the topical administration of certain formulations for skin cancer may not exhibit significant activity, as their bioavailability was shown to be limited. The findings of these research investigations suggest that the application of EGCG topically may have greater potential for skin cancer chemoprevention and chemotherapeutics compared to oral administration.

12.4.13 Silibinin

Silibinin, a phytochemical derived from the seeds of Milk thistle, has shown considerable potential. Silibinin constitutes the principal bioactive constituent inside the silymarin complex. The utilization of silibinin has been limited due to its inadequate bioavailability. Nevertheless, researchers are currently engaged in the endeavour of formulating novel compositions that can be more effectively absorbed, specifically in the form of nanosuspensions (Wang et al., 2010). The findings of a recently conducted investigation pertaining to the application of silymarin-nanostructured lipid carrier (NLC) gel on B16 melanoma cell lines revealed more favourable results in terms of tumour volume reduction and suppression of pro-inflammatory cytokines (Iqbal et al., 2019a). These functions include acting as a scavenger for reactive oxygen species (ROS) and phenylglyoxylic ketyl radicals (Iqbal et al., 2017). Various clinical trials have been conducted to examine the chemotherapeutic effects of silymarin on a diverse array of malignancies, including skin cancer (Deep and Agarwal, 2010). A study has provided confirmation that silymarin exhibits cancer chemotherapeutic efficacy through the inhibition of tumour promotion induced by TPA in mouse skin. A further research investigation has demonstrated that silibinin has a potent anticancer effect by targeting the CDK pathway and inducing cell cycle arrest (Afaq et al., 2003). Moreover, silibinin exhibits antagonistic effects on angiogenesis through its specific targeting of VEGF receptors and iNOS (Deep et al., 2006; Singh et al., 2005). Furthermore, it has been observed that silibinin can also elicit cell death mediated by caspases through both extrinsic and internal signalling pathways (Singh et al., 2006; Kaur et al., 2009).

Additionally, Katiyar et al. (1997) conducted a study in which they found that silymarin demonstrated a protective effect against skin cancer generated by UV radiation in a mouse model of photocarcinogenesis. Furthermore, the injection of silymarin has been found to limit the occurrence of UVB-induced sunburn and skin oedema. It also promotes the production of apoptotic cells, reduces catalase activity, and induces the expression of COX and ornithine decarboxylase. The use of silibinin on UVB-damaged skin has been found to exhibit similar protective effects. Specifically, the levels of proliferating cell nuclear antigen, thymidine dimer-positive cells, and apoptotic sunburn cells were observed to decrease following treatment with silibinin (Vaid and Katiyar, 2010). Silibinin has demonstrated potential efficacy in the context of cancer therapies by inhibiting the MAPK-mediated signalling cascade (Dhanalakshmi et al., 2004; Tong et al., 2018). In a study conducted on SKH-1 mouse skin, it was observed that both dietary consumption and topical application of silibinin effectively suppressed the activation of MAPK, p38, JNK, and Akt pathways that were produced by UV exposure (Singh et al., 2009). Moreover, a study has indicated that silymarin has exhibited a notable reduction in the accumulation of β-catenin in human multiple myeloma (MM) cells, hence impeding the migration of these cells (Mallikarjuna et al., 2004; Vaid et al., 2011, Gu et al., 2007). Taken together, the findings indicate that silymarin/silibinin have demonstrated significant efficacy as chemopreventive and chemotherapeutic drugs in the treatment of skin cancer. Future clinical studies of silymarin in the context of human cancer research may also prioritize the examination of its toxicity and bioavailability, necessitating rigorous scientific inquiry.

12.4.14 Allyl Sulphides (AS)

Allyl sulphides are sulphur metabolites found in garlic oil that have exhibited anticancer capabilities against a range of cancer forms, including skin cancer. The anticarcinogenic properties of allyl sulphides are ascribed to their potent antioxidant, anti-inflammatory, and anti-proliferative attributes. Among the several chemicals examined, diallyl trisulfide (DATS), which falls under the category of allyl sulphides, has demonstrated superior effectiveness in the treatment of skin cancer when compared to mono- and disulfides (Wang et al., 2012). The growth of human melanoma A375 cells and basal cell carcinoma (BCC) cells are impeded by DATS, which operates through various mechanisms including the elevation of intracellular reactive oxygen species (ROS), induction of DNA damage, and induction of G2/M cell cycle arrest. Additionally, it induces endoplasmic reticulum (ER) stress and mitochondria-mediated apoptosis, encompassing both caspase-dependent and caspase-independent mechanisms (Wang et al., 2017).

Numerous investigations have been conducted to elucidate the molecular mechanisms that underlie the preventative actions of allyl sulphides generated from garlic on skin cancer. These studies have demonstrated the remarkable capacity of these compounds to efficiently target numerous signalling pathways [182]. Nevertheless, it is important to highlight that there is a lack of clinical trials conducted on human subjects to assess the efficacy of these chemicals in preventing skin cancer (Zhang et al., 2020).

Moreover, it is worth noting that diallyl sulphide (DAS) has the potential to exhibit chemopreventive and chemotherapeutic properties through the modulation of several signalling pathways. As an example, a scientific investigation found that DAS could decrease the expression of H-ras mRNA, suppress the production of oncogenic p21, increase the levels of p53 and Bax proteins, and reduce the activity of PI3K/Akt and MAPKs (Arora et al., 2006). Recent studies have additionally substantiated the effectiveness of DAS in mice as a preventive measure against skin cancer generated by UV radiation. This is achieved through the modulation of several molecular targets, namely NF-κB, PEG2, COX-2, and p53 (Kalayarasan et al., 2009).

In a mouse model, the topical use of diallyl disulfide (DADS) has demonstrated encouraging outcomes by diminishing the occurrence and propagation of skin cancer. Research findings have demonstrated that DADS (diallyl disulfide) can increase the activity of antioxidant enzymes such as glutathione peroxidase, catalase, and superoxide dismutase. Additionally, DADS has been shown to promote the production of p21 protein and enable the involvement of Nrf2 in the maintenance of cellular redox equilibrium (Gerhauser, 2012). Moreover, diallyl trisulfide (DATS) has demonstrated considerable promise in its ability to decrease the formation and growth of cutaneous papilloma by inhibiting the expression of COX2 through the modulation of JNK or Akt signalling pathways. Additionally, it has been observed that the compound can trigger apoptosis in melanoma (MM) cells by the downregulation of anti-apoptotic proteins, including Bcl-2 and Bcl-xl. The observed effects are consistently observed in human basal cell carcinoma (BCC) cells. Specifically, diallyl trisulfide (DATS) induces cell death by caspase-mediated mechanisms, while simultaneously increasing the expression levels of p53 and Bax and decreasing the expression levels of Bcl-2 and Bcl-xl. The findings of this study suggest that DATS elicits endoplasmic reticulum stress in cancer cells, ultimately resulting in cellular apoptosis. Furthermore, it has been demonstrated that DATS exhibit promises in suppressing the invasion and migration of human MM cells and BCC cells through the downregulation of MMP-2 and MMP-9 expression and the disruption of the integrin signalling pathway (Puccinelli and Stan, 2017).

The comparative analysis of several allyl sulphides demonstrated that diallyl trisulfide (DATS) has a greater potential for reducing the expression of cyclooxygenase-2 (COX-2) in human embryonic kidney cells, as compared to diallyl disulfide (DADS) and diallyl sulphide (DAS). Moreover, it has been observed that DATS exhibits more potent inhibitory effects on human MM cells and BCC cell lines compared to DADS and DAS. Furthermore, it has been demonstrated by this research that allyl sulphides possess the ability to impede the proliferation of cancer cells, induce cell cycle arrest at the G2/M phase, trigger cell death mediated by caspases, and activate the p53 pathway in the presence of oxidative stress. In conclusion, DATS demonstrates considerable potential as a chemopreventive and chemotherapeutic phytochemical when compared to DADS and DAS (Iqbal et al., 2019b). Notwithstanding this promise, it is imperative to acknowledge that yet, there have been no documented human clinical studies, emphasizing the necessity for additional research into its pharmacological characteristics and bioavailability.

12.4.15 Apigenin

Apigenin, known chemically as 4',5,7,-trihydroxyflavone, is a flavone that naturally forms yellow, needle-shaped crystals when it is in its pure state. It is frequently found in oranges (*Citrus sinensis*), tea (*Camellia sinensis*), parsley (*Petroselinum crispum*), celery (*Apium graveolens*), thyme (*Thymus vulgaris*), and onions (*Allium cepa*) (Shukla and Gupta, 2010a). Epidemiological studies have shown that diets containing high levels of flavones, which are bioactive compounds found in plants, are linked to a reduced risk of some types of cancer (Shukla and Gupta, 2010b). This observation has prompted investigations into the anticancer effects of apigenin.

Apigenin has demonstrated anticancer properties in *in vitro* tests conducted on different cell lines, such as head and neck squamous cell carcinoma cells, melanoma cells, and liver cells. The methods described by Chan et al. (2012) involve causing a halt in the cell cycle during the G2/M phase, increasing the expression of tumour necrosis factor receptor (TNF-R) and TNF-related apoptosis-inducing ligand receptor (TRAIL-R) apoptotic pathways, reducing the expression of Bcl-2, and activating caspase-3. The combined influence of these mechanisms contributes to the chemoprotective benefits of apigenin. Apigenin has demonstrated protective properties against UVB radiation on human keratinocytes in laboratory settings and on the skin tissue of mice in live experiments. These effects disrupt cell survival and growth by interfering with the nuclear factor kappa-light-chain-enhancer of activated B cells (NF-κB) and mitogen-activated protein kinase (MAPK) pathways (Ng et al., 2018). Furthermore, apigenin hinders the cancer-causing enzyme Src, resulting in a decrease in the expression of cyclooxygenase 2 (COX-2) induced by UVB radiation. This, in turn, helps alleviate the inflammatory and cancer-causing effects linked to COX-2 (Tong et al., 2007).

In both *in vitro* and *in vivo* experiments, it has been observed that ethosomes containing apigenin exhibit higher levels of skin deposition compared to liposomes (Shen et al., 2014). Das et al. (2013a) investigated the influence of nano-encapsulation on apigenin and found that encapsulated apigenin may offer a more advantageous formulation compared to free apigenin. Apigenin, when encapsulated, showed improved ability to enter the nucleus, leading to increased apoptotic effects. The use of apigenin, which was encapsulated in nanoparticles, on the skin surface effectively prevented the formation of tumours in Swiss albino mice that were subjected to UVB radiation. Remarkably, the mice showed a higher level of inhibition of carcinogenesis when they were given a combination of apigenin through oral and topical administration (Das et al., 2013).

12.4.16 Quercetin

Quercetin, having hydroxyl groups at positions 3, 5, 7, 3', and 4' of its flavonol skeleton (Ay et al., 2021), is a flavonol that is not soluble in cold water, has low solubility in hot water, and is soluble in alcohol. Quercetin, known for its exceptional biological properties, is often regarded as one of the most potent flavonoids (Shah et al., 2016). The main source of its effectiveness against cancer is mostly derived from its antioxidant and anti-inflammatory characteristics. However, further investigation is necessary to fully understand the mechanisms by which quercetin demonstrates its anti-cancer activities (Azeem et al., 2023).

Quercetin, which is found in large quantities in the human diet, can be present in many forms with sugar molecules attached to it, such as galactosides, rhamnosides, arabinosides, or glucosides (Shah et al., 2016). Significant dietary sources of quercetin include apples (*Malus domestica*), tomatoes (*Solanum lycoperscium*), tea (*Camellia sinensis*), grapes (*Vitis vinifera*), Ginkgo (*Ginkgo biloba*), St John's Wort (*Hypericum perforatum*), and even onions (Allium cepa), despite their relatively low intake (Azeem et al., 2023). Other sources that can be added to the list are capers (*Capparis spinosa*), cloves (*Syzygium aromaticum*), black elderberries (*Sambucus nigra*), dark chocolate (*Theobroma cacao*), and oregano (*Origanum vulgare*) (Asensi et al., 2011).

Research has investigated whether diets rich in quercetin provide preventive effects against the onset and progression of cancer, due to their widespread presence in different dietary products. Quercetin has been found to have both cooperative and opposing effects that help defend against chemotherapy and prevent the spread of cancer cells. Quercetin has been shown to have anti-melanoma actions by affecting the survival of cells at low concentrations and causing programmed cell death at higher concentrations (Kim et al., 2019). The studies exploring the role of quercetin conducted by Zhang et al. (2005) and Olson et al. (2010) highlight the mechanisms that involve reducing the expression of B cell lymphoma 2 (Bcl-2), enhancing the activity of caspase-3, and adjusting the expression of the c-fos gene caused by UVB. Moreover, studies have shown that quercetin can prevent UVB-induced oxidative stress, DNA damage, and apoptosis in mouse epidermal cells. It achieves this by decreasing the production of reactive oxygen species (ROS) and restoring the expression of antioxidant enzymes (Zhu et al., 2017). A study conducted by Cao et al. (2014) indicated quercetin's ability to combat melanoma may be attributed to its inhibitory effects on signal transducer and activator of transcription 3 (STAT3), a protein associated with the development of cancer.

Studies conducted on hairless mice exposed to UVB radiation demonstrated that the use of topical formulations containing quercetin, both nonionic and anionic, resulted in a reduction in the increase in myeloperoxidase activity and a drop in glutathione levels. These effects are associated with the damage caused by UVB radiation on the skin (Casagrande et al., 2006). Polymethoxylate derivatives of quercetin and Glycosides have shown promise as potential candidates for topical distribution. The enhanced therapeutic efficacy of quercetin can also be attributed to its metabolic stability as has been reported in a study (Alizadeh and Ebrahimzadeh, 2022). Moreover, studies have shown that quercetin nanoparticles as microemulsion formulations were effectively delivered to the skin under controlled pharmacodynamic settings (Tomou et al., 2023) thereby opening an opportunity for translational potential.

12.5 CONCLUSION

This chapter provides a meta-analysis of the many forms of skin cancer, including their molecular pathways and the multiple factors that contribute to their development. In addition, the potential of different naturally occurring phytochemicals in preventing and treating skin cancer has also been explored within the scope of the present study. A wide spectrum of studies has established that several phytochemicals and their respective formulations possess potent antioxidant, anti-proliferative, anti-inflammatory, and anti-angiogenic properties. These properties allow them to influence important molecular processes involved in the progression of skin cancer. In this respect, the effectiveness of phytochemicals such as resveratrol, indole-3-carbinol, silymarin, curcumin, epigallocatechin-3-gallate, luteolin, genistein, proanthocyanidins, apigenin, and capsaicin, which are derived from various fruits, vegetables, roots, and herbs, particularly in treating advanced skin cancer symptoms have been elucidated in this study. Although not as effective as traditional chemotherapy drugs, their contribution to cancer prevention is apparent. Nevertheless, addressing several challenges and defining future research directions is imperative. As a future direction, more studies need to be directed towards:

Controlled Clinical Trials: Essential to evaluate efficacy, toxicity, and oral bioavailability in humans.

Standardization: Crucial to establish methods for assessing bioavailability, efficacy, and safety meeting global standards.

Large-scale Studies: Required to validate anticancer actions and adverse effects before widespread use.

In essence, while phytochemicals show promise in the holistic management of skin cancer, extensive research and rigorous clinical trials are indispensable to comprehensively comprehend their potential benefits and limitations compared to standard treatments. Critical discussions within this chapter highlight challenges and underscore the promising role of phytochemicals in skin cancer management, emphasizing the need for further investigation and verification.

REFERENCES

Abbasi, B. A., Iqbal, J., Mahmood, T., Khalil, A. T., Ali, B., Kanwal, S., Shah, S. A., & Ahmad, R. 2018. Role of dietary phytochemicals in modulation of miRNA expression: Natural swords combating breast cancer. *Asian Pacific Journal of Tropical Medicine*, 11(9), 501–509.

Afaq, F., Adhami, V. M., Ahmad, N., & Mukhtar, H. 2003. Inhibition of ultraviolet B-mediated activation of nuclear factor κB in normal human epidermal keratinocytes by green tea Constituent (-)-epigallocatechin-3-gallate. *Oncogene*, 22(7), 1035–1044.

Afaq, F., & Katiyar, S. K. 2011. Polyphenols: Skin photoprotection and inhibition of photocarcinogenesis. *Mini Reviews in Medicinal Chemistry*, 11(14), 1200–1215.

Ahmad, N., Gupta, S., & Mukhtar, H. 2000. Green tea polyphenol epigallocatechin-3-gallate differentially modulates nuclear factor κB in cancer cells versus normal cells. *Archives of Biochemistry and Biophysics*, 376(2), 338–346.

Alizadeh, S. R., & Ebrahimzadeh, M. A. 2022. O-Glycoside quercetin derivatives: Biological activities mechanisms of action and structure–activity relationship for drug design a review. *Phytotherapy Research*, 36(2), 778–807.

Arafeh, R., Qutob, N., Emmanuel, R., Keren-Paz, A., Madore, J., Elkahloun, A., Wilmott, J. S., Gartner, J. J., Di Pizio, A., Winograd-Katz, S., & Sindiri, S. 2015. Recurrent inactivating RASA2 mutations in melanoma. *Nature Genetics*, 47(12), 1408–1410.

Aronchik, I., Kundu, A., Quirit, J. G., & Firestone, G. L. 2014. The antiproliferative response of indole-3-carbinol in human melanoma cells is triggered by an interaction with NEDD4-1 and disruption of wild-type PTEN degradation. *Molecular Cancer Research*, 12(11), 1621–1634.

Arora, A., Kalra, N., & Shukla, Y. 2006. Regulation of p21/ras protein expression by diallyl sulfide in DMBA induced neoplastic changes in mouse skin. *Cancer Letters*, 242(1), 28–36.

Asensi, M., Ortega, A., Mena, S., Feddi, F., & Estrela, J. M. 2011. Natural polyphenols in cancer therapy. *Critical Reviews in Clinical Laboratory Sciences*, 48(5–6), 197–216.

Ay, M., Charli, A., Jin, H., Anantharam, V., Kanthasamy, A., & Kanthasamy, A. G. 2021. Quercetin. In *Nutraceuticals* (pp. 749–755). Academic Press.

Azeem, M., Hanif, M., Mahmood, K., Ameer, N., Chughtai, F. R. S., & Abid, U. 2023. An insight into anticancer antioxidant antimicrobial antidiabetic and anti-inflammatory effects of quercetin: A review. *Polymer Bulletin*, 80(1), 241–262.

Aziz, M. H., Reagan-Shaw, S., Wu, J., Longley, B. J., & Ahmad, N. 2005. Chemoprevention of skin cancer by grape constituent resveratrol: Relevance to human disease? *The FASEB Journal*, 19(9), 1193–1195.

Banach, K., Kowalska, J., Rzepka, Z., Beberok, A., Rok, J., & Wrześniok, D. 2021. The role of UVA radiation in ketoprofen-mediated BRAF-mutant amelanotic melanoma cells death–a study at the cellular and molecular level. *Toxicology in Vitro*, 72, 105108.

Bhattacharya, S., Darjatmoko, S. R., & Polans, A. S. 2011. Resveratrol modulates the malignant properties of cutaneous melanoma via changes in the activation and attenuation of the anti-apoptotic proto-oncogenic protein Akt/PKB. *Melanoma Research*, 21(3), 180.

Bode, A. M., & Dong, Z. 2011. The two faces of capsaicin. *Cancer Research*, 71(8), 2809–2814.

Bode, A. M., Ma, W. Y., Surh, Y. J., & Dong, Z. 2001. Inhibition of epidermal growth factor-induced cell transformation and activator protein 1 activation by [6]-gingerol. *Cancer Research*, 61(3), 850–853.

Both, D. M., Goodtzova, K., Yarosh, D. B., & Brown, D. A. 2002. Liposome-encapsulated ursolic acid increases ceramides and collagen in human skin cells. *Archives of Dermatological Research*, 293, 569–575.

Cancer Facts and Figures. 2023. American Cancer Society. www.cancer.org/content/dam/cancer-org/research/cancer-facts-and-statistics/annual-cancer-facts-and-figures/2023/2023-cancer-facts-and-figures.pdf.

Cao, H. H., Tse, A. K. W., Kwan, H. Y., Yu, H., Cheng, C. Y., Su, T., Fong, W. F., & Yu, Z. L. 2014. Quercetin exerts anti-melanoma activities and inhibits STAT3 signaling. *Biochemical Pharmacology*, 87(3), 424–434.

Casagrande, F., & Darbon, J. M. 2000. p21CIP1 is dispensable for the G2 arrest caused by genistein in human melanoma cells. *Experimental Cell Research*, 258(1), 101–108.

Casagrande, R., Georgetti, S. R., Verri Jr, W. A., Dorta, D. J., dos Santos, A. C., & Fonseca, M. J. 2006. Protective effect of topical formulations containing quercetin against UVB-induced oxidative stress in hairless mice. *Journal of Photochemistry and Photobiology B: Biology*, 84(1), 21–27.

Cerami, E., Gao, J., Dogrusoz, U., Gross, B. E., Sumer, S. O., Aksoy, B. A., Jacobsen, A., Byrne, C. J., Heuer, M. L., Larsson, E., & Antipin, Y. 2012. The cBio cancer genomics portal: An open platform for exploring multidimensional cancer genomics data. *Cancer Discovery*, 2(5), 401–404.

Chan, L. P., Chou, T. H., Ding, H. Y., Chen, P. R., Chiang, F. Y., Kuo, P. L., & Liang, C. H. 2012. Apigenin induces apoptosis via tumor necrosis factor receptor-and Bcl-2-mediated pathway and enhances susceptibility of head and neck squamous cell carcinoma to 5-fluorouracil and cisplatin. *Biochimica et Biophysica Acta (BBA)-General Subjects*, 1820(7), 1081–1091.

Chandra Pal, H., Marchiony Hunt, K., Diamond, A. A., Elmets, C. A., & Afaq, F. 2016. Phytochemicals for the management of melanoma. *Mini Reviews in Medicinal Chemistry*, 16(12), 953–979.

Chao, P. C., Hsu, C. C., & Yin, M. C. 2009. Anti-inflammatory and anti-coagulatory activities of caffeic acid and ellagic acid in cardiac tissue of diabetic mice. *Nutrition & Metabolism*, 6(1), 1–8.

Chen, H., & Liu, R. H. 2018. Potential mechanisms of action of dietary phytochemicals for cancer prevention by targeting cellular signaling transduction pathways. *Journal of Agricultural and Food Chemistry*, 66(13), 3260–3276.

Chen, Y. J., Shiao, M. S., Hsu, M. L., Tsai, T. H., & Wang, S. Y. 2001. Effect of caffeic acid phenethyl ester, an antioxidant from propolis, on inducing apoptosis in human leukemic HL-60 cells. *Journal of Agricultural and Food Chemistry*, 49(11), 5615–5619.

Chinembiri, T. N., Du Plessis, L. H., Gerber, M., Hamman, J. H., & Du Plessis, J. 2014. Review of natural compounds for potential skin cancer treatment. *Molecules*, 19(8), 11679–11721.

Choudhari, J., Nimma, R., Nimal, S. K., Venkata, S. K. T., Kundu, G. C., & Gacche, R. N. 2023. Prosopis juliflora (Sw.) DC phytochemicals induce apoptosis and inhibit cell proliferation signaling pathways, EMT, migration, invasion, angiogenesis, and stem cell markers in melanoma cell lines. *Journal of Ethnopharmacology*, 312, 116472.

Christensen, J. G., & LeBlanc, G. A. 1996. Reversal of multidrug resistance in vivo by dietary administration of the phytochemical indole-3-carbinol. *Cancer Research*, 56(3), 574–581.

Cragg, G. M., & Newman, D. J. 2005. Plants as a source of anti-cancer agents. *Journal of Ethnopharmacology*, 100(1–2), 72–79.

Cragg, G. M., & Newman, D. J. 2013. Natural products: A continuing source of novel drug leads. *Biochimica et Biophysica Acta (BBA)-General Subjects*, 1830(6), 3670–3695.

Cui, S., Wang, J., Wu, Q., Qian, J., Yang, C., & Bo, P. 2017. Genistein inhibits the growth and regulates the migration and invasion abilities of melanoma cells via the FAK/paxillin and MAPK pathways. *Oncotarget*, 8(13), 21674.

Dahmke, I. N., Backes, C., Rudzitis-Auth, J., Laschke, M. W., Leidinger, P., Menger, M. D., Meese, E., & Mahlknecht, U. 2013. Curcumin intake affects miRNA signature in murine melanoma with mmu-miR-205-5p most significantly altered. *PLoS ONE*, 8(12), e81122.

Danciu, C., Borcan, F., Bojin, F., Zupko, I., & Dehelean, C. 2013. Effect of the isoflavone genistein on tumor size metastasis potential and melanization in a B16 mouse model of murine melanoma. *Natural Product Communications*, 8(3), 1934578X1300800318.

Das, S., Das, J., Samadder, A., Paul, A., & Khuda-Bukhsh, A. R. 2013a. Strategic formulation of apigenin-loaded PLGA nanoparticles for intracellular trafficking DNA targeting and improved therapeutic effects in skin melanoma in vitro. *Toxicology Letters*, 223(2), 124–138.

Das, S., Das, J., Samadder, A., Paul, A., & Khuda-Bukhsh, A. R. 2013b. Efficacy of PLGA-loaded apigenin nanoparticles in Benzo [a] pyrene and ultraviolet-B induced skin cancer of mice: Mitochondria mediated apoptotic signalling cascades. *Food and Chemical Toxicology*, 62, 670–680.

Deep, G., & Agarwal, R. 2010. Antimetastatic efficacy of silibinin: Molecular mechanisms and therapeutic potential against cancer. *Cancer and Metastasis Reviews*, 29, 447–463.

Deep, G., Singh, R. P., Agarwal, C., Kroll, D. J., & Agarwal, R. 2006. Silymarin and silibinin cause G1 and G2–M cell cycle arrest via distinct circuitries in human prostate cancer PC3 cells: A comparison of flavanone silibinin with flavanolignan mixture silymarin. *Oncogene*, 25(7), 1053–1069.

Dhanalakshmi, S., Mallikarjuna, G. U., Singh, R. P., & Agarwal, R. 2004. Dual efficacy of silibinin in protecting or enhancing ultraviolet B radiation-caused apoptosis in HaCaT human immortalized keratinocytes. *Carcinogenesis*, 25(1), 99–106.

Dong, Z., Ma, W. Y., Huang, C., & Yang, C. S. 1997. Inhibition of tumor promoter-induced activator protein 1 activation and cell transformation by tea polyphenols(-)-epigallocatechin gallate and theaflavins. *Cancer Research*, 57(19), 4414–4419.

Ellis, L. Z., Liu, W., Luo, Y., Okamoto, M., Qu, D., Dunn, J. H., & Fujita, M. 2011. Green tea polyphenol epigallocatechin-3-gallate suppresses melanoma growth by inhibiting inflammasome and IL-1β secretion. *Biochemical and Biophysical Research Communications*, 414(3), 551–556.

Erb, P., Ji, J., Wernli, M., Kump, E., Glaser, A., & Büchner, S. A. 2005. Role of apoptosis in basal cell and squamous cell carcinoma formation. *Immunology Letters*, 100(1), 68–72.

Esmaeili, F., Rajabnejhad, S., Partoazar, A. R., Mehr, S. E., Faridi-Majidi, R., Sahebgharani, M., Syedmoradi, L., Rajabnejhad, M. R., & Amani, A. 2016. Anti-inflammatory effects of eugenol nanoemulsion as a topical delivery system. *Pharmaceutical Development and Technology*, 21(7), 887–893.

Es-Saady, D., Simon, A., Ollier, M., Maurizis, J. C., Chulia, A. J., & Delage, C. 1996. Inhibitory effect of ursolic acid on B16 proliferation through cell cycle arrest. *Cancer Letters*, 106(2), 193–197.

Fang, Y., Bradley, M. J., Cook, K. M., Herrick, E. J., & Nicholl, M. B. 2013. A potential role for resveratrol as a radiation sensitizer for melanoma treatment. *Journal of Surgical Research*, 183(2), 645–653.

Farrar, M. D., Nicolaou, A., Clarke, K. A., Mason, S., Massey, K. A., Dew, T. P., Watson, R. E., Williamson, G., & Rhodes, L. E. 2015. A randomized controlled trial of green tea catechins in protection against ultraviolet radiation–induced cutaneous inflammation. *The American Journal of Clinical Nutrition*, 102(3), 608–615.

Farris, P., Yatskayer, M., Chen, N., Krol, Y., & Oresajo, C. 2014. Evaluation of efficacy and tolerance of a nighttime topical antioxidant containing resveratrol baicalin and vitamin e for treatment of mild to moderately photodamaged skin. *Journal of Drugs in Dermatology*, 13(12), 1467–1472.

Fernandez, M. A., Saenz, M. T., & Garcia, M. D. 1998. Anti-inflammatory activity in rats and mice of phenolic acids isolated from Scrophularia frutescens. *Journal of Pharmacy and Pharmacology*, 50(10), 1183–1186.

Frenkel, K., Wei, H., Bhimani, R., Ye, J., Zadunaisky, J. A., Huang, M. T., Ferraro, T., Conney, A. H., & Grunberger, D. 1993. Inhibition of tumor promoter-mediated processes in mouse skin and bovine lens by caffeic acid phenethyl ester. *Cancer Research*, 53(6), 1255–1261.

Gensler, H. L., Timmermann, B. N., Valcic, S., Wächter, G. A., Dorr, R., Dvorakova, K., & Alberts, D. S. 1996. Prevention of photocarcinogenesis by topical administration of pure epigallocatechin gallate isolated from green tea. *Nutrition and Cancer*, 26(3), 325–335. https://doi.org/10.1080/01635589609514488

George, B. P., Chandran, R., & Abrahamse, H. 2021. Role of phytochemicals in cancer chemoprevention: Insights. *Antioxidants*, 10(9), 1455.

Gerhauser, C. 2012. Cancer chemoprevention and nutri-epigenetics: State of the art and future challenges. In Pezzuto, J., Suh, N. (eds.). *Natural Products in Cancer Prevention and Therapy*. Topics in Current Chemistry, vol 329. Springer, Berlin, Heidelberg. https://doi.org/10.1007/128_2012_360

Ghosh, R., Nadiminty, N., Fitzpatrick, J. E., Alworth, W. L., Slaga, T. J., & Kumar, A. P. 2005. Eugenol causes melanoma growth suppression through inhibition of E2F1 transcriptional activity. *Journal of Biological Chemistry*, 280(7), 5812–5819.

Gu, M., Singh, R. P., Dhanalakshmi, S., Agarwal, C., & Agarwal, R. 2007. Silibinin inhibits inflammatory and angiogenic attributes in photocarcinogenesis in SKH-1 hairless mice. *Cancer Research*, 67(7), 3483–3491.

Gupta, S. C., Patchva, S., Koh, W., & Aggarwal, B. B. 2012. Discovery of curcumin, a component of golden spice, and its miraculous biological activities. *Clinical and Experimental Pharmacology and Physiology*, 39(3), 283–299.

Guthrie, A. R., Chow, H. H. S., & Martinez, J. A. 2017. Effects of resveratrol on drug-and carcinogen-metabolizing enzymes implications for cancer prevention. *Pharmacology Research & Perspectives*, 5(1), e00294.

Hail Jr, N., & Lotan, R. 2002. Examining the role of mitochondrial respiration in vanilloid-induced apoptosis. *Journal of the National Cancer Institute*, 94(17), 1281–1292.

Harmand, P. O., Duval, R., Delage, C., & Simon, A. 2005. Ursolic acid induces apoptosis through mitochondrial intrinsic pathway and caspase-3 activation in M4Beu melanoma cells. *International Journal of Cancer*, 114(1), 1–11.

Harmand, P. O., Duval, R., Liagre, B., Chevolleau, S., & Simon, A. 2003. Ursolic acid induces death receptor-5 (DR5)-related apoptosis, caspase-8-cleavage and mitochondrial release of proapoptotic factors in human HT29 colorectal cancer cells. *Anti-Cancer Drugs*, 14(10), 779–787.

Hoeijmakers, J. H. 2001. Genome maintenance mechanisms for preventing cancer. *Nature*, 411(6835), 366–374.

Horibe, I., Satoh, Y., Shiota, Y., Kumagai, A., Horike, N., Takemori, H., Uesato, S., Sugie, S., Obata, K., Kawahara, H., & Nagaoka, Y. 2013. Induction of melanogenesis by 4′-O-methylated flavonoids in B16F10 melanoma cells. *Journal of Natural Medicines*, 67, 705–710.

Hwang, M. K., Bode, A. M., Byun, S., Song, N. R., Lee, H. J., Lee, K. W., & Dong, Z. 2010. Cocarcinogenic effect of capsaicin involves activation of EGFR signaling but not TRPV1. *Cancer Research*, 70(17), 6859–6869.

Iqbal, B., Ali, J., Ganguli, M., Mishra, S., & Baboota, S. 2019a. Silymarin-loaded nanostructured lipid carrier gel for the treatment of skin cancer. *Nanomedicine*, 14(9), 1077–1093.

Iqbal, J., Abbasi, B. A., Ahmad, R., Batool, R., Mahmood, T., Ali, B., Khalil, A. T., Kanwal, S., Shah, S. A., Alam, M. M., & Bashir, S. 2019b. Potential phytochemicals in the fight against skin cancer: Current landscape and future perspectives. *Biomedicine & Pharmacotherapy*, 109, 1381–1393.

Iqbal, J., Abbasi, B. A., Ahmad, R., Mahmood, T., Kanwal, S., Ali, B., Khalil, A. T., Shah, S. A., Alam, M. M., & Badshah, H. 2018a. Ursolic acid, a promising candidate in the therapeutics of breast cancer: Current status and future implications. *Biomedicine & Pharmacotherapy*, 108, 752–756.

Iqbal, J., Abbasi, B. A., Khalil, A. T., Ali, B., Mahmood, T., Kanwal, S., Shah, S. A., & Ali, W. 2018b. Dietary isoflavones the modulator of breast carcinogenesis: Current landscape and future perspectives. *Asian Pacific Journal of Tropical Medicine*, 11(3), 186–193.

Iqbal, J., Abbasi, B. A., Mahmood, T., Kanwal, S., Ali, B., Shah, S. A., & Khalil, A. T. 2017. Plant-derived anticancer agents: A green anticancer approach. *Asian Pacific Journal of Tropical Biomedicine*, 7(12), 1129–1150.

Iwashita, K., Kobori, M., Yamaki, K., & Tsushida, T. 2000. Flavonoids inhibit cell growth and induce apoptosis in B16 melanoma 4A5 cells. *Bioscience Biotechnology and Biochemistry*, 64(9), 1813–1820.

Jagdeo, J., Adams, L., Lev-Tov, H., Sieminska, J., Michl, J., & Brody, N. 2010. Dose-dependent antioxidant function of resveratrol demonstrated via modulation of reactive oxygen species in normal human skin fibroblasts in vitro. *Journal of Drugs in Dermatology*, 9(12), 1523–1526.

Jang, M., Cai, L., Udeani, G. O., Slowing, K. V., Thomas, C. F., Beecher, C. W., Fong, H. H., Farnsworth, N. R., Kinghorn, A. D., Mehta, R. G., & Moon, R. C. 1997. Cancer chemopreventive activity of resveratrol a natural product derived from grapes. *Science*, 275(5297), 218–220.

Ji, C., Yang, Y. L., He, L., Gu, B., Xia, J. P., Sun, W. L., Su, Z. L., Chen, B., & Bi, Z. G. 2012. Increasing ceramides sensitizes genistein-induced melanoma cell apoptosis and growth inhibition. *Biochemical and Biophysical Research Communications*, 421(3), 462–467.

Jones, V., & Katiyar, S. K. 2013. Emerging phytochemicals for prevention of melanoma invasion. *Cancer Letters*, 335(2), 251–258.

Kalal, B. S., Upadhya, D., & Pai, V. R. 2017. Chemotherapy resistance mechanisms in advanced skin cancer. *Oncology Reviews*, 11(1).

Kalayarasan, S., Prabhu, P. N., Sriram, N., Manikandan, R., Arumugam, M., & Sudhandiran, G. 2009. Diallyl sulfide enhances antioxidants and inhibits inflammation through the activation of Nrf2 against gentamicin-induced nephrotoxicity in Wistar rats. *European Journal of Pharmacology*, 606(1–3), 162–171.

Kang, N. J., Lee, K. W., Shin, B. J., Jung, S. K., Hwang, M. K., Bode, A. M., Heo, Y. S., Lee, H. J., & Dong, Z. 2009. Caffeic acid, a phenolic phytochemical in coffee, directly inhibits Fyn kinase activity and UVB-induced COX-2 expression. *Carcinogenesis*, 30(2), 321–330.

Katiyar, S. K., Afaq, F., Azizuddin, K., & Mukhtar, H. 2001. Inhibition of UVB-induced oxidative stress-mediated phosphorylation of mitogen-activated protein kinase signaling pathways in cultured human epidermal keratinocytes by green tea polyphenol (−)-epigallocatechin-3-gallate. *Toxicology and Applied Pharmacology*, 176(2), 110–117.

Katiyar, S. K., Agarwal, R., Wood, G. S., & Mukhtar, H. 1992. Inhibition of 12-O-tetradecanoylphorbol-13-acetate-caused tumor promotion in 7 12-dimethylbenz [a] anthracene-initiated SENCAR mouse skin by a polyphenolic fraction isolated from green tea. *Cancer Research*, 52(24), 6890–6897.

Katiyar, S. K., Ahmad, N., & Mukhtar, H. 2000. Green tea and skin. *Archives of Dermatology*, 136(8), 989–994.

Katiyar, S. K., Korman, N. J., Mukhtar, H., & Agarwal, R. 1997. Protective effects of silymarin against photocarcinogenesis in a mouse skin model. *Journal of the National Cancer Institute*, 89(8), 556–565.

Kaur, G., Athar, M., & Alam, M. S. 2010. Eugenol precludes cutaneous chemical carcinogenesis in mouse by preventing oxidative stress and inflammation and by inducing apoptosis. *Molecular Carcinogenesis*, 49(3), 290–301.

Kaur, M., Velmurugan, B., Tyagi, A., Deep, G., Katiyar, S., Agarwal, C., & Agarwal, R. 2009. Silibinin suppresses growth and induces apoptotic death of human colorectal carcinoma LoVo cells in culture and tumor xenograft. *Molecular Cancer Therapeutics*, 8(8), 2366–2374.

Keene, M. R., Heslop, I. M., Sabesan, S. S., & Glass, B. D. 2019. Complementary and alternative medicine use in cancer: A systematic review. *Complementary Therapies in Clinical Practice*, 35, 33–47.

Khan, A. Q., Khan, R., Qamar, W., Lateef, A., Ali, F., Tahir, M., & Sultana, S. 2012. Caffeic acid attenuates 12-O-tetradecanoyl-phorbol-13-acetate (TPA)-induced NF-κB and COX-2 expression in mouse skin: Abrogation of oxidative stress, inflammatory responses and proinflammatory cytokine production. *Food and Chemical Toxicology*, 50(2), 175–183.

Kiguchi, K., Constantinou, A. I., & Huberman, E. 1990. Genistein-induced cell differentiation and protein-linked DNA strand breakage in human melanoma cells. *Cancer Communications*, 2(8), 271–278.

Kim, D. S., Jeong, Y. M., Moon, S. I., Kim, S. Y., Kwon, S. B., Park, E. S., Youn, S. W., & Park, K. C. 2006. Indole-3-carbinol enhances ultraviolet B-induced apoptosis by sensitizing human melanoma cells. *Cellular and Molecular Life Sciences CMLS*, 63, 2661–2668.

Kim, J. K., Kim, Y., Na, K. M., Surh, Y. J., & Kim, T. Y. 2007. [6]-Gingerol prevents UVB-induced ROS production and COX-2 expression in vitro and in vivo. *Free Radical Research*, 41(5), 603–614.

Kim, S. H., Yoo, E. S., Woo, J. S., Han, S. H., Lee, J. H., Jung, S. H., Kim, H. J., & Jung, J. Y. 2019. Antitumor and apoptotic effects of quercetin on human melanoma cells involving JNK/P38 MAPK signaling activation. *European Journal of Pharmacology*, 860, 172568.

Kim, S. O., Chun, K. S., Kundu, J. K., & Surh, Y. J. 2004. Inhibitory effects of [6]-gingerol on PMA-induced COX-2 expression and activation of NF-κB and p38 MAPK in mouse skin. *Biofactors*, 21(1–4), 27–31.

Kim, S. Y., Kim, D. S., Jeong, Y. M., Moon, S. I., Kwon, S. B., & Park, K. C. 2011. Indole-3-carbinol and ultraviolet B induce apoptosis of human melanoma cells via down-regulation of MITF. *Die Pharmazie*, 66(12), 982–987.

Kowalczyk, M. C., Kowalczyk, P., Tolstykh, O., Hanausek, M., Walaszek, Z., & Slaga, T. J. 2010. Synergistic effects of combined phytochemicals and skin cancer prevention in SENCAR mice. *Cancer Prevention Research*, 3(2), 170–178.

Kudugunti, S. K., Vad, N. M., Ekogbo, E., & Moridani, M. Y. 2011. Efficacy of caffeic acid phenethyl ester (CAPE) in skin B16-F0 melanoma tumor bearing C57BL/6 mice. *Investigational New Drugs*, 29, 52–62.

Kulkarni, A., Al-Hraishawi, H., Simhadri, S., Hirshfield, K. M., Chen, S., Pine, S., . . . & Ganesan, S. 2017. BRAF fusion as a novel mechanism of acquired resistance to vemurafenib in BRAF V600E mutant melanoma. *Clinical Cancer Research*, 23(18), 5631–5638.

Kundu, J. K., Shin, Y. K., Kim, S. H., & Surh, Y. J. 2006. Resveratrol inhibits phorbol ester-induced expression of COX-2 and activation of NF-κB in mouse skin by blocking IκB kinase activity. *Carcinogenesis*, 27(7), 1465–1474.

Kuttan, R., Sudheeran, P. C., & Joseph, C. D. 1987. Turmeric and curcumin as topical agents in cancer therapy. *Tumori Journal*, 73(1), 29–31.

Lagoa, R., Silva, J., Rodrigues, J. R., & Bishayee, A. 2020. Advances in phytochemical delivery systems for improved anticancer activity. *Biotechnology Advances*, 38, 107382.

Li, Q. S., Li, C. Y., Li, Z. L., & Zhu, H. L. 2012. Genistein and its synthetic analogs as anticancer agents. *Anti-Cancer Agents in Medicinal Chemistry (Formerly Current Medicinal Chemistry-Anti-Cancer Agents)*, 12(3), 271–281.

Liu, P., Du, R., & Yu, X. 2019. Ursolic acid exhibits potent anticancer effects in human metastatic melanoma cancer cells (SK-MEL-24) via apoptosis induction, inhibition of cell migration and invasion, cell cycle arrest, and inhibition of mitogen-activated protein kinase (MAPK)/ERK signaling pathway. *Medical Science Monitor: International Medical Journal of Experimental and Clinical Research*, 25, 1283.

López-Lázaro, M. 2009. Distribution and biological activities of the flavonoid luteolin. *Mini Reviews in Medicinal Chemistry*, 9(1), 31–59.

Luke, J. J., & Schwartz, G. K. 2013. Chemotherapy in the management of advanced cutaneous malignant melanoma. *Clinics in Dermatology*, 31(3), 290–297.

Maisuthisakul, P., Suttajit, M., & Pongsawatmanit, R. 2007. Assessment of phenolic content and free radical-scavenging capacity of some Thai indigenous plants. *Food Chemistry*, 100(4), 1409–1418.

Mallikarjuna, G. U., Dhanalakshmi, S., Singh, R. P., Agarwal, C., & Agarwal, R. 2004. Silibinin protects against photocarcinogenesis via modulation of cell cycle regulators mitogen-activated protein kinases and Akt signaling. *Cancer Research*, 64(17), 6349–6356.

Mangione, C. M., Barry, M. J., Nicholson, W. K., Chelmow, D., Coker, T. R., Davis, E. M., Donahue, K. E., Jaén, C. R., Kubik, M., Li, L., & Ogedegbe, G. 2023. Screening for skin cancer: US preventive services task force recommendation statement. *JAMA*, 329(15), 1290–1295.

Mantena, S. K., Meeran, S. M., Elmets, C. A., & Katiyar, S. K. 2005. Orally administered green tea polyphenols prevent ultraviolet radiation-induced skin cancer in mice through activation of cytotoxic T cells and inhibition of angiogenesis in tumors. *The Journal of Nutrition*, 135(12), 2871–2877.

Manu, K. A., & Kuttan, G. 2008. Ursolic acid induces apoptosis by activating p53 and caspase-3 gene expressions and suppressing NF-κB mediated activation of bcl-2 in B16F-10 melanoma cells. *International Immunopharmacology*, 8(7), 974–981.

Marques, C. M., Dibden, C., Danson, S., Haycock, J. W., & MacNeil, S. 2013. Combined effects of capsaicin and HA14-1 in inducing apoptosis in melanoma cells. *Journal of Cosmetics Dermatological Sciences and Applications*, 3(3), 175.

Mason, L., Moore, R. A., Derry, S., Edwards, J. E., & McQuay, H. J. 2004. Systematic review of topical capsaicin for the treatment of chronic pain. *BMJ*, 328(7446), 991.

McCubrey, J. A., Steelman, L. S., Chappell, W. H., Abrams, S. L., Wong, E. W., Chang, F., Lehmann, B., Terrian, D. M., Milella, M., Tafuri, A., & Stivala, F. 2007. Roles of the Raf/MEK/ERK pathway in cell growth, malignant transformation and drug resistance. *Biochimica et Biophysica Acta (BBA)-Molecular Cell Research*, 1773(8), 1263–1284.

Metz, C. H., Scheulen, M., Bornfeld, N., Lohmann, D., & Zeschnigk, M. 2013. Ultradeep sequencing detects GNAQ and GNA11 mutations in cell-free DNA from plasma of patients with uveal melanoma. *Cancer Medicine*, 2(2), 208–215.

Mologni, L., Costanza, M., Sharma, G. G., Viltadi, M., Massimino, L., Citterio, S., Purgante, S., Raman, H., Pirola, A., Zucchetti, M., & Piazza, R. 2018. Concomitant BCORL1 and BRAF mutations in vemurafenib-resistant melanoma cells. *Neoplasia*, 20(5), 467–477.

Moon, M. K., Lee, Y. J., Kim, J. S., Kang, D. G., & Lee, H. S. 2009. Effect of caffeic acid on tumor necrosis factor-alpha-induced vascular inflammation in human umbilical vein endothelial cells. *Biological and Pharmaceutical Bulletin*, 32(8), 1371–1377.

More, M. P., Pardeshi, S. R., Pardeshi, C. V., Sonawane, G. A., Shinde, M. N., Deshmukh, P. K., Naik, J. B., & Kulkarni, A. D. 2021. Recent advances in phytochemical-based Nano-formulation for drug-resistant Cancer. *Medicine in Drug Discovery*, 10, 100082.

Moyano-Mendez, J. R., Fabbrocini, G., De Stefano, D., Mazzella, C., Mayol, L., Scognamiglio, I., Carnuccio, R., Ayala, F., La Rotonda, M. I., & De Rosa, G. 2014. Enhanced antioxidant effect of trans-resveratrol: Potential of binary systems with polyethylene glycol and cyclodextrin. *Drug Development and Industrial Pharmacy*, 40(10), 1300–1307.

Mukhtar, H., & Ahmad, N. 1999. Green tea in chemoprevention of cancer. *Toxicological Sciences*, 52(suppl_1), 111–117.

Mullenders, L. H., Hazekamp-van Dokkum, A. M., Kalle, W. H., Vrieling, H., Zdzienicka, M. Z., & van Zeeland, A. A. 1993. UV-induced photolesions, their repair, and mutations. *Mutation Research/Genetic Toxicology*, 299(3–4), 271–276.

Nakashima, S., Matsuda, H., Oda, Y., Nakamura, S., Xu, F., & Yoshikawa, M. 2010. Melanogenesis inhibitors from the desert plant Anastatica hierochuntica in B16 melanoma cells. *Bioorganic & Medicinal Chemistry*, 18(6), 2337–2345.

Natarajan, K., Singh, S., Burke Jr, T. R., Grunberger, D., & Aggarwal, B. B. 1996. Caffeic acid phenethyl ester is a potent and specific inhibitor of activation of nuclear transcription factor NF-kappa B. *Proceedings of the National Academy of Sciences*, 93(17), 9090–9095.

Nataren, N., Yamada, M., & Prow, T. 2022. Molecular skin cancer diagnosis: Promise and limitations. *The Journal of Molecular Diagnostics*, 25(1), 17–35.

Ndiaye, M., Philippe, C., Mukhtar, H., & Ahmad, N. 2011. The grape antioxidant resveratrol for skin disorders: Promise prospects and challenges. *Archives of Biochemistry and Biophysics*, 508(2), 164–170.

Ng, C. Y., Yen, H., Hsiao, H. Y., & Su, S. C. 2018. Phytochemicals in skin cancer prevention and treatment: An updated review. *International Journal of Molecular Sciences*, 19(4), p. 941.

Nigam, N., George, J., Srivastava, S., Roy, P., Bhui, K., Singh, M., & Shukla, Y. 2010. Induction of apoptosis by [6]-gingerol associated with the modulation of p53 and involvement of mitochondrial signaling pathway in B [a] P-induced mouse skin tumorigenesis. *Cancer Chemotherapy and Pharmacology*, 65, 687–696.

Nihal, M., Ahsan, H., Siddiqui, I. A., Mukhtar, H., Ahmad, N., & Wood, G. S. 2009. (-)-Epigallocatechin-3-gallate (EGCG) sensitizes melanoma cells to interferon induced growth inhibition in a mouse model of human melanoma. *Cell Cycle*, 8(13), 2057–2063.

Niles, R. M., Cook, C. P., Meadows, G. G., Fu, Y. M., McLaughlin, J. L., & Rankin, G. O. 2006. Resveratrol is rapidly metabolized in athymic (nu/nu) mice and does not inhibit human melanoma xenograft tumor growth. *The Journal of Nutrition*, 136(10), 2542–2546.

Olson, E. R., Melton, T., Dickinson, S. E., Dong, Z., Alberts, D. S., & Bowden, G. T. 2010. Quercetin potentiates UVB-Induced c-Fos expression: Implications for its use as a chemopreventive agent. *Cancer Prevention Research*, 3(7), 876–884.

Onori, P., DeMorrow, S., Gaudio, E., Franchitto, A., Mancinelli, R., Venter, J., Kopriva, S., Ueno, Y., Alvaro, D., Savage, J., & Alpini, G. 2009. Caffeic acid phenethyl ester decreases cholangiocarcinoma growth by inhibition of NF-κB and induction of apoptosis. *International Journal of Cancer*, 125(3), 565–576.

Osmond, G. W., Augustine, C. K., Zipfel, P. A., Padussis, J., & Tyler, D. S. 2012. Enhancing melanoma treatment with resveratrol. *Journal of Surgical Research*, 172(1), 109–115.

Oyagbemi, A. A., Saba, A. B., & Azeez, O. I. 2010. Capsaicin: A novel chemopreventive molecule and its underlying molecular mechanisms of action. *Indian Journal of Cancer*, 47(1), 53–58.

Pal, D., Banerjee, S., Mukherjee, S., Roy, A., Panda, C. K., & Das, S. 2010. Eugenol restricts DMBA croton oil induced skin carcinogenesis in mice: Downregulation of c-Myc and H-ras and activation of p53 dependent apoptotic pathway. *Journal of Dermatological Science*, 59(1), 31–39.

Park, K. K., Chun, K. S., Lee, J. M., Lee, S. S., & Surh, Y. J. 1998. Inhibitory effects of [6]-gingerol a major pungent principle of ginger on phorbol ester-induced inflammation epidermal ornithine decarboxylase activity and skin tumor promotion in ICR mice. *Cancer Letters*, 129(2), 139–144.

Park, K. K., & Surh, Y. J. 1997. Effects of capsaicin on chemically-induced two-stage mouse skin carcinogenesis. *Cancer Letters*, 114(1–2), 183–184.

Pham, D. D. M., Guhan, S., & Tsao, H. 2020. KIT and melanoma: Biological insights and clinical implications. *Yonsei Medical Journal*, 61(7), 562.

Ponti, G., Manfredini, M., Greco, S., Pellacani, G., Depenni, R., Tomasi, A., Maccaferri, M., & Cascinu, S. 2017. BRAF, NRAS, and C-KIT advanced melanoma: Clinico-pathological features, targeted-therapy strategies, and survival. *Anticancer Research*, 37(12), 7043–7048.

Puccinelli, M. T., & Stan, S. D. 2017. Dietary bioactive diallyl trisulfide in cancer prevention and treatment. *International Journal of Molecular Sciences*, 18(8), 1645.

Radi, F. Z., Bouhrim, M., Mechchate, H., Al-Zahrani, M., Qurtam, A. A., Aleissa, A. M., Drioiche, A., Handaq, N., & Zair, T. 2021. Phytochemical analysis, antimicrobial, and antioxidant properties of Thymus zygis L. and Thymus willdenowii Boiss. essential oils. *Plants*, 11(1), 15.

Ramachandran, S., Rajendra Prasad, N., Pugalendi, K. V., & Menon, V. P. 2008. Modulation of UVB-induced oxidative stress by Ursolic acid in human blood lymphocytes. *Asian Journal of Biochemistry*, 3(1), 11–18.

Ratcharin, N., Wongtrakul, P., & Indranupakorn, R. 2012. Preparation of Zingiber officinale extract loaded solid lipid nanoparticles. *Advanced Materials Research*, 506, 389–392.

Rathod, N. B., Elabed, N., Punia, S., Ozogul, F., Kim, S. K., & Rocha, J. M. 2023. Recent developments in polyphenol applications on human health: A review with current knowledge. *Plants*, 12(6), 1217.

Rauth, S., Kichina, J., & Green, A. 1997. Inhibition of growth and induction of differentiation of metastatic melanoma cells in vitro by genistein: Chemosensitivity is regulated by cellular p53. *British Journal of Cancer*, 75(11), 1559–1566.

Rizeq, B., Gupta, I., Ilesanmi, J., AlSafran, M., Rahman, M. M., & Ouhtit, A. 2020. The power of phytochemicals combination in cancer chemoprevention. *Journal of Cancer*, 11(15), 4521.

Ruan, J. S., Liu, Y. P., Zhang, L., Yan, L. G., Fan, F. T., Shen, C. S., Wang, A. Y., Zheng, S. Z., Wang, S. M., & Lu, Y. 2012. Luteolin reduces the invasive potential of malignant melanoma cells by targeting β3 integrin and the epithelial-mesenchymal transition. *Acta Pharmacologica Sinica*, 33(10), 1325–1331.

Rusin, A., Krawczyk, Z., Grynkiewicz, G., Gogler, A., Zawisza-Puchałka, J., & Szeja, W. 2010. Synthetic derivatives of genistein their properties and possible applications. *Acta Biochimica Polonica*, 57(1), 23–34.

Rusin, A., Zawisza-Puchałka, J., Kujawa, K., Gogler-Pigłowska, A., Wietrzyk, J., Świtalska, M., Głowala-Kosińska, M., Gruca, A., Szeja, W., Krawczyk, Z., & Grynkiewicz, G. 2011. Synthetic conjugates of genistein affecting proliferation and mitosis of cancer cells. *Bioorganic & Medicinal Chemistry*, 19(1), 295–305.

Safe, S., Papineni, S., & Chintharlapalli, S. 2008. Cancer chemotherapy with indole-3-carbinol bis (3′-indolyl) methane and synthetic analogs. *Cancer Letters*, 269(2), 326–338.

Saladi, R. N., & Persaud, A. N. 2005. The causes of skin cancer: A comprehensive review. *Drugs of Today*, 41(1), 37–53.

Sarkar, F. H., & Li, Y. 2002. Mechanisms of cancer chemoprevention by soy isoflavone genistein. *Cancer and Metastasis Reviews*, 21, 265–280.

Scherer, D., & Kumar, R. 2010. Genetics of pigmentation in skin cancer – a review. *Mutation Research/Reviews in Mutation Research*, 705(2), 141–153.

Shah, P. M., Priya, V. V., & Gayathri, R. 2016. Quercetin-a flavonoid: A systematic review. *Journal of Pharmaceutical Sciences and Research*, 8(8), 878.

Shankar, E., Kanwal, R., Candamo, M., & Gupta, S. 2016, October. Dietary phytochemicals as epigenetic modifiers in cancer: Promise and challenges. *Seminars in Cancer Biology*, 40, 82–99. Academic Press.

Shen, L. N., Zhang, Y. T., Wang, Q., Xu, L., & Feng, N. P. 2014. Enhanced in vitro and in vivo skin deposition of apigenin delivered using ethosomes. *International Journal of Pharmaceutics*, 460(1–2), 280–288.

Shin, D. H., Kim, O. H., Jun, H. S., & Kang, M. K. 2008. Inhibitory effect of capsaicin on B16-F10 melanoma cell migration via the phosphatidylinositol 3-kinase/Akt/Rac1 signal pathway. *Experimental & Molecular Medicine*, 40(5), 486–494.

Shishodia, S., Majumdar, S., Banerjee, S., & Aggarwal, B. B. 2003. Ursolic acid inhibits nuclear factor-κB activation induced by carcinogenic agents through suppression of IκBα kinase and p65 phosphorylation: Correlation with down-regulation of cyclooxygenase 2 matrix metalloproteinase 9 and cyclin D1. *Cancer Research*, 63(15), 4375–4383.

Shukla, S., & Gupta, S. 2010a. Apigenin and cancer chemoprevention. In *Bioactive Foods in Promoting Health* (pp. 663–689). Academic Press.

Shukla, S., & Gupta, S. 2010b. Apigenin: A promising molecule for cancer prevention. *Pharmaceutical Research*, 27, 962–978.

Simoes, M. F., Sousa, J. S., & Pais, A. C. 2015. Skin cancer and new treatment perspectives: A review. *Cancer Letters*, 357(1), 8–42.

Singh, M., Suman, S., & Shukla, Y. 2014. New enlightenment of skin cancer chemoprevention through phytochemicals: In vitro and in vivo studies and the underlying mechanisms. *BioMed Research International*, 2014.

Singh, R. P., Dhanalakshmi, S., Agarwal, C., & Agarwal, R. 2005. Silibinin strongly inhibits growth and survival of human endothelial cells via cell cycle arrest and downregulation of survivin Akt and NF-κB: Implications for angioprevention and antiangiogenic therapy. *Oncogene*, 24(7), 1188–1202.

Singh, R. P., Dhanalakshmi, S., Mohan, S., Agarwal, C., & Agarwal, R. 2006. Silibinin inhibits UVB-and epidermal growth factor–induced mitogenic and cell survival signaling involving activator protein-1 and nuclear factor-κB in mouse epidermal JB6 cells. *Molecular Cancer Therapeutics*, 5(5), 1145–1153.

Singh, R. P., Raina, K., Deep, G., Chan, D., & Agarwal, R. 2009. Silibinin suppresses growth of human prostate carcinoma PC-3 orthotopic xenograft via activation of extracellular signal-regulated kinase 1/2 and inhibition of signal transducers and activators of transcription signaling. *Clinical Cancer Research*, 15(2), 613–621.

Sjoberg, E. R., Chammas, R., Ozawa, H., Kawashima, I., Khoo, K. H., Morris, H. R., Dell, A., Tai, T., & Varki, A. 1995. Expression of De-N-acetyl-gangliosides in human melanoma cells is induced by genistein or nocodazole (*). *Journal of Biological Chemistry*, 270(7), 2921–2930.

Soll, F., Ternent, C., Berry, I. M., Kumari, D., & Moore, T. C. 2020. Quercetin inhibits proliferation and induces apoptosis of B16 melanoma cells in vitro. *ASSAY and Drug Development Technologies*, 18(6), 261–268.

Somani, S. J., Modi, K. P., Majumdar, A. S., & Sadarani, B. N. 2015. Phytochemicals and their potential usefulness in inflammatory bowel disease. *Phytotherapy Research*, 29(3), 339–350.

Song, H. S., Park, T. W., Sohn, U. D., Shin, Y. K., Choi, B. C., Kim, C. J., & Sim, S. S. 2008. The effect of caffeic acid on wound healing in skin-incised mice. *The Korean Journal of Physiology & Pharmacology*, 12(6), 343–347.

Stary, A., & Sarasin, A. 2002. The genetics of the hereditary xeroderma pigmentosum syndrome. *Biochimie*, 84(1), 49–60.

Subbaramaiah, K., Chung, W. J., Michaluart, P., Telang, N., Tanabe, T., Inoue, H., Jang, M., Pezzuto, J. M., & Dannenberg, A. J. 1998. Resveratrol inhibits cyclooxygenase-2 transcription and activity in phorbol ester-treated human mammary epithelial cells. *Journal of Biological Chemistry*, 273(34), 21875–21882.

Sukumaran, K., Unnikrishnan, M. C., & Kuttan, R. 1994. Inhibition of tumour promotion in mice by eugenol. *Indian Journal of Physiology and Pharmacology*, 38, 306–306.

Sun, C. Y., Zhang, Q. Y., Zheng, G. J., & Feng, B. 2019. Phytochemicals: Current strategy to sensitize cancer cells to cisplatin. *Biomedicine & Pharmacotherapy*, 110, 518–527.

Thakur, V. S., Deb, G., Babcook, M. A., & Gupta, S. 2014. Plant phytochemicals as epigenetic modulators: Role in cancer chemoprevention. *The AAPS Journal*, 16, 151–163.

Tomou, E. M., Papakyriakopoulou, P., Saitani, E. M., Valsami, G., Pippa, N., & Skaltsa, H. 2023. Recent advances in nanoformulations for quercetin delivery. *Pharmaceutics*, 15(6), 1656.

Tong, W. W., Zhang, C., Hong, T., Liu, D. H., Wang, C., Li, J., He, X. K., & Xu, W. D. 2018. Silibinin alleviates inflammation and induces apoptosis in human rheumatoid arthritis fibroblast-like synoviocytes and has a therapeutic effect on arthritis in rats. *Scientific Reports*, 8(1), 3241.

Tong, X., Van Dross, R. T., Abu-Yousif, A., Morrison, A. R., & Pelling, J. C. 2007. Apigenin prevents UVB-induced cyclooxygenase 2 expression: Coupled mRNA stabilization and translational inhibition. *Molecular and Cellular Biology*, 27(1), 283–296.

Trager, M. H., Geskin, L. J., Samie, F. H., & Liu, L. 2022. Biomarkers in melanoma and non-melanoma skin cancer prevention and risk stratification. *Experimental Dermatology*, 31(1), 4–12.

Tsai, S. J., Chao, C. Y., & Yin, M. C. 2011. Preventive and therapeutic effects of caffeic acid against inflammatory injury in striatum of MPTP-treated mice. *European Journal of Pharmacology*, 670(2–3), 441–447.

Vaid, M., & Katiyar, S. K. 2010. Molecular mechanisms of inhibition of photocarcinogenesis by silymarin a phytochemical from milk thistle (Silybum marianum L. Gaertn.). *International Journal of Oncology*, 36(5), 1053–1060.

Vaid, M., Prasad, R., Sun, Q., & Katiyar, S. K. 2011. Silymarin targets β-catenin signaling in blocking migration/invasion of human melanoma cells. *PLoS ONE*, 6(7), e23000.

Vaid, M., Singh, T., Prasad, R., & Katiyar, S. K. 2015. Silymarin inhibits melanoma cell growth both in vitro and in vivo by targeting cell cycle regulators, angiogenic biomarkers, and induction of apoptosis. *Molecular Carcinogenesis*, 54(11), 1328–1339.

Wang, H. C., Chu, Y. L., Hsieh, S. C., & Sheen, L. Y. 2017. Diallyl trisulfide inhibits cell migration and invasion of human melanoma a375 cells via inhibiting integrin/facal adhesion kinase pathway. *Environmental Toxicology*, 32(11), 2352–2359.

Wang, H. C., Pao, J., Lin, S. Y., & Sheen, L. Y. 2012. Molecular mechanisms of garlic-derived allyl sulfides in the inhibition of skin cancer progression. *Annals of the New York Academy of Sciences*, 1271(1), 44–52.

Wang, H. C., Oo Khor, T., Shu, L., Su, Z. Y., Fuentes, F., Lee, J. H., & Tony Kong, A. N. 2012. Plants vs. cancer: A review on natural phytochemicals in preventing and treating cancers and their druggability. *Anti-Cancer Agents in Medicinal Chemistry*, 12(10), 1281–1305.

Wang, Y., Zhang, D., Liu, Z., Liu, G., Duan, C., Jia, L., Feng, F., Zhang, X., Shi, Y., & Zhang, Q. 2010. In vitro and in vivo evaluation of silybin nanosuspensions for oral and intravenous delivery. *Nanotechnology*, 21(15), 155104.

Wang, Z. Y., Huang, M. T., Ho, C. T., Chang, R., Ma, W., Ferraro, T., Reuhl, K. R., Yang, C. S., & Conney, A. H. 1992. Inhibitory effect of green tea on the growth of established skin papillomas in mice. *Cancer Research*, 52(23), 6657–6665.

Warwick, E., Cassidy, A., Hanley, B., Jouni, Z. E., & Bao, Y. 2012. Effect of phytochemicals on phase II enzyme expression in infant human primary skin fibroblast cells. *British Journal of Nutrition*, 108(12), 2158–2165.

Wei, H., Bowen, R., Zhang, X., & Lebwohl, M. 1998. Isoflavone genistein inhibits the initiation and promotion of two-stage skin carcinogenesis in mice. *Carcinogenesis*, 19(8), 1509–1514.

Wei, H., Zhang, X., Wang, Y., & Lebwohl, M. 2002. Inhibition of ultraviolet light-induced oxidative events in the skin and internal organs of hairless mice by isoflavone genistein. *Cancer Letters*, 185(1), 21–29.

Weng, J. R., Tsai, C. H., Kulp, S. K., Wang, D., Lin, C. H., Yang, H. C., Ma, Y., Sargeant, A., Chiu, C. F., Tsai, M. H., & Chen, C. S. 2007. A potent indole-3-carbinol–derived antitumor agent with pleiotropic effects on multiple signaling pathways in prostate cancer cells. *Cancer Research*, 67(16), 7815–7824.

Wu, J., Omene, C., Karkoszka, J., Bosland, M., Eckard, J., Klein, C. B., & Frenkel, K. 2011. Caffeic acid phenethyl ester (CAPE), derived from a honeybee product propolis, exhibits a diversity of anti-tumor effects in pre-clinical models of human breast cancer. *Cancer Letters*, 308(1), 43–53.

Xiang, D., Wang, D., He, Y., Xie, J., Zhong, Z., Li, Z., & Xie, J. 2006. Caffeic acid phenethyl ester induces growth arrest and apoptosis of colon cancer cells via the β-catenin/T-cell factor signaling. *Anti-Cancer Drugs*, 17(7), 753–762.

Xu, D., Yuan, R., Gu, H., Liu, T., Tu, Y., Yang, Z., & He, L. 2013. The effect of ultraviolet radiation on the transforming growth factor beta 1/Smads pathway and p53 in actinic keratosis and normal skin. *Archives of Dermatological Research*, 305(9), 777–786.

Yan, C., & Han, R. 1997. Suppression of adhesion-induced protein tyrosine phosphorylation decreases invasive and metastatic potentials of B16-BL6 melanoma cells by protein tyrosine kinase inhibitor genistein. *Invasion & Metastasis*, 17(4), 189–198.

Yang, S., Irani, K., Heffron, S. E., Jurnak, F., & Meyskens Jr, F. L. 2005. Alterations in the expression of the apurinic/apyrimidinic endonuclease-1/redox factor-1 (APE/Ref-1) in human melanoma and identification of the therapeutic potential of resveratrol as an APE/Ref-1 inhibitor. *Molecular Cancer Therapeutics*, 4(12), 1923–1935.

Yang, Y., Li, Y., Wang, K., Wang, Y., Yin, W., & Li, L. 2013. P38/NF-κB/snail pathway is involved in caffeic acid-induced inhibition of cancer stem cells-like properties and migratory capacity in malignant human keratinocyte. *PLoS ONE*, 8(3), e58915.

Yao, X., Jiang, W., Yu, D., & Yan, Z. 2019. Luteolin inhibits proliferation and induces apoptosis of human melanoma cells in vivo and in vitro by suppressing MMP-2 and MMP-9 through the PI3K/AKT pathway. *Food & Function*, 10(2), 703–712.

Zhang, H., & Rosdahl, I. 2004. Deletion in p16INK4a and loss of p16 expression in human skin primary and metastatic melanoma cells. *International Journal of Oncology*, 24(2), 331–335.

Zhang, J., Lei, Z., Huang, Z., Zhang, X., Zhou, Y., Luo, Z., Zeng, W., Su, J., Peng, C., & Chen, X. 2016. Epigallocatechin-3-gallate (EGCG) suppresses melanoma cell growth and metastasis by targeting TRAF6 activity. *Oncotarget*, 7(48), 79557.

Zhang, X. M., Chen, J., Xia, Y. G., & Xu, Q. 2005. Apoptosis of murine melanoma B16-BL6 cells induced by quercetin targeting mitochondria inhibiting expression of PKC-α and translocating PKC-δ. *Cancer Chemotherapy and Pharmacology*, 55, 251–262.

Zhang, Y. J., Gan, R. Y., Li, S., Zhou, Y., Li, A. N., Xu, D. P., & Li, H. B. 2015. Antioxidant phytochemicals for the prevention and treatment of chronic diseases. *Molecules*, 20(12), 21138–21156.

Zhang, Y. J., Liu, X., Ruan, J., Zhuang, X., Zhang, X., & Li, Z. 2020. Phytochemicals of garlic: Promising candidates for cancer therapy. *Biomedicine & Pharmacotherapy*, 123, 109730.

Zhao, G., Han, X., Cheng, W., Ni, J., Zhang, Y., Lin, J., & Song, Z. 2017. Apigenin inhibits proliferation and invasion and induces apoptosis and cell cycle arrest in human melanoma cells. *Oncology Reports*, 37(4), 2277–2285.

Zhao, G., Han, X., Zheng, S., Li, Z., Sha, Y., Ni, J., Sun, Z., Qiao, S., & Song, Z. 2016. Curcumin induces autophagy, inhibits proliferation and invasion by downregulating AKT/mTOR signaling pathway in human melanoma cells. *Oncology Reports*, 35(2), 1065–1074.

Zhao, S., Tang, L., Chen, W., Su, J., Li, F., Chen, X., & Wu, L. 2021. Resveratrol-induced apoptosis is associated with regulating the miR-492/CD147 pathway in malignant melanoma cells. *Naunyn-Schmiedeberg's Archives of Pharmacology*, 394, 797–807.

Zhu, X., Li, N., Wang, Y., Ding, L., Chen, H., Yu, Y., & Shi, X. 2017. Protective effects of quercetin on UVB irradiation-induced cytotoxicity through ROS clearance in keratinocyte cells. *Oncology Reports*, 37(1), 209–218.

13 Advanced Therapeutics for Wounds and Their Healing Mechanism

Wajid Hussain, Xiaohan Yang, Ishrat Perveen, Manjila Adhikari, Jawad Hussain, Bakhtawar Bukhari, Muhammad Wajid Ullah and Shenqi Wang

13.1 INTRODUCTION

Assessment of wounds has been a problem since ancient times and continues to be so now. Mortality and morbidity associated with wounds depend on various factors, which may include the site and type of the wound, overall medical health, and the promptness and adequacy of medical care provided. Proper wound management, including cleaning, disinfection, and appropriate treatment, is crucial to minimize complications and promote healing (Nguyen et al., 2023). When an injury occurs, various anatomical structures and functions can be disrupted or damaged, leading to a wound. The skin is commonly affected but depending upon the nature and intensity of wounds, other tissues and structures may also be involved. It might be claimed that the skin, the most fragile part of the human body, is most vulnerable to injury since it is readily burnt or wounded by torment or incision (Sharma et al., 2021).

The most complicated process in multicellular animals is wound healing. It comprises multiple phases, including hemostasis/inflammation, proliferation, and remodelling phases. Development of a keloid or a hypertrophic scar may result from the imbalance of either or both of these phases, which can have two very different and harmful effects (Kiritsi and Nyström, 2018; S. Y. Wang et al., 2018). One may assert that the type of damage and any underlying diseases, in addition to the local wound variables and systemic mediators, all influence the healing process (Dissemond and Romanelli, 2022). As the primary technique for wound closure, doctors suggest using stitches, or skin adhesives to close the borders of wounds. When the margins are unreachable, the secondary aim manifests, allowing the incision to be uncovered and the defective region to gradually fill with connective tissue. Slow wound healing and infection risk characterize these wounds. Patients with underlying illnesses including diabetes, vascular, or pressure ulcers like those who have a mechanical strain, post-surgical wound infections, or pressure ulcers frequently experience this condition (Rosen and Manna, 2023). Last but not least, tertiary intention, or delayed primary intention, necessitates the opening of the wound unless the non-viable tissues are removed due to any contamination or infection before approaching the wound borders. The healing ultimately proceeds in a way that is compatible with the original objectives of wound healing (Demmer et al., 2021).

Colonization is the existence of multiplying bacteria irrespective of host reaction, and it's a frequent occurrence. The majority of these organisms belong to the common skin microbiota, such as the Staphylococcus genus and other coagulase-negative organisms, that is, *Propionibacterium acnes*, *Brevibacterium sp.*, *Pityrosporum sp.*, *Staphylococcus Corynebacterium sp.* (Sandström

DOI: 10.1201/9781003394846-13

et al., 2023). Growing bacteria now invade the wound site and initiate the host's reaction. Proteus, pseudomonas, Acinetobacter, Stenotrophomonas (Xanthomonas), and Klebsiella, are examples of bacteria or common pathogens discovered at the site of the wound (Vivcharenko et al., 2023).

The natural skin microbiome is said to predominate in early acute wounds. Beta-hemolytic Streptococcus will shortly be followed by *S. aureus*. These are typical microorganisms detected in ulcers of a diabetic foot. Proteus, *E. coli*, and Klebsiella, facultative anaerobic gram-negative rods will colonies the site after around four weeks. Whenever the wound becomes worse, deeper structures are affected. Multi-microbial infections are prevalent in humans (Mojica et al., 2022). Anaerobes are frequently more prevalent than aerobes in long-term chronic wounds. Pseudomonas, Acinetobacter, and Stenotrophomonas (Xanthomonas), for instance, are frequently discovered on fetal cuts (Sandström et al., 2023).

13.2 TYPES OF WOUNDS

Acute wounds, which include wounds from surgery, crunches, smouldering, minor cuts, and chafing, like severe slashes caused by trauma or injuries brought on by being crushed or being shot with a pistol, are those that result in the exterior degradation of undamaged skin. Regardless of the type of cutaneous injury, it is generally known that severe abrasions should repair in the expected extent of time. However, the chosen healing method will change based on the kind, location, and extent of the wound (Dissemond and Romanelli, 2022).

However, in the case of most serious traumatic wounds, which include an injury or shotgun deformity, the occurrence of muscle that has devitalized and also contaminated with feasible and infeasible foreign substances necessitates clinical removal along with the antimicrobial (ciprofloxacin and linezolid) treatment. This characteristic enables remedial to proceed naturally, via irritation and granulate process, for eventual re-epithelization and reshaping (Kolimi et al., 2022).

Chronic wounds can develop for a number of causes, including decreased vascular supply, insufficient venous drainage, and metabolic illnesses like diabetes mellitus. These explanations include leg and foot sores, weighted wounds, and chronic wounds (Han and Ceilley, 2017). Chronic ulceration may become worse due to ageing, obesity, smoking, poor diet, immunosuppression brought on by an ailment (like AIDS), or treatment (like radiotherapy or chemotherapy) (Furlan et al., 2020). On the other hand, continuous outward skin force, most frequently on the bums, resoles and sacrum leads to the development of pressure or decubitus tissue ulcers. However, the underlying illness is frequently the cause of hypertension and in that case, pressure lesions heal slowly and unexpectedly (Zaidi and Sharma, 2022). Figure 13.1 displays all the phases of the wound-healing process.

13.3 ACUTE WOUND HEALING PROCESS

13.3.1 Haemostasis

The typical response of a vascular injury is haemostasis, which involves clot formation and prevention of bleeding. Tissue hypoxia as well as acidity are brought on by the reduced blood flow caused by arteriolar constriction in a short time. Additionally, mast cells release histamine at the same time. It may increase vascular permeability and blood vessel dilation, which will aid in the intrusion of inflammatory cells in the extracellular spaces, surrounding the region of the wound (Okur et al., 2020). Clotting is brought on by the activation and accumulation of platelets and thrombocytes in the fibrin network (Kolimi et al., 2022).

13.3.2 Inflammation

A provocative reaction is quickly elicited when neutrophils passively enter the damaged blood arteries. The invasion of macrophages and eventually active neutrophils from nearby blood arteries maintains the seditious response. This process is regulated by growth factor signals from adjacent

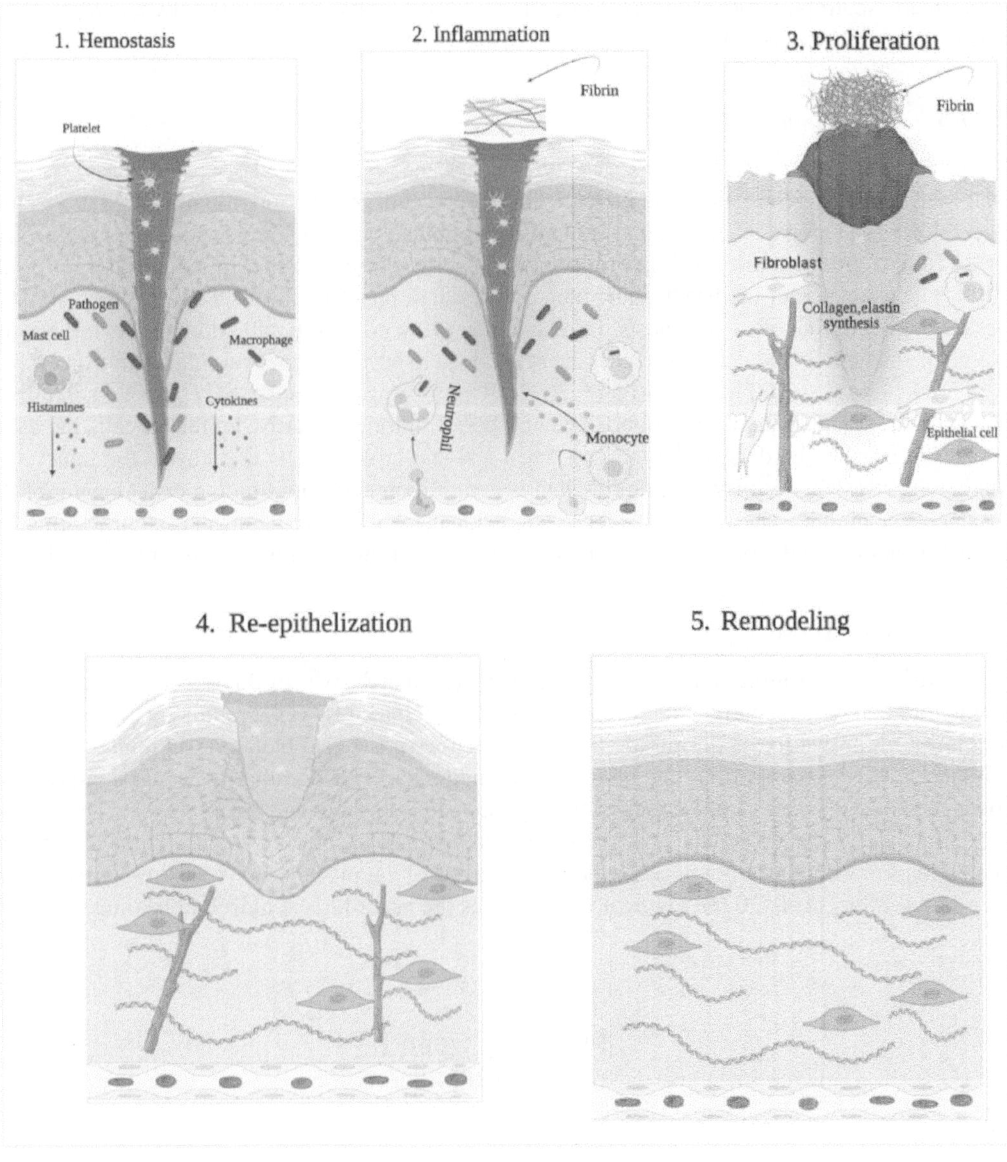

FIGURE 13.1 Different stages involved in the wound healing process.

cells and the serum in addition to external epitopes such as the lipo-polysaccharides (LPS) of encroaching bacteria. It is widely known that neutrophils around the site of the lesion release pro-inflammatory cytokines as well as active antimicrobial compounds like reactive oxygen species (ROS), proteases, and cationic peptides. The inflammatory response is maintained by the immigration of active neutrophils in response to platelet degranulation, complement system activation, and products of bacterial dilapidation (Okur et al., 2020). Then, neutrophils are initially activated and gather cells that help to maintain uncontaminated tissues and also aid in the loss of harmful infections. Pro-inflammatory cytokines cause a considerable amount of neutrophils to infiltrate through blood capillary walls shortly after the formation of lesions like necrosis factor alpha (TNF-), interferon-gamma (IFN-), and tumour necrosis factor beta (IL-1), which are present where the lesion has formed. Neutrophils either undergo programmed cell death or phagocytized by macrophages or may leak from the outside of the wound (Okur et al., 2020).

13.3.3 Proliferation

The proliferative phase lasts for four to twenty-one days and is categorized by angiogenesis, extracellular matrix (ECM) production, and epithelization. Platelet degranulation initiates ECM production since PDGF is a recognized activator of proteoglycan and collagen synthesis. A vascular connective tissue called "granulation tissue" is produced during the phase of proliferation (Rodrigues et al., 2019). The formation of the haemostatic plug as a result of platelet excretes TGF-, PDGF, and FGF signals the beginning of angiogenesis. In reaction to hypoxia, VEGF and additional cytokines are excreted, and endothelial cells are successively initiated to encourage neovascularization and repair of injured blood vessels (Diller and Tabor, 2022). After damage has occurred, epithelial cell migration starts at the margins of the wound. The number of mitotically active epithelial cells increases noticeably near the borders of the incision when a single layer of cells first covers the injury site. Epithelial cells that are migrating collide, halting migration and initiating the development of the basement membrane. These growth factors are critical for inducing angiogenesis, promoting the development of granulation tissue, and controlling the inflammatory response (Okur et al., 2020).

13.3.4 Remodelling

The third stage of healing, remodelling, begins approximately two to three weeks following the lesion initially and can continue for a year or more. The collagen III-rich granulation tissue replaces the clot of fibrin that forms during the first inflammatory phase and blood vessel remodelling that takes place during the proliferative phase. It then undergoes reintegration by a scar comprised of collagenous tissues and type I collagen that is supplied with younger blood vessels. The growth agents (polypeptides) are produced at the wound site by multiple activated cells. These are the primary agents in wound healing because they encourage cellular proliferation and attract new cells to the wound (Mony et al., 2023). Figure 13.2 illustrates the procedures required in healing a wound.

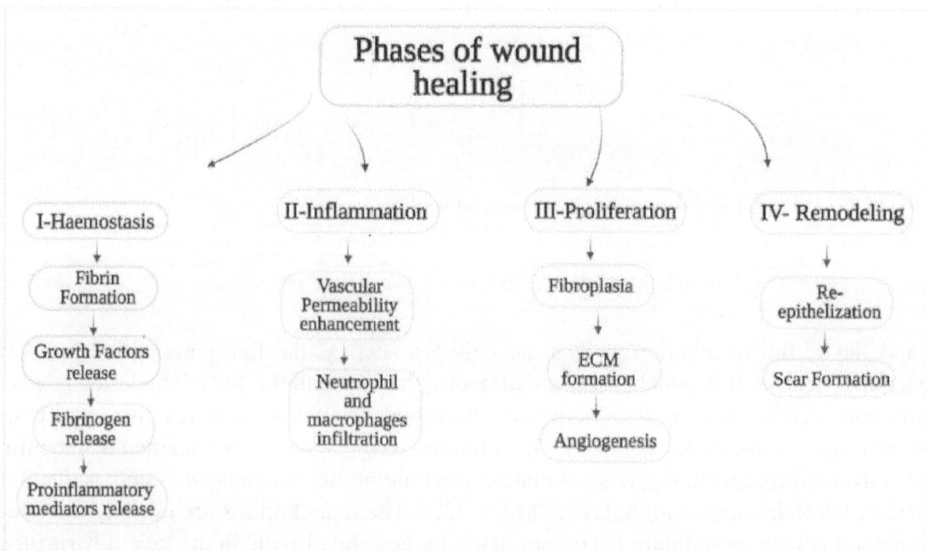

FIGURE 13.2 Stages involved in the process of wound healing.

13.3.5 Complications Associated with Chronic Wound Healing

Many wounds may heal quickly, but some wound types are known to be chronic and stuck at the inflammatory stage. Numerous variables contribute to this compromise. High levels of inflammatory mediators caused the extracellular matrix of cells as well as growth factors, are essential for healing, to initially suffer damage (Campos et al., 2008). The risk of a wound becoming swiftly populated by microorganisms is a significant issue during the wound-healing process. This occurs as a result of wound infections that impair vital fibrin for healing and decrease growth factors. Furthermore, because microorganisms or pathogens produce inflammation, chronic wounds usually contain a staggering amount of microorganisms. Hence, it can be concluded that a considerable decrease in the number of germs will accelerate the healing process (Dissemond and Romanelli, 2022). Furthermore, the complex material known as biofilm, which may stick to surfaces, contributes to the emergence and maintenance of certain conditions in the environment around chronic wounds (Goswami et al., 2023). In addition to other effects, hypoxia inhibits wound healing, encourages the growth of pathogenic organisms like bacteria, and limits the production of collagen and fibroblasts (Molnar et al., 2014). It seems that prevalent chronic conditions like cardiovascular illness (Okur et al., 2018), smoking, lung diseases, and peripheral vascular disease (PVD) contribute to hypoxia (Koike et al., 2020). Smoking also hinders wound healing because it interferes with chemotaxis, migratory function, and oxidative bactericidal procedures throughout the inflammatory stages (Liu et al., 2022).

Diabetes is one metabolic condition that impairs wound healing. A smaller epidermal layer, slower inflammatory, migratory, and proliferative responses, and a higher likelihood of chronic illness make older people more prone to have chronic wounds or their consequences (Burgess et al., 2021). Malnutrition with inadequate protein can also delay the healing of wounds. These elements, therefore, affect how well wounds heal, and eliminate any barriers to guarantee a speedy and painless wound closure (Okur et al., 2020).

13.4 BIOLOGICAL FACTORS ASSOCIATED WITH WOUND HEALING

Some wounds heal fast, while others tend to persist and stay in the inflammation phase. The outcome is influenced by several factors. The promoters and ECM are firstly harmed by high quantities of inflammatory mediators, which are crucial for the healing process (Ryan et al., 2021). A significant issue throughout the healing process is the potential for bacteria or fungus to quickly colonize a wound. This is because infection of the wound reduces growth hormone levels and annihilates fibrin, which is essential for healing. Furthermore, non-healing wounds usually have higher bacterial counts because bacteria generate inflammation. The process of healing will be visibly improved by reducing bacterial contamination (Okur et al., 2020).

Some biological factors for wound healing are:

13.4.1 Endogenous Electrical Field

The percentage of epithelized wound healing is controlled by the electric field produced by a wound. Intrinsic factors generated at epithelioid wounds are a key trait of all transporting epithelial tissues that segregate ions and sustain transcellular potential contrast. An endogenous electrical field that reaches up to one millimetre from the abrasion edge is produced by epithelial cells as they grow and move in connection to close a wound (Jia et al., 2021). The organism itself generates this electrical field. During the epithelial wound healing process, physiologically strong EFs dominate existing guidance signals and regulate cell migration. By organizing cells into coordinated tissues, these *in vivo* EFs have the potential to serve as an integrator in the development, healing, and regeneration of wound tissues (Li et al., 2023).

An endogenous electric field (EF) is measured near the margin of a freshly produced wound. It ranges from 42 to 100 mV/mm (Li et al., 2023). Keratinocyte directionality and velocity *in vitro* are positively linked by EF resilience between 25–200 mV/mm range. Keratinocytes moving with EF strength of about 25 mV/mm are oriented towards the cathode. This is termed as electro taxis. Modern sophisticated electrostimulation techniques involve releasing low-level electrical current stimulation through silver-infused nylon to prevent infection, lessen vasoconstriction in an abrasion, improve blood flow, and hasten wound healing (Ji et al., 2020). The bioelectric field, however, differs significantly from the electrical stimulation techniques. Tissue is stimulated by a strong current utilized in electrostimulation treatment, which differs from the bioelectric field in both direction and intensity. The recovery of granulation tissue is the most important stage of wound healing (Luo et al., 2023). During the first stages of wound healing, the bioelectric field that encourages keratin cells to migrate towards the wound's core is quite powerful. The exposed wounds are thus closed (Liang et al., 2020).

13.4.2 Adipose-Derived Stem Cells (ASCs)

Progenitor cells found in a stromal-vascular portion of hydrolyzed adipose tissue are known as ASCs. ASCs are acquired via a process known as filtration, collagenase breakdown, in addition to centrifugation of fat tissue to attain cellular fractionation. They are extracted from entire fat or lipo-aspirated fat (Hassanshahi et al., 2019). They are capable of differentiating into cells with lineages that produce tissues including fat, bone, and cartilage. ASCs have an additional advantage over MSCs in that they may be frequently and easily extracted via a procedure that is minimally invasive, such as liposuction, producing little to no damage and a higher output of MSCs (Schneider et al., 2023).

ASCs can directly affect the method of wound repair. The principal mechanism is perceived to be its paracrine release of substances which encourage the distinction and propagation of surrounding cells and stem cells (D. Chang et al., 2021; Mokry and Pisal, 2020). Potential mechanisms or regenerative abilities of ASCs for accelerating the wound healing process have been shown in Figure 13.3. ASCs have the potential to encourage the mobilization of intrinsic cells and hasten

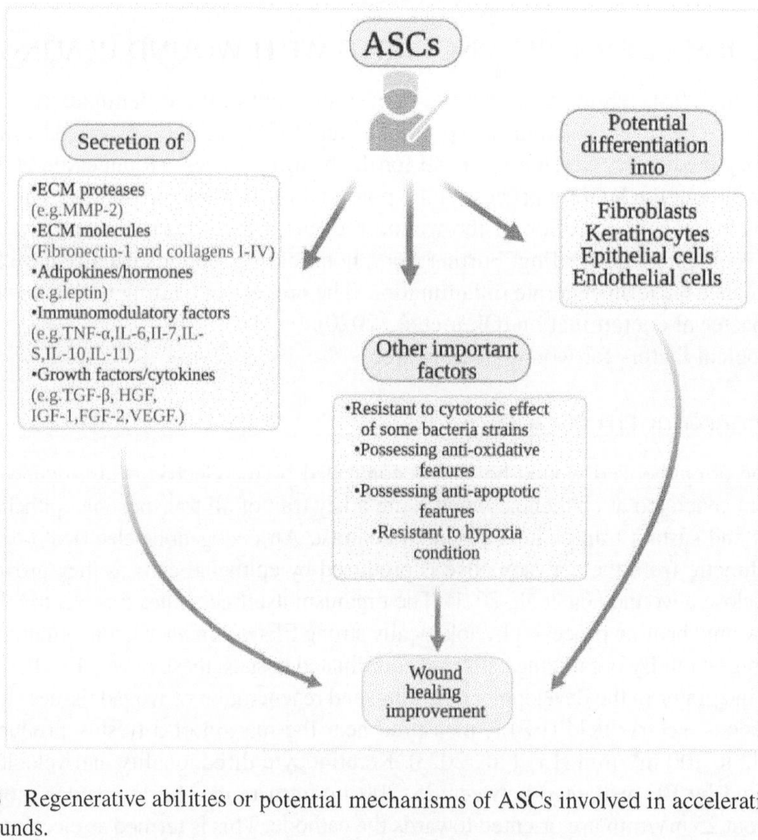

FIGURE 13.3 Regenerative abilities or potential mechanisms of ASCs involved in accelerating the healing process of wounds.

wound healing by secreting a number of growth factors, including fibroblast growth factor 2 (FGF2), insulin-like growth factor 1 (IGF-1), hepatocyte growth factor (HGF), vascular endothelial growth factor (VEGF), and transforming growth factor 1 (TGF-1) (Elsharawy et al., 2012; Rai et al., 2022).

Anti-inflammatory cytokines present at the injury site also encourage neovascularization. This paracrinicity of ASCs is also demonstrated in the repairing of radiation incisions, where the stem cells trigger the formation of the growth factors, that is, Keratinocyte (KGF) and VEGF in surrounding cells, which leads to neovascularization in the wound bed. ASCs' conditioned media, which contains EGF, PDGF-AA, VEGF, and FGF-2, was demonstrated to dramatically improve the migration and propagation of dermal fibroblasts (Li and Guo, 2018; Lu et al., 2021).

Related research found that treatment with a 50% human ASC-conditioned medium improved the function of fibroblasts and keratinocytes, leading to increased fibroblast, keratinocyte proliferation and the advancement of human keratinocytes in the healing of wounds. Interestingly, conditioned media made from ASCs boosted the creation of fibronectin, collagen types I and III, signifying that soluble factors made from ASCs show a substantial role in the enhancement of wound repair. A hypoxic environment was also found to induce the expression of hypoxia-induced factor-1, which in turn produces proangiogenic factors like VEGF, HGF, and FGF-2. In contrast, an existing hypoxic environment in the wound bed (1.0–5.0% oxygen absorption) is desirable for ASCs to migrate, multiply, and discriminate. ASCs' ability to bind to cell adhesion molecules CAMs like vascular cell adhesion protein 1 and intercellular adhesion molecule 1 is interestingly increased by hypoxia, which increases ASC numbers and FGF2 production, which promotes neovascularization (Li and Guo, 2018; Schneider et al., 2023).

Due to their strong resilience against the cytotoxic effects produced by bacterial strains, ASCs may also be an effective alternative to conventional cellular therapies for the treatment of infected wounds like Staphylococcus, *Streptococcus pyogenes*, and *Escherichia coli* (Chen et al., 2021). The acquaintance of the above-mentioned bacterial strains and their constituent parts did not affect the development of ASCs. ASCs have a variety of potential uses, including the capability to discriminate into diverse cell kinds, release growth factors and cytokines which tend to promote vascularization, lower levels of programmed cell death, and oxidative trauma, and exhibit antimicrobial capabilities to hasten the healing of wounds.

13.4.3 Wound Associated Conditions

The primary risk factors for wound healing are thought to be age, immunological function, malnutrition, pollution, hypoxia or perfusion, infections, smoking, medications, radiation, and chemotherapy. The development of medicines to speed up wound healing and treat wound impairment may result from a greater understanding of how these factors affect repair (Su et al., 2019). Vasculature disruption, which results in hypoxemia, is a major factor that slows down wound therapeutics. Giving extra oxygen (O2) to treat hypoxemia in the perioperative and patient settings can significantly speed up the healing of wounds. O_2 also assists critical procedures like vascularization, cell mobility, and ECM synthesis in addition to its roles as nutrition and antibiotic (M. Wang et al., 2021). Protein-energy malnutrition leads to immunodeficiency, which can worsen infection severity and frequency. It can also cause the thymus gland to atrophy and the peripheral lymphoid tissue to waste away, which impairs cell-mediated immune responses (He et al., 2023).

Macrophage activation (essential for the phagocytosis of dead cells and antimicrobial activity), the excretion of cytokines-like growth factors, and the formation of T-lymphocytes (cytotoxic) which are antigen-specific are all necessary to ensure wound healing (Jin et al., 2023). Potential mechanisms or regenerative abilities of ASCs for accelerating wound healing processes have been given in Figure 13.3. The cellular ingestion, the production of ROS, and degranulation of microbicidal components, which typically comprised of cathelicidin, elastase, defensins, lactoferrin, and cathepsin G, are all part of neutrophils' (polys) antibacterial and cytotoxic function. Neutrophils may now synthesize vasculogenic and fibrotic factors, MMPs, pro-inflammatory agents, anti-inflammatory

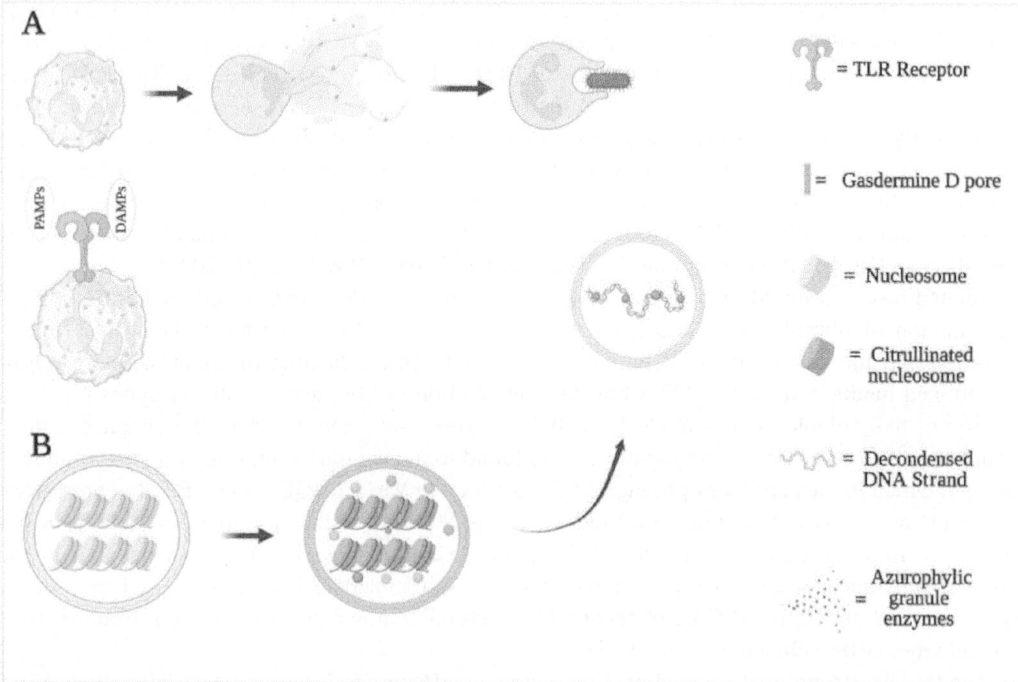

FIGURE 13.4 Mechanistic pathways involved in the production of extracellular neutrophil traps (NET).

agents, and immunoregulatory cytokines (Capó et al., 2023). Neutrophil extracellular traps (NETs), which are composed of non-histone and histone proteins, microbic mediators, and chromatin, move outside of the cell by a sequence of cytoplasmatic or nuclear events which are termed as NETosis. Formerly believed to be merely a protective/apoptotic process, NETosis is now understood to be an overly protective response that, in certain cases, has significant detrimental effects on tissue physiology and may even cause or exacerbate diseases (Sabbatini et al., 2021).

Figure 13.4 illustrates that both M1 and M2 pro-inflammatory and anti-inflammatory types of macrophages play an equal role in tissue clearance when it comes to the removal of NETs. The persistence of autoimmunity-triggering substances may be caused by delayed macrophage-dependent clearance (Sabbatini et al., 2021). Normal skin experiences typical ageing-related changes that affect how well wounds heal. Additionally, as people age, the healing process for wounds changes. Although older people's recovery was once thought to be flawed, current research shows that although older people's healing takes longer, it nevertheless produces results that are qualitatively comparable to those of younger patients (Zhu et al., 2021). Some additional wound healing agents and their respective effects have been depicted in Table 13.1.

13.5 RECENT ADVANCES IN BIOMATERIAL-BASED WOUND HEALING THERAPY

13.5.1 NANOPARTICLES

Wound healing is a difficult condition that necessitates wound management. Nanofabrication provides numerous innovative restorative healthcare methods. Nanomaterials improve wound curative and blister therapy. Nanomaterials have a size of 10–9 m (1 nm). Silver, gold, and zinc nanoparticles have exceptional features, which may include micro-biostatic, antibacterial activity, and reduced *in vivo* toxicity.

TABLE 13.1
Additional Wound-Healing Agents and Their Respective Effects

Factors	Effect on Wound Healing
Inflammation	Triggers immune response removes, debris, and fights infection.
Blood Clotting	Stops bleeding and creates a healing scaffold in the form of a clot.
Angiogenesis	Formation of cutting-edge blood vessels to supply the wound with nutrition and oxygen.
Granulation Tissue Formation	Creates a foundation for new tissue growth and provides structural support.
Collagen Production	Production of collagen, the main protein in wound healing, provides strength to the wound.
Matrix Remodeling	Reorganization and remodelling of the newly formed tissue for optimal strength and function.
Growth Factors	Stimulate cellular activities, including proliferation, migration, and differentiation.
Nutrition	Adequate intake of nutrients, vitamins, and minerals supports overall wound healing.
Oxygenation	Sufficient oxygen supply promotes cell metabolism and tissue. regeneration
Chronic Conditions	Underlying health issues (e.g., diabetes, vascular disease) may impair wound healing.
Medications	Certain medications (e.g., immunosuppressants) can affect the healing process.
Age	Wound healing tends to be slower in older individuals due to decreased cellular function.
Mechanical Factors	Excessive tension, pressure, or movement can impede wound healing.

Source: From El-Sherbeni and Negm, 2023; Guo and DiPietro, 2010; Rybka et al., 2022a.

13.5.1.1 Silver Nanoparticle

Silver nanoparticles are frequently applied in the healing of blisters, infection of wounds, and ulcer formations. $AgNO_3$, for instance, is utilized in the treatment of persistent wounds (Butler et al., 2015). For effective drug distribution, which is essential for the treatment of persistent wounds, many types of silvery coverings in wound bandages are now readily accessible (Cameron et al., 2016). Even when applied for an extended period, silver nanoparticles (AgNPs)-based dressings did not pose any challenge or difficulty. When AgNPs and collagen are assorted, they demonstrate high antiseptic activity, forming a perfect constituent of wound bandage (Navani et al., 2015). Wound healing properties of metal nanostructures as mono conjugates have been demonstrated (Kaba and Egorova, 2015). When various wound dressing materials are coupled with metal nanoparticles, they quickly remove germs from the site of inflammation or wound site (GhavamiNejad et al., 2015). This has been considered as the main reason in protracting the inflammation phase of chronic wounds (Jayakumar et al., 2011). To effectively prevent infection, AgNPs are commonly utilized in the formation of burnt balms, compression ulcer wound bandages, and medical devices. AgNPs are widely known for their broad antibacterial action, and for killing microbes, fungi, viruses, and protozoans (Rybka et al., 2022).

13.5.1.2 Gold Nano Particles

AuNPs (gold nanoparticles) are nontoxic and widely utilized in tissue repair, medication, and wound therapeutics (Krishani et al., 2023; Akturk et al., 2016). Au nanoparticles can combine with pathogen-specific antibodies for light-induced hyperthermia (Qin et al., 2023) or via phototoxic compounds for photodynamic therapy (PDT) to generate an antibacterial effect (Qin et al., 2023). AuNP-conjugated photosensitizer had better antifungal efficacy in *Candida albicans*-infested injuries in mice (Roma-Rodrigues et al., 2020). A recent study found that external treatment of AuNPs to superficial injuries improved regeneration by increasing granulation tissue development, re-epithelialization process, collagen fibre material and ECM deposition. The total healing time is shortened by these discrepancies, especially in the early phases. Rats were used in an *in vitro* wound healing study (Xiu et al., 2012). AgNP monotherapy outperformed AuNP monotherapy in terms of its ability to eliminate free radicals and wound recovery augmentation (Naraginti et al., 2016; Roma-Rodrigues et al., 2020; Rybka et al., 2022).

13.5.1.3 Zinc Oxide (ZnO) Nanoparticles (NPs)

ZnO nanoparticles (NPs) hold considerable therapeutic and nutraceutical significance. They are used in the manufacturing of beauty products like make-up, creams, and lotions for the skin because they have antiseptic, anti-inflammatory, and uncontaminated properties (Garg et al., 2015). Zinc was reportedly found to lessen swelling, promote re-epithelialization, lower blood pressure, and limit the growth of germs in persistent wounds. Zn is a crucial activator for the enzyme metalloproteinase and aids healing of the ECM (Kolarikova et al., 2023). Dimension and abundance of ZnO nanoparticles affect how well they can heal wounds. ZnO exhibits significant antibacterial activity when combined with chitosan hydrogel, making it an ideal component for wound dressing materials (Dkhil et al., 2020; Păunica-Panea et al., 2016; Gheorghiță et al., 2023). ZnO can be more effective at lower doses when it is used in a nanoformulation where it functions as a biocompatible polymer. Additionally, zinc controls keratinocyte migration and auto phagocytosis, two processes necessary for the best possible wound healing. Zn ions are released from ZnO because of a biphasic process that the nanomaterial goes through. Zn ions quickly hydrate in the presence of biological fluids for the formation of hydrated ZnO, which is the bacteriostatic mediator (Roma-Rodrigues et al., 2020).

13.5.2 Natural Product

13.5.2.1 Aloe Vera

Aloe vera has a long history of usage in traditional medicine for a variety of ailments, including digestion, cleaning, and antibacterial and anti-inflammatory effects. The leaves of A. vera encompass an ointment in the middle that is largely employed in wound-healing remedies (Albayrak et al., n.d.). Aloe vera has a variety of potent elements, including polysaccharides, that contribute to its wound-healing ability. Acemannan is a polysaccharide that has been linked to fibroblast proliferation and type I collagen formation, the skin of A. vera plants should be used with caution for therapeutic purposes because it has been reported to have carcinogenic characteristics, which are assumed to the incidence of anthracene diones inside the skin. It has been proven that whole-leaf cuttings of A. vera and maximum dosages cause intestinal tumors in mice (Boudreau et al., 2013; Chelu et al., 2023).

Gel and whole-leaf material induced positive wound restoration response. Though utilizing an in situ wound examination by dermal keratinocytes (HaCaTs), aloe vera gel therapy outperformed whole-leaf material to initiate wound repair (Fox et al., 2017). The re-epithelialization process depends on encouraging the recruitment of dermal keratinocytes onto the wound site since preserving the skin's outer protective layer is so important. This aids the developing epidermis in controlling the outflow of water from the tissue and serves as a defence against some bacteriological organisms (Fox et al., 2017; Massoud et al., 2022).

The growth of cells from connective tissue around the wound followed by the creation of EMX, which replaces the transient fibrinogen framework, is another important aspect of wound healing. Fibroblast viability was shown not to be affected by the accumulation of aloe vera gel, and it even seemed to boost proliferation, especially after one day. Aloe vera gel enhanced fibroblast migration toward a bared section within 24 hours, according to an *in situ* scrab wound study; yet some of its better migrant response may be attributable to the induced proliferative capability. However, because of their role in fibroblast migration, integrins 1 and 1 gene expression are suggestively amplified after dealing with aloe vera gel, supporting the wound regeneration effect experiential in scrape assay (Shafaie et al., 2020). As a result, A. vera gel could independently encourage fibroblast migration and proliferation, producing an even stronger reaction.

TABLE 13.2
Applications of Nanofibers and Microfibers for the Healing of Wounds

Fabrication Method	Animal Type (model)	Cell Type	Components	Electrical Stimuli	Conductivity	References
Electrospinning	Mouse (III)	fibroblast cells (embryonic), skin cells	doxorubicin PCL/BP/Gel/PEG	nill	nill	(Xue et al., 2021)
Electrospinning	Human (nill)	fibroblast cells	Col/CS/PPy/PEO	nill	59.564× 10–3–164.274×10–3Sm–	(Xue et al., 2021)
Electrospinning	Mouse (I)	fibroblast cell (L929)	QCS/PCL/PANI	nill	nill	(X. Huang et al., 2020)
Coating through a plasma process/Electrospinning	Human (nill)	keratinocyte cell	PVP/iodine/PPy	nill	nill	(Román-Doval et al., 2019)
Glutaraldehyde/Electrospinning/Crosslinking	nill	nill	CS/Mxene	nill	nill	(Román-Doval et al., 2019)
Electrospinning	V	nill	PAA/rGO	nill	nill	(Altinbasak et al., 2018)
Electrospinning	Mouse (I)	fibroblast cell (L929)	arginine/NPs/GO/PCL/Ag	nill	nill	(Shahmoradi et al., 2018)
Chemical polymerization, coating	Human (nill)	keratinocyte cells	SF/PANI/PPy	nill	2.2×10^{-5}–1.6×10^{-4} S cm^{-1}	(GH et al., 2017)
Electrospinning	Human (nill)	dermal fibroblasts	chitin/CSA/PANI	nill	4.2×10^{-4} S cm^{-1}	(Gu et al., 2018)
Chemical polymerization	Human (nill)	primary dermal fibroblast	PET/PPy	period 600 s, width 300 s, 50–100 mV mm^{-1}, or period 1200 s, width 10 s	nill	(Y. Wang et al., 2017)
Electrospinning	Mouse (nill)	fibroblast cell (L929)	PVA/GO	nill	nill	(GH et al., 2017)
Template-assisted polymerization/electrospinning	Mouse (nill)	fibroblast cell	PNIPMA/MWCNT/PANI	n/a	nill	(Ruiz-Velasco et al., 2017)
Chemical polymerization, coating, Electrospinning	Human (nill)	skin fibroblast	PHBV/PEDOT	n/a	1.45 ± 0.03 μS m^{-1}	(Patra et al., 2016)
Chemical polymerization	Human (nill)	dermal fibroblasts	PLLA/PEDOT	6 h, 50 mV mm^{-1}	0.1 kΩ sq^{-1}	(Y. Wang et al., 2013)

(Continued)

TABLE 13.2 *(Continued)*
Applications of Nanofibers and Microfibers for the Healing of Wounds

Fabrication Method	Animal Type (model)	Cell Type	Components	Electrical Stimuli	Conductivity	References
Chemical polymerization, coating	Human (n/a)	dermal fibroblasts	PET/PPy	period 1200 s, width 10 s, 100 mV mm^{-1}, or period 600 s, width 300 s	nill	(H. C. Chang et al., 2016)
Electrospinning	Human (nill)	skin fibroblasts	graphene/PET/PPy/CS/PVA	n/a period 1200 s, width 10 s, 5 V,	nill 63.4 ± 0.9 kΩ sq^{-1}	(Y. Wang et al., 2013) (Niu et al., 2015)
Electrospinning coating Chemical polymerization	nill	fibroblast cells (COS-1)	3ABAPANI/PPy/PLA/SF	nill	2.4–8.1 mS cm^{-1} nill	(Gizdavic-Nikolaidis et al., 2010)
Polymerization, coating	Adult Human (I, IV)	fibroblasts, mesenchymal cells (stem cells)	SF/PPY	nill		(Aznar-Cervantes et al., 2012)
Electrospinning	Mouse (nill)	NIH-3T3 fibroblasts, skeletal muscles	PLCL/CSA / PANI	20 mA, 200 mA	0.0138 S cm^{-1}	(Jeong et al., 2008)
Plasma polymerization	Human (nill)	embryonic kidney cells, dermal fibroblasts, tenocytes, and keratinocyte	PGA/PLLA/PPy-I	nill	nill	(Ruiz-Velasco et al., 2017)

nill = n/a or not available; V: Fully thick wound (*S. aureus* infected); IV: Fully thick wound in rabbits.

13.5.2.2 Caesalpinia Ferrea

Tree saps have been demonstrated to hasten the healing of wounds alongside having beneficial wound-healing properties. *In vitro* evaluation of *Lepiniopsis ternatensis* indicated that fibroblast proliferation stimulates a favourable reaction to chronic wounds by decreasing its proliferative ability. TNF-secretion was also decreased in M1 macrophages after treatment with the highest concentration (1.0% v/v) of L. ternatensis sap, which had been activated through granulocyte-macrophage colony-stimulating factor (GM-CSF) (Moses et al., 2023). TNF is associated with high MMP expression, which causes the ECM to break down in chronic wounds; suppressing TNF may help to reduce the ulcers' chronicity (Moses et al., 2020, 2023; Raziyeva et al., 2021).

When human neutrophils and macrophages were exposed to this exudate, they released more IL-6, similar to how they did when they were stimulated with GM-CSF or lipopolysaccharide (LPS), respectively. Neutrophils and macrophages were both stimulated to produce TNF in a dose-dependent manner (Moses et al., 2020, 2023; Nguyen et al., 2023). A pictorial depiction of the programmed neutrophil death has been given in Figure 13.5.

Exudation of *Ficus septica*, which includes alkaloid ficuseptine, exposed notable uncontaminated and anti-inflammatory activities *in situ* when compared to the prior plant exudates. Expression of TNF and IL-6 from neutrophils and M1 macrophages were markedly dose-dependently reduced. Papua New Guineans use F. septica latex to topically treat infected skin lesions. The fact that IL-6 increased noticeably even at the lowest dose suggests that, depending on the concentration used, this exudate may have pro- and/or anti-inflammatory effects (Deli et al., 2022).

F. septica exudation as an innovative therapy was found as effective as Savlon antiseptic lotion and just washing skin sores with cleanser and water in a clinical study. With 50 patients in each therapy arm, the cluster randomized study included 150 patients. Three dermatologists performed an unbiased evaluation of ulcer photos taken on days seven, eight, and 14. On day 14, it was discovered that the ficus exudate's capacity to treat/improve the ulcer was on par with that of Savlon cream and soap and water (Deli et al., 2022).

13.5.2.3 Epoxy-Tiglianes

The Australian rainforest plant *Fontaine picrosperma* included epoxy-taglines, a unique class of diterpene esters, which QBiotics Group was able to identify and extract (Moses et al., 2023). EBC-46, the prototype epoxy-tagline, is currently being evaluated in Phase II humanoid medical tribunals as a limited therapy for a variety of malignancies. It has also been recently accepted as a veterinarian pharmacological worldwide. PKC activation contributes to the efficacy of EBC-46, a strong cellular gesturing element (Reddell et al., 2021).

13.5.3 Therapeutic Drugs

13.5.3.1 Growth Factors

Growth factor usage in chronic wounds has been thoroughly researched for a number of years in order to promote angiogenesis and improve wound healing. Growth factors (endogenous) are necessary for efficient wound healing and are important in wound healing (Veith et al., 2019).

13.5.3.2 Platelet-Derived Growth Factors (PDGFs)

Platelet-derived growth factors, often known as PDGFs, are crucial chemotactic substances in the healing of wounds. The only growth factor therapy for wound healing currently approved by the FDA is becaplermin (100 g/g PDGF-BB) (Deptuła et al., 2020). Becaplermin has been proven to hasten wound curative in intravenous limb ulcers in several medical investigations (Parella et al., 2023). At a dose of 2.20 g/cm^2 of ulcer area, 48% of patients (29/61) in a medical test using becaplermin to treat foot ulcers (diabetic) had been found to heal ulcers which is found opposed to 25% (14/57) in the placebo group (Peng and Lavigne, 2023).

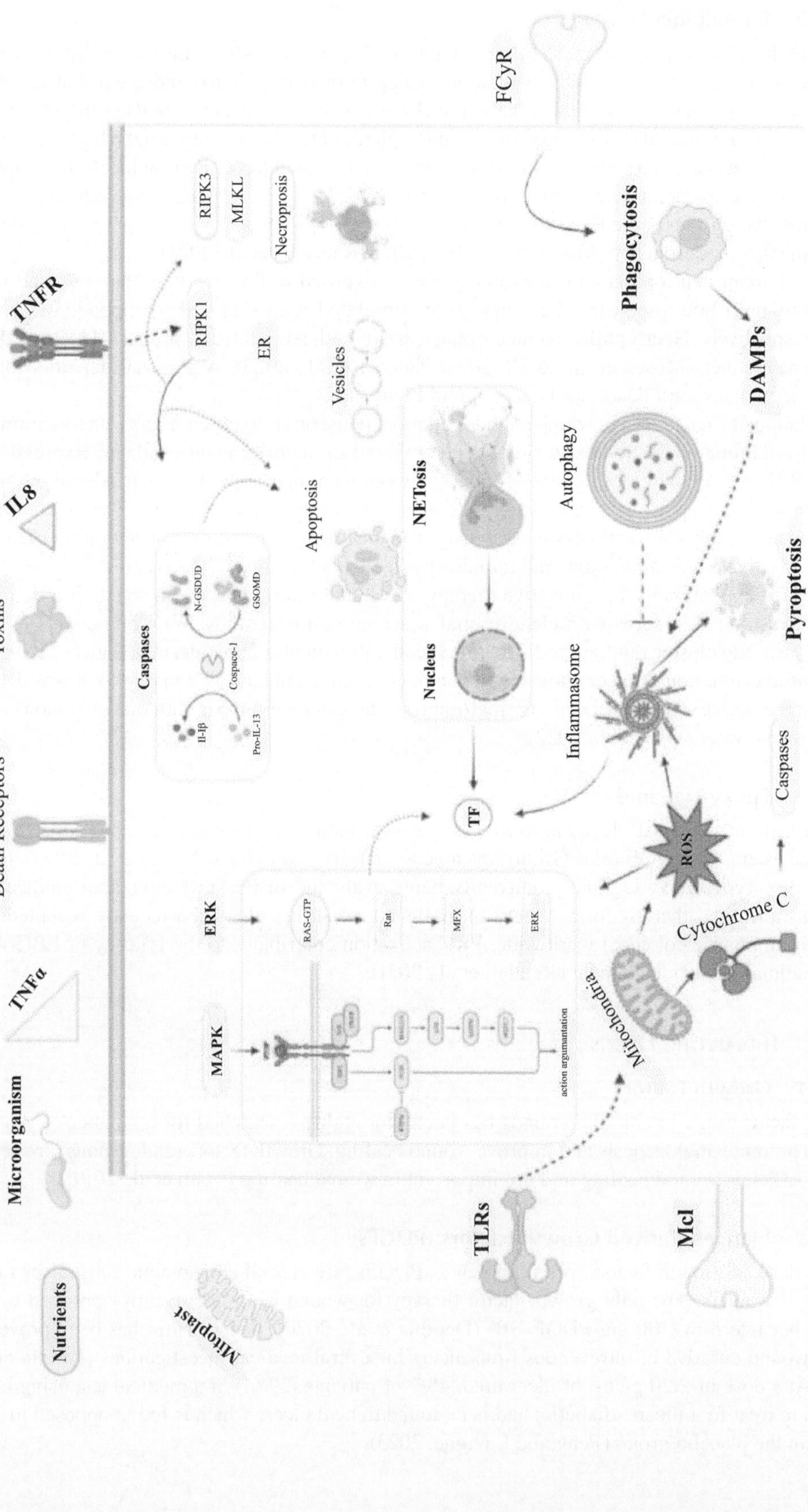

FIGURE 13.5 Mechanistic pathway indicating programmed neutrophil death in cell.

Becaplermin treatment resulted in statistically noteworthy advances in this kind of enduring wound in medical trials directed in 1998 and 1999 (Moses et al., 2023). Diabetic foot ulcers were reduced in groups that received becaplermin once daily. Wound healing rates were quicker in comparison to placebo controls. These encouraging findings could be related to PDGF-BB's dimensions to increase cell migration to wound spots and rouse angiogenic response. Additional medical trials, however, had revealed no improvement in wound therapeutic through becaplermin therapy (Parella et al., 2023).

The effectiveness of PDGF-BB therapies may be inhibited by therapeutic resistance, requiring higher dosages and raising the possibility of side effects related to growth treatment using variables. The FDA also allotted a black box warning for becaplermin because subjects who received three tubes of the medication and had cancer in the past were five times more likely to die (Deptuła et al., 2020). This study highlights the risks of utilizing pro-growth and angiogenic medications in individuals who previously had malignancy, even while the risk of emerging another type of carcinoma did not increase in those who didn't have cancer. New delivery strategies are necessary for expansion.

PDGF-BB effectiveness without enhancing dosage and its cost-effectiveness. The additional advantage required for medicines to be utilized in a clinical environment may be provided by the instantaneous use of manifold growth factors and proteins. Pericyte recruitment is directed when PDGF-BB and epidermal growth factor (EGF) are administered simultaneously, which helps to stabilize blood vessels (Rieg et al., 2022; Yang et al., 2020). Combining PDGF-BB with FGF-2 enhances vascular firmness in rats and rabbits with hind limb ischemia, according to other multiple growth factor studies (Bae et al., 2016; Yang et al., 2020). The introduction of PDGF-BB and VEGF-A from electrically spun nanofibre frames concurrently improved wound healing in mice and promoted angiogenesis at the site of the injury (Yang et al., 2020).

13.5.3.3 Epidermal Growth Factor

Epidermal growth factor (EGF) is a fascinating factor for angiogenesis therapy and identified angiogenesis modifier. Primary medical investigations show that wound re-epithelialization and healing times were accelerated. However, worries concerning the involvement of recombinant EGF in tumours have diminished interest in this therapy in the United States (Morbidelli et al., 2021). Despite this, other regions of the world have expressed a strong interest in using EGF therapy to hasten the healing of chronic wounds. Recombinant human EGF is marketed outside of the United States in several different forms, such as Easyef (Nepidermin), Regen-D 150, and Heberprot-P. In a prospective, open-label crossover study including 89 patients, the commercial Easyef spray or ointment proved effective in treating diabetic foot ulcers (Tottoli et al., 2020). In those with persistent diabetic foot ulcers, Regen-D 150, an EGF-having gel formulation, shows an acceleration of wound repair and lessens the peril of confiscation. The lyophilized EGF product known as *Heberprot-P* has been studied by more than 100,000 persons (Viswanathan et al., 2020).

It has been demonstrated to improve chronic ulcer cure and prevent diabetes-related amputations in Cuban patients. Recombinant human EGF has also been proven to aid in the healing of mild to severe diabetic foot ulcers in clinical studies conducted in Vietnam. EGF has been shown to stimulate angiogenesis through VEGF-mediated pathways that involve phosphatidylinositol 3-kinase/protein kinase B (PI3K/Akt) and extracellular signal-regulated kinase-1/2 (ERK1/2) (Viswanathan et al., 2020). EGF can induce VEGF-A production in cancer cells, hence promoting tumour angiogenesis. It is a potent chemo-attractant for keratinocyte relocation to aid in wound re-epithelialization with heparin-bound EGF being one of the most efficient kinds. EGF is a regulator as well (Morbidelli et al., 2021).

13.5.3.4 Fibroblast Growth Factor

Fibroblast growth factors (FGFs) are a class of growth factors that have an impact on a variety of physiological functions, which makes them perfect for use in the healing of cutaneous wounds

(Franková et al., 2016). Multiple FGF families exist. Among the family members that are expressed throughout the repairing procedure are FGF-1, FGF-2, FGF-7, FGF-10, and FGF-22. FGF ligands typically bind to receptors when heparin or heparan sulfate is present, and these indications cause angiogenesis and endothelial cell propagation and movement (Farooq et al., 2021). Due to its important connection to wound angiogenesis, FGF-2 is the FGF family member used in wound healing applications that have been the subject of the most research. The encouragement of re-epithelization is more typically associated with other FGFs, most notably FGF-7 (Hui et al., 2018).

It has been shown that FGF-7 is necessary for keratinocyte migration to the wound site and that animals with a deficit in FGF-7 have slower wound healing. Furthermore, it has been demonstrated that the simultaneous administration of FGF-7 and FGF-2 promotes skin wound repair with supplementary advanced vascular networks. Several revisions found that delivering FGF-2 to wounds improves healing. FGF-2 administration enhanced angiogenesis and hastened wound repair in freeze-damaged skin implantations. Current investigations confirmed similar indications in a variety of wound types. Skin covers to incisional herniations are examples of models (Liu et al., 2021). Collagen and cellulose nanocrystals (CNCs) were used to create a scaffold, which was then filled with FGF-2-impregnated microspheres to strengthen the collagen network mechanically (Norouzkhani et al., 2023; Rieg et al., 2022). Compared to FGF-2 from the scaffold alone or a control group, FGF-2 released by gelatin nanospheres dramatically amplified several new-fangled blood vessels in a subcutaneous implantation model (Takaya et al., 2022).

13.5.3.5 Vascular Endothelial Growth Factors

Five different growth factors make up the Vascular Endothelial Growth Factor (VEGF) family: VEGF-A, -B, -C, -D, and Placenta Growth Factor (PlGF). Both adult and embryonic angiogenesis and lymphangiogenesis are primarily influenced by these variables. The VEGF family has also been discovered to regulate the creation of new blood vessels during wound healing and is implicated in blood vessel permeability and vasodilation. However, due in part to the treatment's increased vasopermeability, the use of VEGF therapy in medical situations has been constrained. To enhance angiogenesis in critical limb or heart ischemia, recombinant VEGF-A has been used in several clinical trials. However, the therapeutic effects of VEGF therapy were limited, and further study was not explored (Brogowska et al., 2023).

Serum VEGF-A and FGF-2 concentrations are negatively correlated with one another during wound healing, with high concentrations of FGF-2 being used to attract endothelial cells initially and high concentrations of VEGF-A being utilized to stabilize new vessel growth and extend the angiogenic signalling window (Yu et al., 2009). Rats with ischemic injury had better blood circulation in their hind limbs after receiving simultaneous infusions of FGF-2 and VEGF-A, according to research by Losi et al. in a hind limb ischemia model (Rieg et al., 2022).

Additionally, the level of VEGF-A generated by macrophages was increased after the administration of granulocyte-macrophage colony-stimulating factor (GM-CSF) to diabetic patients with limb ulcers. Applying VEGF-A to wounds in db/db mice encouraged angiogenesis and wound healing In wounds handled with VEGF-A, there are more endothelial progenitor cells, which results in the creation of new blood vasculature, and more pro-angiogenic soluble substances (Hui et al., 2018; Takaya et al., 2022).

13.5.4 Mesenchymal Stem-Cells

Adult nonhematopoietic pluripotent stem cells are known as MSC. Due to their application in regenerative medicine, they have attracted a lot of attention lately (Cherubino et al., 2011). Navel string blood, endometrial tissue, myeloid tissues, adipose tissues, Wharton's jelly, liver, amnionic liquid, placenta, and tooth tissue are among the tissues from which MSC can be extracted. Additionally, they are plastic adherent cells that do not display the surface markers CD14, CD34, CD45, or

HLA-DR but do express CD73, CD90, and CD105. Unexpectedly, MSC can differentiate into both stomal and non-mesenchymal cell lineages, including nerve cells, neuroglial cells, and liver cells. Mesenchymal cellular ancestry includes bone-forming cells, cartilage cells, fat cells, endothelial cells, and cardiomyocytes. They possess qualities that are immunosuppressive, anti-inflammatory, tissue-restorative, neuroprotective, and differentiation. These characteristics make them a desirable therapeutic approach in cellular therapy and regenerative medicine. A great deal of clinical research on their use for various illnesses has been published. Acute and chronic heart failure, stroke, tumours, multiple sclerosis, amyotrophic lateral sclerosis, Crohn's disease, chronic kidney and liver disease, diabetes, osteoarthritis, rheumatoid arthritis, osteonecrosis, lumbar intervertebral disc degeneration, spinal cord contusions, sepsis, and critical limb ischemia are a few of them. Since MSC can grow *in vitro* and be maintained in liquid nitrogen while still being active, express medium and low levels of MHC class I and II molecules, and can be administered directly, they have enormous therapeutic potential (Ghaneialvar et al., 2018).

MSC takes part in the four stages of wound recovery when harm occurs. They are capable of integrating into damaged tissue, growing into functional cells, inducing anti-inflammatory reactions, and endorsing neovascularization for tissue repair. But in addition to MSC, trophic chemicals they create that have a paracrine function are also credited with this repair. The secretome is made up of proteins, extracellular vesicles, extracellular matrix elements, chemokines, growth factors, and inflammatory mediators that can mediate intercellular communication, diversity of cells, chemotaxis, immunomodulation, inhibition of fibrosis and apoptosis, and secretion of pro-angiogenic factors (Eddy et al., 2018).

The physiological reaction of MSC to a skin abrasion is as follows: In response to the local inflammation, neutrophils release TNF and IL-1, and MSC infiltrates the lesion. MSCs create pro-inflammatory mediators such CCL2, CCL3, CSF2, IL-6, IL-8, CCL2, and CCL3 when they reach the area. These are responsible for triggering the immune system's adaptive response and initiating the destruction of injured tissue. Additionally, MSC recruits plasmacytoid Antigen-presenting cells APCs and restricts the amount of stimulated T cells along with further distinctive immune cells to aid the transition to the following stage. EGF, HGF, VEGF, ANGPT1, PDGF-B, and FGF are a few of the proteins and chemicals that MSC release to aid in angiogenesis and healing during the developmental stage (Ghaneialvar et al., 2018). Additionally, they can cause local phagocytes to the transformation of their phenotype from seditious (M1) to a wound-healing (M2). In the end, MSC excretes matrix metalloproteinases during the remodelling phase, which causes matrix formation and tissue metalloproteinase blockers. By doing this, they prevent the sedimentation of extracellular matrix proteins and prevent the creation of wounds (Morbidelli et al., 2021).

13.5.5 Electroconductive Biomaterials

13.5.5.1 2d Conductive Materials

13.5.5.1.1 Film

Thin film is semipermeable, permitting the passage of oxygen and water vapour but obstructing a sizable concentration of odours and bacteria. Film-based wound bandage is appropriate for sensitive skin and thin wounds with less transude because of its elasticity, featherweight and hydration capability. Meanwhile, the film shouldn't be too dense because it would have poor oxygen and vapour transport, which would worsen wounds from maceration of the skin and moisture accumulation. Thin films, such as conductive materials, can also be employed as a matrix for bioactive compounds. Because of their affluence of manipulating both *in situ* and *ex vivo*, as well as their subsequent employment in other more complex structures, conducting nanofilms are broadly used to examine the process of by what method conducting constituents modify cellular processes. As a result, conductive film-based wound bandages have fascinated the focus of several researchers, with many of these studies concentrating on the mechanism through which conductive materials promote wound healing (Alven et al., 2022). Vapour transmission and oxygen absorbency were overlooked because cells were cultivated on the surfaces of these films [98–100]. The most popular method for making films is solvent casting. Thus, CPs synthesized by polymerization via solvent-soluble monomers demonstrate enormous advantages in terms of performance conductive film design as opposed to carbon nanomaterials, which tend to congregate and need the help of other amphipathic polymers or technologies to form monodispersing (Nguyen et al., 2023). Polypyrrole (PPy), poly(3,4-ethylenedioxythiophene) (PEDOT), and polyaniline (PANI) and its oligomeric compounds were frequently employed due to their biocompatibility, ease of fabrication, and adjustable conductance created via different doping agents. Even while regulated electrochemical polymerization can make pure CP films, their characteristic delicateness, motorized inflexibility, and indestructibility prevent them from being used as wound bandages (Table 13.2). Basic mixing and coupling with supplementary flexible bioinert polymers is a popular technique for getting beyond these restrictions (Shi et al., 2020).

13.6 CHALLENGES ASSOCIATED WITH WOUND HEALING

Identifying treatment goals at any particular time is a very important component that the doctor must take into account when assessing wounds. It is critical, to comprehend the distinction between wound healing and wound management. It is true that actual granulation tissue formation usually gets carried through the patient's body and is far more dependent on nutrients than dressings. Due to this, it is difficult to find the "Ideal Dressing," but the doctor should choose it after considering the wound's condition, the desired course of therapy, and the patient's preferences based on prior experiences. The key component of wound care is to establish a warm, moist, toxin-free environment to encourage spontaneous wound healing. The doctor should address all of the patient's health issues since, among other things, coexisting illnesses might impair the healing of a wound. To offset these effects, wound management must be carried out concurrently with the treatment of any underlying medical conditions and evaluated accordingly. A wound dressing's desired qualities include being secure, non-irritating, easy to use, painless, affordable, and absorbent (Sadati et al., 2019).

13.7 CONCLUSION AND FUTURE PROSPECTIVE

Hemostasis, inflammation, proliferation, and remodelling are the steps that constitute the natural process of wound healing. Complex and controlled molecular and cellular paths are used in this procedure. Wound healing is significantly influenced by biological elements such as the endogenous electric field, adipose-derived stem cells, wound-related circumstances, etc. Various therapeutic modalities, including natural materials (aloe vera, honey, tree saps etc.), therapeutic drugs (growth factors i.e., PDGFs, Fibroblast, and epidermal, etc.), nanoparticles (silver, gold, zinc oxide, etc.), multifunctional synthetic patches, electrical stimulation, and combined therapies have also been used for wound healing applications. Furthermore, by recognizing the requirement of explaining wound healing materials that target the wound healing mechanism, future studies should also be conducted on the recent developments in wound healing and the general difficulties or future challenges involved in converting the research into workable practical treatments for wound therapy.

ACKNOWLEDGMENTS

This work was supported by the National Key Research and Development Program of China (Grant#: 2017YFC1104402) and Metasequoia Teacher Research Start-up Fund, Nanjing Forestry University, Nanjing 210037, PR China (163105998).

REFERENCES

Akturk, O., Kismet, K., Yasti, A. C., Kuru, S., Duymus, M. E., Kaya, F., Caydere, M., Hucumenoglu, S., & Keskin, D. (2016). Collagen/gold nanoparticle nanocomposites: A potential skin wound healing biomaterial. *Journal of Biomaterials Applications*, *31*(2), 283–301. https://doi.org/10.1177/0885328216644536

Albayrak, S., Aksoy, A., Sagdic, O., & Hamzaoglu, E. (n.d.). Compositions, antioxidant and antimicrobial activities of Helichrysum (Asteraceae) species collected from Turkey. *Food Chemistry*. www.sciencedirect.com/science/article/pii/S030881460900795X

Altinbasak, I., Jijie, R., Barras, A., Golba, B., Sanyal, R., Bouckaert, J., Drider, D., Bilyy, R., Dumych, T., Paryzhak, S., Vovk, V., Boukherroub, R., Sanyal, A., & Szunerits, S. (2018). Reduced graphene-oxide-embedded polymeric nanofiber mats: An "on-demand" photothermally triggered antibiotic release platform. *ACS Applied Materials & Interfaces*, *10*(48), 41098–41106. https://doi.org/10.1021/acsami.8b14784

Alven, S., Peter, S., Mbese, Z., & Aderibigbe, B. A. (2022). Polymer-based wound dressing materials loaded with bioactive agents: Potential materials for the treatment of diabetic wounds. *Polymers*, *14*(4), 724. https://doi.org/10.3390/polym14040724

Aznar-Cervantes, S., Roca, M. I., Martinez, J. G., Meseguer-Olmo, L., Cenis, J. L., Moraleda, J. M., & Otero, T. F. (2012). Fabrication of conductive electrospun silk fibroin scaffolds by coating with polypyrrole for biomedical applications. *Bioelectrochemistry*, *85*, 36–43. https://doi.org/10.1016/j.bioelechem.2011.11.008

Bae, Y. J., Cho, C. H., Lee, W. J., Huh, J. S., & Lim, J. O. (2016). Optimization of recombinant human platelet-derived growth factor-BB encapsulated in Poly (lactic-co-glycolic acid) microspheres for applications in wound healing. *Tissue Engineering and Regenerative Medicine*, *13*(1), 13–20. https://doi.org/10.1007/s13770-015-0029-z

Boudreau, M. D., Mellick, P. W., Olson, G. R., Felton, R. P., Thorn, B. T., & Beland, F. A. (2013). Clear evidence of carcinogenic activity by a whole-leaf extract of aloe barbadensis miller (Aloe vera) in F344/N rats. *Toxicological Sciences*, *131*(1), 26–39. https://doi.org/10.1093/toxsci/kfs275

Brogowska, K. K., Zajkowska, M., & Mroczko, B. (2023). Vascular endothelial growth factor ligands and receptors in breast cancer. *Journal of Clinical Medicine*, *12*(6), 2412. https://doi.org/10.3390/jcm12062412

Burgess, Jamie L., Austin Wyant, W., Abujamra, B. A., Kirsner, R. S., & Jozic, I. (2021). Diabetic wound-healing science. *Medicina, 57*(10), 1072. https://doi.org/10.3390/medicina57101072

Butler, K. S., Peeler, D. J., Casey, B. J., Dair, B. J., & Elespuru, R. K. (2015). Silver nanoparticles: Correlating nanoparticle size and cellular uptake with genotoxicity. *Mutagenesis*, *30*(4), 577–591. https://doi.org/10.1093/mutage/gev020

Cameron, P., Gaiser, B. K., Bhandari, B., Bartley, P. M., Katzer, F., & Bridle, H. (2016). Silver nanoparticles decrease the viability of cryptosporidium parvum oocysts. *Applied and Environmental Microbiology*, *82*(2), 431–437. https://doi.org/10.1128/AEM.02806-15

Campos, A. C., Groth, A. K., & Branco, A. B. (2008). Assessment and nutritional aspects of wound healing. *Current Opinion in Clinical Nutrition and Metabolic Care*, *11*(3), 281–288. https://doi.org/10.1097/MCO.0b013e3282fbd35a

Capó, X., Monserrat-Mesquida, M., Quetglas-Llabrés, M., Batle, J. M., Tur, J. A., Pons, A., Sureda, A., & Tejada, S. (2023). Hyperbaric oxygen therapy reduces oxidative stress and inflammation, and increases growth factors favouring the healing process of diabetic wounds. *International Journal of Molecular Sciences*, *24*(8), 7040. https://doi.org/10.3390/ijms24087040

Chang, D., Fan, T., Gao, S., Jin, Y., Zhang, M., & Ono, M. (2021). Application of mesenchymal stem cell sheet to treatment of ischemic heart disease. *Stem Cell Research & Therapy*, *12*(1), 384. https://doi.org/10.1186/s13287-021-02451-1

Chang, H. C., Sun, T., Sultana, N., Lim, M. M., Khan, T. H., & Ismail, A. F. (2016). Conductive PEDOT:PSS coated polylactide (PLA) and poly(3-hydroxybutyrate-co-3-hydroxyvalerate) (PHBV) electrospun membranes: Fabrication and characterization. *Materials Science and Engineering: C*, *61*, 396–410. https://doi.org/10.1016/j.msec.2015.12.074

Chelu, M., Popa, M., Ozon, E. A., Pandele Cusu, J., Anastasescu, M., Surdu, V. A., Calderon Moreno, J., & Musuc, A. M. (2023). High-content aloe vera based hydrogels: physicochemical and pharmaceutical properties. *Polymers*, *15*(5), 1312. https://doi.org/10.3390/polym15051312

Chen, M. G., Wu, M. Y., Zou, M. L., & Zeng, M. Y. (2021). Effect of MicroRNA-146a modified adipose-derived stem cell exosomes on rat back wound healing. *The International Journal of Lower Extremity Wounds*, 153473462110380. https://doi.org/10.1177/15347346211038092

Cherubino, M., Rubin, J. P., Miljkovic, N., Kelmendi-Doko, A., & Marra, K. G. (2011). Adipose-derived stem cells for wound healing applications. *Annals of Plastic Surgery*, *66*(2), 210–215.

Chin, J. D., Zhao, L., Mayberry, T. G., Cowan, B. C., Wakefield, M. R., & Fang, Y. (2023). Photodynamic therapy, probiotics, acetic acid, and essential oil in the treatment of chronic wounds infected with Pseudomonas aeruginosa. *Pharmaceutics*, *15*(6), 1721. https://doi.org/10.3390/pharmaceutics15061721

Deli, J., González-Beiras, C., Guldan, G. S., Moses, R. L., Dally, J., Moseley, R., Lundy, F. T., Corbacho-Monne, M., Walker, S. L., Cazorla, M. U., Ouchi, D., Fang, R., Briggs, M., Kiapranis, R., Yahimbu, M., Mitjà, O., & Prescott, T. A. K. (2022). Ficus septica exudate, a traditional medicine used in Papua New Guinea for treating infected cutaneous ulcers: In vitro evaluation and clinical efficacy assessment by cluster randomised trial. *Phytomedicine*, *99*, 154026. https://doi.org/10.1016/j.phymed.2022.154026

Demmer, W., Sorg, H., Steiert, A., Hauser, J., & Tilkorn, D. J. (2021). Wound healing and therapy in soft tissue defects of the hand and foot from a surgical point of view. *Medical Sciences*, *9*(4), 71. https://doi.org/10.3390/medsci9040071

Deptuła, M., Karpowicz, P., Wardowska, A., Sass, P., Sosnowski, P., Mieczkowska, A., Filipowicz, N., Dzierżyńska, M., Sawicka, J., Nowicka, E., Langa, P., Schumacher, A., Cichorek, M., Zieliński, J., Kondej, K., Kasprzykowski, F., Czupryn, A., Janus, Ł., Mucha, P., . . . Pikuła, M. (2020). Development of a peptide derived from platelet-derived growth factor (PDGF-BB) into a potential drug candidate for the treatment of wounds. *Advances in Wound Care*, *9*(12), 657–675. https://doi.org/10.1089/wound.2019.1051

Diller, R. B., & Tabor, A. J. (2022). The role of the extracellular matrix (ECM) in wound healing: A review. *Biomimetics*, *7*(3), 87. https://doi.org/10.3390/biomimetics7030087

Dissemond, J., & Romanelli, M. (2022). Inflammatory skin diseases and wounds. *British Journal of Dermatology*, *187*(2), 167–177. https://doi.org/10.1111/bjd.21619

Dkhil, M. A., Diab, M. S. M., Aljawdah, H. M. A., Murshed, M., Hafiz, T. A., Al-Quraishy, S., & Bauomy, A. A. (2020). Neuro-biochemical changes induced by zinc oxide nanoparticles. *Saudi Journal of Biological Sciences*, *27*(10), 2863–2867. https://doi.org/10.1016/j.sjbs.2020.07.009

Eddy, A. C., Bidwell, G. L., & George, E. M. (2018). Pro-angiogenic therapeutics for preeclampsia. *Biology of Sex Differences*, *9*(1), 36. https://doi.org/10.1186/s13293-018-0195-5

Elsharawy, M. A., Naim, M., & Greish, S. (2012). Human CD34+ stem cells promote healing of diabetic foot ulcers in rats. *Interactive CardioVascular and Thoracic Surgery*, *14*(3), 288–293. https://doi.org/10.1093/icvts/ivr068

El-Sherbeni, S. A., & Negm, W. A. (2023). The wound healing effect of botanicals and pure natural substances used in in vivo models. *Inflammopharmacology*, *31*(2), 755–772. https://doi.org/10.1007/s10787-023-01157-5

Farooq, M., Khan, A. W., Kim, M. S., & Choi, S. (2021). The role of fibroblast growth factor (FGF) signaling in tissue repair and regeneration. *Cells*, *10*(11), 3242. https://doi.org/10.3390/cells10113242

Fox, L. T., Mazumder, A., Dwivedi, A., Gerber, M., du Plessis, J., & Hamman, J. H. (2017). In vitro wound healing and cytotoxic activity of the gel and whole-leaf materials from selected aloe species. *Journal of Ethnopharmacology*, *200*, 1–7. https://doi.org/10.1016/j.jep.2017.02.017

Franková, J., Pivodová, V., Vágnerová, H., Juránová, J., & Ulrichová, J. (2016). Effects of silver nanoparticles on primary cell cultures of fibroblasts and keratinocytes in a wound-healing model. *Journal of Applied Biomaterials & Functional Materials*, *14*(2). https://doi.org/10.5301/jabfm.5000268

Furlan, C., Berenbeim, J. A., & Dessent, C. E. H. (2020). Photoproducts of the photodynamic therapy agent verteporfin identified via laser interfaced mass spectrometry. *Molecules*, *25*(22), 5280. https://doi.org/10.3390/molecules25225280

Garg, T., Rath, G., & Goyal, A. K. (2015). Comprehensive review on additives of topical dosage forms for drug delivery. *Drug Delivery*, *22*(8), 969–987. https://doi.org/10.3109/10717544.2013.879355

GH, D., Kong, D., Gautrot, J., & Vootla, S. K. (2017). Fabrication and characterization of conductive conjugated polymer-coated *Antheraea mylitta* silk fibroin fibers for biomedical applications. *Macromolecular Bioscience*, *17*(7), 1600443. https://doi.org/10.1002/mabi.201600443

Ghaneialvar, H., Soltani, L., Rahmani, H. R., Lotfi, A. S., & Soleimani, M. (2018). Characterization and classification of mesenchymal stem cells in several species using surface markers for cell therapy purposes. *Indian Journal of Clinical Biochemistry*, *33*(1), 46–52. https://doi.org/10.1007/s12291-017-0641-x

GhavamiNejad, A., Rajan Unnithan, A., Ramachandra Kurup Sasikala, A., Samarikhalaj, M., Thomas, R. G., Jeong, Y. Y., Nasseri, S., Murugesan, P., Wu, D., Hee Park, C., & Kim, C. S. (2015). Mussel-inspired electrospun nanofibers functionalized with size-controlled silver nanoparticles for wound dressing application. *ACS Applied Materials & Interfaces*, *7*(22), 12176–12183. https://doi.org/10.1021/acsami.5b02542

Gheorghiță, D., Moldovan, H., Robu, A., Bița, A.-I., Grosu, E., Antoniac, A., Corneschi, I., Antoniac, I., Bodog, A. D., & Băcilă, C. I. (2023). Chitosan-based biomaterials for hemostatic applications: A review of recent advances. *International Journal of Molecular Sciences*, *24*(13), 10540. https://doi.org/10.3390/ijms241310540

Gizdavic-Nikolaidis, M., Ray, S., Bennett, J. R., Easteal, A. J., & Cooney, R. P. (2010). Electrospun functionalized polyaniline copolymer-based nanofibers with potential application in tissue engineering. *Macromolecular Bioscience*, *10*(12), 1424–1431. https://doi.org/10.1002/mabi.201000237

Goswami, A. G., Basu, S., Banerjee, T., & Shukla, V. K. (2023). Biofilm and wound healing: From bench to bedside. *European Journal of Medical Research*, *28*(1), 157. https://doi.org/10.1186/s40001-023-01121-7

Gu, B. K., Park, S. J., & Kim, C.-H. (2018). Beneficial effect of aligned nanofiber scaffolds with electrical conductivity for the directional guide of cells. *Journal of Biomaterials Science, Polymer Edition*, *29*(7–9), 1053–1065. https://doi.org/10.1080/09205063.2017.1364097

Guo, S., & Dipietro L. A. (2010). Factors affecting wound healing. *Journal of Dental Research*, *89*(3), 219–229. doi: 10.1177/0022034509359125. Epub 2010 Feb 5. PMID: 20139336; PMCID: PMC2903966.

Han, G., & Ceilley, R. (2017). Chronic wound healing: A review of current management and treatments. *Advances in Therapy*, *34*(3), 599–610. https://doi.org/10.1007/s12325-017-0478-y

Hassanshahi, A., Hassanshahi, M., Khabbazi, S., Hosseini-Khah, Z., Peymanfar, Y., Ghalamkari, S., Su, Y., & Xian, C. J. (2019). Adipose-derived stem cells for wound healing. *Journal of Cellular Physiology*, *234*(6), 7903–7914. https://doi.org/10.1002/jcp.27922

He, Y., Chang, Q., & Lu, F. (2023). Oxygen-releasing biomaterials for chronic wounds breathing: From theoretical mechanism to application prospect. *Materials Today Bio*, *20*, 100687. https://doi.org/10.1016/j.mtbio.2023.100687

Herman, A., & Herman, A. P. (2023). Antimicrobial peptides for diabetic wound healing: Preclinical and clinical studies. *Advances in Wound Care*. https://doi.org/10.1089/wound.2023.0011

Huang, H., Liu, R., Yang, J., Dai, J., Fan, S., Pi, J., Wei, Y., & Guo, X. (2023). Gold nanoparticles: Construction for drug delivery and application in cancer immunotherapy. *Pharmaceutics*, *15*(7), 1868. https://doi.org/10.3390/pharmaceutics15071868

Huang, X., Liang, P., Jiang, B., Zhang, P., Yu, W., Duan, M., Guo, L., Cui, X., Huang, M., & Huang, X. (2020). Hyperbaric oxygen potentiates diabetic wound healing by promoting fibroblast cell proliferation and endothelial cell angiogenesis. *Life Sciences*, *259*, 118246. https://doi.org/10.1016/j.lfs.2020.118246

Hui, Q., Jin, Z., Li, X., Liu, C., & Wang, X. (2018). FGF family: From drug development to clinical application. *International Journal of Molecular Sciences*, *19*(7), 1875. https://doi.org/10.3390/ijms19071875

Jayakumar, R., Prabaharan, M., Sudheesh Kumar, P. T., Nair, S. V., & Tamura, H. (2011). Biomaterials based on chitin and chitosan in wound dressing applications. *Biotechnology Advances*, *29*(3), 322–337. https://doi.org/10.1016/j.biotechadv.2011.01.005

Jeong, S. I., Jun, I. D., Choi, M. J., Nho, Y. C., Lee, Y. M., & Shin, H. (2008). Development of electroactive and elastic nanofibers that contain polyaniline and poly(L-lactide- $co-\varepsilon$ -caprolactone) for the control of cell adhesion. *Macromolecular Bioscience*, *8*(7), 627–637. https://doi.org/10.1002/mabi.200800005

Ji, R., Teng, M., Zhang, Z., Wang, W., Zhang, Q., Lv, Y., Zhang, J., & Jiang, X. (2020). Electric field downregulates CD9 to promote keratinocytes migration through AMPK pathway. *International Journal of Medical Sciences*, *17*(7), 865–873. https://doi.org/10.7150/ijms.42840

Jia, N., Yang, J., Liu, J., & Zhang, J. (2021). Electric field: A key signal in wound healing. *Chinese Journal of Plastic and Reconstructive Surgery*, *3*(2), 95–102. https://doi.org/10.1016/S2096-6911(21)00090-X

Jin, J., Zhu, X., Deng, Z., Zhang, P., Xiao, Y., Han, H., Li, Y., & Li, H. (2023). Protein-energy malnutrition and worse outcomes after major cancer surgery: A nationwide analysis. *Frontiers in Oncology*, *13*. https://doi.org/10.3389/fonc.2023.970187

Kaba, S., & Egorova, E. (2015). In vitro studies of the toxic effects of silver nanoparticles on HeLa and U937 cells. *Nanotechnology, Science and Applications*, *19*. https://doi.org/10.2147/NSA.S78134

Kiritsi, D., & Nyström, A. (2018). The role of TGFβ in wound healing pathologies. *Mechanisms of Ageing and Development*, *172*, 51–58. https://doi.org/10.1016/j.mad.2017.11.004

Koike, Y., Yozaki, M., Utani, A. et al. (2020). Fibroblast growth factor 2 accelerates the epithelial–mesenchymal transition in keratinocytes during wound healing process. *Scientific Reports, 10*, 18545. https://doi.org/10.1038/s41598-020-75584-7

Kolarikova, M., Hosikova, B., Dilenko, H., Barton-Tomankova, K., Valkova, L., Bajgar, R., Malina, L., & Kolarova, H. (2023). Photodynamic therapy: Innovative approaches for antibacterial and anticancer treatments. *Medicinal Research Reviews*, *43*(4), 717–774. https://doi.org/10.1002/med.21935

Kolimi, P., Narala, S., Nyavanandi, D., Youssef, A. A. A., & Dudhipala, N. (2022). Innovative treatment strategies to accelerate wound healing: Trajectory and recent advancements. *Cells*, *11*(15), 2439. https://doi.org/10.3390/cells11152439

Krishani, M., Shin, W. Y., Suhaimi, H., & Sambudi, N. S. (2023). Development of scaffolds from bio-based natural materials for tissue regeneration applications: A review. *Gels*, *9*(2), 100. doi: 10.3390/gels9020100. PMID: 36826270; PMCID: PMC9957409.

Li, P., & Guo, X. (2018). A review: Therapeutic potential of adipose-derived stem cells in cutaneous wound healing and regeneration. *Stem Cell Research & Therapy*, *9*(1), 302. https://doi.org/10.1186/s13287-018-1044-5

Li, P., Xu, J., Shi, Q., Wang, J., Zhang, W., Zheng, L., Wang, M., & Fan, Y. (2023). Pulse capacitive coupling electric field regulates cell migration, proliferation, polarization, and vascularization to accelerate wound healing. *Advances in Wound Care*, *12*(9), 498–512. https://doi.org/10.1089/wound.2021.0194

Liang, Y., Tian, H., Liu, J., Lv, Y., Wang, Y., Zhang, J., & Huang, Y. (2020). Application of stable continuous external electric field promotes wound healing in pig wound model. *Bioelectrochemistry*, *135*, 107578. https://doi.org/10.1016/j.bioelechem.2020.107578

Liu, D., Zhu, L., & Yang, C. (2022). The effect of preoperative smoking and smoke cessation on wound healing and infection in post-surgery subjects: A meta-analysis. *International Wound Journal*, *19*(8), 2101–2106. https://doi.org/10.1111/iwj.13815

Liu, Y., Liu, Y., Deng, J., Li, W., & Nie, X. (2021). Fibroblast growth factor in diabetic foot ulcer: Progress and therapeutic prospects. *Frontiers in Endocrinology*, *12*. https://doi.org/10.3389/fendo.2021.744868

Lu, C., Kolbenschlag, J., Nüssler, A. K., Ehnert, S., McCaig, C. D., Čebron, U., Daigeler, A., & Prahm, C. (2021). Direct current electrical fields improve experimental wound healing by activation of cytokine secretion and Erk1/2 pathway stimulation. *Life (Basel, Switzerland)*, *11*(11). https://doi.org/10.3390/life11111195

Luo, R., Liang, Y., Yang, J., Feng, H., Chen, Y., Jiang, X., Zhang, Z., Liu, J., Bai, Y., Xue, J., Chao, S., Xi, Y., Liu, X., Wang, E., Luo, D., Li, Z., & Zhang, J. (2023). Reshaping the endogenous electric field to boost wound repair via electrogenerative dressing. *Advanced Materials*, *35*(16). https://doi.org/10.1002/adma.202208395

Massoud, D., Alrashdi, B. M., Fouda, M. M. A., El-kott, A., Soliman, S. A., & Abd-Elhafeez, H. H. (2022). Aloe vera and wound healing: A brief review. *Brazilian Journal of Pharmaceutical Sciences*, *58*. https://doi.org/10.1590/s2175-97902022e20837

Mojica, M. F., Humphries, R., Lipuma, J. J., Mathers, A. J., Rao, G. G., Shelburne, S. A., Fouts, D. E., Van Duin, D., & Bonomo, R. A. (2022). Clinical challenges treating *Stenotrophomonas maltophilia* infections: An update. *JAC-Antimicrobial Resistance*, *4*(3). https://doi.org/10.1093/jacamr/dlac040

Mokry, J., & Pisal, R. (2020). Development and maintenance of epidermal stem cells in skin adnexa. *International Journal of Molecular Sciences*, *21*(24), 9736. https://doi.org/10.3390/ijms21249736

Molnar, J. A., Underdown, M. J., & Clark, W. A. (2014). Nutrition and chronic wounds. *Advances in Wound Care*, *3*(11), 663–681. https://doi.org/10.1089/wound.2014.0530

Mony, M. P., Harmon, K. A., Hess, R., Dorafshar, A. H., & Shafikhani, S. H. (2023). An updated review of hypertrophic scarring. *Cells*, *12*(5), 678. doi: 10.3390/cells12050678

Morbidelli, L., Genah, S., & Cialdai, F. (2021). Effect of microgravity on endothelial cell function, angiogenesis, and vessel remodeling during wound healing. *Frontiers in Bioengineering and Biotechnology*, *9*. https://doi.org/10.3389/fbioe.2021.720091

Moses, R. L., Dally, J., Lundy, F. T., Langat, M., Kiapranis, R., Tsolaki, A. G., Moseley, R., & Prescott, T. A. K. (2020). Lepiniopsis ternatensis sap stimulates fibroblast proliferation and down regulates macrophage TNF-α secretion. *Fitoterapia*, *141*, 104478. https://doi.org/10.1016/j.fitote.2020.104478

Moses, R. L., Prescott, T. A. K., Mas-Claret, E., Steadman, R., Moseley, R., & Sloan, A. J. (2023). Evidence for natural products as alternative wound-healing therapies. *Biomolecules*, *13*(3), 444. https://doi.org/10.3390/biom13030444

Naraginti, S., Kumari, P. L., Das, R. K., Sivakumar, A., Patil, S. H., & Andhalkar, V. V. (2016). Amelioration of excision wounds by topical application of green synthesized, formulated silver and gold nanoparticles in albino Wistar rats. *Materials Science and Engineering: C*, *62*, 293–300. https://doi.org/10.1016/j.msec.2016.01.069

Navani, N. K., Ramulu Lambadi, P., Kumar Sharma, T., Kumar, P., Vasnani, P., Mouli Thalluri, S., Bisht, N., & Pathania, R. (2015). Facile biofunctionalization of silver nanoparticles for enhanced antibacterial properties, endotoxin removal, and biofilm control. *International Journal of Nanomedicine*, 2155. https://doi.org/10.2147/IJN.S72923

Nguyen, H. M., Ngoc Le, T. T., Nguyen, A. T., Thien Le, H. N., & Pham, T. T. (2023). Biomedical materials for wound dressing: Recent advances and applications. *RSC Advances*, *13*(8), 5509–5528. https://doi.org/10.1039/D2RA07673J

Niu, X., Rouabhia, M., Chiffot, N., King, M. W., & Zhang, Z. (2015). An electrically conductive 3D scaffold based on a nonwoven web of poly(l-lactic acid) and conductive poly(3,4-ethylenedioxythiophene). *Journal of Biomedical Materials Research Part A*, *103*(8), 2635–2644. https://doi.org/10.1002/jbm.a.35408

Norouzkhani, N., Ghazanfari, M. J., Falakdami, A., Takasi, P., Mollaei, A., Ghorbani Vajargah, P., Mehrabi, H., Firooz, M., Hosseini, S. J., Mobayen, M., & Karkhah, S. (2023). Implementation of telemedicine for burns management: Challenges and opportunities. *Burns*, *49*(2), 482–484. https://doi.org/10.1016/j.burns.2022.10.011

Okur, M. E., Karantas, I. D., Okur, N. U., & Siafaka, P. I. (2018). Hypertension in 2017: Update in treatment and pharmaceutical innovations. *Current Pharmaceutical Design*, *23*(44), 6795–6814. https://doi.org/10.2174/1381612823666170927123454

Okur, M. E., Karantas, I. D., Şenyiğit, Z., Üstündağ Okur, N., & Siafaka, P. I. (2020). Recent trends on wound management: New therapeutic choices based on polymeric carriers. *Asian Journal of Pharmaceutical Sciences*, *15*(6), 661–684. https://doi.org/10.1016/j.ajps.2019.11.008

Okur, M. E., Karantas, I. D., & Siafaka, P. I. (2017). Diabetes mellitus: A review on pathophysiology, current status of oral pathophysiology, current status of oral medications and future perspectives. *ACTA Pharmaceutica Sciencia*, *55*(1), 61. https://doi.org/10.23893/1307-2080.APS.0555

Parella, K., Moody, K., Wortel, D., Colegrove, H., & Elser, J. A. (2023). HOXA3 accelerates wound healing in diabetic and aged non-diabetic mammals. *Scientific Reports*, *13*(1), 9923. https://doi.org/10.1038/s41598-023-36933-4

Patra, H. K., Sharma, Y., Islam, M. M., Jafari, M. J., Murugan, N. A., Kobayashi, H., Turner, A. P. F., & Tiwari, A. (2016). Inflammation-sensitive in situ smart scaffolding for regenerative medicine. *Nanoscale*, *8*(39), 17213–17222. https://doi.org/10.1039/C6NR06157E

Păunica-Panea, G., Ficai, A., Marin, M. M., Marin, Ştefania, Albu, M. G., Constantin, V. D., Dinu-Pîrvu, C., Vuluga, Z., Corobea, M. C., & Ghica, M. V. (2016). New collagen-dextran-zinc oxide composites for wound dressing. *Journal of Nanomaterials*, *2016*, 1–7. https://doi.org/10.1155/2016/5805034

Peng, Z., & Lavigne, J.-P. (2023). Editorial: Pathogenesis of diabetic foot ulcers. *Frontiers in Pharmacology*, *14*. https://doi.org/10.3389/fphar.2023.1218201

Qin, H., Chen, Y., Wang, Z. et al. (2023). Biosynthesized gold nanoparticles that activate Toll-like receptors and elicit localized light-converting hyperthermia for pleiotropic tumor immunoregulation. *Nature Communications*, *14*, 5178. https://doi.org/10.1038/s41467-023-40851-4

Rai, V., Moellmer, R., & Agrawal, D. K. (2022). Stem cells and angiogenesis: Implications and limitations in enhancing chronic diabetic foot ulcer healing. *Cells*, *11*(15), 2287. https://doi.org/10.3390/cells11152287

Raziyeva, K., Kim, Y., Zharkinbekov, Z., Kassymbek, K., Jimi, S., & Saparov, A. (2021). Immunology of acute and chronic wound healing. *Biomolecules*, *11*(5), 700. https://doi.org/10.3390/biom11050700

Reddell, P., De Ridder, T. R., Morton, J. M., Jones, P. D., Campbell, J. E., Brown, G., Johannes, C. M., Schmidt, P. F., & Gordon, V. (2021). Wound formation, wound size, and progression of wound healing after intratumoral treatment of mast cell tumors in dogs with tigilanol tiglate. *Journal of Veterinary Internal Medicine*, *35*(1), 430–441. https://doi.org/10.1111/jvim.16009

Rieg, A. D., Suleiman, S., Anker, C., Bünting, N. A., Verjans, E., Spillner, J., Kalverkamp, S., von Stillfried, S., Braunschweig, T., Uhlig, S., & Martin, C. (2022). Platelet-derived growth factor (PDGF)-BB regulates the airway tone via activation of MAP2K, thromboxane, actin polymerisation and Ca2+-sensitisation. *Respiratory Research*, *23*(1), 189. https://doi.org/10.1186/s12931-022-02101-x

Rodrigues, M., Kosaric, N., Bonham, C. A., & Gurtner, G. C. (2019). Wound healing: A cellular perspective. *Physiological Reviews*, *99*(1), 665–706. https://doi.org/10.1152/physrev.00067.2017

Román-Doval, R., Tellez-Cruz, M. M., Rojas-Chávez, H., Cruz-Martínez, H., Carrasco-Torres, G., & Vásquez-Garzón, V. R. (2019). Enhancing electrospun scaffolds of PVP with polypyrrole/iodine for tissue engineering of skin regeneration by coating via a plasma process. *Journal of Materials Science*, *54*(4), 3342–3353. https://doi.org/10.1007/s10853-018-3024-7

Roma-Rodrigues, C., Pombo, I., Fernandes, A. R., & Baptista, P. V. (2020). Hyperthermia induced by gold nanoparticles and visible light photothermy combined with chemotherapy to tackle doxorubicin sensitive and resistant colorectal tumor 3D spheroids. *International Journal of Molecular Sciences*, *21*(21), 8017. https://doi.org/10.3390/ijms21218017

Rosen, R. D., & Manna, B. (2023, May 1). Wound Dehiscence. In: *StatPearls* [Internet]. Treasure Island, FL: StatPearls Publishing.

Ruiz-Velasco, G., Martínez-Flores, F., Morales-Corona, J., Olayo-Valles, R., & Olayo, R. (2017). Polymeric scaffolds for skin. *Macromolecular Symposia*, *374*(1), 1600133. https://doi.org/10.1002/masy.201600133

Ryan, C. N. M., Doulgkeroglou, M. N., & Zeugolis, D. I. (2021). Electric field stimulation for tissue engineering applications. *BMC Biomedical Engineering*, *3*(1), 1. https://doi.org/10.1186/s42490-020-00046-0

Rybka, M., Mazurek, Ł., & Konop, M. (2022). Beneficial effect of wound dressings containing silver and silver nanoparticles in wound healing – from experimental studies to clinical practice. *Life*, *13*(1), 69. https://doi.org/10.3390/life13010069

Sabbatini, M., Magnelli, V., & Renò, F. (2021). NETosis in wound healing: When enough is enough. *Cells*, *10*(3), 494. https://doi.org/10.3390/cells10030494

Sadati, L., Froozesh, R., Beyrami, A., Khaneghah, Z. N., Elahi, S. A., Asl, M. F., & Salehi, T. (2019). A comparison of three dressing methods for pilonidal sinus surgery wound healing. *Advances in Skin & Wound Care*, *32*(7), 1–5. https://doi.org/10.1097/01.ASW.0000558268.59745.d2

Sandström, N., Söderquist, B., Wistrand, C., & Friberg, Ö. (2023). The presence of skin bacteria in the sternal wound and contamination of implantation materials during cardiac surgery. *Journal of Hospital Infection*, *135*, 145–151. https://doi.org/10.1016/j.jhin.2023.03.018

Schneider, I., Calcagni, M., & Buschmann, J. (2023). Adipose-derived stem cells applied in skin diseases, wound healing and skin defects: A review. *Cytotherapy*, *25*(2), 105–119. https://doi.org/10.1016/j.jcyt.2022.08.005

Shafaie, S., Andalib, S., Shafaei, H., Montaseri, A., & Tavakolizadeh, M. (2020). Differential biological behavior of fibroblasts and endothelial cells under aloe vera gel culturing. *International Journal of Molecular and Cellular Medicine*, *9*(3), 234–246.

Shahmoradi, S., Golzar, H., Hashemi, M., Mansouri, V., Omidi, M., Yazdian, F., Yadegari, A., & Tayebi, L. (2018). Optimizing the nanostructure of graphene oxide/silver/arginine for effective wound healing. *Nanotechnology*, *29*(47), 475101. https://doi.org/10.1088/1361-6528/aadedc

Sharma, A., Khanna, S., Kaur, G., & Singh, I. (2021). Medicinal plants and their components for wound healing applications. *Future Journal of Pharmaceutical Sciences*, *7*(1), 53. https://doi.org/10.1186/s43094-021-00202-w

Shi, C., Wang, C., Liu, H., Li, Q., Li, R., Zhang, Y., Liu, Y., Shao, Y., & Wang, J. (2020). Selection of appropriate wound dressing for various wounds. *Frontiers in Bioengineering and Biotechnology*, *8*. https://doi.org/10.3389/fbioe.2020.00182

Su, L., Zheng, J., Wang, Y., Zhang, W., & Hu, D. (2019). Emerging progress on the mechanism and technology in wound repair. *Biomedicine & Pharmacotherapy*, *117*, 109191. https://doi.org/10.1016/j.biopha.2019.109191

Takaya, K., Aramaki-Hattori, N., Sakai, S., Okabe, K., Asou, T., & Kishi, K. (2022). Fibroblast growth factor 7 suppresses fibrosis and promotes epithelialization during wound healing in mouse fetuses. *International Journal of Molecular Sciences*, *23*(13), 7087. https://doi.org/10.3390/ijms23137087

Tottoli, E. M., Dorati, R., Genta, I., Chiesa, E., Pisani, S., & Conti, B. (2020). Skin wound healing process and new emerging technologies for skin wound care and regeneration. *Pharmaceutics*, *12*(8), 735. https://doi.org/10.3390/pharmaceutics12080735

Veith, A. P., Henderson, K., Spencer, A., Sligar, A. D., & Baker, A. B. (2019). Therapeutic strategies for enhancing angiogenesis in wound healing. *Advanced Drug Delivery Reviews*, *146*, 97–125. https://doi.org/10.1016/j.addr.2018.09.010

Viswanathan, V., Juttada, U., & Babu, M. (2020). Efficacy of recombinant human epidermal growth factor (Regen-D 150) in healing diabetic foot ulcers: A hospital-based randomized controlled trial. *The International Journal of Lower Extremity Wounds*, *19*(2),158–164. doi:10.1177/1534734619892791

Vivcharenko, V., Trzaskowska, M., & Przekora, A. (2023). Wound dressing modifications for accelerated healing of infected wounds. *International Journal of Molecular Sciences*, *24*(8), 7193. https://doi.org/10.3390/ijms24087193

Wang, M., Xu, X., Lei, X., Tan, J., & Xie, H. (2021). Mesenchymal stem cell-based therapy for burn wound healing. *Burns & Trauma*, *9*. https://doi.org/10.1093/burnst/tkab002

Wang, S. T., Neo, B. H., & Betts, R. J. (2021). Glycosaminoglycans: Sweet as sugar targets for topical skin anti-aging. *Clinical, Cosmetic and Investigational Dermatology*, *14*, 1227–1246. https://doi.org/10.2147/CCID.S328671

Wang, S. Y., Kim, H., Kwak, G., Yoon, H. Y., Jo, S. D., Lee, J. E., Cho, D., Kwon, I. C., & Kim, S. H. (2018). Development of biocompatible HA hydrogels embedded with a new synthetic peptide promoting cellular migration for advanced wound care management. *Advanced Science*, *5*(11), 1800852. https://doi.org/10.1002/advs.201800852

Wang, Y., Rouabhia, M., Lavertu, D., & Zhang, Z. (2017). Pulsed electrical stimulation modulates fibroblasts' behaviour through the Smad signalling pathway. *Journal of Tissue Engineering and Regenerative Medicine*, *11*(4), 1110–1121. https://doi.org/10.1002/term.2014

Wang, Y., Rouabhia, M., & Zhang, Z. (2013). PPy-coated PET fabrics and electric pulse-stimulated fibroblasts. *Journal of Materials Chemistry B*, *1*(31), 3789. https://doi.org/10.1039/c3tb20257g

Xiu, Z., Zhang, Q., Puppala, H. L., Colvin, V. L., & Alvarez, P. J. J. (2012). Negligible particle-specific antibacterial activity of silver nanoparticles. *Nano Letters*, *12*(8), 4271–4275. https://doi.org/10.1021/nl301934w

Xue, C., Sutrisno, L., Li, M., Zhu, W., Fei, Y., Liu, C., Wang, X., Cai, K., Hu, Y., & Luo, Z. (2021). Implantable multifunctional black phosphorus nanoformulation-deposited biodegradable scaffold for combinational photothermal/chemotherapy and wound healing. *Biomaterials*, *269*, 120623. https://doi.org/10.1016/j.biomaterials.2020.120623

Yang, X., Zhan, P., Wang, X., Zhang, Q., Zhang, Y., Fan, H., Li, R., & Zhang, M. (2020). Polydopamine-assisted PDGF-BB immobilization on PLGA fibrous substrate enhances wound healing via regulating anti-inflammatory and cytokine secretion. *PLoS ONE*, *15*(9), e0239366. https://doi.org/10.1371/journal.pone.0239366

Yu, Y., Wan, Y., & Huang, C. (2009). The biological functions of NF-κB1 (p) and its potential as an anti-cancer target. *Current Cancer Drug Targets*, *9*(4), 566–571. https://doi.org/10.2174/156800909788486759

Zaidi, S. R. H., & Sharma, S. (2022, August 9). Pressure ulcer. In: *StatPearls* [Internet]. Treasure Island, FL: StatPearls Publishing; 2023. Available from: www.Ncbi.Nlm.Nih.Gov/Books/NBK553107/.

Zhu, S., Yu, Y., Ren, Y., Xu, L., Wang, H., Ling, X., Jin, L., Hu, Y., Zhang, H., Miao, C., & Guo, K. (2021). The emerging roles of neutrophil extracellular traps in wound healing. *Cell Death & Disease*, *12*(11), 984.

14 Medicinal Plants and Natural Compounds for the Treatment of Dermatological Disorders

Zozan Guleken

14.1 INTRODUCTION

14.1.1 Overview of Skin Physiology

The skin (cutis) covers the entire external surface of the body. Its primary function is the prevention of the loss of physiologically important components of the body and protection of the body from harsh environmental conditions (Proksch et al., 2008). It protects the body from pathogens, ultraviolet (UV) light, chemicals, and mechanical injury and acts as a physical barrier. It also regulates body temperature and perspiration (Tobina, 2006). It regulates and prevents water and electrolyte loss. In addition, the skin has a sensory function. The receptors it contains allow us to feel touch, pain, temperature, pressure, and vibration. In addition to providing a strong barrier against external agents, antimicrobial peptides, and lipids in the skin form a biomolecular barrier against bacterial membranes (Kwiecien et al., 2019). The skin is also a complex organ involved in many inflammatory processes such as tumour immunity, autoimmunity, and allergy (Matejuk, 2018). Depending on the thickness of the epidermis and dermis, the distribution of epidermal extensions, the content of melanocytes (pigment-containing cells) and their structure and function, the skin layers have different characteristics (Lin and Fisher, 2007). On the palm of the hand and the soles of the foot, there is another layer between the horny layer and the granular layer. This is the thickest layer of tissue and is called the lucid layer. The skin, from the outside to the inside; the epidermis (upper skin), the dermis (skin) and the hypodermis (subcutaneous tissue) consist of the following structural layers. The epidermis, the outermost layer of the skin, serves as a vital barrier providing basic protection against mechanical, chemical, and microbial threats from the external environment. Its thickness varies across different body regions, with the palms and soles exhibiting the thickest epidermal layers (Wickett and Visscher, 2006). Comprised predominantly of keratinocytes, the epidermis undergoes continuous renewal, with cells originating from the basal cell layer and progressing through distinct sublayers, including the spiny, granular, and cornified layers (Suter et al., 2009). Melanocytes, responsible for melanin production, are also present in the basal layer, providing protection against UV radiation (Lin and Fisher, 2007). The epidermis contains four different sublayers: stratum corneum (horn-like layer), stratum lucidum (inhibitory layer), stratum granulosum (granular layer), stratum spinosum (spiny layer), and stratum basale (basal cell layer). The stratum corneum, the outermost layer, consists of anucleate squamous cells rich in protein and lipid, serving as a barrier against dehydration and microbial invasion (Wickett and Visscher, 2006). Understanding the complex structure and functions of the epidermis is essential for comprehending skin physiology and its role in maintaining overall health and homeostasis. In this introduction, we explore the intricate organization and functions of the epidermis, laying the foundation for a comprehensive understanding of skin physiology.

Epidermis (Upper Skin) The epidermis is the outermost layer of the skin that shows basic protection. It is the first layer of skin to encounter mechanical and chemical damage or infections from outside. Therefore, it is the layer that shows sensitivity to damage caused by bumps, allergies, and

DOI: 10.1201/9781003394846-14

chemical substance irritations. The thickness of the epidermis varies depending on the body region. The thickness of the epidermis is about 1 mm on the palms and soles of the feet, which consist of a hairless structure. This thickness thins down to 0.1 mm in other parts of the body. The forehead, temples, and around the eyes are even thinner (Fuchs and Byrne, 1994). The thickness of the epidermis varies depending on gender, genetic makeup and climatic conditions. The main structure of the epidermis is keratinocytes originating from the ectodermal layer. Keratinocytes make up about 95 per cent of the cells of the epidermis. Keratin cells form morphologically different epidermis layers as they move from the basement membrane to the surface of the skin. The epidermis contains four different sublayers: stratum corneum (horn-like layer), stratum lucidum (inhibitory layer), stratum granulosum (granular layer), stratum spinosum (spiny layer), and stratum basale (basal cell layer) (Lefèvre-Utile et al., 2021).

Basal Cell Layer (Stratum basale): Contains mitotically active stem cells responsible for continuous production of keratinocytes. Houses melanocytes that produce melanin, providing protection against UV radiation (Yousef and Sharma, 2018).

Spiny Layer (Stratum spinosum): Provides structural support and integrity to the epidermis. Contains dendritic cells involved in immune responses. Allows for communication between neighbouring cells through desmosomes (Nguyen and Soulika, 2019).

Granular Layer (Stratum granulosum): Synthesizes and accumulates keratohyaline granules which contain keratin precursors and aid in keratinization. Produces lamellar granules containing glycolipids that contribute to the formation of the skin's barrier function (*Anatomy, Skin (Integument), Epidermis – StatPearls – NCBI Bookshelf*, n.d.).

Stratum Corneum: Forms the outermost layer of the epidermis, serving as a protective barrier against dehydration, microbial invasion, and environmental insults. Composed of multiple layers of anuclear squamous cells derived from keratinocytes. Rich in proteins and lipids, providing structural integrity and water resistance (G.K. Menon et al., 2012). Lacks vasculature, relying on diffusion from the underlying dermis for nutrient supply. Contains cold-sensing receptors and secretes antimicrobial peptides known as defensins as part of the immune response (Murphrey and Zito, 2018). Undergoes constant shedding and renewal to maintain skin integrity and homeostasis.

Each layer of the epidermis contributes uniquely to the overall function and health of the skin, collectively ensuring its protective and regulatory roles in maintaining bodily homeostasis.

14.1.2 Cellular Components of the Epidermis

Keratinocytes, the primary cell type in the epidermis, play a pivotal role in skin barrier function by forming tight junctions, producing keratin, and secreting lipids, while also regulating calcium absorption and contributing to vitamin D synthesis; moreover, they can express HLA-II and secrete cytokines, thereby influencing inflammatory processes (Jiang et al., 2020).

Melanocytes, situated in the stratum basale, are responsible for producing melanin, the pigment that determines skin colour, and for transferring melanin granules to adjacent keratinocytes to shield DNA from UV radiation; deficiency of melanocytes can result in vitiligo, an autoimmune disorder characterized by depigmented patches on the skin (Cichorek et al., 2013).

Merkel cells, specialized oval-shaped epidermal cells located in the stratum basale, communicate with cutaneous neurons through mechanoreceptors, contributing to tactile sensation in regions such as fingertips, palms, soles of the feet, and oral and genital mucosa; however, their exact role in cutaneous immunity remains incompletely understood (Abraham and Mathew, 2019).

Langerhans cells, dendritic cells situated in the epidermis, are pivotal in both innate and acquired immune responses by presenting antigens to T cells; originating from bone marrow CD34 positive stem cells, they express MHC I and MHC II molecules to uptake antigens and ferry them to lymph nodes, contributing to immune surveillance and tolerance mechanisms against environmental antigens (Romani et al., 2010).

Memory T cells, which persist in the epidermis for prolonged durations, are associated with a range of skin disorders including atopic and contact dermatitis, cutaneous T cell lymphoma, psoriasis, and vitiligo, highlighting their significant role in mediating immune responses within the skin (Chen and Shen, 2020).

Neutrophils and T cells have the capacity to infiltrate the epidermis during specific skin disorders, yet are typically absent from the epidermal layer under normal physiological conditions (Orsmond et al., 2021).

14.1.3 Dermis (Skin)

The dermis comprises extracellular matrix proteins providing structural integrity and elasticity to the skin and allows free circulation of cell populations. Separated from the epidermis by the basement membrane, it supports the epidermis and is divided into the superficial papillary layer (stratum papillare) and deep reticular layer (stratum reticulare) (Watt and Fujiwara, 2011). The papillary layer forms dermal papillae, composed of loose connective tissue and vascularized networks, while the reticular layer consists of dense, irregular connective tissue supporting sensory and sympathetic nerves. Both layers contain blood vessels, lymphatics, sweat glands, hair follicles, and other structures (Kim and Dao, 2020). The cellular components include fibroblasts, endothelial cells, and neurones, facilitating immune responses and sensory functions. Additionally, innate immune cells such as dermal dendritic cells, plasmacytoid dendritic cells, and macrophages, along with adaptive immune cells like T lymphocytes and B cells, contribute to immune surveillance and response within the dermis (Elbe-Bürger, 2012; Nguyen and Soulika, 2019).

14.1.4 Hypodermis (Subcutis) and Blood Supply

The hypodermis, or subcutaneous tissue, lies beneath the dermis and connects the skin to the underlying fascia, serving as a fat storage layer and providing insulation and cushioning. Blood supply to the skin is facilitated by two plexuses within the dermis and subcutaneous tissues, regulating temperature through vasodilation and vasoconstriction controlled by the autonomic nervous system (Kim and Dao, 2020; *Hypodermis – Concept | Anatomy and Physiology | JoVe*, n.d.).

14.1.5 Nerves in the Skin and Dermatomes

Somatic and autonomic sensory nerves in the skin convey pain, temperature, touch, and proprioception. Mechanoreceptors and free nerve endings enable sensory perception, while autonomic innervation regulates vascular tone, pilomotor stimulation, and sweating. Dermatomes, anatomical regions innervated by specific spinal nerve roots, facilitate sensory signal transmission, aiding in diagnosing nerve damage and spinal lesions (Archer and Eedy, 2010).

14.1.6 Skin Flora

The skin surface maintains a pH range of 4.0 to 7.0, with the acidic pH of the stratum corneum (SC) crucial for skin health, microbial control, lipid barrier formation, and SC homeostasis. Metabolic processes in the epidermis, including passive formation of lactic acid and active production of free fatty acids (FFA), cholesterol sulfate, urocanic acid, and pyrrolidone carboxylic acid (PCA), contribute to this acidity (Boer et al., 2016). Proton pumps and sodium/hydrogen exchange proteins in cell membranes regulate extracellular acidity in the SC layers, establishing a pH gradient essential for normal flora preservation and prevention of pathogenic colonization (Lefèvre-Utile et al., 2021). The epidermal layer harbours most of the skin's normal flora, including Staphylococcus aureus, Cutibacterium acnes, Corynebacterium, Streptococcus, Candida, and Clostridium perfringens, fostering a mutually symbiotic relationship that fortifies the skin's defences and maintains acidification (Fournière et al., 2020).

14.2 SKIN DISORDERS

Skin disorders can be classified into several categories based on various criteria such as aetiology (cause), clinical presentation, affected skin layers, and systemic involvement (Soltani, 2002; Rustin, 1990).

Skin disorders encompass a wide array of conditions that affect the largest organ of the human body. Understanding their classification is pivotal in diagnosis, treatment, and management. The classification of skin disorders is presented in Figure 14.1.

14.2.1 Infectious Skin Disorders

Caused by microorganisms like bacteria, viruses, fungi, or parasites, encompass a variety of conditions including bacterial infections such as cellulitis, impetigo, and folliculitis; viral infections like herpes simplex virus (HSV), human papillomavirus (HPV), and varicella-zoster virus (VZV); fungal infections such as tinea (ringworm), candidiasis, and dermatophytosis; and parasitic infections including scabies and lice infestations (Chiller et al., 2001).

14.2.2 Inflammatory Skin Disorders

Inflammation is a complex process by which the body attempts to repair tissue damage and defend itself against harmful agents and is triggered by infectious agents, noxious stimuli (mechanical and chemical) and autoimmune responses. Inflammatory signs include reddening, swelling, itching, heat, and pain. (Ikeda et al., 2008; Chakraborty and Burns, 2023; Dawid-Pać, 2013). Some intracellular biochemicals are secreted by cells when an inflammatory factor is present. Cytokines (regulatory glycoproteins of the immune system) are secreted from monocytes and macrophages. They play a major role in inflammation by stimulating the activation of cells involved in inflammation (neutrophils, macrophages, and mast cells), allowing communication among these cells, inducing the synthesis of prostaglandins, and influencing the synthesis of C-reactive protein. The cytokine family is divided into proinflammatory (interleukins IL-1, IL-6, IL-8, IL-17, IL-18, alpha and beta interferons and TNFs) and anti-inflammatory (e.g. IL-4, IL-10, IL-13) cytokines. An overabundance of proinflammatory cytokines leads to a systemic inflammatory response, whereas an overabundance of anti-inflammatory cytokines leads to an anti-inflammatory response. (Dawid-Pać, 2013).

Eicosanoids are also involved in the regulation of inflammatory processes and cellular response: prostaglandins (PGD, PGE and PGF), prostacyclins (PGI), thromboxanes (TXA), leukotrienes (LTB, LTC, LTD), mediators produced in response to stimuli from arachidonic acid (AA) associated with the cell membrane in the form of phospholipids (G. Bhat, 2022).

Inflammatory skin disorders marked by inflammation of the skin and surrounding tissues comprise various conditions such as dermatitis (including atopic dermatitis, contact dermatitis, and seborrheic dermatitis), psoriasis (a chronic autoimmune condition with red, scaly patches of skin), rosacea (a chronic inflammatory condition affecting the face, leading to redness and visible

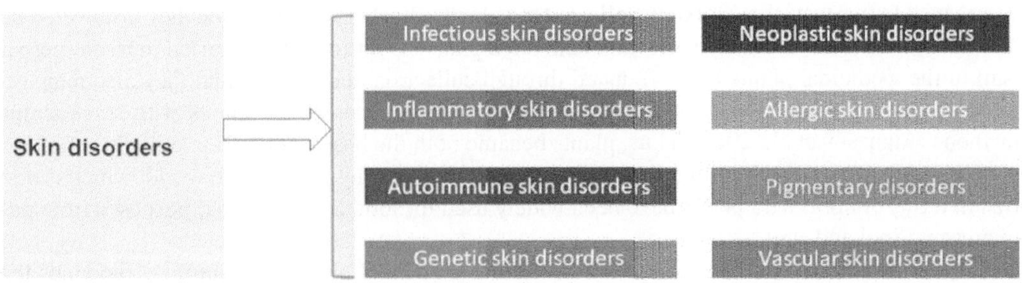

FIGURE 14.1 The classification of skin disorders is presented.

blood vessels), and acne (a common inflammatory skin condition involving the hair follicles and oil glands).

Autoimmune skin disorders arise from an aberrant immune response targeting the body's own tissues and include conditions such as systemic lupus erythematosus (SLE), a multi-organ autoimmune disease affecting the skin among other organs; pemphigus vulgaris, a rare autoimmune disorder leading to blistering of the skin and mucous membranes; and dermatomyositis, a connective tissue disease distinguished by muscle weakness and skin rash (Das et al., 2019).

Genetic skin disorders stem from inherited genetic mutations impacting the skin's structure or function and encompass conditions like epidermolysis bullosa, marked by fragile skin and blister formation; ichthyosis, leading to dry, scaly skin; and albinism, characterized by the absence of melanin pigment in the skin, hair and eyes (Parker et al., 2021).

Neoplastic skin disorders encompass the emergence of abnormal growths or tumours on the skin, including basal cell carcinoma (BCC), squamous cell carcinoma (SCC), melanoma, and cutaneous lymphoma (Manuelyan et al., 2017).

Allergic skin disorders arise from allergic reactions to specific substances or allergens, leading to conditions such as urticaria (hives), allergic contact dermatitis, and atopic dermatitis (eczema) (Katoh, 2018).

Pigmentary skin disorders encompass changes in skin colouration resulting from alterations in melanin production or distribution, including conditions such as vitiligo, hyperpigmentation (e.g., melasma, post-inflammatory hyperpigmentation), and hypopigmentation (e.g., albinism) (Thawabteh et al., 2023).

Vascular skin disorders are characterized by abnormalities in blood vessels, vascular dysregulation within the skin and include conditions such as rosacea, telangiectasia, and purpura (Del Rosso, 2013; Zawodny et al., 2023).

14.3 MEDICINAL PLANTS AND NATURAL COMPOUNDS

The use of plants in treatment began with the history of humanity, and many drugs used in modern medicine are obtained from plants (F.S. Li and Weng, 2017; Fowler, 2006). Throughout history, skin diseases have had a serious impact on the quality of life of human beings. Skin diseases are still mainly treated by traditional healers in rural areas using plant species, although attempts have been made to cure skin diseases with modern medicines (Almoshari, 2022).

Medicinal and aromatic plants were perhaps not the first plants to be cultivated. However, just like today, thousands of years ago people were collecting these plants wild from nature. Because they needed not only food, clothing, and shelter but also treatment. It did not take long for ancient people to realize that such plants had healing powers. While feeding on them, they perhaps saw that their pains subsided, their discomforts disappeared, and somehow the wound they applied to them healed quickly (Schmidt et al., 2008). By trial and error, they chewed different organs and parts such as flowers, fruits, seeds, stems, leaves, roots and barks, and tried to understand which ones were medicinal or which ones were poisonous by their taste and smell (Petrovska, 2012). By observing nature well in this way, they not only collected the plants whose medicinal value they discovered but also cultivated them like other important cultivated plants. Over time, they started to trade (herbalism) in the medicinal plants they produced through collection and culture, and they also managed to obtain the first medicines containing the bioactive substances of the plants with some simple methods (Atanasov et al., 2015). Thus, plants became both the basic food and livelihood sources of people and the first sources of medicine (G. Bhat, 2022). In many parts of the world, which is very rich in terms of flora, wild plants have been widely used for food and medicinal purposes from past to present (Sõukand et al., 2017).

Table 14.1 provides a detailed overview of various medicinal plants and natural compounds that have shown promise in treating inflammatory skin disorders. It's fascinating to present the diverse range of bioactive substances found in these plants, like essential oils, flavonoids, tannins, and more.

TABLE 14.1
Medicinal Plants and Natural Compounds for the Treatment of Inflammatory Skin Disorders

0	Medicinal plant	Part used	Metabolites	Uses	Reference
1	Matricaria recutita L. (German chamomile)	Flower	Essential oil (α-bisabolol, oxides A, B, C, matricin, chamazulene), flavone derivatives (apigenin, luteolin, apigenin-7-glucoside)	Anti-inflammatory activity by inhibiting prostaglandins and leukotrienes synthesis; inhibition of cyclooxygenase and 5-lipooxygenase activity; bactericidal and fungicidal activity; treatment of skin inflammations, irritations, bacterial skin diseases, eczema, wounds, abscesses, frostbite, and insect bites	(Mailänder et al., 2022; Dawid-Pać, 2013; Dorina Shengjergji Dervishi et al., 2022; Chandrashekhar et al., 2011).
2	Calendula officinalis L.	Flower	Triterpene saponins (oleanolic acid glycosides), triterpene alcohols (α-, β-amyrins, faradiol), flavonoids (quercetin and isorhamnetin)	Anti-inflammatory effects related to flavonoids and triterpene derivatives; inhibition of lipoxygenase; treatment of poorly healing wounds, bruises, rashes, boils, and dermatitis	(Silva et al., 2021)
3	Hamamelis virginiana L. (witch hazel)	Barks	Tannins (hamamelitannins and proanthocyanidins)	Anti-inflammatory effects; inhibition of leukotriene synthesis; treatment of skin inflammation	(Piazza, Martinelli, Magnavacca, et al., 2022)
4	Hamamelis virginiana L.	Leaves	Tannins (proanthocyanidins, small amount of hamamelitannins)	Treatment of atopic eczema, atopic xeroderma, and other eczemas; compresses or washes for local inflammations of the skin	(Piazza, Martinelli, Vrhovsek, et al., 2022)
5	Aloe barbadensis Mill. syn. A. vera (L.) Burm. f.	Leaves (fresh leaves)	Carbohydrates (mannose-6-phosphate, acemannan), glycoproteins, sterols (lupeol, β-sitosterol), enzymes (bradykinase)	External treatment of minor wounds, inflammatory skin disorders, skin irritations including burns, bruises, abrasions	(Tunçay and Kaya, 2021)
6	Althaea officinalis L.	Roots	Mucilage polysaccharides (arabinogalactans, galacturonorhamnans, glucans, arabinans)	Treatment of furunculosis, eczema, dermatitis	(Agency, 2008)
7	Avena sativa L.	Fruits	Mucilage polysaccharides (β-glucan), proteins (glutelin, avenin), flavonoids	Symptomatic treatment of minor skin inflammations, aid in healing of minor wounds	(Veerasubramanian et al., 2018; EL Hosary et al., 2020)
8	Avena sativa L.	Straw	Polysaccharides (β-glucan), silicon dioxide (soluble esters of silicic acid with polyphenols, monosaccharides, oligosaccharides)	Treatment of inflammatory and seborrheic skin diseases	(Reynertson et al., 2015)

(Continued)

TABLE 14.1 *(Continued)*
Medicinal Plants and Natural Compounds for the Treatment of Inflammatory Skin Disorders

0	Medicinal plant	Part used	Metabolites	Uses	Reference
9	Echinacea purpurea (L.) Moench	Fresh herb	Alkamides, caffeic acid derivatives (cichoric acid, caftaric acid, chlorogenic acid)	Treatment of small superficial wounds, inflammation of the skin	(Burlou-Nagy et al., 2022)
10	Symphytum officinale L.	Roots	Amine compound (allantoin), phenolic acids (rosmarinic acid, salicylic acid, caffeic acid), mucilage polysaccharides, glycopeptides	Treatment of skin inflammation, leg ulcerations	(Salehi et al., 2019)
11	Achillea millefolium L.s.l.	Herb	Essential oil (sesquiterpenes, chamazulene, α-, β-pinene), tannins, sesquiterpene lactones (matricine), flavonoids (luteolin, apigenin, isorhamnetin, rutin)	Supportive treatment of small wounds, mild inflammations of the skin	(Farasati Far et al., 2023)
12	Plantago lanceolata L.	Leaves /Herb	Phenylethanoids (acteoside, plantamajoside), mucilage polysaccharides, iridoids (aucubin, catalpol), tannins	Supportive therapy of poorly healing wounds, skin inflammation	(De Cássia Israel Cardoso et al., 2019; Kuranel et al., 2016)
13	Salvia officinalis L.	Leaves	Phenolic acids (rosmarinic acid), triterpenoids acids (ursolic acid, oleanolic acid), essentials oil (monoterpenes), catechin-type tannins (salviatannin)	Relief of minor skin inflammations, bacterial infections of the skin	(Agency, 2008)
14	Hypericum perforatum L.	Herb	Phloroglucinol derivates (hyperforin, adhyperforin), naphtodianthrones (hypericin), flavonoids (amentoflavone, hyperoside)	Treatment of minor cuts, burns, skin ulcers	(Altan et al., 2019)
15	Oenothera biennis L.	Seed	Triglycerides of fatty acids (γ-linolenic acid, linoleic acid)	Symptomatic treatment of atopic dermatitis, improvement of skin condition	(Vieira et al., 2016; Lee and Bielory, 2010)
16	Quercus spp.	Barks	Tannins	Symptomatic treatment of minor inflammations of the skin, purulent skin diseases, exuding eczema, frostbite	(Dawid-Pać, 2013; ESCOP, 2003)

(Continued)

TABLE 14.1 *(Continued)*
Medicinal Plants and Natural Compounds for the Treatment of Inflammatory Skin Disorders

0	Medicinal plant	Part used	Metabolites	Uses	Reference
17	Juglans regia L.	Leaves	Tannins	Treatment of mild, superficial inflammations of the skin, excessive perspiration of hands and feet, dandruff, scalp itching	(Taheri et al., 2020)
18	Linum usitatissimum L.	Seed	Triglycerides of fatty acids (α-linolenic acid, linoleic acid), mucilage polysaccharides (neutral and acidic polysaccharides composed mainly of galacturonic acid)	Local inflammations of the skin, furunculosis	(Cunnane et al., 1993)
19	Trigonella foenum-graecum L.	Seed	Mucilage polysaccharides (galactomannan), steroidal saponins (diosgenin, yamogenin, tigogenin)	Treatment of furunculosis, eczema, ulcers, local inflammations of the skin	[64, 67]
20	Aquilaria agallocha Roxb. (Arachis hypogaea L., Fabaceae)	Barks, roots	Agarospirol, jinkoh-eremol and hinesol	Treatments of the skin thickness, ear weight, reduced oxidative stress, and pro-inflammatory cytokines production	(Yadav et al., 2013)
21	Arachis hypogaea L. (Fabaceae)	Peanut skin	Proanthocyanidin	Treatment of inflammation and melanogenesis; suppressive activities against melanogenesis and cytokine production	(Tatsuno et al., 2012; Kyei et al., 2021)
22	Chromolaena odorata L. (Asteraceae)	Leaves	Chromomoric acid	Treatment of inflammation and skin lesions. Modulating nuclear factor erythroid 2-related factor 2 (Nrf2)	(Heiss et al., 2014) ref
23	Citrus unshiu (Swingle) Marcow. (Rutaceae)	Fruits	Citrusina XI	Anti-inflammatory properties inhibiting pro-inflammatory mediator release. Reduces NO production by decreasing iNOS expression and NF-κB activation associated with IκBα degradation and NF-κB phosphorylation in LPS-induced raw cells	(Noh et al., 2015)
24	Curcuma longa L. (Zingiberaceae)	Rhizomes	Curcumin	Treatment of skin disruption and inflammation	(Fang et al., 2010)
25	Mangifera indica Wall. (Anacardiaceae)	Fruits	Mangiferin	Treatment of in-vivo inflammatory allergy-mediated skin disorder (inhibition of IgE production)	(Rivera et al., 2010)

(Continued)

TABLE 14.1 *(Continued)*
Medicinal Plants and Natural Compounds for the Treatment of Inflammatory Skin Disorders

0	Medicinal plant	Part used	Metabolites	Uses	Reference
26	Panax ginseng C.A. Mey. (Araliaceae)	Roots	Ginsenoside Rb1	Treatment of contact dermatitis or psoriasis by the regulation of COX-2 produced by macrophage cells	(Lorz et al., 2020; Shin et al., 2005)
27	Punica granatum L. (Punicaceae)	Fruits	Punicalagins	Modulation of melanin production inhibits melanin production, potential for use as a cosmetic skin-lightening agent	(Rana et al., 2013)
28	Juniperus cedrus Webb & Berthel (Cupressaceae)	Woods	Hinokitiol	Treatment of hair loss by suppressing factors that promote follicular apoptosis, such as TNF-α, in addition to stimulating new hair growth	(Byeon et al., 2008)
29	Zanthoxylum piperitum Hook. & Arn. (Rutaceae)	Fruits	ZPDC glycoprotein	Prevent dysfunction of the immune system caused by allergic responses and modulation of mast cell degranulation	(Maione et al., 2016)
30	Andrographis paniculata Nees (Acanthaceae)	Whole plant	Andrograpanin	Treatment of infection and inflammation via inhibition of iNOS, TNF-α and IL-6 production	(Lee and Lim, 2010)

These natural compounds have a variety of effects, from reducing inflammation and fighting microbes to promoting skin healing. What's particularly interesting is how each plant extract works differently. Some inhibit inflammatory pathways, while others support the immune system or aid in tissue repair. And the conditions they can address range from eczema and dermatitis to wounds and rashes.

Medicinal plants offer a glimpse into the potential of natural remedies for addressing skin inflammation.

The table highlights a variety of medicinal plants and their corresponding natural compounds that have demonstrated effectiveness in managing inflammatory skin disorders. Here's a closer look at some of the key ingredients mentioned in the table:

Essential Oils: Extracted from various plant parts such as flowers, leaves, or bark, essential oils contain potent bioactive compounds like α-bisabolol, chamazulene, and oxides, known for their anti-inflammatory and antimicrobial properties. They are commonly used in aromatherapy and topical applications for skin health.

***Flavonoids*:** These are a diverse group of plant metabolites with antioxidant and anti-inflammatory properties. Examples include apigenin, luteolin, and quercetin, which help reduce inflammation, neutralize free radicals, and promote tissue repair in the skin.

Tannins: Found in plant parts like bark and leaves, tannins have astringent properties and can help soothe and protect the skin. They are known for their ability to reduce inflammation, tighten skin tissue, and alleviate skin irritation.

Polysaccharides: These complex carbohydrates, such as arabinogalactans and glucans, are known for their moisturizing and soothing properties. They form a protective barrier on the skin, helping to retain moisture and reduce inflammation.

Terpenoids: These are aromatic compounds found in essential oils and plant resins, known for their anti-inflammatory, antimicrobial, and analgesic properties. Examples include α- and β-amyrins, which can help alleviate skin inflammation and promote wound healing.

***Phenolic Compounds*:** This diverse group of plant metabolites includes phenolic acids like rosmarinic acid and caffeic acid, which exhibit antioxidant and anti-inflammatory effects. They help protect the skin from oxidative stress and reduce inflammation.

Alkaloids: These nitrogen-containing compounds, such as matrine and oxymatrine, have diverse pharmacological activities, including anti-inflammatory and antimicrobial effects. They can help alleviate skin inflammation and promote wound healing.

These natural compounds, derived from medicinal plants, offer a holistic approach to skincare by addressing inflammation, promoting tissue repair, and supporting overall skin health. Their therapeutic effects make them valuable ingredients in topical preparations for managing various inflammatory skin conditions.

Table 14.2 presents a comprehensive summary of various medicinal plants and their metabolites known for their therapeutic potential in treating infectious skin disorders. Each entry in the table provides valuable information about the medicinal plant, the specific part used, key metabolites

TABLE 14.2
Plants and Natural Compounds for the Treatment of Infectious Skin Disorders

S. no.	Medicinal plant	Part used	Metabolites	Uses	Reference
1	Artemisia annua L.	Whole plant	Artemisinin, flavonoids, terpenoids	Anti-Inflammatory and immunoregulatory functions	(Shi et al., 2015)
2	Bupleurum smithii var. parvifolium R.H.S.	Whole plant	Saikosaponins, flavonoids, polysaccharides	Immunomodulatory function on macrophages	(Cheng et al., 2010)
3	Camellia sinensis (L.) Kuntze	Leaves	Catechins (e.g., EGCG), caffeine, theanine	JAK inhibitor, antioxidant	(Camargo et al., 2016)
4	Curcuma longa L.	Rhizomes	Curcuminoids (e.g., curcumin), turmerones	Anti-Inflammatory, antioxidant. Mediated through its ability to inhibit cyclooxygenase-2 (COX-2), lipoxygenase (LOX), and inducible nitric oxide synthase (iNOS)	(V. P. Menon and Sudheer, 2007)
5	Fagopyrum tataricum (L.) Gaertn.	Seeds	Rutin, quercetin, phenolic acids	Reduction in serum antibodies and CD4+ T cell activation	(W. Li et al., 2016)
6	Epimedium alpinum L.	Aerial parts	Icariin, flavonoids, polysaccharides	Inhibition of CD4 (+) T cell overactivity, enhancement of Treg cell suppression	(Shen and Wang, 2018)

(Continued)

TABLE 14.2 *(Continued)*
Plants and Natural Compounds for the Treatment of Infectious Skin Disorders

S. no.	Medicinal plant	Part used	Metabolites	Uses	Reference
7	Garcinia mangostana L.	Fruit rinds	Xanthones (e.g., alpha-mangostin), tannins	Reduction in proteinuria and serum antibodies	(W. Li et al., 2015)
8	Hypericum perforatum L.	Aerial parts	Hypericin, hyperforin, flavonoids	Reduction in serum levels of inflammatory cytokines	(Mahmoudi et al., 2022)
9	Illicium verum Hook.f.	Fruits	Anethole, lignans, terpenes	Inhibition of IFN-γ-induced mRNA and protein expression	(Sung et al., 2012)
10	Isodon serra (Maxim.) Kudô	Leaves	Diterpenoids (e.g., enmein), flavonoids	Suppression of murine splenic lymphocyte overproduction	(Y. Zhang et al., 2005)
11	Nelumbo nucifera Gaertn.	Rhizomes	Alkaloids (e.g., armepavine), flavonoids	Suppression of T cell proliferation and cytokine expression	(Liu et al., 2004)
12	Olea europaea L.	Leaves	Oleuropein, hydroxytyrosol, flavonoids	Anti-Inflammatory, immunomodulatory, antioxidant	(Silvestrini et al., 2023)
13	Paeonia lactiflora Pall.	Roots	Paeoniflorin, flavonoids	Suppression of T cell activation and cytokine production	(Z. Zhang et al., 2022)
14	Pinus pinaster Aiton	Barks	Proanthocyanidins (e.g., pycnogenol), catechins	Inhibition of pro-inflammatory cytokine expression	(Fan et al., 2015)
15	Mentha × Piperita L. + Resveratrol	Leaves	Menthol, piperine, resveratrol	Reduction in lupus morbidity measures and oxidative stress	(Pannu and Bhatnagar, 2020)
16	Sophora flavescens Aiton	Roots	Matrine, oxymatrine, flavonoids	Reduction in proteinuria, autoantibodies, and IFN-γ expression	(Wang et al., 2022)
17	Tripterygium wilfordii Hook. f.	Roots	Triptolide, celastrol, diterpenoids	Inhibition of neutrophil extracellular trap formation	(Guan et al., 2021) Ref

present, and their respective therapeutic uses in combating infectious skin conditions. The summarized information highlights the diverse range of natural compounds found in these plants, including flavonoids, terpenoids, alkaloids and phenolic acids, among others. These bioactive compounds exhibit a wide array of pharmacological activities such as anti-inflammatory, antioxidant, immunomodulatory and antimicrobial properties. Moreover, Table 14.2 emphasizes the specific mechanisms of action associated with these natural compounds, ranging from the inhibition of inflammatory cytokine expression to the suppression of T cell activation and cytokine production. Additionally, several plant extracts have been shown to reduce proteinuria, autoantibodies, and

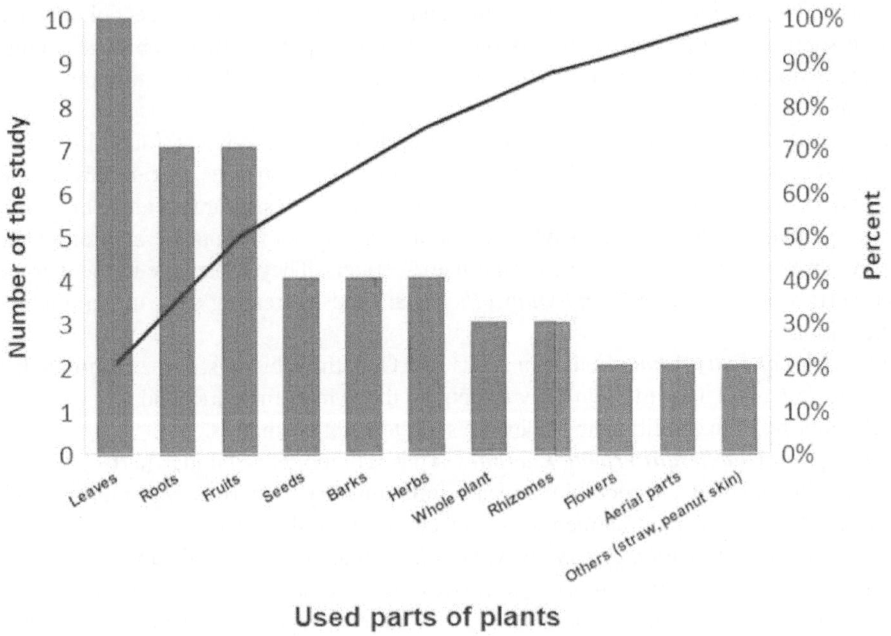

FIGURE 14.2 The specific plant parts used for skin diseases and the corresponding number or percentage of selected studies are presented.

interferon-gamma (IFN-γ) expression, indicating their potential to modulate immune responses and alleviate symptoms of infectious skin disorders. Overall, the comprehensive compilation of medicinal plants and their bioactive constituents underscores the rich diversity of natural resources available for the management of infectious skin conditions. The parts of the plants and number of the studies used for the skin diseases are presented in Figure 14.2.

A comprehensive analysis of studies on the utilization of various plant parts reveals that leaves were the focus in ten instances (37.04%), followed by roots and fruits, each accounting for seven studies (25.93%), respectively, with seeds, barks, and herbs each represented in four studies (14.81%). Additionally, three studies (11.11%) investigated the use of both whole plants and rhizomes, while flowers and aerial parts each garnered attention in two studies (7.41%). Furthermore, two studies (7.41%) explored other plant parts such as straw and peanut skin.

Essential Oils: Extracted from plant parts like flowers, leaves, or fruits, essential oils contain bioactive compounds known for their antimicrobial properties. Ingredients such as α-bisabolol and chamazulene are effective against bacteria and fungi, making them valuable in treating infectious skin conditions.

Triterpenoids: These compounds, found in plants like Calendula officinalis, have antimicrobial and anti-inflammatory properties. Triterpene saponins and alcohols, along with flavonoids, contribute to the plant's ability to combat infections and reduce inflammation in the skin.

Tannins: Derived from plant bark or leaves, tannins have astringent and antimicrobial properties. Compounds like hamamelitannins and proanthocyanidins, found in Hamamelis virginiana, help inhibit the growth of bacteria and fungi, making them effective in treating skin infections.

Carbohydrates and Glycoproteins: These compounds, found in Aloe vera, have moisturizing and wound-healing properties. Mannose-6-phosphate, acemannan, and glycoproteins like bradykinase promote tissue repair and provide relief from inflammatory skin conditions.

Mucilage Polysaccharides: Derived from plant roots or seeds, mucilage polysaccharides have soothing and protective effects on the skin. Compounds like arabinogalactans and glucans, found in Althaea officinalis and Avena sativa, help alleviate inflammation and promote wound healing.

Alkamides and Caffeic Acid Derivatives: These bioactive compounds, found in Echinacea purpurea, possess antimicrobial and immunomodulatory properties. Alkamides and caffeic acid derivatives help combat infections and support the skin's natural defence mechanisms.

Phenolic Compounds: Plant-derived phenolic acids, such as rosmarinic acid and salicylic acid, have antimicrobial and anti-inflammatory effects. They contribute to the therapeutic properties of plants like *Symphytum officinale*, used in treating skin inflammation and ulcerations.

Triglycerides of Fatty Acids: Found in seeds like Oenothera biennis, these compounds have moisturizing and anti-inflammatory effects on the skin. γ-linolenic acid and linoleic acid help improve skin condition and alleviate symptoms of dermatitis.

Sesquiterpenes and Sesquiterpene Lactones: These compounds, found in herbs like Achillea millefolium, exhibit antimicrobial and anti-inflammatory properties. Matricine, along with flavonoids, supports the treatment of skin wounds and inflammations.

Triterpenoids and Steroidal Saponins: Compounds like diosgenin and yamogenin, found in seeds like Trigonella foenum-graecum, have antimicrobial and anti-inflammatory effects. They help alleviate skin infections and inflammations, supporting overall skin health.

These natural compounds, derived from medicinal plants, offer a promising avenue for the treatment of infectious skin disorders. Their antimicrobial, anti-inflammatory, and wound-healing properties make them valuable ingredients in topical formulas aimed at combating skin infections and promoting skin health.

14.4 CONCLUSION AND FUTURE REMARKS

The present review on medicinal plants and natural compounds for the treatment of dermatological disorders. Alternative methods for skin disorders have a greater impact on every household or even in society. A better understanding of skin disorders among patients as public representatives is also desirable in the current world. Throughout history, various cultures have relied on botanical remedies to address skin ailments, and modern research continues to uncover their therapeutic potential. From soothing inflammation to combating microbial infections, these natural agents offer diverse mechanisms of action that can complement conventional dermatological treatments or serve as standalone therapies. One of the key advantages of medicinal plants and natural compounds is their typically lower risk of adverse effects compared to synthetic pharmaceuticals. Many botanical extracts and essential oils possess potent anti-inflammatory, antimicrobial, antioxidant, and wound-healing properties without causing harmful side effects. Additionally, their abundance in bioactive compounds provides a rich source of novel therapeutic agents that can be further explored and developed for dermatological applications. Moreover, the trend towards eco-friendly and sustainable healthcare practices has spurred renewed interest in botanical medicine. With growing concerns about environmental impact and antibiotic resistance associated with conventional treatments, natural alternatives offer a compelling solution. By harnessing the healing power of plants, we can reduce reliance on synthetic drugs while promoting environmental conservation and biodiversity preservation. However, despite the promising attributes of medicinal plants and natural compounds, several challenges remain. Standardization of herbal extracts, quality control, and regulatory issues pose significant hurdles in ensuring the safety and efficacy of botanical preparations. Additionally, further research is needed to elucidate the mechanisms of action, optimize formulations, and conduct rigorous clinical trials to validate their therapeutic benefits.

Looking ahead, continued interdisciplinary collaboration between traditional knowledge systems, modern science, and clinical practice will be essential for advancing the field of botanical dermatology. Integrating traditional wisdom with contemporary research methodologies can facilitate the discovery of novel bioactive compounds and innovative formulations for dermatological care. Furthermore, the development of evidence-based guidelines and protocols for the use of medicinal plants in dermatology is imperative to ensure safe and effective therapeutic outcomes. This involves conducting well-designed clinical trials, establishing pharmacovigilance systems, and fostering education and awareness among healthcare professionals and consumers. Moreover, the advent of advanced technologies such as metabolomics, genomics, and nanotechnology hold great promise for unlocking the full potential of medicinal plants in dermatological applications. These cutting-edge tools enable precise characterization of plant-derived compounds, targeted delivery systems, and personalized treatment approaches tailored to individual patient needs. In conclusion, while challenges persist, the integration of medicinal plants and natural compounds into dermatological practice represents a holistic and sustainable approach to skin health. By embracing the wisdom of nature and leveraging scientific innovation, we can usher in a new era of dermatological care that is safe, effective, and environmentally conscious.

REFERENCES

Abraham, J., & Mathew, S. (2019). Merkel cells: A collective review of current concepts. *International Journal of Applied and Basic Medical Research*, *9*(1), 9. https://doi.org/10.4103/ijabmr.ijabmr_34_18

Agency, E. E. M. (2008). Echinaceae purpureae herba – herbal medicinal product. *European Medicines Agency*. Avena Sativa L., Herba and Avena Sativa L., Fructus. https://www.ema.europa.eu/en/medicines/herbal/echinaceae-purpureae-herba

Almoshari, Y. (2022). Medicinal plants used for dermatological disorders among the people of the kingdom of Saudi Arabia: A narrative review. *Saudi Journal of Biological Sciences*, *29*(6). https://doi.org/10.1016/j.sjbs.2022.103303

Altan, A., Aras, M. H., Damlar, I., Gokce, H., Ozcan, O., & Alpaslan, C. (2019). The effect of Hypericum Perforatum on wound healing of oral mucosa in diabetic rats. *European Oral Research*, *52*(3), 143–149. https://doi.org/10.26650/eor.2018.505

Anatomy, Skin (Integument), Epidermis – StatPearls – NCBI Bookshelf. (n.d.). Retrieved November 1, 2020, from https://www.ncbi.nlm.nih.gov/books/NBK470464/

Archer, C. B., & Eedy, D. J. (2010). The skin and the nervous system. *Rook's Textbook of Dermatology: Eighth Edition*, *4*, 3192–3216. https://doi.org/10.1002/9781444317633.ch63

Atanasov, A. G., Waltenberger, B., Pferschy-Wenzig, E. M., Linder, T., Wawrosch, C., Uhrin, P., Temml, V., Wang, L., Schwaiger, S., Heiss, E. H., Rollinger, J. M., Schuster, D., Breuss, J. M., Bochkov, V., Mihovilovic, M. D., Kopp, B., Bauer, R., Dirsch, V. M., & Stuppner, H. (2015). Discovery and resupply of pharmacologically active plant-derived natural products: A review. *Biotechnology Advances*, *33*(8), 1582–1614. https://doi.org/10.1016/j.biotechadv.2015.08.001

Bhat, S. G. (2022). Medicinal plants and its pharmacological values. *Natural Medicinal Plants*. https://doi.org/10.5772/intechopen.99848

Boer, M., Duchnik, E., Maleszka, R., & Marchlewicz, M. (2016). Structural and biophysical characteristics of human skin in maintaining proper epidermal barrier function. *Postepy Dermatologii i Alergologii*, *33*(1), 1–5. https://doi.org/10.5114/pdia.2015.48037

Burlou-Nagy, C., Bănică, F., Jurca, T., Vicaş, L. G., Marian, E., Muresan, M. E., Bácskay, I., Kiss, R., Fehér, P., & Pallag, A. (2022). Echinacea purpurea (L.) Moench: Biological and pharmacological properties—a review. *Plants*, *11*(9), 1244. https://doi.org/10.3390/plants11091244

Byeon, S. E., Lee, Y. C., Kim, J. C., Han, J. G., Lee, H. Y., & Cho, J. Y. (2008). Hinokitiol, a natural tropolone derivative, inhibits TNF-α production in LPS-activated macrophages via suppression of NF-κB. *Planta Medica*, *74*(8), 828–833. https://doi.org/10.1055/s-2008-1074548

Camargo, L. E. A., Pedroso, L. S., Vendrame, S. C., Mainardes, R. M., & Khalil, N. M. (2016). Atividade antioxidante e antifúngica das folhas de Camellia sinensis (L.) Kuntze, obtidas por diferentes formas de produção. *Brazilian Journal of Biology*, *76*(2), 428–434. https://doi.org/10.1590/1519-6984.18814

Chakraborty, R. K., & Burns, B. (2023). Systemic inflammatory response syndrome. *StatPearls*. https://www.ncbi.nlm.nih.gov/books/NBK547669/

Chandrashekhar, V. M., Halagali, K. S., Nidavani, R. B., Shalavadi, M. H., Biradar, B. S., Biswas, D., & Muchchandi, I. S. (2011). Anti-allergic activity of German chamomile (Matricaria recutita L.) in mast cell mediated allergy model. *Journal of Ethnopharmacology*, *137*(1), 336–340. https://doi.org/10.1016/j.jep.2011.05.029

Chen, L., & Shen, Z. (2020). Tissue-resident memory T cells and their biological characteristics in the recurrence of inflammatory skin disorders. *Cellular and Molecular Immunology*, *17*(1), 64–75. https://doi.org/10.1038/s41423-019-0291-4

Cheng, X. Q., Li, H., Yue, X. L., Xie, J. Y., Zhang, Y. Y., Di, H. Y., & Chen, D. F. (2010). Macrophage immunomodulatory activity of the polysaccharides from the roots of Bupleurum smithii var. parvifolium. *Journal of Ethnopharmacology*, *130*(2), 363–368. https://doi.org/10.1016/j.jep.2010.05.019

Chiller, K., Selkin, B. A., & Murakawa, G. J. (2001). Skin microflora and bacterial infections of the skin. *Journal of Investigative Dermatology Symposium Proceedings*, *6*(3), 170–174. https://doi.org/10.1046/j.0022-202x.2001.00043.x

Cichorek, M., Wachulska, M., Stasiewicz, A., & Tymińska, A. (2013). Skin melanocytes: Biology and development. *Postepy Dermatologii i Alergologii*, *30*(1), 30–41. https://doi.org/10.5114/pdia.2013.33376

Cunnane, S. C., Ganguli, S., Menard, C., Liede, A. C., Hamadeh, M. J., Chen, Z.-Y., Wolever, T. M. S., & Jenkins, D. J. A. (1993). High α-linolenic acid flaxseed (Linum usitatissimum): Some nutritional properties in humans. *British Journal of Nutrition*, *69*(2), 443–453. https://doi.org/10.1079/bjn19930046

Das, D., Akhtar, S., Kurra, S., Gupta, S., & Sharma, A. (2019). Emerging role of immune cell network in autoimmune skin disorders: An update on pemphigus, vitiligo and psoriasis. *Cytokine and Growth Factor Reviews*, *45*, 35–44. https://doi.org/10.1016/j.cytogfr.2019.01.001

Dawid-Pać, R. (2013). Medicinal plants used in treatment of inflammatory skin diseases. *Postepy Dermatologii i Alergologii*, *30*(3), 170–177. https://doi.org/10.5114/pdia.2013.35620

De Cássia Israel Cardoso, F., Peruzzo Apolinário, P., Da Silva Cunha Breder, J., Paranhos, T., Ceretta Oliveira, H., Dini Polidoro, A., Souza Oliveira Kumakura, A. R., & Melo Lima, M. H. (2019). A protocol for systematic review of Plantago major L. effectiveness in accelerating wound-healing in animal models. *Systematic Reviews*, *8*(1). https://doi.org/10.1186/s13643-019-1255-6

Del Rosso, J. Q. (2013). Management of facial erythema of rosacea: What is the role of topical α-adrenergic receptor agonist therapy? *Journal of the American Academy of Dermatology*, *69*(6 Suppl. 1). https://doi.org/10.1016/j.jaad.2013.06.009

Dervishi, D. S., Panariti, E., & Nuna, A. (2022). Trend of using herbal based preparations in pediatric skin inflammation disorders. *GSC Biological and Pharmaceutical Sciences*, *21*(3), 190–195. https://doi.org/10.30574/gscbps.2022.21.3.0485

Elbe-Bürger, A. (2012). Skin architecture and function. In *Handbook of burns: Reconstruction and rehabilitation* (Vol. 2, pp. 29–46). https://doi.org/10.1007/978-3-7091-0315-9_3

EL Hosary, R., El-Mancy, S. M. S., El Deeb, K. S., Eid, H. H., EL Tantawy, M. E., Shams, M. M., Samir, R., Assar, N. H., & Sleem, A. A. (2020). Efficient wound healing composite hydrogel using Egyptian Avena sativa L. polysaccharide containing β-glucan. *International Journal of Biological Macromolecules*, *149*, 1331–1338. https://doi.org/10.1016/j.ijbiomac.2019.11.046

ESCOP. (2003). *2nd edition of the ESCOP monographs: Two volume – ESCOP*. European Scientific Cooperative On Phytotheraphy. ESCOP Monographs The Scientific Foundation for Herbal Medicinal Products (2nd ed.). https://www.escop.com/escop-products/publications/

Fan, B., Dun, S. H., Gu, J. Q., Guo, Y., & Ikuyama, S. (2015). Pycnogenol attenuates the release of proinflammatory cytokines and expression of perilipin 2 in lipopolysaccharide-stimulated microglia in part via inhibition of NF-κB and AP-1 activation. *PLoS ONE*, *10*(9). https://doi.org/10.1371/journal.pone.0137837

Fang, J.-Y., Hung, C.-F., Chiu, H.-C., Wang, J.-J., & Chan, T.-F. (2010). Efficacy and irritancy of enhancers on the in-vitro and in-vivo percutaneous absorption of curcumin. *Journal of Pharmacy and Pharmacology*, *55*(5), 593–601. https://doi.org/10.1211/002235703765344496

Farasati Far, B., Behzad, G., & Khalili, H. (2023). Achillea millefolium: Mechanism of action, pharmacokinetic, clinical drug-drug interactions and tolerability. *Heliyon*, *9*(12), e22841. https://doi.org/10.1016/j.heliyon.2023.e22841

Fournière, M., Latire, T., Souak, D., Feuilloley, M. G. J., & Bedoux, G. (2020). Staphylococcus epidermidis and cutibacterium acnes: Two major sentinels of skin microbiota and the influence of cosmetics. *Microorganisms*, *8*(11), 1–31. https://doi.org/10.3390/microorganisms8111752

Fowler, M. W. (2006). Plants, medicines and man. *Journal of the Science of Food and Agriculture*, *86*(12), 1797–1804. https://doi.org/10.1002/jsfa.2598

Fuchs, E., & Byrne, C. (1994). The epidermis: Rising to the surface. *Current Opinion in Genetics and Development*, *4*(5), 725–736. https://doi.org/10.1016/0959-437X(94)90140-X

Guan, H., Xie, L., Ji, Z., Song, R., Qi, J., & Nie, X. (2021). Triptolide inhibits neutrophil extracellular trap formation. *Annals of Translational Medicine*, *9*(17), 1384–1384. https://doi.org/10.21037/atm-21-3522

Heiss, E. H., Tran, T. V. A., Zimmermann, K., Schwaiger, S., Vouk, C., Mayerhofer, B., Malainer, C., Atanasov, A. G., Stuppner, H., & Dirsch, V. M. (2014). Identification of chromomoric acid C-I as an Nrf2 activator in Chromolaena odorata. *Journal of Natural Products*, *77*(3), 503–508. https://doi.org/10.1021/np400778m

Hypodermis – concept | Anatomy and Physiology | JoVe. (n.d.). Retrieved March 21, 2024, from https://app.jove.com/science-education/v/13994/hypodermis?language=Turkish

Ikeda, Y., Murakami, A., & Ohigashi, H. (2008). Ursolic acid: An anti- and pro-inflammatory triterpenoid. *Molecular Nutrition and Food Research*, *52*(1), 26–42. https://doi.org/10.1002/mnfr.200700389

Jiang, Y., Tsoi, L. C., Billi, A. C., Ward, N. L., Harms, P. W., Zeng, C., Maverakis, E., Michelle Kahlenberg, J., & Gudjonsson, J. E. (2020). Cytokinocytes: The diverse contribution of keratinocytes to immune responses in skin. *JCI Insight*, *5*(20). https://doi.org/10.1172/jci.insight.142067

Katoh, N. (2018). Allergic skin disorders. *Japanese Journal of Allergology*, *67*(9), 1243–1247. https://doi.org/10.15036/arerugi.67.1243

Kim, J. Y., & Dao, H. (2020). Physiology, integument. *StatPearls*. http://www.ncbi.nlm.nih.gov/pubmed/32119273

Kuranel, E., Küpeli Akkol, E., Süntar, I., Gürsoy, Ş., Keleş, H., & Aktay, G. (2016). Investigating biological activity potential of Plantago lanceolata L. In healing of skin wounds by a preclinical research. *Turkish Journal of Pharmaceutical Sciences*, *13*(2), 135–144. https://doi.org/10.5505/tjps.2016.08370

Kwiecien, K., Zegar, A., Jung, J., Brzoza, P., Kwitniewski, M., Godlewska, U., Grygier, B., Kwiecinska, P., Morytko, A., & Cichy, J. (2019). Architecture of antimicrobial skin defense. *Cytokine and Growth Factor Reviews*, *49*, 70–84. https://doi.org/10.1016/j.cytogfr.2019.08.001

Kyei, S. K., Eke, W. I., Abdul-Karim, H., Darko, G., & Akaranta, O. (2021). Phytochemicals from peanut (Arachis hypogaea L.) skin extract with potential for pharmacological activity. *Current Bioactive Compounds*, *17*(9). https://doi.org/10.2174/1573407217666210202092052

Lee, J., & Bielory, L. (2010). Complementary and alternative interventions in atopic dermatitis. *Immunology and Allergy Clinics of North America*, *30*(3), 411–424. https://doi.org/10.1016/j.iac.2010.06.006

Lee, J., & Lim, K. T. (2010). Inhibitory effect of phytoglycoprotein (24 kDa) on allergy-related factors in compound 48/80-induced mast cells in vivo and in vitro. *International Immunopharmacology*, *10*(5), 591–599. https://doi.org/10.1016/J.INTIMP.2010.02.009

Lefèvre-Utile, A., Braun, C., Haftek, M., & Aubin, F. (2021). Five functional aspects of the epidermal barrier. *International Journal of Molecular Sciences*, *22*(21). https://doi.org/10.3390/ijms222111676

Li, F. S., & Weng, J. K. (2017). Demystifying traditional herbal medicine with modern approaches. *Nature Plants*, *3*. https://doi.org/10.1038/nplants.2017.109

Li, W., Li, H., Zhang, M., Wang, M., Zhong, Y., Wu, H., Yang, Y., Morel, L., & Wei, Q. (2016). Quercitrin ameliorates the development of systemic lupus erythematosus-like disease in a chronic graft-versus-host murine model. *American Journal of Physiology – Renal Physiology*, *311*(1), F217–F226. https://doi.org/10.1152/ajprenal.00249.2015

Li, W., Li, H., Zhang, M., Zhong, Y., Wang, M., Cen, J., Wu, H., Yang, Y., & Wei, Q. (2015). Isogarcinol extracted from Garcinia mangostana L. Ameliorates systemic lupus erythematosus-like disease in a murine model. *Journal of Agricultural and Food Chemistry*, *63*(38), 8452–8459. https://doi.org/10.1021/acs.jafc.5b03425

Lin, J. Y., & Fisher, D. E. (2007). Melanocyte biology and skin pigmentation. *Nature*, *445*(7130), 843–850. https://doi.org/10.1038/nature05660

Liu, C. P., Tsai, W. J., Lin, Y. L., Liao, J. F., Chen, C. F., & Kuo, Y. C. (2004). The extracts from Nelumbo nucifera suppress cell cycle progression, cytokine genes expression, and cell proliferation in human peripheral blood mononuclear cells. *Life Sciences*, *75*(6), 699–716. https://doi.org/10.1016/j.lfs.2004.01.019

Lorz, L. R., Kim, M. Y., & Cho, J. Y. (2020). Medicinal potential of Panax ginseng and its ginsenosides in atopic dermatitis treatment. *Journal of Ginseng Research*, *44*(1), 8–13. https://doi.org/10.1016/j.jgr.2018.12.012

Mahmoudi, M., Rastin, M., Arababadi, M. K., Anaeigoudari, A., & Nosratabadi, R. (2022). Enhancing the efficacy of Hypericum perforatum in the treatment of an experimental model of multiple sclerosis using gold nanoparticles: An in vivo study. *Avicenna Journal of Phytomedicine*, *12*(3), 325–336. https://doi.org/10.22038/AJP.2022.19574

Mailänder, L. K., Lorenz, P., Bitterling, H., Stintzing, F. C., Daniels, R., & Kammerer, D. R. (2022). Phytochemical characterization of chamomile (Matricaria recutita L.) roots and evaluation of their antioxidant and antibacterial potential. *Molecules*, *27*(23). https://doi.org/10.3390/molecules27238508

Maione, F., Russo, R., Khan, H., & Mascolo, N. (2016). Medicinal plants with anti-inflammatory activities. *Natural Product Research*, *30*(12), 1343–1352. https://doi.org/10.1080/14786419.2015.1062761

Manuelyan, K. L., Bogdanov, I. A., & Darlenski, R. B. (2017). Skin signs of systemic infections and neoplastic diseases. *Giornale Italiano Di Dermatologia e Venereologia*, *152*(5), 489–499. https://doi.org/10.23736/S0392-0488.17.05719-4

Matejuk, A. (2018). Skin immunity. *Archivum Immunologiae et Therapiae Experimentalis*, *66*(1), 45–54. https://doi.org/10.1007/s00005-017-0477-3

Menon, G. K., Cleary, G. W., & Lane, M. E. (2012). The structure and function of the stratum corneum. *International Journal of Pharmaceutics*, *435*(1), 3–9. https://doi.org/10.1016/j.ijpharm.2012.06.005

Menon, V. P., & Sudheer, A. R. (2007). Antioxidant and anti-inflammatory properties of curcumin. *Advances in Experimental Medicine and Biology*, *595*, 105–125. https://doi.org/10.1007/978-0-387-46401-5_3

Murphrey, M. B., & Zito, P. M. (2018). Histology, stratum corneum. *StatPearls*. http://www.ncbi.nlm.nih.gov/pubmed/30020671

Nguyen, A. V., & Soulika, A. M. (2019). The dynamics of the skin's immune system. *International Journal of Molecular Sciences*, *20*(8), 1811. https://doi.org/10.3390/ijms20081811

Noh, H. J., Hwang, D., Lee, E. S., Hyun, J. W., Yi, P. H., Kim, G. S., Lee, S. E., Pang, C., Park, Y. J., Chung, K. H., Kim, G. D., & Kim, K. H. (2015). Anti-inflammatory activity of a new cyclic peptide, citrusin XI, isolated from the fruits of Citrus unshiu. *Journal of Ethnopharmacology*, *163*, 106–112. https://doi.org/10.1016/j.jep.2015.01.024

Orsmond, A., Bereza-Malcolm, L., Lynch, T., March, L., & Xue, M. (2021). Skin barrier dysregulation in psoriasis. *International Journal of Molecular Sciences*, *22*(19). https://doi.org/10.3390/ijms221910841

Pannu, N., & Bhatnagar, A. (2020). Combinatorial therapeutic effect of resveratrol and piperine on murine model of systemic lupus erythematosus. *Inflammopharmacology*, *28*(2), 401–424. https://doi.org/10.1007/s10787-019-00662-w

Parker, J. C., Rangu, S., Grand, K. L., Bhoj, E. J., Castelo-Soccio, L., & Sheppard, S. E. (2021). Genetic skin disorders: The value of a multidisciplinary clinic. *American Journal of Medical Genetics, Part A*, *185*(4), 1159–1167. https://doi.org/10.1002/ajmg.a.62095

Petrovska, B. B. (2012). Historical review of medicinal plants' usage. *Pharmacognosy Reviews*, *6*(11), 1–5. https://doi.org/10.4103/0973-7847.95849

Piazza, S., Martinelli, G., Magnavacca, A., Fumagalli, M., Pozzoli, C., Terno, M., Canilli, L., Angarano, M., Maranta, N., Dell'Agli, M., & Sangiovanni, E. (2022). Unveiling the ability of witch hazel (Hamamelis virginiana L.) bark extract to impair keratinocyte inflammatory cascade typical of atopic eczema. *International Journal of Molecular Sciences*, *23*(16). https://doi.org/10.3390/ijms23169279

Piazza, S., Martinelli, G., Vrhovsek, U., Masuero, D., Fumagalli, M., Magnavacca, A., Pozzoli, C., Canilli, L., Terno, M., Angarano, M., Dell'Agli, M., & Sangiovanni, E. (2022). Anti-inflammatory and anti-acne effects of hamamelis virginiana bark in human keratinocytes. *Antioxidants*, *11*(6), 1119. https://doi.org/10.3390/antiox11061119

Proksch, E., Brandner, J. M., & Jensen, J. M. (2008). The skin: An indispensable barrier. *Experimental Dermatology*, *17*(12), 1063–1072. https://doi.org/10.1111/j.1600-0625.2008.00786.x

Rana, J., Diwakar, G., Saito, L., Scholten, J. D., & Mulder, T. (2013). Inhibition of melanin content by punicalagins in the super fruit pomegranate (Punica granatum). *Journal of Cosmetic Science*. https://pubmed.ncbi.nlm.nih.gov/24397882/

Reynertson, K. A., Garay, M., Nebus, J., Chon, S., Kaur, S., Mahmood, K., Kizoulis, M., & Southall, M. D. (2015). Anti-inflammatory activities of colloidal oatmeal (Avena sativa) contribute to the effectiveness of oats in treatment of itch associated with dry, irritated skin. *Journal of Drugs in Dermatology*. https://pubmed.ncbi.nlm.nih.gov/25607907/

Rivera, D. G., Balmaseda, I. H., León, A. Á., Hernández, B. C., Montiel, L. M., Garrido, G. G., Hernández, R. D., & Cuzzocrea, S. (2010). Anti-allergic properties of Mangifera indica L. extract (Vimang) and contribution of its glucosylxanthone mangiferin. *Journal of Pharmacy and Pharmacology*, *58*(3), 385–392. https://doi.org/10.1211/JPP.58.3.0014

Romani, N., Clausen, B. E., & Stoitzner, P. (2010). Langerhans cells and more: Langerin-expressing dendritic cell subsets in the skin. *Immunological Reviews*, *234*(1), 120–141. https://doi.org/10.1111/j.0105-2896.2009.00886.x

Rustin, M. H. A. (1990). Andrews' diseases of the skin – clinical dermatology. *Postgraduate Medical Journal*, *66*(781), 984–984. https://doi.org/10.1136/pgmj.66.781.984

Salehi, B., Sharopov, F., Tumer, T. B., Ozleyen, A., Rodríguez-Pérez, C., Ezzat, S. M., Azzini, E., Hosseinabadi, T., Butnariu, M., Sarac, I., Bostan, C., Acharya, K., Sen, S., Kasapoglu, K. N., Daşkaya-Dikmen, C., Özçelik, B., Baghalpour, N., Sharifi-Rad, J., Fokou, P. V. T., . . . Martins, N. (2019). Symphytum species: A comprehensive review on chemical composition, food applications and phytopharmacology. *Molecules*, *24*(12), 2272. https://doi.org/10.3390/molecules24122272

Schmidt, B., Ribnicky, D. M., Poulev, A., Logendra, S., Cefalu, W. T., & Raskin, I. (2008). A natural history of botanical therapeutics. *Metabolism: Clinical and Experimental*, *57*(Suppl. 1), S3. https://doi.org/10.1016/j.metabol.2008.03.001

Shen, R., & Wang, J.-H. (2018). The effect of icariin on immunity and its potential application. *American Journal of Clinical and Experimental Immunology*, *7*(3), 50–56. http://www.ncbi.nlm.nih.gov/pubmed/30038846%0Ahttp://www.pubmedcentral.nih.gov/articlerender.fcgi?artid=PMC6055068

Shi, C., Li, H., Yang, Y., & Hou, L. (2015). Anti-inflammatory and immunoregulatory functions of artemisinin and its derivatives. *Mediators of Inflammation*, *2015*. https://doi.org/10.1155/2015/435713

Shin, Y. W., Bae, E. A., Kim, S. S., Lee, Y. C., & Kim, D. H. (2005). Effect of ginsenoside Rb1 and compound K in chronic oxazolone-induced mouse dermatitis. *International Immunopharmacology*, *5*(7–8), 1183–1191. https://doi.org/10.1016/J.INTIMP.2005.02.016

Silva, D., Ferreira, M. S., Sousa-Lobo, J. M., Cruz, M. T., & Almeida, I. F. (2021). Anti-inflammatory activity of calendula officinalis l. Flower extract. *Cosmetics*, *8*(2), 31. https://doi.org/10.3390/cosmetics8020031

Silvestrini, A., Giordani, C., Bonacci, S., Giuliani, A., Ramini, D., Matacchione, G., Sabbatinelli, J., Di Valerio, S., Pacetti, D., Procopio, A. D., Procopio, A., & Rippo, M. R. (2023). Anti-inflammatory effects of olive leaf extract and its bioactive compounds oleacin and oleuropein-aglycone on senescent endothelial and small airway epithelial cells. *Antioxidants*, *12*(8). https://doi.org/10.3390/antiox12081509

Soltani, K. (2002). Skin disease: Diagnosis and treatment. *Annals of Internal Medicine*, *136*(9), 712. https://doi.org/10.7326/0003-4819-136-9-200205070-00027

Sõukand, R., Hrynevich, Y., Vasilyeva, I., Prakofjewa, J., Vnukovich, Y., Paciupa, J., Hlushko, A., Knureva, Y., Litvinava, Y., Vyskvarka, S., Silivonchyk, H., Paulava, A., Kõiva, M., & Kalle, R. (2017). Multifunctionality of the few: Current and past uses of wild plants for food and healing in Liubań region, Belarus. *Journal of Ethnobiology and Ethnomedicine*, *13*(1), 1–42. https://doi.org/10.1186/s13002-017-0139-x

Sung, Y. Y., Kim, Y. S., & Kim, H. K. (2012). Illicium verum extract inhibits TNF-α- and IFN-γ-induced expression of chemokines and cytokines in human keratinocytes. *Journal of Ethnopharmacology*, *144*(1), 182–189. https://doi.org/10.1016/j.jep.2012.08.049

Suter, M. M., Schulze, K., Bergman, W., Welle, M., Roosje, P., & Müller, E. J. (2009). The keratinocyte in epidermal renewal and defence. *Veterinary Dermatology*, *20*(5–6), 515–532. https://doi.org/10.1111/j.1365-3164.2009.00819.x

Taheri, A., Mirghazanfari, S. M., & Dadpay, M. (2020). Wound healing effects of Persian walnut (Juglans regia L.) green husk on the incision wound model in rats. *European Journal of Translational Myology*, *30*(1), 210–218. https://doi.org/10.4081/ejtm.2019.8671

Tatsuno, T., Jinno, M., Arima, Y., Kawabata, T., Hasegawa, T., Yahagi, N., Takano, F., & Ohta, T. (2012). Anti-inflammatory and anti-melanogenic proanthocyanidin oligomers from peanut skin. *Biological and Pharmaceutical Bulletin*, *35*(6), 909–916. https://doi.org/10.1248/bpb.35.909

Thawabteh, A. M., Jibreen, A., Karaman, D., Thawabteh, A., & Karaman, R. (2023). Skin pigmentation types, causes and treatment—a review. *Molecules*, *28*(12). https://doi.org/10.3390/molecules28124839

Tobina, D. J. (2006). Biochemistry of human skin—our brain on the outside. *Chemical Society Reviews*, *35*(1), 52–67. https://doi.org/10.1039/b505793k

Tunçay, H. A., & Kaya, G. İ. (2021). (2021). Ankara Üniversitesi Eczacılık Fakültesi Dergisi » Makale » ALOE VERA (L.) BURM. F. (SARISABIR) BİTKİSİNİN FİTOTERAPİDE KULLANIMI. *Journal of Faculty of Pharmacy of Ankara University*. https://dergipark.org.tr/tr/pub/jfpanu/issue/65218/834326

Veerasubramanian, P. K., Thangavel, P., Kannan, R., Chakraborty, S., Ramachandran, B., Suguna, L., & Muthuvijayan, V. (2018). An investigation of konjac glucomannan-keratin hydrogel scaffold loaded with Avena sativa extracts for diabetic wound healing. *Colloids and Surfaces B: Biointerfaces*, *165*, 92–102. https://doi.org/10.1016/j.colsurfb.2018.02.022

Vieira, B. L., Lim, N. R., Lohman, M. E., & Lio, P. A. (2016). Complementary and alternative medicine for atopic dermatitis: An evidence-based review. *American Journal of Clinical Dermatology*, *17*(6), 557–581. https://doi.org/10.1007/s40257-016-0209-1

Wang, M. C., Huang, W. C., Chen, L. C., Yeh, K. W., Lin, C. F., & Liou, C. J. (2022). Sophoraflavanone G from Sophora flavescens ameliorates allergic airway inflammation by suppressing Th2 response and oxidative stress in a murine asthma model. *International Journal of Molecular Sciences*, *23*(11). https://doi.org/10.3390/ijms23116104

Watt, F. M., & Fujiwara, H. (2011). Cell-extracellular matrix interactions in normal and diseased skin. *Cold Spring Harbor Perspectives in Biology*, *3*(4), 1–14. https://doi.org/10.1101/cshperspect.a005124

Wickett, R. R., & Visscher, M. O. (2006). Structure and function of the epidermal barrier. *American Journal of Infection Control*, *34*(10 Suppl.), S98–S110. https://doi.org/10.1016/j.ajic.2006.05.295

Yadav, D., Mudgal, V., Agrawal, J., Maurya, A., Bawankule, D., Chanotiya, C., Khan, F., & Thul, S. (2013). Molecular docking and ADME studies of natural compounds of agarwood oil for topical anti-inflammatory activity. *Current Computer Aided-Drug Design*, *9*(3), 360–370. https://doi.org/10.2174/1573409911309030012

Yousef, H., & Sharma, S. (2018). Anatomy, skin (integument), epidermis. In *StatPearls*. StatPearls Publishing. http://www.ncbi.nlm.nih.gov/pubmed/29262154

Zawodny, P., Malec, W., Gill, K., Skonieczna-Żydecka, K., & Sieńko, J. (2023). Assessment of the effectiveness of treatment of vascular lesions within the facial skin with a laser with a wavelength of 532 nm based on photographic diagnostics with the use of polarized light. *Sensors*, *23*(2). https://doi.org/10.3390/s23021010

Zhang, Y., Liu, J., Jia, W., Zhao, A. H., & Li, T. (2005). Distinct immunosuppressive effect by Isodon serra extracts. *International Immunopharmacology*, *5*(13–14), 1957–1965.

Zhang, Z., Zhang, Y., Zhao, Z., Li, P., Chen, D., Wang, W., Han, Y., Zou, S., Jin, X., Zhao, J., Liu, H., Wang, X., & Zhu, W. (2022). Paeoniflorin drives the immunomodulatory effects of mesenchymal stem cells by regulating Th1/Th2 cytokines in oral lichen planus. *Scientific Reports*, *12*(1), 18678. https://doi.org/10.1038/s41598-022-23158-0

Index

Note: Page numbers in *italics* indicate a figure and page numbers in **bold** indicate a table on the corresponding page.

D

E

J

K

L

M

N

O

P

T